Plant Development and Organogenesis

Plant Development and Organogenesis

From Basic Principles to Applied Research

Special Issue Editor
Giovanna Frugis

MDPI • Basel • Beijing • Wuhan • Barcelona • Belgrade

Special Issue Editor
Giovanna Frugis
CNR, Istituto di Biologia e Biotecnologia Agraria (IBBA)
Italy

Editorial Office
MDPI
St. Alban-Anlage 66
4052 Basel, Switzerland

This is a reprint of articles from the Special Issue published online in the open access journal *Plants* (ISSN 2223-7747) from 2018 to 2019 (available at: https://www.mdpi.com/journal/plants/special_issues/plant_dev?authAll=true).

For citation purposes, cite each article independently as indicated on the article page online and as indicated below:

LastName, A.A.; LastName, B.B.; LastName, C.C. Article Title. *Journal Name* **Year**, *Article Number*, Page Range.

ISBN 978-3-03928-126-8 (Pbk)
ISBN 978-3-03928-127-5 (PDF)

© 2020 by the authors. Articles in this book are Open Access and distributed under the Creative Commons Attribution (CC BY) license, which allows users to download, copy and build upon published articles, as long as the author and publisher are properly credited, which ensures maximum dissemination and a wider impact of our publications.

The book as a whole is distributed by MDPI under the terms and conditions of the Creative Commons license CC BY-NC-ND.

Contents

About the Special Issue Editor vii

Giovanna Frugis
Plant Development and Organogenesis: From Basic Principles to Applied Research
Reprinted from: *Plants* **2019**, *8*, 299, doi:10.3390/plants8090299 1

Angelo De Paolis, Giovanna Frugis, Donato Giannino, Maria Adelaide Iannelli, Giovanni Mele, Eddo Rugini, Cristian Silvestri, Francesca Sparvoli, Giulio Testone, Maria Luisa Mauro, Chiara Nicolodi and Sofia Caretto
Plant Cellular and Molecular Biotechnology: Following Mariotti's Steps
Reprinted from: *Plants* **2019**, *8*, 18, doi:10.3390/plants8010018 7

Riccardo Di Mambro, Sabrina Sabatini and Raffaele Dello Ioio
Patterning the Axes: A Lesson from the Root
Reprinted from: *Plants* **2019**, *8*, 8, doi:10.3390/plants8010008 34

Emanuela Pierdonati, Simon Josef Unterholzner, Elena Salvi, Noemi Svolacchia, Gaia Bertolotti, Raffaele Dello Ioio, Sabrina Sabatini and Riccardo Di Mambro
Cytokinin-Dependent Control of *GH3* Group II Family Genes in the *Arabidopsis* Root
Reprinted from: *Plants* **2019**, *8*, 94, doi:10.3390/plants8040094 47

Ilaria Fraudentali, Renato Alberto Rodrigues-Pousada, Alessandro Volpini, Paraskevi Tavladoraki, Riccardo Angelini and Alessandra Cona
Stress-Triggered Long-Distance Communication Leads to Phenotypic Plasticity: The Case of the Early Root Protoxylem Maturation Induced by Leaf Wounding in Arabidopsis
Reprinted from: *Plants* **2018**, *7*, 107, doi:10.3390/plants7040107 56

Maurizio Trovato, Roberto Mattioli and Paolo Costantino
From *A. rhizogenes* RolD to Plant P5CS: Exploiting Proline to Control Plant Development
Reprinted from: *Plants* **2018**, *7*, 108, doi:10.3390/plants7040108 66

Jennifer C. Fletcher
The CLV-WUS Stem Cell Signaling Pathway: A Roadmap to Crop Yield Optimization
Reprinted from: *Plants* **2018**, *7*, 87, doi:10.3390/plants7040087 79

Jan Traas
Organogenesis at the Shoot Apical Meristem
Reprinted from: *Plants* **2019**, *8*, 6, doi:10.3390/plants8010006 90

Annis E Richardson and Sarah Hake
Drawing a Line: Grasses and Boundaries
Reprinted from: *Plants* **2019**, *8*, 4, doi:10.3390/plants8010004 99

Eva Hellmann, Donghwi Ko, Raili Ruonala and Ykä Helariutta
Plant Vascular Tissues—Connecting Tissue Comes in All Shapes
Reprinted from: *Plants* **2018**, *7*, 109, doi:10.3390/plants7040109 119

Giovanna Sessa, Monica Carabelli, Marco Possenti, Giorgio Morelli and Ida Ruberti
Multiple Pathways in the Control of the Shade Avoidance Response
Reprinted from: *Plants* **2018**, *7*, 102, doi:10.3390/plants7040102 137

Willeke Leijten, Ronald Koes, Ilja Roobeek and Giovanna Frugis
Translating Flowering Time from *Arabidopsis thaliana* to Brassicaceae and Asteraceae Crop Species
Reprinted from: *Plants* **2018**, *7*, 111, doi:10.3390/plants7040111 . . . . **160**

Bill Gordon-Kamm, Nagesh Sardesai, Maren Arling, Keith Lowe, George Hoerster, Scott Betts and Todd Jones
Using Morphogenic Genes to Improve Recovery and Regeneration of Transgenic Plants
Reprinted from: *Plants* **2019**, *8*, 38, doi:10.3390/plants8020038 . . . . **202**

Nathan P. Grant, Amita Mohan, Devinder Sandhu and Kulvinder S. Gill
Inheritance and Genetic Mapping of the Reduced Height (*Rht18*) Gene in Wheat
Reprinted from: *Plants* **2018**, *7*, 58, doi:10.3390/plants7030058 . . . . **220**

Natalia V. Tsvetkova, Natalia D. Tikhenko, Bernd Hackauf and Anatoly V. Voylokov
Two Rye Genes Responsible for Abnormal Development of Wheat–Rye Hybrids Are Linked in the Vicinity of an Evolutionary Translocation on Chromosome 6R
Reprinted from: *Plants* **2018**, *7*, 55, doi:10.3390/plants7030055 . . . . **229**

About the Special Issue Editor

Giovanna Frugis is a Senior Scientist at the Institute of Agricultural Biology and Biotechnology of the National Research Council of Italy (CNR), where she has been since 2001 after her postdoc in the Prof. Nam-Hai Chua Laboratory of Plant Molecular Biology at the Rockefeller University, New York (US). Her main interests span from the identification of genes that control morphogenesis and the differentiation of higher plants to their possible use for crop genetic improvement. Much of her work has been on translating research from plant models to crops, and vice-versa, from cultivated species to model plants.

Editorial

Plant Development and Organogenesis: From Basic Principles to Applied Research

Giovanna Frugis

Istituto di Biologia e Biotecnologia Agraria (IBBA), Unit of Rome, Consiglio Nazionale delle Ricerche (CNR), Via Salaria Km. 29,300, Monterotondo Scalo, 00015 Roma, Italy; giovanna.frugis@cnr.it

Received: 1 August 2019; Accepted: 20 August 2019; Published: 24 August 2019

Abstract: The way plants grow and develop organs significantly impacts the overall performance and yield of crop plants. The basic knowledge now available in plant development has the potential to help breeders in generating plants with defined architectural features to improve productivity. Plant translational research effort has steadily increased over the last decade, due to the huge increase in the availability of crop genomic resources and *Arabidopsis*-based sequence annotation systems. However, a consistent gap between fundamental and applied science has yet to be filled. One critical point is often the unreadiness of developmental biologists on one side, to foresee agricultural applications for their discoveries, and of the breeders on the other, to exploit gene function studies to apply candidate gene approaches when advantageous. In this Special Issue, developmental biologists and breeders make a special effort to reconcile research on basic principles of plant development and organogenesis with its applications to crop production and genetic improvement. Fundamental and applied science contributions interwine and chase each other, giving the reader different but complementary perpectives from only apparently distant corners of the same world.

Keywords: plant development and organogenesis; translational research; crop productivity; genetic improvement; *Arabidopsis thaliana*; regulatory networks; phytohormones; *rol* genes; plant cell and tissue culture

---

I am very pleased to introduce this Special Issue, which aims at reconciling research on basic principles of plant development and organogenesis with its applications to crop production and genetic improvement. This issue is published in honor of Domenico Mariotti, who significantly contributed to building up the Italian research community in Agricultural Genetics and Biotechnology and carried out the first experiments of *Agrobacterium*-mediated plant genetic transformation and regeneration in Italy during the 1980s. Domenico never believed in a clear distinction between fundamental and applied science; this is shown by his many scientific contributions to the field of cellular and molecular biotechnology in plants of agricultural interest spanning from basic to applied research. The review from De Paolis et al. [1] is dedicated to him, and summarizes the recent advances obtained in plant biotechnology and fundamental research following Mariotti scientific interests as guiding principles. Most of these themes recur throughout the Special Issue, where specific papers deepen into basic principles of developmental transitions and organogenesis, giving them a perspective in applied research and crop genetic improvement.

When we called for this Special Issue we were not prepared to such a prompt and enthusiastic response from the many friends/colleagues working in basic or applied research. We received many excellent manuscripts that made a major effort in forecasting translational solutions to improve crop production while addressing and reviewing fundamental knowledge of key plant developmental processes in model species [2–6]. Important contributions also came from researchers working on crop species [7,8] and plant breeding companies [6,9] that decided to openly share their strategies with the scientific community.

1. Key Questions in Root Developmental Biology and Target Genes for Root Crop Design

Di Mambro et al. [10] addressed the central question in developmental biology on how the body plan is established and maintained in multicellular organisms using *Arabidopsis* root as a simple model to study the molecular mechanisms of proximodistal and radial axes formation. The review describes all the main pathways and genes involved in establishing the two axes of growth in *Arabidopsis*, highlighting the involvement of some common players in controlling both axes and calling for more research in crop species in which root development shows higher levels of complexity [10]. Radial axis patterning is established by a finely regulated mechanism that controls the biosynthesis and activity of the phytohormone cytokinin, which in turn regulates auxin distribution and signaling. In another recent article, Di Mambro et al. have shown that cytokinin/auxin (CK/AUX) crosstalk is also involved in the regulation of root meristem size [11]. Cytokinins shape an auxin gradient by promoting the expression of *GH3.17*, which encodes an auxin-conjugating enzyme, in the most external layer of the root to position an auxin minimum in the last meristematic cells of the root to trigger cell differentiation [11]. In this Special Issue, Pierdonati et al. [12] from the same research group demonstrated that two additional *GH3* genes are expressed in the root, and also contribute to cytokinin-dependent positioning of the auxin minimum for root meristem size regulation. Fraudentali et al. showed how the CK/AUX-driven basic developmental frame can be taken over by reactive oxygen species (ROS) and other hormones signaling under stress conditions in the *Arabidopsis* plant model [13]. Leaf wounding triggers leaf to root long-distance communication resulting in early root xylem differentiation independent from root growth or meristem size. Root architecture and phenotypic plasticity influence crop productivity by affecting water and nutrient uptake, especially under environmental stress. These studies pave the way to unravel how long-distance communication may mediate phenotypic plasticity to adapt to changing environmental and stress conditions through the modification of the basic pathways of development [13].

The basic principles of root vascular development, provascular tissue formation and xylem differentiation, are described in the article from Hellmann et al. [4] where the key genetic pathways of primary and secondary development of *Arabidopsis thaliana* root are extensively reviewed, together with vascular development in shoot and hypocotyls. In this work the authors also focus on how this knowledge can and has been applied to agronomically important plants for production of wood and edible tubers as storage organs, providing important strategies and ideas to improve cambial activity in these processes [4].

The many regulatory candidate genes and pathways that are currently available in the *Arabidopsis* model are ready to be tested in crop biology and represent a valuable tool to be explored in breeding programs for root architectural traits.

2. Highjacking Plant Developmental Plans: The Case of the *Agrobacterium Rhizogenes Rol* Genes

In the review from De Paolis et al. [1] two sections are dedicated to the "hairy root" syndrome induced by *Agrobacterium rhizogenes*, characterized by the emergence of adventitious roots at the wound site of infected plants, and application of *A. rhizogenes* rooting locus (*rol*) genes to fruit tree propagation and transformation. How these *rol* genes act to highjack somatic plant cells to induce root meristem initiation and maintain indeterminate adventitious root growth is still a fascinating "enigma" after more than 30 years since their identification. However, evidence exists that they may act through the modification of as-of-yet unknown enzymatic reactions in the metabolism/signaling of cytokinins, auxin, and gibberellins as well as in ROS signaling [1]. In light of the current deep knowledge on root meristem formation and maintenance in *Arabidopsis*, it would be interesting to study the effect of *rol* genes in this model system to eventually identify their candidate target genes and pathways and understand their mode of action.

Trovato et al. [14] present a brief historical survey on the *rol* genes focusing on *rolD*, the only well characterized *rol* gene encoding an ornithine cyclodeaminase, which converts ornithine into proline. This type of enzyme is not present in plants, which synthesize proline through a more complex two-step reaction. The review illustrates how converging studies on *rolD* and proline function allowed to assess proline involvement in different plant developmental processes such as root elongation, flowering time, embryo formation, and pollen fertility. These studies corroborate the idea that different *rol* genes may act by interfering with plant metabolic pathways by encoding enzymes that bypass or redirect basic biochemical pathways. Since proline also acts as redox buffer and ROS scavenger, different *rol* genes may share a common role in the homeostasis of reactive oxygen species that can act as signaling molecules to regulate cellular processes underlying development [14].

3. Know the Old SAM: The Shoot Apical Meristem as the Key Developmental Switch in the Roadmap to Crop Yield Optimization

Three fascinating reviews guide the readers into the shoot apical meristem (SAM) world, where cells have to decide whether to keep on staying indeterminate (stem cells) or start the cell differentiation journey leading to the formation of complex organs such as leaves, flowers, and fruits. Several developmental features of plants, such as overall plant architecture, leaf shape, and vasculature architecture, that are major agricultural traits, depend on the activity of the SAM. The optimization of such developmental traits thus has great potential to increase biomass and crop yield. The failure of organizing a proper SAM in the embryo was also suggested to be involved in the post-zygotic incompatibility of wheat–rye hybrids [8].

The review of Fletcher [2] clearly summarizes the molecular mechanisms involved in stem cell maintenance in shoot and floral meristems through the molecular negative feedback loop called the CLAVATA (CLV)–WUSCHEL (WUS) pathway (CLV–WUS), both in the *Arabidopsis* model plant and crop species such as tomato, rice, and maize, highlighting similarities and specificities. Fletcher also illustrates the several examples of increased yield traits due to CLV–WUS pathway modulation in crop domestication, and foresees the great opportunity of using genome editing to enhance yield traits in a wide variety of agricultural plant species by fine-tuning the highly conserved CLV–WUS system [2].

The review of Traas [15] focuses on the basic principles guiding lateral organ formation at the shoot apical meristem, particularly on how auxin-dependent pathways can modulate wall structure to set particular growth rates and growth directions. How the molecular activity is translated into changes in geometry for oriented growth of organs and tissues is still unknown. The author brings the readers at the intersection of transcriptional regulation, mechanical forces and complex feedbacks from the cytoskeleton and the cell wall on gene expression, critically discussing the many questions that remain open in the field [15].

Richardson and Hake [3] consider another fascinating aspect of organogenesis at the shoot apical meristem, the formation of boundaries between pluripotent meristematic cells and differentiating organs. Their review critically summarizes the current understanding of boundary specification during vegetative development in grass crops in comparison with eudicot models. Gene regulatory networks (GRNs) underlying meristem/organ boundaries, as well as genetic modules that have been co-opted to specify within-organ boundaries to generate morphological diversity, are deeply analyzed in both eudicots and grass crops [3]. These GRNs are driven by different classes of transcription factors, the most important of which are NAC domain (NAM/ATAF/CUC), LBD (lateral organ boundaries domain), and KNOTTED1-like homeobox (KNOX) transcription factors (TFs). A specific section in De Paolis et al. [1] is also dedicated to KNOX TFs. Since boundary specification have a profound effect on leaf shape and plant productivity, GRN-based strategies to exploit this knowledge for crop genetic improvement are suggested. Also, the authors highlight the importance of translational research to develop accurate computational models of crop growth and development to help predict the effects of a changing climate on crop productivity [3].

4. Heading to the Sun: Vascular Growth and Developmental Changes in Shoot Architecture

Vascular development underlies every organogenesis and morphogenesis process to ensure resource delivery and mechanical support to any tissue and organ. Hellmann et al. [4] provide a comprehensive overview of the research on *Arabidopsis thaliana* vascular development and then focus on how this knowledge has been applied and expanded in research on the wood of trees and storage organs of crop plants. Basic principles of vascular development in roots, hypocotyl, leaves, and stems are reviewed, and gene regulatory networks involved are dissected and compared amongst models, woody species and Brassica crops, providing important hints on how to modulate cambial activity to improve productivity [4].

Translational biology from *Arabidopsis* to Brassica species is also the subject of the review from Leijten et al. [6] where the genetic networks involved in flowering time regulation in *Arabidopsis* are compared with related crop species in the Brassicaceae and with more distant vegetable crops within the Asteraceae family. Flowering time diversity has adaptive value in natural populations and plays a major role in agricultural production. In particular, it represents a crucial breeding trait for yield and nutritive quality of vegetable crops. This review is a collaboration among two public Institutions (the Italian CNR and the University of Amsterdam) involved in basic research, with the Research and Development group of Enza Zaden, an international vegetable-breeding company which develops new vegetable varieties that are grown and consumed all over the world. As a result, fundamental and applied science views on flowering time regulation intertwine, providing a comprehensive overview of basic genetic principles, available alleles and quantitative trait loci (QTL) and new perspectives for breeding strategies [6]. An overview of the molecular mechanisms of the shoot transition from juvenility to adult phases and flowering in fruit tree species can be found in the last section of De Paolis et al. [1].

A useful allele that can be used for wheat breeding programs to develop semi-dwarf cultivars is described in an article by Grant et al. [7]. The introduction of semi-dwarf varieties, that are more responsive to changing agriculture practices, was important during the green revolution in the mid-twentieth century to increase cereal production. Grant et al. report the inheritance and genetic mapping of the *Reduced Height 18 (Rht18)* gene in wheat and the selection of a semi-dwarf line with superior agronomic characteristics that could be utilized in breeding programs [7].

The genetic pathways that plants activate to sense and react to the presence of neighboring plants in the shade avoidance response is reviewed in Sessa et al. [5]. The authors critically summarize the current knowledge on the multiple pathways and regulators involved in this adaptive process, that can result in phenotypes with a high relative fitness in individual plants growing within dense vegetation. Recent advances in the molecular description of the shade avoidance response in crops, such as maize and tomato, and their similarities and differences with *Arabidopsis*, are discussed together with strategies to attenuate shade avoidance at defined developmental stages and/or in specific organs in high-density crop plantings [5].

5. Plant Cell Culture: Powerful Tools for Biotechnology

Most crops are recalcitrant to genetic transformation and/or regeneration; this represents a bottleneck in applying genome editing (GE) technologies to enhance crop productivity. In their review, Gordon-Kamm et al. [9] from the Agriculture Division of DowDuPonts (Corteva Agriscience company, Dupont Pioneer) provide an overview on how ectopic overexpression of genes involved in morphogenesis could and have been used to improve transformation efficiencies of recalcitrant crops. These genes are mainly regulators of embryo and meristem formation, or involved in hormonal pathways, and are discussed by the authors based on their practical or potential benefit when used for transformation. Due to their important function in plant growth and development, constitutive or strong expression of these genes often cause undesired pleiotropic effects. Gordon-Kamm et al. share with the readers the many possible strategies to limit/overcome pleiotropic deleterious problems, providing examples from the literature and from their own in-house experience in cereal crops [9]. These strategies might be applied to most recalcitrant crop species, including crop legume species that are mainly recalcitrant to in vitro culture and for which high throughput genetic transformation systems are yet to be developed. This is highlighted in the section dedicated to the genetic transformation of legumes in De Paolis et al. [1], where the power of in vitro plant cell and tissue cultures for applied biotechnology is also reviewed in the first section.

6. Conclusions

The knowledge acquired so far on the genetic basis of plant development, and its great potential in crop science and breeding to improve the yield and quality of agricultural products, are summarized in this Special Issue. Several target genes and pathways for root and shoot design are available for application in precision breeding to improve performance and productivity of crops, and more will come in the near future with the increase of translational research in plants. The readers will find several hints, molecular tools, and strategies to translate plant development basic research into crop productivity traits.

Funding: This research received no external funding.

Acknowledgments: I would like to deeply thank all the colleagues that contributed to this Special Issue, giving their perspectives and ideas to increase translation of fundamental knowledge from model plants to crops. Among those, a special thought goes to Ida Ruberti, who recently left us. She enthusiastically accepted to contribute with her recognized expertise on the molecular mechanisms of shade avoidance, and will be always remembered for the many contributions to the field of plant science and for her profound devotion to science.

Conflicts of Interest: The author declares no conflict of interest.

References

1. De Paolis, A.; Frugis, G.; Giannino, D.; Iannelli, M.A.; Mele, G.; Rugini, E.; Silvestri, C.; Sparvoli, F.; Testone, G.; Mauro, M.L.; et al. Plant Cellular and Molecular Biotechnology: Following Mariotti's Steps. *Plants* **2019**, *8*, 18. [CrossRef] [PubMed]
2. Fletcher, J.C. The CLV-WUS Stem Cell Signaling Pathway: A Roadmap to Crop Yield Optimization. *Plants* **2018**, *7*, 87. [CrossRef] [PubMed]
3. Richardson, A.E.; Hake, S. Drawing a Line: Grasses and Boundaries. *Plants* **2019**, *8*, 4. [CrossRef] [PubMed]
4. Hellmann, E.; Ko, D.; Ruonala, R.; Helariutta, Y. Plant Vascular Tissues-Connecting Tissue Comes in All Shapes. *Plants* **2018**, *7*, 109. [CrossRef] [PubMed]
5. Sessa, G.; Carabelli, M.; Possenti, M.; Morelli, G.; Ruberti, I. Multiple Pathways in the Control of the Shade Avoidance Response. *Plants* **2018**, *7*, 102. [CrossRef] [PubMed]
6. Leijten, W.; Koes, R.; Roobeek, I.; Frugis, G. Translating Flowering Time From *Arabidopsis thaliana* to Brassicaceae and Asteraceae Crop Species. *Plants* **2018**, *7*, 111. [CrossRef] [PubMed]
7. Grant, N.P.; Mohan, A.; Sandhu, D.; Gill, K.S. Inheritance and Genetic Mapping of the *Reduced Height (Rht18)* Gene in Wheat. *Plants* **2018**, *7*, 58. [CrossRef] [PubMed]
8. Tsvetkova, N.V.; Tikhenko, N.D.; Hackauf, B.; Voylokov, A.V. Two Rye Genes Responsible for Abnormal Development of Wheat-Rye Hybrids Are Linked in the Vicinity of an Evolutionary Translocation on Chromosome 6R. *Plants* **2018**, *7*, 55. [CrossRef] [PubMed]
9. Gordon-Kamm, B.; Sardesai, N.; Arling, M.; Lowe, K.; Hoerster, G.; Betts, S.; Jones, A.T. Using Morphogenic Genes to Improve Recovery and Regeneration of Transgenic Plants. *Plants* **2019**, *8*, 38. [CrossRef] [PubMed]
10. Di Mambro, R.; Sabatini, S.; Dello Ioio, R. Patterning the Axes: A Lesson from the Root. *Plants* **2019**, *8*, 8. [CrossRef] [PubMed]
11. Di Mambro, R.; Svolacchia, N.; Dello Ioio, R.; Pierdonati, E.; Salvi, E.; Pedrazzini, E.; Vitale, A.; Perilli, S.; Sozzani, R.; Benfey, P.N.; et al. The Lateral Root Cap Acts as an Auxin Sink that Controls Meristem Size. *Curr. Biol.* **2019**, *29*, 1199–1205. [CrossRef] [PubMed]
12. Pierdonati, E.; Unterholzner, S.J.; Salvi, E.; Svolacchia, N.; Bertolotti, G.; Dello Ioio, R.; Sabatini, S.; Di Mambro, R. Cytokinin-Dependent Control of GH3 Group II Family Genes in the Arabidopsis Root. *Plants* **2019**, *8*, 94. [CrossRef] [PubMed]
13. Fraudentali, I.; Rodrigues-Pousada, R.A.; Volpini, A.; Tavladoraki, P.; Angelini, R.; Cona, A. Stress-Triggered Long-Distance Communication Leads to Phenotypic Plasticity: The Case of the Early Root Protoxylem Maturation Induced by Leaf Wounding in Arabidopsis. *Plants* **2018**, *7*, 107. [CrossRef] [PubMed]

14. Trovato, M.; Mattioli, R.; Costantino, P. From *A. rhizogenes RolD* to Plant *P5CS*: Exploiting Proline to Control Plant Development. *Plants* **2018**, *7*, 108. [CrossRef]
15. Traas, J. Organogenesis at the Shoot Apical Meristem. *Plants* **2019**, *8*, 6. [CrossRef] [PubMed]

© 2019 by the author. Licensee MDPI, Basel, Switzerland. This article is an open access article distributed under the terms and conditions of the Creative Commons Attribution (CC BY) license (http://creativecommons.org/licenses/by/4.0/).

Review

Plant Cellular and Molecular Biotechnology: Following Mariotti's Steps

Angelo De Paolis [1,†], Giovanna Frugis [2,†], Donato Giannino [2,†], Maria Adelaide Iannelli [2,†], Giovanni Mele [2,†], Eddo Rugini [3,†], Cristian Silvestri [3,†], Francesca Sparvoli [4,†], Giulio Testone [2,†], Maria Luisa Mauro [5], Chiara Nicolodi [2] and Sofia Caretto [1,*]

1. Istituto di Scienze delle Produzioni Alimentari (ISPA), Consiglio Nazionale delle Ricerche (CNR), Via Monteroni, 73100 Lecce, Italy; angelo.depaolis@ispa.cnr.it
2. Istituto di Biologia e Biotecnologia Agraria (IBBA), UOS Roma, Consiglio Nazionale delle Ricerche (CNR), Via Salaria Km. 29,300, Monterotondo Scalo, 00015 Roma, Italy; giovanna.frugis@cnr.it (G.F.); donato.giannino@cnr.it (D.G.); mariaadelaide.iannelli@cnr.it (M.A.I.); giovanni.mele@cnr.it (G.M.); giulio.testone@cnr.it (G.T.); chiara.nicolodi@cnr.it (C.N.)
3. Dipartimento di Scienze Agrarie e Forestali (DAFNE), Università degli Studi della Tuscia, Via San Camillo De Lellis S.N.C., 01100 Viterbo, Italy; rugini@unitus.it (E.R.); silvestri.c@unitus.it (C.S.)
4. Istituto di Biologia e Biotecnologia Agraria (IBBA), Consiglio Nazionale delle Ricerche (CNR), Via Bassini 15, 20133 Milano, Italy; sparvoli@ibba.cnr.it
5. Dipartimento di Biologia e Biotecnologie, Sapienza Università di Roma, P.le A. Moro 5, 00185 Roma, Italy; marialuisa.mauro@uniroma1.it
* Correspondence: sofia.caretto@ispa.cnr.it; Tel.: +39-0832-422605
† These authors equally contributed to this work.

Received: 31 October 2018; Accepted: 7 January 2019; Published: 10 January 2019

Abstract: This review is dedicated to the memory of Prof. Domenico Mariotti, who significantly contributed to establishing the Italian research community in Agricultural Genetics and carried out the first experiments of *Agrobacterium*-mediated plant genetic transformation and regeneration in Italy during the 1980s. Following his scientific interests as guiding principles, this review summarizes the recent advances obtained in plant biotechnology and fundamental research aiming to: (i) Exploit in vitro plant cell and tissue cultures to induce genetic variability and to produce useful metabolites; (ii) gain new insights into the biochemical function of *Agrobacterium rhizogenes rol* genes and their application to metabolite production, fruit tree transformation, and reverse genetics; (iii) improve genetic transformation in legume species, most of them recalcitrant to regeneration; (iv) untangle the potential of KNOTTED1-like homeobox (KNOX) transcription factors in plant morphogenesis as key regulators of hormonal homeostasis; and (v) elucidate the molecular mechanisms of the transition from juvenility to the adult phase in *Prunus* tree species.

Keywords: Plant in vitro cultures; somatic cell selection; hairy roots; *rol* genes; *Agrobacterium rhizogenes*; genetic transformation; recalcitrant species; KNOX transcription factors; plant development; tree phase change

1. Introduction

In the 1990's, plant biotechnology experienced a remarkable development, exerting a significant impact on genetics for crop improvement in agricultural sciences. The scientific interests of Domenico Mariotti were very much influenced by this trend, focusing on in vitro plant cell and tissue cultures of important crop species, as valuable starting tools for genetic improvement, by selecting or inducing plant genome changes. This promising scientific approach let him foresee significant achievements for applied research, as well as the possibility to add relevant new knowledge to the molecular mechanisms of plant cell development. This review, dedicated to his memory, reports on the research progress accomplished in the last 10 years, following the scientific lines drawn by his many contributions to the field of cellular and molecular biotechnology in plants of agricultural interest. His biotechnological approach will be highlighted, starting from the induction of new in vitro variability and identification of useful genetic traits for applied research (Figure 1). The study of "hairy root" syndrome induced by *Agrobacterium rhizogenes* will then be considered, in terms of new insights in the function of *rol* genes and their biotechnological application for plant genetic transformation. A specific focus regards the progress in the genetic transformation of tree species and recalcitrant legume species. As for plant development, the last two paragraphs focus on the advances on KNOX transcription factors as key regulators of hormonal homeostasis in morphogenesis, and on the study of the transition from juvenility to the adult phase in fruit trees of the *Prunus* species.

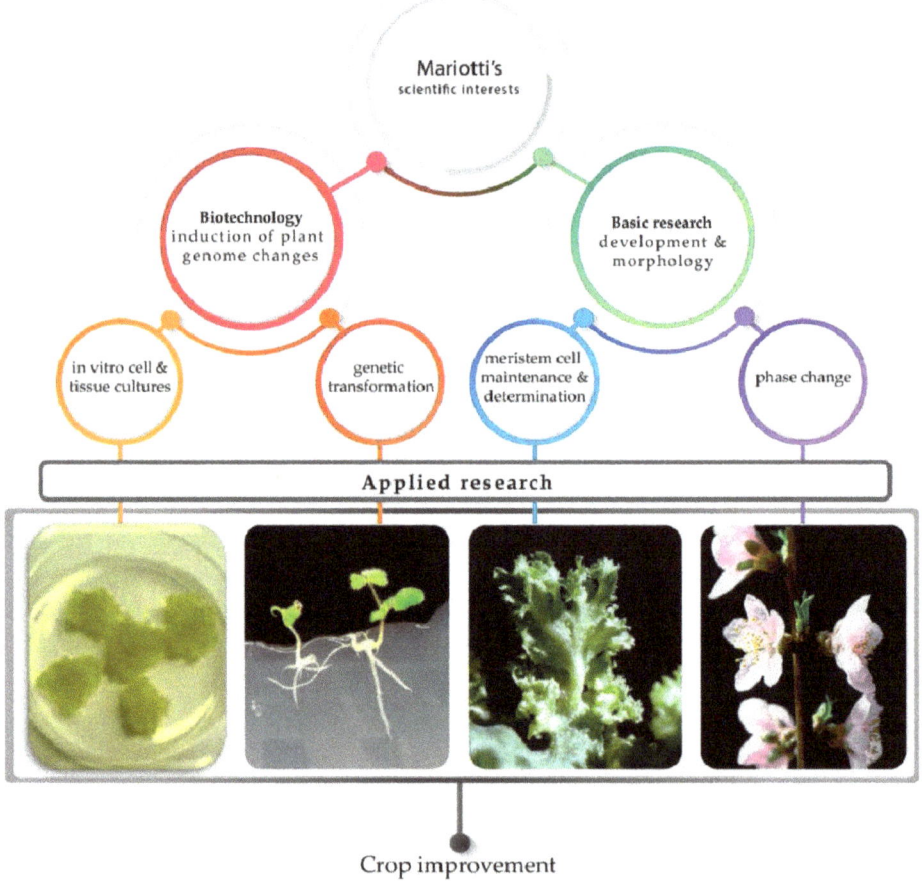

Figure 1. Outline of the main fields explored in this review following Mariotti's scientific interests. His research spanned from basic research to applied biotechnology, foreseeing the great potential of in vitro cell and tissue culture for plant transformation and crop genetic improvement. All photographs in the figure have been taken by the authors of the paper.

2. In Vitro Plant Cell and Tissue Cultures for Applied Biotechnology

In the last decades, based on the totipotency of most plant cells, many achievements have been accomplished by exploiting plant cell and tissue cultures of either model or crop species. One great potential for plant biotechnology is due to the genetic variability detectable in plant in vitro tissues, known as 'somaclonal variation' [1]. The exposure of plant cells to stressful in vitro conditions can enhance natural variability, which can be exploited for identifying novel useful variants. A proper selection strategy can help in identifying specific traits. To this regard, Mariotti's group contributed to gain insight into herbicide resistance in crop species achieved by somatic cell selection, being one of the successful applications of plant biotechnology as an alternative to gene transfer. On the other hand, the use of transgenic plants has encountered several regulatory restrictions in many countries. A stepwise selection, by applying increasing concentrations of herbicide, led to the identification of carrot cell lines as resistant to the sulfonylurea herbicide, chlorsulfuron (CS). Such resistance was due to gene amplification of the target enzyme, acetohydroxyacid synthase (AHAS) [2]. Alternatively, one-step selection, by applying a single toxic concentration of the herbicide, led to the isolation of mutant forms of the AHAS enzyme in resistant tobacco and sugarbeet cells [3–5]. In several cases, the resistance was maintained in the plants regenerated from the resistant cell lines [6]. Since then, herbicide resistance in crops for better weed management has been widely accomplished by genetically modified plants. In particular, in the United States, glyphosate resistant crop species have been largely developed and cultivated [7]. Nevertheless, somatic cell selection has continued to be applied for crop improvement. Very recently, two variants of potato cell cultures and regenerated plants resistant to CS were identified by somatic cell selection and the resistance in both cases was due to mutant AHAS genes, confirming the effectiveness of crop cell selection for this purpose. Moreover, the identified mutant genes can be useful as selectable marker genes in potato transformation [8].

The potential of in vitro variability of plant cell cultures can be of wide interest in many fields of applied research. Recently, plant cell cultures have been investigated as sources of metabolites, which can be used as food additives, pharmaceuticals, cosmetic ingredients, and as an alternative to the extraction of metabolites from field grown plants. To obtain an efficient plant cell culture process for metabolite production, it is necessary to establish cell lines by optimizing growth rate/product yields and enhancing the desired products using elicitors, precursors, or abiotic stress (Figure 2). Plant metabolite production by cell cultures can offer the advantage of a continuous supply, independent of environmental and seasonal changes, and using small spaces; moreover, it often ensures the obtainment of natural compounds that can hardly be produced in the same quality or specificity by chemical synthesis [9].

Vitamin E from plant sources comprises two groups of important antioxidant molecules, tocopherols and tocotrienols, that are differently distributed in the plant tissues [10]. The major natural vitamin E form is α-tocopherol, which can be extracted from the tissues of several food plant species [11]. Synthetic α-tocopherol, being a racemic mixture of eight different stereoisomers, is always less effective than the natural form, (R,R,R) α-tocopherol. For this reason, it is important to obtain vitamin E from natural sources, such as in vitro cell and tissue cultures [11]. Cell cultures of two oil plants, safflower and sunflower, were successfully established, producing the natural α-form as the main tocopherol [12,13]. Moreover, the sunflower in the in vitro production system confirmed that a certain degree of variability, often characterizing plant cell cultures, could be useful to identify highly productive cell lines. Two sunflower cell lines were identified and characterized for producing different amounts of α-tocopherol in cell suspension cultures' screening. In spite of the different content of α-tocopherol (almost threefold higher in the high producing cell line, HT, than in the low producing one, LT), these cell lines had very similar growth curves. It is interesting to note that HT cells also produced higher levels of vitamin C and glutathione. On the other hand, LT cells had higher activities of antioxidant enzymes, such as ascorbate peroxidase and catalase, compared to HT [14]. Recently, suspension cell cultures of mung bean were shown to be valuable for an in vitro system for producing both antioxidant tocopherols and phytosterols [15].

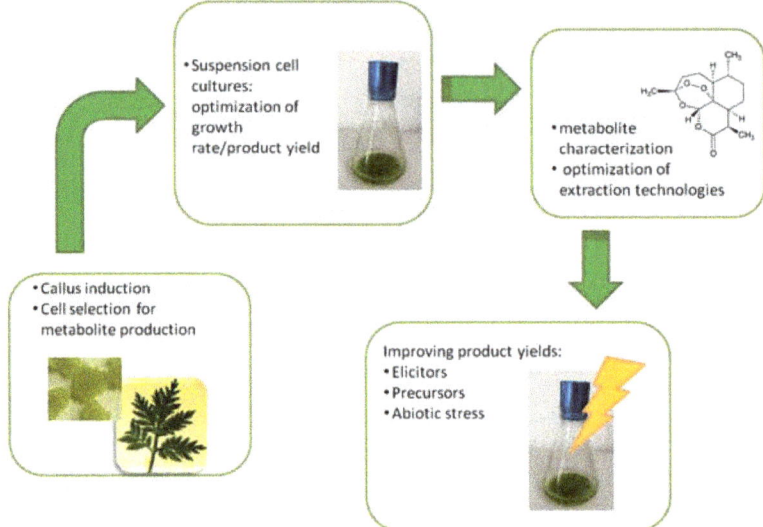

Figure 2. Schematic framework for the production of bioactive compounds by plant cell cultures.

Besides antioxidants, many phytochemicals belonging to the class of secondary metabolites are known to exert biological activities, which can be beneficial for human health and are of pharmaceutical interest. Human demand for these compounds has been growing along with the preference for natural products. Plant cell cultures for the production of these bioactive compounds can have significant advantages as supply sources, mainly when the desired compounds occur in very small amounts and/or are accumulated in specific tissues of the plant [16]. The apocarotenoid crocin is a main component of the yellow spice, saffron, known as a precious food ingredient with valuable pharmaceutical properties and found only in the stigma of *Crocus sativus* L. flowers [17,18]. Efforts have been made to establish crocin in in vitro production systems as an alternative to production from saffron plants, which is expensive and time-consuming. Although the induction of saffron callus cultures from stigma is very difficult to achieve, callus cultures induced from style explants were established and revealed to be more efficient in terms of the growth rate and crocin production compared to corm-derived calli, when the plant growth regulator, thidiazuron, was used [19].

As for pharmaceuticals, a successful example of efficient in vitro systems is represented by the anticancer drug, taxol, produced by cell suspension cultures of *Taxus* spp. The drug is intensively used for the treatment of different types of cancer and the cell culture technology avoids sacrificing yew trees. Such an in vitro production process has been extensively investigated and this has led to significant yield improvements. The availability of plant cell suspension cultures acting as "bio-factories" of specific compounds offers the possibility of scaling up to large volumes for industrial production. This is the case of *Taxus* cell cultures, nowadays used for industrial-scale biotechnological production to the commercialization of the anticancer drug, paclitaxel (taxol) [20].

Another plant metabolite of pharmaceutical interest is the sesquiterpene, artemisinin. It is an antimalarial compound, produced at low levels by the aerial parts, leaves, and inflorescences, of the plant, *Artemisia annua* L., an annual herb native to Asia. Due to its efficacy, it is strongly recommended by the World Health Organization as the first choice in therapeutic protocols against malaria, but unfortunately the concentration in field grown plants is quite low, being 0.1–1% dry weight, thus its worldwide supply is insufficient. Although many efforts have been made to obtain new *A. annua* genotypes characterized by enhanced yields through breeding strategies, a certain degree of variability in field grown plants was also observed [21,22]. Metabolic engineering was applied using transgenic plants of both *Artemisia* and tobacco; however, the obtained content increases of artemisinin or its precursors were not sufficient to overcome the drug shortage [23,24]. In addition, an engineered microbial system was established, however, it led to the production of the precursor, artemisinic acid, to be chemically converted to artemisinin [25]. Due to the complexity of the artemisinin molecule, chemical synthesis requires a laborious and costly process. Furthermore, it was reported that pure artemisinin was less effective than intact dried leaves in treating malaria [26], thus there is the need to explore other supply sources, such as in vitro cell culture technologies. *A. annua* in vitro cell cultures were established by optimizing the use of plant growth regulators and culture conditions. Different strategies were applied to improve artemisinin production, such as the elicitation by methyl jasmonate, which was successful for improving yields in both suspension cell cultures and hairy root cultures of *A. annua* [27,28]. The availability of suspension cell cultures has the advantage of scaling up for possible industrial production. Interestingly, *A. annua* suspension cell cultures were characterized by the ability to exudate artemisinin into the culture medium, making it easier to recover the desired native product [27]. Recently, cyclic oligosaccharides have been used in different cell culture systems for enhancing metabolite production. Resveratrol from grape cell cultures was reported to be increased by the application of β-cyclodextrins (β-CD), which acted as true elicitors [29]. Moreover, artemisinin production was significantly improved by applying different types of CD to *A. annua* cell cultures. In particular, dimethylated β-CD induced a 300-fold increase of artemisinin, most likely by reducing the negative feedback as a consequence of artemisinin-CD complex formation [30].

3. The "Hairy Root" Syndrome Induced by *Agrobacterium rhizogenes*

The "hairy root" syndrome, characterized by the emergence of adventitious roots at the wound site of infected plants, was first described in the 1930s–1960s as an indicator of pathogen attack in horticultural plants. The responsible bacterial agent, *Agrobacterium rhizogenes*, was identified and the role of gene transfer from the resident bacterial plasmid to the plant genome was revealed [31]. *A. rhizogenes*, as the related *Agrobacterium tumefaciens* species, are well known for the capacity to transfer part of their DNA (Ri, root-inducing; Ti, tumor-inducing) to the plant genome during a natural infection process, leading to abnormal roots (hairy roots) or tumors (crown galls), respectively [32,33]. The expression of transfer DNA (T-DNA) causes abnormal growth and leads to the production of characteristic amino acid and sugar derivatives (opines), which can be used by the bacteria for their own growth. Being natural plant genetic engineers, in the 1980s, *A. tumefaciens* started to be exploited in biotechnology for plant genetic transformation [34]. Modified Ti plasmids, which lacked T-DNA genes related to the syndrome (disarmed), though retaining the entire *vir* (virulence) region, were used for the introduction and integration of foreign DNA in the plant cells and subsequent regeneration of transgenic plants. *A. rhizogenes* raised additional interest as Ri T-DNA transformed roots could be regenerated into whole plants with a characteristic "hairy root" phenotype. Hairy root plants have reduced apical dominance, shortened internodes, wrinkled and wider leaves, adventitious root formation, altered flower morphology, and reduced content of pollen and seeds [35], indicating a role of the T-DNA genes in modulating various developmental processes. The major *A. rhizogenes* genes involved in the hairy root syndrome were identified in 1985 among the 18 open reading frames in the T-DNA [36], and named *rol* genes (*A*, *B*, *C*, and *D*) after "rooting locus" or oncogenes for their capacity to alter plant cell programs [37]. The laboratory of Domenico Mariotti contributed to the characterization of the *rol* genes' function [32,38–40], although most work was addressed to *rol* genes' applications to induce adventitious root formation in recalcitrant species for micropropagation, and to modify developmental traits in crops [41–46]. Studies from several independent laboratories have contributed to suggest biochemical functions for the different *rol* genes [47]. The phenotype of plants transformed with either *rolA*, *rolB*, or *rolC*, and biochemical in vitro assays suggested their involvement in phytohormone homeostasis, such as gibberellins, auxin, and cytokinin metabolism and/or signaling, respectively (Figure 3a). However, conflicting results were produced, from which no definitive conclusions can be drawn. Contradictory indications were also published on the involvement of *rol* genes in reactive oxygen species (ROS) homeostasis, heading to a possible function of *rolB* in either increasing or decreasing ROS signaling [48,49], and to *rolC* as an ROS suppressor [50]. Differently, rolD was shown to act as an ornithine cyclodeaminase, which converts ornithine into proline, thus inducing acceleration and stimulation of flowering in both plants and tissue cultures [51].

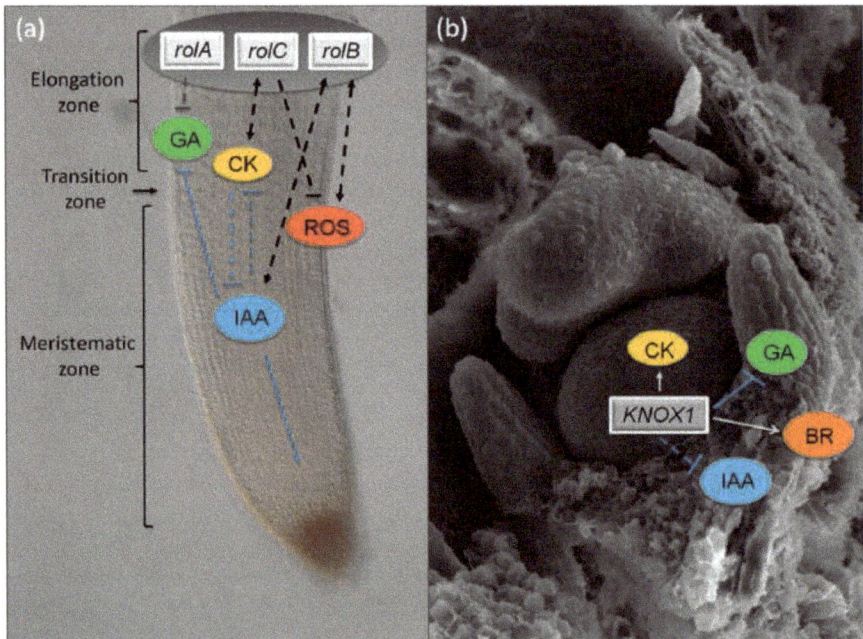

Figure 3. A simplified view of the involvement of *A. rhizogenes rol* genes and plant class 1 KNOX transcription factors in hormonal homeostasis in the root (left panel) or shoot (right panel) apical meristem. (**a**) *rolA, rolB*, and *rolC* may control hairy roots formation and their indefinite growth by hijacking some as-of-yet unknown components of the gibberellin (GA), auxin (IAA), and cytokinin (CK) metabolism, respectively; (**b**) class 1 KNOX control boundaries between undifferentiated cells and differentiating organs through the regulation of hormone metabolism and signaling. KNOX expression in the shoot apical meristem establishes a regime of high CK, low GA, and a gradient of auxin and brassinosteroids (BR) to keep the indeterminacy of the SAM and setting boundaries for proper organ separation during plant development.

Levesque et al. [52] coined the term "*plast*" genes, standing for "developmental plasticity", to describe those *Agrobacterium* genes able to change the development when introduced into wild-type plants. According to this study, "*plast*" genes encode a family of 11 proteins (from both *A. rhizogenes* and *A. tumefaciens*), with sequence similarity values ranging between 13% and 34%, which may share similar functions, and whose diversification could result from a process of coevolution between different *Agrobacterium* species/strains and plant species. This family of ca. 70 proteins includes rolB and rolC [53] and proteins from plant species (e.g., *Nicotiana*, *Linaria*, and *Ipomoea*) that contain T-DNA genes (cellular, *cT-DNAs*) from *A. rhizogenes* in their genomes [53]. This is a very interesting example of horizontal gene transfer, which likely occurred by sparse events of spontaneous regeneration of transformed plants from *A. rhizogenes*-induced hairy roots in the natural environment. Some *Agrobacterium*-derived *cT-DNA* genes, such as *rolC*, *orf13*, and *orf14*, or some involved in opine production, are frequently intact and expressed in natural transformants, potentially able to influence plant growth and the microbiome root environment. Indeed, overexpression studies in plants suggest that "*plast*" genes have growth-modifying properties similar to their *A. rhizogenes* equivalents [54,55].

It was hypothesized that the effect of T-DNA on the regenerative capacity and the interaction with microorganism communities might have affected the evolution of natural transformant plants [56]. However, loss-of-function studies of expressed *cT-DNA* genes should be performed to assess their possible adaptive roles in plants.

Although the biochemical features of *rol* genes remain poorly understood, they have been proven to be powerful tools in plant biotechnology and functional biology research. The peculiar features displayed by hairy roots, such as a high growth rate in hormone-free liquid media, unlimited branching, and biochemical and genetic stability, make them a promising tool for metabolic engineering and large-scale metabolite production [57]. Potential applications of *rolC* and *rolD* genes in floriculture have been suggested for their effects on plant architecture and flowering promotion, respectively. Also, *rol* genes were shown to activate secondary metabolism in transformed cells from the *Solanaceae*, *Araliaceae*, *Rubiaceae*, *Vitaceae*, and *Rosaceae* families, paving the way for their possible exploitation for secondary metabolite production [57–59]. As an example, more than a 100-fold increase in resveratrol production was also obtained in *Vitis amurensis* cells transformed with the *rolB* bacterial gene from *A. rhizogenes* [60]. Fruits of transgenic tomato plants that overexpress *rolB* exhibited higher nutritional quality and foliar tolerance to two fungal pathogens [61], improved photosynthetic processes, and a more effective protection against oxidative damage and excess energy [62]. As *rolB* is the major activator of the secondary metabolism, its mechanism of action was further investigated, revealing a possible rolB function in activating specific MYB transcription factors to accelerate secondary metabolite production [63].

Besides biotechnological uses, an interesting application of hairy roots in fundamental biology studies exploits the ability of *A. rhizogenes* to elicit adventitious roots to obtain the so-called "composite plants", which comprise a transgenic hairy root system attached to non-transformed shoots and leaves [64]. Initially used for micropropagation purposes, the obtainment of composite plants has become a powerful tool in gene function studies of root biology, especially those involving legume-*rhizobium* symbiosis [65]. The T-DNA harboring the transgene of interest in a disarmed binary vector is generally used to co-transform *A. rhizogenes* containing the complete Ri T-DNA, the latter allowing fast growth of transgenic roots. For these studies, relatively low virulence *A. rhizogenes* strains, such as Arqua-1 and K599, are used, which elicit a limited number of transformed roots, with growth and morphology comparable to normal roots. Transformation of *Medicago truncatula* with *A. rhizogenes* Arqua-1 allows the production of composite plants with transgenic roots that are suitable for studies of root-specific interactions because they can be nodulated by *Sinorhizobium meliloti*, efficiently colonised by endomycorrhizal fungi, and infected by pathogenic/parasitic organisms [65]. *A. rhizogenes*-transformed composite plants were achieved in different plant genera (i.e., tomato, potato, poplar) [66–68], including those species that are usually recalcitrant to *A. tumefaciens* transformation, providing alternative solutions in gene function studies.

Despite the huge effort made over the last three decades of research, the biochemical and cellular functions of *rol* genes, with the exception of *rolD*, remain elusive. Due to the coevolution process that occurred between *A. rhizogenes* and dicot species, *rol* genes have typical eukaryotic *cis*-regulatory motives in their promoters, but likely encode proteins of bacterial origins. Proteins encoded by *rol* genes do not display any clear sequence homology with known plant or bacterial proteins, but different and contrasting enzymatic properties have been attributed without further confirmation. Additional research to solve this "enigma" should consider that *rol* genes evolved to highjack somatic plant cells to induce root meristem initiation and maintain indeterminate adventitious root growth independently of the aerial part of the plant. Hence, the possible targets of *rol* genes should be searched amongst the main pathways involved in these root biology processes. In the past decade, most aspects of root patterning and function have been extensively explored, and the role of auxin, cytokinin, and gibberellin in root development were assessed [69], although several biochemical steps of hormone homeostasis are still unclear. Proteins encoded by *rolB* and *rolC* may be involved in as-of-yet unknown enzymatic reactions in the metabolism/signaling of these hormones in the root. This may occur either directly via already existing plant biochemical functions, or indirectly through interference with specific substrate availability, thus shifting the biochemical equilibrium. The root system of *Arabidopsis thaliana* has been established as a powerful tool to study genetic networks and signaling underlying root development [70]. It would be very interesting to study the effect of *rol* genes in the *Arabidopsis* system in light of the current knowledge on root meristem formation and maintenance. This would allow identification of candidate target genes and pathways regulated by *rol* genes at the cellular level. Moreover, the availability of complete genome information of both plants and agrobacteria, including Ri and Ti plasmids [71,72], and the possibility to run transcriptome analysis of plant-*Agrobacterium* interactions may help to integrate previous knowledge with novel molecular data to unravel *rol* genes' mechanism of action.

4. Application of *A. rhizogenes rol* Genes to Fruit Tree Transformation

In the early 1980s, the *Agrobacterium rhizogenes* wt was used in fruit trees to improve propagation of difficult-to-root varieties and rootstocks. At that time, gene transfer represented a pioneeristic work in woody plants because regeneration methods were poorly available or not developed yet, considering the usual recalcitrance of these species to in vitro manipulation, as well as molecular techniques. However, after many efforts and with many initial failures, the work was rewarded with many positive results, which consisted of chimeric or fully transformed plants; the former was achieved by bacterial direct inoculum through a wound at the base of the shoot, while the latter was produced by whole plant regeneration (shoot organogenesis or somatic embryogenesis) from "hairy roots". Later, transgenic whole plants were obtained for one or few *rol* genes of the *riT-DNA* plasmid of *Agrobacterium tumefaciens*. Several traits of fruit species were successfully modified by genes of *A. rhizogenes* and the major results are summarized in Table 1. The first woody plants modified with *A. rhizogenes* NCPPB pRi1855, using in vitro micro-shoots, were almond cv Tuono [73] and, later, olive cv Moraiolo [74,75]. Both species showed abundant rooting in auxin free medium or in very low auxin concentration, while in almond, the detached roots continued to grow in vitro even in hormone-free medium and to produce opines, and those of olive plants rarely expressed these abilities. The reason could be ascribed to transient gene expression or to the organogenesis of non-transformed cells, after stimuli from the adjacent transgenic ones or the bacterium diffusible exudates [76]. Olive plants showed less vigor than those rooted with auxin, similarly to plum MrS2/5, cherry F12/1, and cherry rootstocks Colt in field conditions [77]. Subsequently, the *A. rhizogenes* gene transfer technology to induce in vitro rooting spread throughout several fruit species (Table 1).

Table 1. Main results in woody fruit species obtained by the use of *riT-DNA* and *rol* genes of *Agrobacterium rhizogenes*.

Species	Gene(s)	Results	Ref.
Olive, Almond, Walnut, F12/I, MrS/5, Colt, apple	*riT-DNA*	Chimeric plants (better rooting)	[73,78–80]
Papaya (*Carica papaya*)	*riT-DNA*	Reduced growth habit	[81]
Colt rootstock (*P. avium* × *P. pseudocerasus*)	*riT-DNA*	Reduced growth habit	[79]
Kiwifruit (*Actinidia deliciosa*), cv Hayward	*rolB*	bigger fruits, drought tolerance	
Kiwifruit (*A. deliciosa*), cv Hayward and GTH	*rolABC*	reduced plant size, flower set, increased drought tolerance	[43,44,81]
Citrange Troyer (*Citru sinensis* × *P. trifoliata*)	*rolABC*	drought tolerance	[82]
Olive (*Olea europaea* L.) cv Canino	*rolABC*	Reduced growth habit, increased drought tolerance	[83–85]
Apple rootstock	*rolA*	Reduced growth habit	[86]
Apple rootstock	*rolB*	Reduced growth habit	[87]
Pear (*P. communis* L.)	*rolB*	Increased rooting ability	[88]
Strawberry (*Fragraria* × *ananassa*)	*rolC*	Higher fruit set and resistance to *Phytophtora cactorum*	[89]
Pear rootstock	*rolB*	Increased rooting ability	[88]
Richter 110 (*Vitis berlandieri* × *V. rupestris*)	*rolB*	better rooting	[80]

While many species are easily induced to in vitro rooting by *A. rhizogenes* wt, in vivo experiments proved difficult or impossible. Rinallo and Mariotti [45], after unsuccessful experiments with *A. rhizogenes* wt, obtained abundant rooting in chestnut cuttings using *A. tumefaciens* harboring the *rolB* gene, in combination with etiolation and auxin treatments. Later, it has been demonstrated that auxins and putrescine play an important co-adjuvant role in *A. rhizogenes*-mediated root induction [75]. Only cuttings from seedlings of *Asimina triloba* L. were responsive to *A. rhizogenes* treatment; therefore, juvenility should be considered a key factor for successful transformation [90]. According to Sutter and Luza [91], plant response to *A. rhizogenes* involves auxins through either hormone increased concentration or increased sensitivity of the infected cells, based on the analogies of the morphological response of plant tissues treated with auxins.

Whole transformed plants with *riT-DNA* were achieved following the regeneration from "hairy roots" in papaya [81], cherry rootstock Colt [92], and kiwifruit [93], which showed the typical hairy root syndrome. Plant regeneration of fully transgenic plants is feasible in vitro and in vivo (in the pot or in the field) from spontaneous regeneration of hairy roots, particularly in species (e.g., *Prunus* spp.) that show high efficiency of regeneration from roots [79]. However, the "hairy root" phenotype is exhibited not only by fully transformed plants, but also by chimeric plants (having only transformed roots). This phenomenon limits the use of *A. rhizogenes* wt to overcome the difficulties encountered in the rooting of hard-to-root species, since sole transgenic roots also modify the canopy morphology. Nonetheless, a large scale selection of *Prunus* spp. regenerated form hairy cultures was effective to produce *riT-DNA* dwarfing rootstocks that did not alter the fruit quality of grafted conventional sweet cherry scions [77]. These novel approaches have the advantage of shortening the time required for selection and escape the stringent regulations on genetically modified organisms, because no recombinant vector is used. The idea of producing *riT-DNA* transgenic plants with a high rooting ability of (mature) cuttings is still challenging as seen in *riT-DNA* Colt rootstocks, which showed rooting recalcitrance by hardwood and semi-hardwood cuttings, and also by layering in the field [77,79], while the explants easily rooted in vitro, even without auxin supply.

To avoid the strong "hairy root" phenotype, *rol* genes from the *riT-DNA* were cloned into *A. tumefaciens* to produce several transgenic fruit plants. Specifically, through induced shoot organogenesis from leaves, male *rolABC* "GTH" [44,74] and female "Hayward" kiwifruits were produced [43] together with many offsprings (*rolABC* "GTH" × "Hayward" control), and, subsequently, *rolABC* "Canino" olive tree, through cyclic somatic embryogenesis of maternal tissue [84,85], and 10 years of field trials were also conducted. Overall, the transgenic *rolABC* phenotype is characterized by pleiotropic effects; they include: Internode and shoot shortening; reduction of trunk, leaf lamina, and petioles; reduced number of total flowers and increased number of single flower per bud; delay of vegetative growth in autumn; increased rooting ability in vitro and in vivo; increased tolerance to drought and decreased transpiration rate; increase of putrescine levels; enhanced *Pseudomonas syringae* susceptibility [94]; and fruit shape alteration and dwarfing properties of rootstocks [44,95]. Several of these traits also occurred in other *rolABC* transgenic fruit trees, including cherry 'Inmil' (*P. incisa* × *serrala*) and Damil (*P. dawyckensis*) [96] and walnut hybrid [97], whereas in transgenic *Citrus* spp. plants, a higher photosynthetic efficiency, better development of root systems, and higher tolerance to oxidative stress were reported [98]. Furthermore, the soils underneath transgenic plants did not change in its composition of microbial populations [82]. The same behavior has been observed in other species, such as *rolABC* olive cv Canino, in field trials, where the plants showed a strong reduction of apical dominance with a short internode length, with a tendency to axillary buds' outgrowth and prolonged vegetative growth in late autumn with a high risk of frost damage in winter [83]. Regarding the single *rol* transformation, *rolB* female kiwifruit appeared morphologically similar to the controls, with a slight increase in fruit size and a normal shape; nevertheless, a reduction in the number of triple flowers per bud (the triple flowers is a negative phenomenon in the female cultivar, Hayward), a higher drought tolerance, and self-rooting were scored [77,99]. In apples, *rolB* induced the typical hairy root phenotype and transgenic rootstocks affected the internode length, canopy size, flowering, and fruiting of the conventional scion, whilst the fruit quality was preserved [86,87,100,101]. RT-PCR analysis revealed that neither the *rolB* gene nor its mRNA were detectable in the scion, indicating no translocation from the rootstock to scion. Similar results have been observed in the pear rootstock [88] and in grapes [80]. *RolC* gene insertion into kiwifruit (*A. deliciosa* A. Chev) generated yellow leaves, stunted growth, and reduction of fruit size and flower number, thus was unsuitable for commercial uses [99]. *RolC* plants have been produced also in *A. kolomikta* [102], in *Fragraria × ananassa*, cv Calipso, and raspberry [103]. In the latter species, the increase of cytokinins' metabolism was accompanied with increased yield and fruit downsizing, enhanced sugar content and tolerance to *Phytophthora cactorum* [103], boosted rooting ability, and precocious flowering [89]. *RolC* overexpression reduced the vigor in pear rootstocks [104] and in *Poncirus trifoliatae*, together with the internode shortening, enhanced rooting ability [105].

Overall, the whole *riT-DNA* of *A. rhizogenes* and *rol* genes, singly or in association, merit further investigation, since the results so far obtained suggest a favorable use for improving different fruit tree species, both varieties and rootstocks, to be used in modern agriculture, suitable for mechanization and for adverse soil and climate conditions. In addition, the use of wild type bacterium could also allow the stringent rules of genetically modified organism regulations to be overcome.

5. Genetic Transformation of Legumes

In her review on "Advances in development of transgenic pulse crops" published in 2008, Susan Eapen wrote: 'To date, genetic transformation has been reported in all the major pulse crops like *Vigna* species, *Cicer arietinum*, *Cajanus cajan*, *Phaseolus* spp., *Lupinus* spp., *Vicia* spp. and *Pisum sativum*, but transgenic pulse crops have not yet been commercially released. The reason for lack of commercialization of transgenic pulse crops can be attributed to the difficulty in developing transgenics with reproducibility, which in turn is due to lack of competent totipotent cells for transformation, long periods required for developing transgenics and lack of coordinated research efforts by the scientific community and long term funding' [106].

One of the main interests of Domenico Mariotti was the genetic transformation of crop plants, in particular grain legumes, mediated by *Agrobacterium*. These crops are recalcitrant to in vitro culture and this makes it more difficult to achieve genetic transformation. Mariotti was very clear that the key toward success was to be able to reach the meristematic areas and then stimulate organ regeneration, avoiding the callus phase. With this in mind, he contributed to establishing protocols for chickpea and common bean transformation [107,108].

Nowadays, 10 years later, things have not gone very far. Few transgenic legume crops have been approved and registered for commercialization, most of which have been produced in soybean [109], alfalfa [110], and only one is in a common bean, the EMBRAPA EMB-PVØ51-1 variety, resistant to Bean Golden Mosaic Virus [111]; however, only GM soybean and alfalfa are currently cultivated.

Compared to other crops, progress in legume transformation is still very poor. Besides technical problems, this may be due to the lower economical relevance of some of these crops compared to cereals, despite the increasing interest that is arising for legumes in the last years, and to the fact that most of them are mainly cultivated and consumed in developing countries of Asia (*Cajanus cajan*, *Cicer arietinum*, *Lens culinaris*, *Vigna radiate*, and *Vigna mungo*), Africa (*Vigna unguiculata*, *Phaseolus vulgaris*), and Central and South America (*Phaseolus vulgaris*) [112]. Furthermore, the strict regulations imposed by several European countries on GM crops cultivation have strongly limited the economic interest as well as the technical advancements in recalcitrant crops, such as legumes. Therefore, despite the importance of pulse legumes to both human and agroecosystem health, these crop species still lack a high throughput genetic transformation system. Main limiting technical factors regard the recalcitrance of pulses for regeneration, low competency of regenerating cells for transformation, and lack of a reproducible in planta transformation system [106,113]. *Agrobacterium tumefaciens*-mediated gene transfer is still the most commonly used procedure for legume transformation. Consistent attempts for high-frequency recovery of transgenic events with *Agrobacterium*-mediated transformation in major grain legumes have resulted in marginal success, despite optimization of several crucial parameters [114,115]. Some good results have been obtained with direct gene transfer using particle gun bombardment, a technique that is mostly genotype independent and that may overcome problems related to plant regeneration [116]. In fact, legume in vitro regeneration is still a challenge for plant researchers; however, the extensive use of the model legume plants, *Medicago truncatula* and *Lotus japonicas*, for molecular studies has favored the development of efficient regeneration and *Agrobacterium*-mediated transformation protocols for these two species [114].

Root transformation using *A. rhizogenes* has emerged as an alternative to traditional transformation and is gaining importance as an effective tool for reverse genetics studies in plants, especially legumes in which studies have focused on genes involved in root biology and root–microbe interactions [114,115]. For example, transgenic adventitious roots have been proven to be a good system to investigate the role of genes involved in symbiosis [116].

In vitro regeneration of legumes is based on direct organogenesis, indirect organogenesis, or somatic embryogenesis from different explant types. The determination of species-specific parameters, like the explant source, plant genotype, and media components, are key to gain successful regeneration. When possible, the somatic embryogenesis approach is favored, as each event of regeneration is supposed to be derived from one cell and chromosomal rearrangements are less frequent, however, this system may increase the frequency of unwanted traits arising from somaclonal variation. In many cases, the regeneration of shoots from the cotyledonary node or from other meristematic explants after *Agrobacterium* infection has been proven to be a rapid and relatively efficient method in a number of legume species [113]. Mariotti's group contributed to this field, proposing a method to obtain common bean plant regeneration from different genotypes, through meristematic organogenesis [117]. However, the pioneering work of Domenico Mariotti and co-workers started before, when in 1989, they published a first study reporting the development of transgenic common bean and runner bean (*P. vulgaris* and *P. coccineus*, respectively) plants based on a rapid and efficient plant regeneration system, which reduced the in vitro culture and avoided the callus phase [108]. The transformation method was based on *A. tumefaciens* infection of the primary node of young explants deprived of both apical meristem and the upper part of axillary buds. They obtained good percentages (15–20%) of shoot regenerations on the selective media for both species, and among these, about 60% were positive to GUS staining [114]. Unfortunately, in the paper, no data were presented on the stability through generation of the transformants, so it remains to be demonstrated that the efficacy of the method can produce stable transformed T1 and T2 plants. A few years later, Domenico Mariotti and his coworkers reported the first transformed chickpea plantlets obtained after co-cultivation of embryonic axis [107].

Subsequently, several reports were made of chickpea transformation using the embryonic axis or parts thereof. Indeed, frequent common features of legume crop transformation protocols include the use of cotyledonary nodes or embryonic axes as explants for genetic transformation, the use of grafting to overcome problems related to organogenesis, and the addition of thiols compounds to improve the transformation efficiency [114,118–121].

Although we are still far from efficient and high throughput transformation systems, for some legume crops (chickpea, cowpea, lupin, common bean, peanut), a number of successful transformation events have been reported in the last 10 years, underlying the development of robust transformation methods, although very often still poorly efficient and genotype dependent. Chickpea has been transformed for resistance against target pests, bruchids and aphids, as well as for traits conferring tolerance to drought and salinity [122]. In all these works, transgenic chickpea plants were always obtained by *Agrobacterium*-mediated methods, with only one exception, in which the method used was based on particle gun bombardment [123]. Some progress has been gained also with the transformation of *Vigna* species (*V. unguiculata*, *V. radiate*, and *V. mungo*) and transgenic plants have obtained resistance to biotic stresses, abiotic stresses, or herbicides [124–127]. Only cowpea lines tolerant to a herbicide from the imidazoline class (imazapyr) were obtained by means of particle gun bombardment [127]; in all other cases, transformation was achieved by the use of *Agrobacterium tumefaciens*. Improved protocols, based on the method set up by Pigeaire et al. [128], are also available for lupin species' (*Lupinus angustifolius*, *L. luteus*) transformation [129,130] and have been applied to develop plants that are resistant to fungal disease [131] or to improve the seed sulphur amino acid content [132]. Common bean, the only food legume crop for which a GM variety has been approved, was transformed by the use of the biolistic method [133,134]; however, a recent paper reported the possibility to transform this crop by *Agrobacterium*-mediated transformation using indirect organogenesis [135]. Successful genetic transformation protocols have been reported in the peanut both via *Agrobacterium tumefaciens* [136,137] and biolistic/particle bombardment [138]. Moreover, several papers report examples of genetic transformation of peanuts to improve traits related to abiotic and biotic stresses and for the production of oral vaccines [139]. Very few reports are available for other legume crops, such as the lentil [140] and faba bean [141].

In the last years, the emergence of genome-editing technologies has revolutionized plant research, and it is now possible to create specific and precise genetic modification as well as modulate the function of DNA sequences in their endogenous genomic context [142]. The power of this new technology has been accompanied with a burst of edited crops to speed up breeding. In the near future, we can expect that increasing efforts will be put into advancing knowledge and technical skills to improve genetic transformation of legumes and hopefully gaps with other crops will be reduced.

6. KNOX Transcription Factors as Key Regulators of Hormonal Homeostasis in Plant Morphogenesis

KNOTTED1-like homeobox (KNOX) transcription factors (TF) belong to the Three Amino acid Loop Extension (TALE) ancestral superclass of homeodomain transcription factors conserved in animals, plants, and fungi [143], and are subdivided into three phylogenetic classes (class 1, 2, and M) [144]. Functional studies of class 1 *KNOX* genes in the 1990s assigned a prominent role of KNOX transcription factors in regulating cell fate determination at the shoot apical meristem (SAM) and in leaf morphogenesis and architecture [145–147]. However, at that time, neither direct nor indirect relationships between the expression of *KNOX* genes and the modification of plant biochemical functions were known. In the late 1990s, a few laboratories started to hypothesize that KNOX may act through modification of hormonal homeostasis, mainly cytokinins (CKs) and gibberellins (GAs) [148]. Among these, Mariotti's laboratory first established the occurrence of a strict correlation among *KNAT1* (an *Arabidopsis* class 1 *KNOX*), overproduction of specific cytokinins in the leaves, and leaf architecture through *KNAT1* overexpression in the crop species, *Lactuca sativa* [149]. Accumulation of cytokinins in the vascular bundles at the leaf margins suggested that KNAT1 might change the determinate state of the leaves to indeterminate by increasing cytokinins' biosynthesis [150]. This let them hypothesize a leading role of cytokinins in leaf development and morphology, and a possible role of KNOX in the regulation of cytokinin production, though the plant genes for the cytokinin biosynthesis had not been identified yet. The discovery of plant *ISOPENTENYL TRANSFERASE* genes (*IPTs*) encoding the cytokinin biosynthetic enzymes [151,152] paved the way to establish a direct regulatory link between KNOX TFs and cytokinin biosynthesis. Independent studies in model species provided molecular evidence for the positive regulation of CK biosynthesis by KNOX in the SAM through the activation of some *IPT* genes [153–155], and positioned cytokinins both upstream and downstream of class 1 KNOX. Further studies on compound-leafed species confirmed a major role of cytokinins in leaf architecture by regulating morphogenetic activity in leaf margins. Shani et al. elegantly demonstrated that expression of class 1 KNOXs during leaf primordia development correlated to the maintenance of an indeterminate state that would prompt the leaf to undertake morphological processes for leaflet production [156], and that CK mediates this function in the regulation of leaf shape [157].

Gibberellins homeostasis was also placed downstream of class 1 KNOX, which were shown to directly repress GA biosynthesis and up-regulate GA catabolism [158–160]. These and further studies identified a key role of class 1 KNOX in maintaining high levels of CK and low levels of GAs to keep the indeterminacy of the SAM and to set boundaries for proper organ separation during plant development [161].

Indications that KNOX action may also involve modulation of the auxin pathway came from genome-wide studies in maize [162]. ChIP-seq analysis showed a direct binding of the maize KNOX KN1 to auxin-related genes, including those involved in auxin signaling and transport, and some of them showed differential expression in *Kn1-N* (gain of function mutant) leaves. Moreover, KN1 can bind genes involved in the synthesis of auxin and its precursor, tryptophan, suggesting that KN1 may directly control the auxin pathway at all levels. Several genes involved in auxin biosynthesis and transport, in GA biosynthesis and in CK catabolism, signaling, and response were also identified in a recent work as modulated by the class 1 KNOX *Arabidopsis* protein, SHOOT MERISTEMLESS (STM), using STMoe and STM-RNAi time-course data and meta-analysis [163].

In addition to cytokinins, gibberellins, and auxin, class 1 KNOXs were also shown to regulate the brassinosteroids (BRs) pathway. BRs are growth-promoting phytohormones involved in diverse aspects of plant growth and development [164]. They promote differentiation through activation of a large number of genes related to cell elongation and cell wall modification [165]. In rice, a class 1 *KNOX* gene, *OSH1*, was shown to negatively regulate the BR pathway and in particular, the genes involved in the BR catabolism [166]. The regulation of the BR catabolism is evolutionarily conserved in maize and is important for SAM function and organ boundary formation in leaves [167]. Among the different functions of the BRs, the regulation of vascular bundles' formation and lignin deposition appears to be relevant [168]. Although a direct link among class 1 *KNOX* genes, BRs and lignin deposition is still to be determined, and the *Arabidopsis KNAT1* mutant, *brevipedicellus* (*bp*), shows increased lignin deposition in the stems [169]. Lignin mislocalization and inappropriate cell differentiation in discrete regions of *bp* stems suggests a role of KNAT1 in regulating cell wall properties, particularly lignin deposition and quality, to prevent premature cell differentiation. Characterization of a *KNAT1* ortholog in *Prunus persica* tree species, *KNOPE1*, confirmed this role in preventing lignin deposition as *KNOPE1* expression was inversely correlated with that of lignin genes and lignin deposition along the peach shoot stems and was down-regulated in lignifying vascular tissues [170].

In contrast to class 1 *KNOX* genes, which are expressed primarily in meristematic tissues, class 2 *KNOX* gene expression occurs in differentiating organs [161,171,172]. The function of class 2 KNOX proteins, as well as potential connections with hormonal pathways, has long remained unknown. Recently, the *Arabidopsis KNAT3/4/5* class 2 *KNOX* genes were shown to act redundantly to promote differentiation of aerial organs, antagonistically to the action of class 1 *KNOX* genes [173]. In *Arabidopsis*, *KNAT3/4/5* loss-of-function phenotypes were reminiscent of a gain-of-function of class 1 KNOX phenotypes, and produced leaves with altered leaf margins and shape. In the compound-leafed species, *Cardamine irsuta*, a reduction or increase in class 2 KNOX activity led to an increase or decrease in leaf complexity, respectively, confirming the antagonistic relationship between class 1 and class 2 KNOX transcription factors [173]. However, no connection with specific hormonal pathways has been described so far for class 2 KNOX in leaf development.

Evidence that class 2 KNOX TFs may act through the inhibition of the cytokinin pathway, antagonistically to class 1 KNOX proteins, came from studies on the role of *KNOX* genes in legume root nodule organogenesis. Functional studies of the *Medicago truncatula KNAT3/4/5* class 2 *KNOX* genes [174] suggested that class 2 KNOX TFs regulate legume nodule development through a cytokinin regulatory module, involving a type-A cytokinin response regulator, to control nodule organ boundaries and shape like the class 2 KNOX function in leaf development [175]. It is tempting to speculate that *KNAT3/4/5-like* genes may constitute a regulatory pathway acting in shoot and aerial organ development, which are recruited for the morphogenetic process that underlies plant-rhizobia symbiosis.

Further investigations are needed to fully comprehend the role of *KNOX* genes in developmental processes underlying plant morphogenesis. Despite their pivotal roles in controlling multiple hormonal pathways, KNOX of class 1 can directly regulate key transcription factors of important developmental processes. These TFs, which are overrepresented among target genes [163], include *CUP SHAPED COTYLEDON (CUC)* transcription factors involved in the specification of the meristem-organ boundary zone, the *TEOSINTE BRANCHED1/CYCLOIDEA/PCF1 (TCP)* family of bHLH that also control cell differentiation, and *AINTEGUMENTA-like (AIL) AP2* transcription factors *PLETHORA (PLT) (AIL/PLT)* that regulate pluripotency and phyllotaxis. To fully comprehend regulatory networks controlled by TALEs, studies on KNOX should be reconciled and integrated with those on BEL1-like homeobox (BLH or BELL) TFs, the other subgroup of the TALE protein family, which form functional heterodimers with KNOXs. So far, it is not known if specific KNOX-BLH complexes have a different affinity for the same targets or diversified target specificity, neither if they act as transcriptional activators of repressors in different developmental contexts. Moreover, class 1, class 2, and class M interplays need further studies to untangle the proposed antagonistic function in cell differentiation, likely mediated by different hormonal pathways, including possible regulation of common targets in an opposite way.

7. Phase Change in Fruit Trees: Advances and Perspectives in Peach and *Prunus* Species

Plant post-embryonic development encompasses the juvenile, adult vegetative, and reproductive phases. In tree species, the end of juvenility and the first flower appearance may not coincide, implying the occurrence of an adult vegetative phase [176]; all these transitions occur gradually along the shoot so that intermediate patterns are evident [176]. The adult vegetative-reproductive switch of meristems encompasses the perception of the flowering signal (flower induction), the meristem re-organization (flower initiation), and flower organ morphogenesis (differentiation). Tree flower buds can undergo dormancy, a growth slowdown that is abandoned after response to specific environmental conditions [177]. Rejuvenation is a reversible shift of all or part of the tree from an older to a younger phase; e.g., explants from mature trees may reverse to juvenile traits, such as enhanced rooting during tissue culture [178]. The explant age is crucial for the success of in vitro technologies. Mariotti's group conducted research to develop phase-specific markers at the morphological, histological, cytological [179], and gene expression levels using *P. persica* as a model. Specifically, they identified differentially transcribed genes putatively subtending differences in organs of juvenile, juvenile-like, and mature shoots [180,181]. Peach juvenility spans 3–5 years and is affected by proper seedling management [182]. Juvenile and adult vegetative traits can differ in leaf size, growth vigor, and photosynthetic activity. In mature plants, the one-year branch has a major role in flowering; leaf axillary meristems produce single or clustered buds bearing single flowers or shoots in multiple combinations. These processes are under the control of the shoot growth speed, node length, and expansion grade of subtending leaves [178]. Flower induction is poorly investigated in the peach; vegetative to reproductive meristem transition and flower initiation mostly occur in summer as studied in three-bud clusters (a central vegetative plus two side flower buds). During dormancy, organ development is continuous in both vegetative and flower buds [183,184]; flower bud dormancy release is regulated by chill and heat requirements, water and nutrient conditions, and hormonal equilibria [185].

Extensive research in annual and perennial model species has unraveled gene networks of phase changes, addressing functional conservation in trees [186], and providing tools to favor allele introgression and enhance micropropagation. The juvenile to adult vegetative shift is coordinated by the decreased expression of two microRNAs, *miR156* and *miR157*, which repress the protein synthesis of SQUAMOSA PROMOTER BINDING PROTEIN-LIKE (SBP/SPL) family transcription factors. These latter are upstream regulators of *APETALA1 (AP1)*, *LEAFY (LFY)*, and *FRUITFULL (FUL)*, key MADS-box transcription factors that confer floral identity to meristems. The *SPL* genes can also control vegetative organs in adulthood, providing models that explain the co-existence of the vegetative phase change and adult vegetative-reproductive changes along the tree shoot. The *miR156/miR172* abundance levels can mirror the leaf stage in various species; higher contents of *miR156* vs. *miR172* mark juvenility, while the opposite typifies vegetative adulthood. As for rejuvenation, in vitro culture causes the appearance of juvenile traits accompanied by high *miR156* levels [176]. Finally, the upstream regulation of *miR156/SPLs* module includes gibberellin-mediated stimuli, glucose levels, several biotic and abiotic cues, the biogenesis process, and epigenetic control [187]. Regarding trees, the *miR156* ectopic expression in poplars reduces the *SPL* and *miR172* expression and prolongs juvenility, confirming evolutionary conservation [188]. In apples, two *miR156* precursors and mature forms decrease during the juvenile-adult vegetative transition; the ectopic expression of pre-*miR156* in tobacco represses the endogenous *SPL* levels and triggers adventitious rooting [189,190]. Moreover, *miR156* levels are elevated in in vitro rejuvenated explants of *Prunus* spp. [191] and peach seedlings and in vitro plants showed higher levels of *miR156* and lower expression of *SPL* and *miR172* than the adult ones [192]. Finally, in a work to which Mariotti contributed, DNA methylation was shown to be lower in meristems of young/juvenile-like shoots vs. adult ones, supporting epigenetic mechanisms being associated to phase maintenance [179].

Flowering initiation involves interactions of inner and outer stimuli able to trigger the adult vegetative-reproductive transition in the shoot apical meristem (SAM) [193]. In the *Arabidopsis* annual model, the pathways responding to internal (autonomous, gibberellin, circadian clock, age, and sugar balance) and external signals (vernalization, temperature, and photoperiod) converge towards floral integrators, which can act in the SAM as floral transition promoters or repressors that cross-interact. Major promoters are SUPPRESSOR OF OVEREXPRESSION OF CONSTANS 1 (SOC1), FLOWERING LOCUS T and D (FT and FD), and AGAMOUS-LIKE24 (AGL24) that activate meristem identity factors, such as LFY, AP1, SEPALLATA3 (SEP3), and FUL, which set the irreversible transition. Repressors are necessary to modulate the floral transition by ensuring the appropriate time-space expression of flowering promoters; key actors are FLOWERING LOCUS C (FLC), SHORT VEGETATIVE PHASE (SVP), and TERMINAL FLOWER 1 (TFL1). Focusing on the FT product, it moves from leaves to SAM, where it is bound to FD, to establish meristem re-programming/flower initiation via *SOC1* triggering. As for TFL, it represses flowering by competing vs FT in FD binding. Moreover, age-related and vernalization events share the control mechanisms based on the *miR156/SPL* and *miR172/AP2*-like modules in perennial models. Finally, the *miR172/AP2* module controls the floral destiny of axillary buds [186]. The equivalence of floral integrators/meristem genes between *Arabidopsis* and fruit trees was reported in various studies [177]. Table 2 includes some key functional studies of *Prunus spp.* genes. Models from perennials have been crucial to unravel the mechanisms of seasonal flower induction of *Prunus* trees. Contextually, Mariotti and colleagues found that the message localization patterns of a maintenance DNA-methyltransferase gene differed in vegetative vs reproductive buds during flower initiation, suggesting a role of methylation in re-programming bud fates [180].

Organ identity genes guide flower piece growth and the *Arabidopsis* ABCDE model proposes five classes of activities that act alone or in combination (A: *AP1* and *AP2* specify sepals and petals; B: *AP3* and *PISTILLATA*, petals and stamens; C: *AGAMOUS*, stamen and carpels; D: *SHATTERPROOF1* and *2* and *SEEDSTICK*, ovules; E: *SEP1-4*, redundant function). These genes encode MADS-box transcription factors and peach putative orthologues have been characterized [194]. Flower differentiation and development are under miRNA specific control [195] and many peach miRNAs have been sequenced, though they are functionally undefined [196]. AGL24-like factors (peach *DAM1-6*) control seasonal dormancy in the peach evergreen mutant and integrate day-length and temperature signals to regulate endo-dormancy [194].

Modern peach breeding exploits marker-assisted selection; the facts that juvenility length is inherited [182] and that a juvenile quantitative trait loci (QTL) was found in *P. mume* offer tools to shorten unproductive stages [197]. As for maturity in *Prunus* trees, the term "flowering time", which should properly refer to SAM adult vegetative-reproductive transition, usually measures the number of disclosed flowers (a.k.a. blooming date). The blooming date is controlled by several QTLs that are spread over eight linkage groups and affect seasonal distribution and production. Apricot, sweet cherry, and peach maintain QTL locations though peach specific ones' reside on group 6. These QTLs were associated to flowering genes, including *LFY* and *TFL1* [198], whereas the chilling requirement and blooming date QTLs co-localization [199] support shared determinism. QTL detections and genome wide association studies have received great benefit from peach genome sequence allowing high throughput genotyping of *Prunus* spp. and the assignment of novel QTL of flowering [200].

Biotechnology approaches can shorten the vegetative stages of both scion and rootstocks [182]. Recalcitrance to the genetic transformation of peach has been a long-lasting drawback for the low tissue regeneration efficiency [178]. Cultivar-independent protocols are still necessary and, so far, there have been no transgenic peaches with modified traits. Rootstock genetic engineering was successful and effective to control scion traits [201]. Peach gene function is currently addressed by transient RNA-interference technologies [202,203] and new approaches exploit development genes to enhance in vitro regeneration [204]. Potentially, reproductive maturity can be achieved by finely tuning the *Prunus* flowering gene expression (Table 2). Namely, the *FT* gene overexpression, which causes early and continuous flowering in the plum [205], has led to "FasTrack" breeding strategies for the rapid incorporation of important traits into desired cultivars. The system uses multiple backcrosses and molecular marker selections to produce improved and non-transgenic varieties in five years [206]. The use of recombinant viral vectors (*Apple latent spherical virus*) was effective to induce precocious flowering; the *Arabidopsis FT* delivered into apple seedlings caused the endogenous *TFL1* silencing and precocious anthesis, reducing the breeding cycle to one year [207]. Other virus-based vectors were effective to silence genes in the peach [202,203], offering tools to phase shift manipulation. Finally, recent strategies of gene editing exploit the delivery of guide RNA and Cas9 protein mixture to apple protoplasts from which non-transgenic edited lines were regenerated [208], further paving the way in *Prunus* trees. As for peach micropropagation, monitoring of miRNA expression levels can be useful to assess the maturity status and regenerative/rooting potential of explants. Hence, tuning the miRNA levels by induction can be useful to control rejuvenation, embryogenesis, and somaclonal variation associated to in vitro cultivation [209].

Table 2. Ectopic expression of some flowering genes from/into *Prunus* species.

Gene [1]	Donor	Receiver	Assay [2]	Phenotypic Effect	Ref.
AP1	*Prunus avium*	*Arabidopsis thaliana*	oe	early flowering	[210]
CO	*Prunus persica*	*Arabidopsis thaliana*	co	flowering promotion	[211]
FT	*Prunus avium*	*Arabidopsis thaliana*	oe	early flowering	[212]
	Prunus persica	*Arabidopsis thaliana*	co	flowering promotion	[211]
	Populus trichocarpa	*Prunus domestica*	oe	early flowering	[205]
MADS5	*Prunus persica*	*Arabidopsis thaliana*	oe	early flowering	[213]
MADS7	*Prunus persica*	*Arabidopsis thaliana*	oe	early flowering	[213]
SOC1	*Prunus mume*	*Arabidopsis thaliana*	oe	early flowering	[214]
CBF	*Prunus persica*	*Malus domestica*	oe	cold-induced dormancy	[215]
DAM6	*Prunus mume*	*Populus tremula×P. tremuloides*	oe	dormancy promotion	[216]
SVP1	*Prunus mume*	*Arabidopsis thaliana*	oe	flowering delay	[189]
TFL1	*Prunus persica*	*Arabidopsis thaliana*	oe	flowering delay	[217]

[1], AP1, APETALA1; CO, CONSTANS; FT, FLOWERING LOCUS T; MADS5 and MADS7, SEPALLATA-like; SOC1, SUPPRESSOR OF OVEREXPRESSION OF CONSTANS 1; CBF, C-REPEAT BINDING FACTOR; DAM6, DORMANCY-ASSOCIATED MADS box6; SVP1, SHORT VEGETATIVE PHASE 1; TFL1, TERMINAL FLOWER1.
[2], Functional assay: oe, overexpression; co, complementation.

8. Conclusions

Research updates of the topics, which were in the scientific interests of Domenico Mariotti, dealing with plant genetics for crop improvement in agricultural sciences, were focused on. In the last decade, following the routes of his insights, many goals were reached from plant biotechnology applications of in vitro cell cultures to the genetic transformation of relevant crops, including new knowledge on plant organogenesis of model plants, as well as phase transition in fruit tree crops. However, despite the achieved progress, further efforts are needed to shed more light on the genetic basis of key developmental processes in model and crop species. By identifying new useful genetic traits, it will be possible to further exploit the high potential of plant cells for improving crop production.

Author Contributions: S.C. and G.F. designed the review; A.D.P. and S.C. wrote Section 2; G.F. and M.A.I. wrote Section 3; E.R. and C.S. wrote Section 4; F.S. wrote Section 5; G.F. and G.M. wrote Section 6; D.G. and G.T. wrote Section 7 and provided careful revision of the whole manuscript; M.L.M. co-wrote Section 3; C.N. co-wrote Section 6; S.C. supervised the manuscript. All authors read and approved the final version of the manuscript.

Funding: This research received no external funding.

Acknowledgments: The authors thank Eleanna Longo and Elisabetta Di Giacomo for illustrations used in Figures 2 and 3, respectively.

Conflicts of Interest: The authors declare no conflict of interest.

References

1. Krishna, H.; Alizadeh, M.; Singh, D.; Singh, U.; Chauhan, N.; Eftekhari, M.; Sadh, R.K. Somaclonal variations and their applications in horticultural crops improvement. *3 Biotech* **2016**, *6*, 54. [CrossRef] [PubMed]
2. Caretto, S.; Giardina, M.C.; Nicolodi, C.; Mariotti, D. Chlorsulfuron resistance in *Daucus carota* cell lines and plants: Involvement of gene amplification. *Theor. Appl. Genet.* **1994**, *88*, 520–524. [CrossRef] [PubMed]
3. Chaleff, R.; Ray, T. Herbicide-resistant mutants from tobacco cell cultures. *Science* **1994**, *223*, 1148–1151. [CrossRef] [PubMed]
4. Caretto, S.; Giardina, M.; Nicolodi, C.; Mariotti, D. In-vitro cell selection: Production and characterization of tobacco cell lines and plants resistant to the herbicide chlorsulfuron. *J. Genet. Breed.* **1993**, *47*, 115–120.
5. Terry, R.W.; Newell, F.B.; Stephen, F.S.; Donald, P. Biochemical mechanism and molecular basis for ALS-inhibiting herbicide resistance in sugarbeet (*Beta vulgaris*) somatic cell selections. *Weed Sci.* **1998**, *46*, 13–23.
6. Caretto, S.; Giardina, M.C.; Nicolodi, C.; Mariotti, D. Acetohydroxyacid synthase gene amplification induces clorsulfuron resistance in Daucus Carota L. In *Current Issues in Plant Molecular and Cellular Biology, Proceedings of the VIIIth International Congress on Plant Tissue and Cell Culture, Florence, Italy, 12–17 June 1994*; Terzi, M., Cella, R., Falavigna, A., Eds.; Springer: Dordrecht, The Netherlands, 1995; pp. 235–240.
7. Duke, S.O. Perspectives on transgenic, herbicide-resistant crops in the United States almost 20 years after introduction. *Pest Manag. Sci.* **2015**, *71*, 652–657. [CrossRef] [PubMed]
8. Barrell, P.J.; Latimer, J.M.; Baldwin, S.J.; Thompson, M.L.; Jacobs, J.M.E.; Conner, A.J. Somatic cell selection for chlorsulfuron-resistant mutants in potato: Identification of point mutations in the *acetohydroxyacid synthase* gene. *BMC Biotechnol.* **2017**, *17*, 49. [CrossRef] [PubMed]
9. Smetanska, I. Production of secondary metabolites using plant cell cultures. *Adv. Biochem. Eng. Biotechnol.* **2008**, *111*, 187–228. [PubMed]
10. Mene-Saffrane, L. Vitamin E biosynthesis and its regulation in plants. *Antioxidants* **2017**, *7*, 2. [CrossRef]
11. Caretto, S.; Nisi, R.; Paradiso, A.; De Gara, L. Tocopherol production in plant cell cultures. *Mol. Nutr. Food Res.* **2010**, *54*, 726–730. [CrossRef] [PubMed]
12. Furuya, T.; Yoshikawa, T.; Kimura, T.; Kaneko, H. Production of tocopherols by cell culture of safflower. *Phytochemistry* **1987**, *26*, 2741–2747. [CrossRef]
13. Caretto, S.; Bray Speth, E.; Fachechi, C.; Gala, R.; Zacheo, G.; Giovinazzo, G. Enhancement of vitamin E production in sunflower cell cultures. *Plant Cell Rep.* **2004**, *23*, 174–179. [CrossRef]
14. Caretto, S.; Paradiso, A.; D'Amico, L.; De Gara, L. Ascorbate and glutathione metabolism in two sunflower cell lines of differing α-tocopherol biosynthetic capability. *Plant Physiol. Biochem.* **2002**, *40*, 509–513. [CrossRef]
15. Almagro, L.; Raquel Tudela, L.; Belén Sabater-Jara, A.; Miras-Moreno, B.; Pedreño, M.A. Cyclodextrins increase phytosterol and tocopherol levels in suspension cultured cells obtained from mung beans and safflower. *Biotechnol. Prog.* **2017**, *33*, 1662–1665. [CrossRef] [PubMed]
16. Yue, W.; Ming, Q.L.; Lin, B.; Rahman, K.; Zheng, C.J.; Han, T.; Qin, L.P. Medicinal plant cell suspension cultures: Pharmaceutical applications and high-yielding strategies for the desired secondary metabolites. *Crit. Rev. Biotechnol.* **2016**, *36*, 215–232. [CrossRef] [PubMed]
17. José Bagur, M.; Alonso Salinas, G.; Jiménez-Monreal, A.; Chaouqi, S.; Llorens, S.; Martínez-Tomé, M.; Alonso, G. Saffron: An old medicinal plant and a potential novel functional food. *Molecules* **2018**, *23*, 30. [CrossRef] [PubMed]

18. Winterhalter, P.; Straubinger, M. Saffron—Renewed interest in an ancient spice. *Food Rev. Int.* **2000**, *16*, 39–59. [CrossRef]
19. Moradi, A.; Zarinkamar, F.; Caretto, S.; Azadi, P. Influence of thidiazuron on callus induction and crocin production in corm and style explants of *Crocus sativus* L. *Acta Physiol. Plant* **2018**, *40*, 185. [CrossRef]
20. Malik, S.; Cusidó, R.M.; Mirjalili, M.H.; Moyano, E.; Palazón, J.; Bonfill, M. Production of the anticancer drug taxol in *Taxus baccata* suspension cultures: A review. *Process Biochem.* **2011**, *46*, 23–34. [CrossRef]
21. Graham, I.A.; Besser, K.; Blumer, S.; Branigan, C.A.; Czechowski, T.; Elias, L.; Guterman, I.; Harvey, D.; Isaac, P.G.; Khan, A.M.; et al. The genetic map of *Artemisia annua* L. identifies loci affecting yield of the antimalarial drug artemisinin. *Science* **2010**, *327*, 328–331. [CrossRef]
22. Townsend, T.; Segura, V.; Chigeza, G.; Penfield, T.; Rae, A.; Harvey, D.; Bowles, D.; Graham, I.A. The use of combining ability analysis to identify elite parents for *Artemisia annua* F1 hybrid production. *PLoS ONE* **2013**, *8*, e61989. [CrossRef]
23. Tang, K.; Shen, Q.; Yan, T.; Fu, X. Transgenic approach to increase artemisinin content in *Artemisia annua* L. *Plant Cell Rep.* **2014**, *33*, 605–615. [CrossRef] [PubMed]
24. Zhang, Y.; Nowak, G.; Reed, D.W.; Covello, P.S. The production of artemisinin precursors in tobacco. *Plant Biotechnol. J.* **2011**, *9*, 445–454. [CrossRef] [PubMed]
25. Paddon, C.J.; Keasling, J.D. Semi-synthetic artemisinin: A model for the use of synthetic biology in pharmaceutical development. *Nat. Rev. Microbiol.* **2014**, *12*, 355–367. [CrossRef]
26. Weathers, P.J.; Jordan, N.J.; Lasin, P.; Towler, M.J. Simulated digestion of dried leaves of *Artemisia annua* consumed as a treatment (pACT) for malaria. *J. Ethnopharmacol.* **2014**, *151*, 858–863. [CrossRef] [PubMed]
27. Caretto, S.; Quarta, A.; Durante, M.; Nisi, R.; De Paolis, A.; Blando, F.; Mita, G. Methyl jasmonate and miconazole differently affect arteminisin production and gene expression in *Artemisia annua* suspension cultures. *Plant Biol.* **2011**, *13*, 51–58. [CrossRef]
28. Ahlawat, S.; Saxena, P.; Alam, P.; Wajid, S.; Abdin, M.Z. Modulation of artemisinin biosynthesis by elicitors, inhibitor, and precursor in hairy root cultures of *Artemisia annua* L. *J. Plant Interact.* **2014**, *9*, 811–824. [CrossRef]
29. Zamboni, A.; Vrhovsek, U.; Kassemeyer, H.H.; Mattivi, F.; Velasco, R. Elicitor-induced resveratrol production in cell cultures of different grape genotypes (*Vitis* spp.). *Vitis* **2006**, *45*, 63–68.
30. Durante, M.; Caretto, S.; Quarta, A.; De Paolis, A.; Nisi, R.; Mita, G. beta-Cyclodextrins enhance artemisinin production in *Artemisia annua* suspension cell cultures. *Appl. Microbiol. Biotechnol.* **2011**, *90*, 1905–1913. [CrossRef]
31. Chilton, M.D.; Tepfer, D.A.; Petit, A.; David, C.; Cassedelbart, F.; Tempe, J. *Agrobacterium rhizogenes* inserts T-DNA into the genomes of the host plant-root cells. *Nature* **1982**, *295*, 432–434. [CrossRef]
32. Cardarelli, M.; Mariotti, D.; Pomponi, M.; Spano, L.; Capone, I.; Costantino, P. *Agrobacterium rhizogenes* T-DNA genes capable of inducing hairy root phenotype. *Mol. Gen. Genet.* **1987**, *209*, 475–480. [CrossRef] [PubMed]
33. Zambryski, P.; Tempe, J.; Schell, J. Transfer and function of T-DNA genes from *agrobacterium* Ti and Ri plasmids in plants. *Cell* **1989**, *56*, 193–201. [CrossRef]
34. Zambryski, P.; Joos, H.; Genetello, C.; Leemans, J.; Montagu, M.V.; Schell, J. Ti plasmid vector for the introduction of DNA into plant cells without alteration of their normal regeneration capacity. *EMBO J.* **1983**, *2*, 2143–2150. [CrossRef] [PubMed]
35. Tepfer, D. Transformation of several species of higher plants by *Agrobacterium rhizogenes*: Sexual transmission of the transformed genotype and phenotype. *Cell* **1984**, *37*, 959–967. [CrossRef]
36. White, F.F.; Taylor, B.H.; Huffman, G.A.; Gordon, M.P.; Nester, E.W. Molecular and genetic analysis of the transferred DNA regions of the root-inducing plasmid of *Agrobacterium rhizogenes*. *J. Bacteriol.* **1985**, *164*, 33–44. [PubMed]
37. Costantino, P.; Capone, I.; Cardarelli, M.; De Paolis, A.; Mauro, M.L.; Trovato, M. Bacterial plant oncogenes: The *rol* genes' saga. *Genetica* **1994**, *94*, 203–211. [CrossRef] [PubMed]
38. Spano, L.; Mariotti, D.; Cardarelli, M.; Branca, C.; Costantino, P. Morphogenesis and auxin sensitivity of transgenic tobacco with different complements of Ri T-DNA. *Plant Physiol. Biochem.* **1988**, *87*, 479–483. [CrossRef]

39. Capone, I.; Cardarelli, M.; Mariotti, D.; Pomponi, M.; De Paolis, A.; Costantino, P. Different promoter regions control level and tissue specificity of expression of *Agrobacterium rhizogenes rolB* gene in plants. *Plant Mol. Biol.* **1991**, *16*, 427–436. [CrossRef] [PubMed]
40. Cardarelli, M.; Spanò, L.; Mariotti, D.; Mauro, M.L.; Van Sluys, M.A.; Costantino, P. The role of auxin in hairy root induction. *Mol. Genet. Genom.* **1987**, *208*, 457–463. [CrossRef]
41. Spano, L.; Mariotti, D.; Pezzotti, M.; Damiani, F.; Arcioni, S. Hairy root transformation in alfalfa (*Medicago sativa* L.). *Theor. Appl. Genet.* **1987**, *73*, 523–530. [CrossRef] [PubMed]
42. Frugis, G.; Caretto, S.; Santini, L.; Mariotti, D. *Agrobacterium rhizogenes rol* genes induce productivity-related phenotypical modifications in "creeping-rooted" alfalfa types. *Plant Cell Rep.* **1995**, *14*, 488–492. [CrossRef]
43. Rugini, E.; Mariotti, D. *Agrobacterium rhizogenes* T-DNA genes and rooting in woody species. *Acta Hortic.* **1992**, *300*, 301–308. [CrossRef]
44. Rugini, E.; Pellegrineschi, A.; Mencuccini, M.; Mariotti, D. Increase of rooting ability in the woody species kiwi (*Actinidia deliciosa* A. Chev.) by transformation with *Agrobacterium rhizogenes rol* genes. *Plant Cell Rep.* **1991**, *10*, 291–295. [CrossRef]
45. Rinallo, C.; Mariotti, D. Rooting of *Castanea sativa* Mill. shoots: Effect of *Agrobacterium rhizogenes* T-DNA genes. *J. Hortic. Sci.* **1993**, *68*, 399–407. [CrossRef]
46. Mariotti, D.; Fontana, G.; Santini, L.; Costantino, P. Evaluation under field conditions of the morphological alterations ("hairy root phenotype") induced on *Nicotiana tabacum* by different Ri plasmid T-DNA genes. *J. Genet. Breed.* **1989**, *43*, 157–164.
47. Mauro, M.L.; Costantino, P.; Bettini, P.P. The never ending story of *rol* genes: A century after. *Plant Cell Tissue Organ Cult.* **2017**, *131*, 201–212. [CrossRef]
48. Bulgakov, V.P.; Gorpenchenko, T.Y.; Veremeichik, G.N.; Shkryl, Y.N.; Tchernoded, G.K.; Bulgakov, D.V.; Aminin, D.L.; Zhuravlev, Y.N. The *rolB* gene suppresses reactive oxygen species in transformed plant cells through the sustained activation of antioxidant defense. *Plant Physiol.* **2012**, *158*, 1371–1381. [CrossRef]
49. Wang, Y.; Peng, W.; Zhou, X.; Huang, F.; Shao, L.; Luo, M. The putative *Agrobacterium* transcriptional activator-like virulence protein VirD5 may target T-complex to prevent the degradation of coat proteins in the plant cell nucleus. *New Phytol.* **2014**, *203*, 1266–1281. [CrossRef]
50. Bulgakov, V.P.; Aminin, D.L.; Shkryl, Y.N.; Gorpenchenko, T.Y.; Veremeichik, G.N.; Dmitrenok, P.S.; Zhuravlev, Y.N. Suppression of reactive oxygen species and enhanced stress tolerance in *Rubia cordifolia* cells expressing the *rolC* oncogene. *Mol. Plant Microbe Interact.* **2008**, *21*, 1561–1570. [CrossRef]
51. Trovato, M.; Maras, B.; Linhares, F.; Costantino, P. The plant oncogene *rolD* encodes a functional ornithine cyclodeaminase. *Proc. Natl. Acad. Sci. USA* **2001**, *98*, 13449–13453. [CrossRef]
52. Levesque, H.; Delepelaire, P.; Rouze, P.; Slightom, J.; Tepfer, D. Common evolutionary origin of the central portions of the Ri TL-DNA of *Agrobacterium rhizogenes* and the Ti T-DNAs of *Agrobacterium tumefaciens*. *Plant Mol. Biol.* **1988**, *11*, 731–744. [CrossRef]
53. Otten, L. The *Agrobacterium* phenotypic plasticity (*Plast*) genes. *Curr. Top. Microbiol. Immunol.* **2018**.
54. Mohajjel-Shoja, H.; Clement, B.; Perot, J.; Alioua, M.; Otten, L. Biological activity of the *Agrobacterium rhizogenes*-derived *trolC* gene of *Nicotiana tabacum* and its functional relation to other plast genes. *Mol. Plant Microbe Interact.* **2011**, *24*, 44–53. [CrossRef]
55. Chen, K.; de Borne, F.D.; Sierro, N.; Ivanov, N.V.; Alouia, M.; Koechler, S.; Otten, L. Organization of the TC and TE cellular T-DNA regions in *Nicotiana otophora* and functional analysis of three diverged TE-6b genes. *Plant J.* **2018**, *94*, 274–287. [CrossRef]
56. Matveeva, T.V. *Agrobacterium*-mediated transformation in the evolution of plants. In *Agrobacterium Biology*; Gelvin, S., Ed.; Springer: Cham, Switzerland, 2018; Volume 418, pp. 421–441.
57. Doran, P.M. Biotechnology of hairy root systems. *Adv. Biochem. Eng. Biotechnol.* **2013**, *134*, V–Vi.
58. Bulgakov, V.P. Functions of *rol* genes in plant secondary metabolism. *Biotechnol. Adv.* **2008**, *26*, 318–324. [CrossRef]
59. Chandra, S. Natural plant genetic engineer *Agrobacterium rhizogenes*: Role of T-DNA in plant secondary metabolism. *Biotechnol. Lett.* **2012**, *34*, 407–415. [CrossRef]
60. Kiselev, K.V.; Dubrovina, A.S.; Veselova, M.V.; Bulgakov, V.P.; Fedoreyev, S.A.; Zhuravlev, Y.N. The *rolB* gene-induced overproduction of resveratrol in *Vitis amurensis* transformed cells. *J. Biotechnol.* **2007**, *128*, 681–692. [CrossRef]

61. Arshad, W.; Haq, I.U.; Waheed, M.T.; Mysore, K.S.; Mirza, B. *Agrobacterium*-mediated transformation of tomato with *rolB* gene results in enhancement of fruit quality and foliar resistance against fungal pathogens. *PLoS ONE* **2014**, *9*, e96979. [CrossRef]
62. Bettini, P.P.; Marvasi, M.; Fani, F.; Lazzara, L.; Cosi, E.; Melani, L.; Mauro, M.L. Agrobacterium rhizogenes *rolB* gene affects photosynthesis and chlorophyll content in transgenic tomato (*Solanum lycopersicum* L.) plants. *J. Plant Physiol.* **2016**, *204*, 27–35. [CrossRef]
63. Bulgakov, V.P.; Veremeichik, G.N.; Grigorchuk, V.P.; Rybin, V.G.; Shkryl, Y.N. The *rolB* gene activates secondary metabolism in Arabidopsis calli via selective activation of genes encoding MYB and bHLH transcription factors. *Plant Physiol. Biochem. PPB* **2016**, *102*, 70–79. [CrossRef] [PubMed]
64. Taylor, C.G.; Fuchs, B.; Collier, R.; Lutke, W.K. Generation of composite plants using *Agrobacterium rhizogenes*. *Methods Mol. Biol.* **2006**, *343*, 155–167.
65. Boisson-Dernier, A.; Chabaud, M.; Garcia, F.; Becard, G.; Rosenberg, C.; Barker, D.G. *Agrobacterium rhizogenes*-transformed roots of *Medicago truncatula* for the study of nitrogen-fixing and endomycorrhizal symbiotic associations. *Mol. Plant Microbe Interact.* **2001**, *14*, 695–700. [CrossRef] [PubMed]
66. Ho-Plagaro, T.; Huertas, R.; Tamayo-Navarrete, M.I.; Ocampo, J.A.; Garcia-Garrido, J.M. An improved method for *Agrobacterium rhizogenes*-mediated transformation of tomato suitable for the study of arbuscular mycorrhizal symbiosis. *Plant Methods* **2018**, *14*, 34. [CrossRef] [PubMed]
67. Horn, P.; Santala, J.; Nielsen, S.L.; Huhns, M.; Broer, I.; Valkonen, J.P. Composite potato plants with transgenic roots on non-transgenic shoots: A model system for studying gene silencing in roots. *Plant Cell Rep.* **2014**, *33*, 1977–1992. [CrossRef] [PubMed]
68. Neb, D.; Das, A.; Hintelmann, A.; Nehls, U. Composite poplars: A novel tool for ectomycorrhizal research. *Plant Cell Rep.* **2017**, *36*, 1959–1970. [CrossRef]
69. Pacifici, E.; Polverari, L.; Sabatini, S. Plant hormone cross-talk: The pivot of root growth. *J. Exp. Bot.* **2015**, *66*, 1113–1121. [CrossRef]
70. Benfey, P.N.; Bennett, M.; Schiefelbein, J. Getting to the root of plant biology: Impact of the Arabidopsis genome sequence on root research. *Plant J.* **2010**, *61*, 992–1000. [CrossRef]
71. Moriguchi, K.; Maeda, Y.; Satou, M.; Hardayani, N.S.; Kataoka, M.; Tanaka, N.; Yoshida, K. The complete nucleotide sequence of a plant root-inducing (Ri) plasmid indicates its chimeric structure and evolutionary relationship between tumor-inducing (Ti) and symbiotic (Sym) plasmids in *Rhizobiaceae*. *J. Mol. Biol.* **2001**, *307*, 771–784. [CrossRef]
72. Suzuki, K.; Hattori, Y.; Uraji, M.; Ohta, N.; Iwata, K.; Murata, K.; Kato, A.; Yoshida, K. Complete nucleotide sequence of a plant tumor-inducing Ti plasmid. *Gene* **2000**, *242*, 331–336. [CrossRef]
73. Rugini, E. Progress in studies on in vitro culture of Almonds. In Proceedings of the 41° Conference on Plant Tissue Culture and Its Agricultural Applications, Nothingam, UK, 17–21 September 1984; p. 73.
74. Rugini, E.; Fedeli, E. Olive (*Olea europaea* L.) as an oilseed crop. In *Legumes and Oilseed Crops I*; Bajaj, Y.P.S., Ed.; Springer: New York, NY, USA, 1990; pp. 593–641.
75. Rugini, E. Involvement of polyamines in auxin and *Agrobacterium rhizogenes*-induced rooting of fruit trees in vitro. *Am. J. Hortic. Sci.* **1992**, *117*, 532–536.
76. Rugini, E. Piante da frutto transgeniche e considerazioni sulle conseguenze dei divieti impost alla ricerca in Italia. *Italus Hortus* **2015**, *12*, 79–92.
77. Rugini, E.; Silvestri, C.; Cristofori, V.; Brunori, E.; Biasi, R. Ten years field trial observations of ri-TDNA cherry Colt rootstocks and their effect on grafted sweet cherry cv Lapins. *Plant Cell Tissue Organ Cult.* **2015**, *123*, 557–568. [CrossRef]
78. Damiano, C.; Archilletti, T.; Caboni, E.; Lauri, P.; Falasca, G.; Mariotti, D.; Ferraiolo, G. *Agrobacterium* mediated transformation of almond: In vitro rooting through localized infection of *A. rhizogenes* w.t. *Acta Hortic.* **1995**, *392*, 161–170. [CrossRef]
79. Rugini, E.; Gutierrez-Pesce, P. Transformation in *Prunus* species. In *Biotechnology in Agriculture and Forestry*; Bajaj, Y.P.S., Ed.; Springer: Berlin, Germany, 1999; Volume 44, pp. 245–262.
80. Geier, T.; Eimert, K.; Scherer, R.; Nickel, C. Production and rooting behaviour of *rolB*-transgenic plants of grape rootstock 'Richter 110' (*Vitis berlandieri* × *V. rupestris*). *Plant Cell Tissues Organ Cult.* **2008**, *94*, 269–280. [CrossRef]

81. Rugini, E. Trasformation of kiwi, cherry and papaya with *rol* genes. In Proceedings of the V Congress on University and Biotechnology Innovation, Brescia, IT, USA, 20–21 June 1994; University of Brescia: Brescia, IT, USA, 1994; pp. 68–69.
82. La Malfa, S.; Distefano, G.; Domina, F.; Nicolosi, E.; Toscano, V.; Gentile, A. Evaluation of *Citrus* rootstock transgenic for *rolABC* gene. *Acta Hortic.* **2011**, *892*, 131–140. [CrossRef]
83. Rugini, E.; Rita, B.; Muleo, R. Olive (*Olea europaea* var. *sativa*) transformation. In *Molecular Biology of Woody Plants*; Jain, S., Minocha, S., Eds.; Kluwer Academic Publishers: Dordrecht, The Netherlands, 2000; Volume 2, pp. 245–279.
84. Rugini, E.; Gutierrez-Pesce, P.; Spampinato, P.L.; Ciarmiello, A.; D'Ambrosio, C. New perspective for biotechnologies in olive breeding: Morphogenesis, in vitro selection and gene transformation. *Acta Hortic.* **1999**, *474*, 107–110. [CrossRef]
85. Rugini, E.; Cristofori, V.; Silvestri, C. Genetic improvement of olive (*Olea europaea* L.) by conventional and in vitro biotechnology methods. *Biotechnol. Adv.* **2016**, *34*, 687–696. [CrossRef]
86. Zhu, L.; Holefors, A.; Ahlman, A.; Xue, Z.; Welander, M. Transformation of the apple rootstock M.9/29 with the *rolB* gene and its influence on rooting and growth. *Plant Sci.* **2001**, *160*, 433–439. [CrossRef]
87. Smolka, A.; Li, X.Y.; Heikelt, C.; Welander, M.; Zhu, L.H. Effects of transgenic rootstocks on growth and development of non-transgenic scion cultivars in apple. *Transgenic Res.* **2010**, *19*, 933–948. [CrossRef]
88. Zhu, L.-H.; Li, X.-Y.; Ahlman, A.; Welander, M. The rooting ability of the dwarfing pear rootstock BP10030 (*Pyrus communis*) was significantly increased by introduction of the *rolB* gene. *Plant Sci.* **2003**, *165*, 829–835. [CrossRef]
89. Landi, L.; Capocasa, F.; Costantini, E.; Mezzetti, B. *ROLC* strawberry plant adaptability, productivity, and tolerance to soil-borne disease and mycorrhizal interactions. *Transgenic Res.* **2009**, *18*, 933–942. [CrossRef] [PubMed]
90. Ayala-Silva, T.; Beyl, C.A.; Dortch, G. *Agrobacterium rhizogenes* mediated-transformation of *Asimina triloba* L. cuttings. *Pak. J. Biol. Sci.* **2007**, *10*, 132–136. [CrossRef] [PubMed]
91. Sutter, E.G.; Luza, J. Development anatomy of roots induced by *Agrobacterium rhizogenes* in *Malus pumila* 'M.26' shoots grown in vitro. *Int. J. Plant Sci.* **1993**, *154*, 59–67. [CrossRef]
92. Gutierrez-Pesce, P.; Taylor, K.; Muleo, R.; Rugini, E. Somatic embryogenesis and shoot regeneration from transgenic roots of the cherry rootstock Colt (*Prunus avium* × *P. pseudocerasus*) mediated by pRi 1855 T-DNA of *Agrobacterium rhizogenes*. *Plant Cell Rep.* **1998**, *17*, 574–580. [CrossRef]
93. Yazawa, M.; Suginuma, C.; Ichikawa, K.; Kamada, H.; Akihama, T. Regeneration of transgenic plants from hairy root of kiwi fruit (*Actinidia deliciosa*) induced by *Agrobacterium rhizogenes*. *Jpn. J. Breed.* **1995**, *45*, 241–244. [CrossRef]
94. Balestra, G.M.; Rugini, E.; Varvaro, L. Increased susceptibility to *Pseudomonas syringae* pv. *Syringae* and *Pseudomonas viridiflava* of kiwi plants having transgenic *rolABC* genes and its inheritance in the T1 offspring. *J. Phytopathol.* **2001**, *149*, 189–194. [CrossRef]
95. Rugini, E. Risultati preliminari sulla caratterizzazione morfo-fisiologica di cultivar di actinidia (*Actinidia deliciosa* A. Chev.) transgenica con geni *rol* per modificare l'architettura e la capacità rizogena della pianta. In *Atti Giornate Scientifiche S.O.I*; Istituto Sperimentale Frutticoltura: Rome, Italy, 1992; pp. 142–143.
96. Druart, P.; Delporte, F.; Brazda, M.; Ugarte-Ballon, C.; da Câmara Machado, A.; Laimer da Câmara Machado, M.; Jacquemin, J.; Watillon, B. Genetic transformation of cherry trees. *Acta Hortic.* **1998**, *468*, 71–76. [CrossRef]
97. Vahdati, K.; McKenna, J.R.; Dandekar, A.M.; Uratsu, S.L.; Hackett, W.P.; Negrei, P.; McGranahan, G.H. Rooting and other characteristics of a transgenic walnut hybrid (*Juglans hindsii* × *J. regia*) rootstock expressing *rolABC*. *J. Am. Soc. Sci.* **2002**, *127*, 724–728.
98. Gentile, A.; Deng, Z.N.; La Malfa, S.; Domina, F.; Germanà, C.; Tribulato, E. Morphological and physiological effects of *rolABC* genes into *Citrus* genome. *Acta Hortic.* **2004**, *632*, 235–242. [CrossRef]
99. Rugini, E. Risultati preliminari di una sperimentazione di campo di miglioramento genetico dell'*Actinidia* con tecniche biotecnologiche per tolleranza a stress idrico, funghi patogeni, e modifica dell'architettura della chioma. *Kiwi Inf.* **2012**, 4–18.
100. Welander, M.; Pawlicki, N.; Holefors, A.; Wilson, F. Genetic transformation of the apple rootstock M26 with the *RolB* gene and its influence on rooting. *J. Plant Physiol.* **1998**, *153*, 371–380. [CrossRef]

101. Zhu, L.-H.; Welander, M. Growth characteristics of apple cultivar Gravenstein plants grafted onto the transformed rootstock M26 with *rolA* and *rolB* genes under non-limiting nutrient conditions. *Plant Sci.* **1999**, *147*, 75–80. [CrossRef]
102. Firson, A.; Dolgov, S. *Agrobacterium* transformation of *Actinidia kolomikta*. *Acta Hortic.* **1997**, *447*, 323–327. [CrossRef]
103. Mezzetti, B.; Costantini, E.; Chionchetti, F.; Landi, L.; Pandolfini, T.; Spena, A. Genetic transformation in strawberry and raspberry for improving plant productivity and fruit quality. *Acta Hortic.* **2004**, *649*, 107–110. [CrossRef]
104. Bell, R.L.; Scorza, R.; Srinivasan, C.; Webb, K. Transformation of "Beurre Bosc" pear with the *rolC* gene. *J. Am. Soc. Hortic. Sci.* **1999**, *124*, 570–574.
105. Kaneyoshi, J.; Kobayashi, S. Characteristics of transgenic trifoliate orange (*Poncirus trifoliata* Raf.) possessing the *rolc* gene of *Agrobacterium rhizogenes* Ri plasmid. *Hortic. J.* **1999**, *68*, 734–738. [CrossRef]
106. Eapen, S. Advances in development of transgenic pulse crops. *Biotechnol. Adv.* **2008**, *26*, 162–168. [CrossRef]
107. Fontana, G.S.; Santini, L.; Caretto, S.; Frugis, G.; Mariotti, D. Genetic transformation in the grain legume *Cicer arietinum* L. (chickpea). *Plant Cell Rep.* **1993**, *12*, 194–198. [CrossRef] [PubMed]
108. Mariotti, D.; Fontana, G.S.; Santini, L. Genetic transformation of grain legumes: *Phaseolus vulgaris* L. and *Phaseolus coccineus* L. *J. Genet. Breed.* **1989**, *43*, 77–82.
109. International Service for the Acquisition of Agri-Biotech Applications, I. Soybean (*Glycine max* L.) GM Events (40 Events). Available online: http://www.isaaa.org/gmapprovaldatabase/crop/default.asp?CropID=19&Crop=Soybean (accessed on 28 November 2018).
110. International Service for the Acquisition of Agri-Biotech Applications, I. Alfalfa (*Medicago sativa*) GM Events (5 Events). Available online: http://www.isaaa.org/gmapprovaldatabase/crop/default.asp?CropID=1&Crop=Alfalfa (accessed on 28 November 2018).
111. International Service for the Acquisition of Agri-Biotech Applications, I. Bean (*Phaseolus vulgaris*) GM Events (1 Event). Available online: http://www.isaaa.org/gmapprovaldatabase/crop/default.asp?CropID=3&Crop=Bean (accessed on 28 November 2018).
112. Food and Agriculture Organization of the United Nations, F. FAOSTAT. Available online: http://www.fao.org/faostat/en/#home (accessed on 28 November 2018).
113. Atif, R.M.; Patat-Ochatt, E.M.; Svabova, L.; Ondrej, V.; Klenoticova, H.; Jacas, L.; Griga, M.; Ochatt, S.J. Gene transfer in legumes. In *Progress in Botany*; Luttge, U., Beyschlag, W., Francis, D., Cushman, J., Eds.; Springer: Berlin, Germany, 2013; Volume 74, pp. 37–100.
114. Iantcheva, A.; Mysore, K.S.; Ratet, P. Transformation of leguminous plants to study symbiotic interactions. *Int. J. Dev. Biol.* **2013**, *57*, 577–586. [CrossRef] [PubMed]
115. Estrada-Navarrete, G.; Alvarado-Affantranger, X.; Olivares, J.-E.; Guillén, G.; Díaz-Camino, C.; Campos, F.; Quinto, C.; Gresshoff, P.M.; Sanchez, F. Fast, efficient and reproducible genetic transformation of *Phaseolus* spp. by *Agrobacterium rhizogenes*. *Nat. Protoc.* **2007**, *2*, 1819. [CrossRef] [PubMed]
116. Nova-Franco, B.; Íñiguez, L.P.; Valdés-López, O.; Alvarado-Affantranger, X.; Leija, A.; Fuentes, S.I.; Ramírez, M.; Paul, S.; Reyes, J.L.; Girard, L.; et al. The Micro-*RNA172c-APETALA2-1* node as a key regulator of the common bean-*Rhizobium etli* nitrogen fixation symbiosis. *Plant Physiol.* **2015**, *168*, 273. [CrossRef] [PubMed]
117. Moda-Crinò, V.; Nicolodi, C.; Chichriccò, G.; Mariotti, D. In vitro meristematic organogenesis and plant regeneration in bean (*Phaseolus vulgaris* L.) cultivars. *J. Genet. Breed.* **1995**, *49*, 133–138.
118. Citadin, C.T.; Ibrahim, A.B.; Aragao, F.J. Genetic engineering in Cowpea (*Vigna unguiculata*): History, status and prospects. *GM Crops* **2011**, *2*, 144–149. [CrossRef] [PubMed]
119. Kumar, M.; Yusuf, M.A.; Nigam, M.; Kumar, M. An update on genetic modification of Chickpea for increased yield and stress tolerance. *Mol. Biotechnol.* **2018**, *60*, 651–663. [CrossRef] [PubMed]
120. Hnatuszko-Konka, K.; Kowalczyk, T.; Gerszberg, A.; Wiktorek-Smagur, A.; Kononowicz, A.K. *Phaseolus vulgaris*-recalcitrant potential. *Biotechnol. Adv.* **2014**, *32*, 1205–1215. [CrossRef]
121. Nguyen, A.H.; Hodgson, L.M.; Erskine, W.; Barker, S.J. An approach to overcoming regeneration recalcitrance in genetic transformation of lupins and other legumes. *Plant Cell Tissues Organ Cult.* **2016**, *127*, 623–635. [CrossRef]
122. Das, A.; Parida, S.K. Advances in biotechnological applications in three important food legumes. *Plant Biotechnol. Rep.* **2014**, *8*, 83–99. [CrossRef]

123. Indurker, S.; Misra, H.S.; Eapen, S. Genetic transformation of chickpea (*Cicer arietinum* L.) with insecticidal crystal protein gene using particle gun bombardment. *Plant Cell Rep.* **2007**, *26*, 755–763. [CrossRef]
124. Sahoo, D.P.; Kumar, S.; Mishra, S.; Kobayashi, Y.; Panda, S.K.; Sahoo, L. Enhanced salinity tolerance in transgenic mungbean overexpressing Arabidopsis antiporter (*NHX1*) gene. *Mol. Breed.* **2016**, *36*, 144. [CrossRef]
125. Singh, P.; Kumar, D.; Sarin, N.B. Multiple abiotic stress tolerance in *Vigna mungo* is altered by overexpression of *ALDRXV4* gene via reactive carbonyl detoxification. *Plant Mol. Biol.* **2016**, *91*, 257–273. [CrossRef]
126. Mishra, S.; Behura, R.; Awasthi, J.P.; Dey, M.; Sahoo, D.; Das Bhowmik, S.S.; Panda, S.K.; Sahoo, L. Ectopic overexpression of a mungbean vacuolar Na+/H+ antiporter gene (*VrNHX1*) leads to increased salinity stress tolerance in transgenic *Vigna unguiculata* L. Walp. *Mol. Breed.* **2014**, *34*, 1345–1359. [CrossRef]
127. Citadin, C.T.; Cruz, A.R.; Aragao, F.J. Development of transgenic imazapyr-tolerant cowpea (*Vigna unguiculata*). *Plant Cell Rep.* **2013**, *32*, 537–543. [CrossRef] [PubMed]
128. Pigeaire, A.; Abernethy, D.; Smith, P.M.; Simpson, K.; Fletcher, N.; Lu, C.-Y.; Atkins, C.A.; Cornish, E. Transformation of a grain legume (*Lupinus angustifolius* L.) via *Agrobacterium tumefaciens*-mediated gene transfer to shoot apices. *Mol. Breed.* **1997**, *3*, 341–349. [CrossRef]
129. Atkins, C.A.; Emery, R.J.; Smith, P.M. Consequences of transforming narrow leafed lupin (*Lupinus angustifolius* [L.]) with an *ipt* gene under control of a flower-specific promoter. *Transgenic Res.* **2011**, *20*, 1321–1332. [CrossRef] [PubMed]
130. Barker, S.J.; Si, P.; Hodgson, L.; Ferguson-Hunt, M.; Khentry, Y.; Krishnamurthy, P.; Averis, S.; Mebus, K.; O'Lone, C.; Dalugoda, D.; et al. Regeneration selection improves transformation efficiency in narrow-leaf lupin. *Plant Cell Tissue Organ Cult. (PCTOC)* **2016**, *126*, 219–228. [CrossRef]
131. Wijayanto, T.; Barker, S.J.; Wylie, S.J.; Gilchrist, D.G.; Cowling, W.A. Significant reduction of fungal disease symptoms in transgenic lupin (*Lupinus angustifolius*) expressing the anti-apoptotic *baculovirus* gene *p35*. *Plant Biotechnol. J.* **2009**, *7*, 778–790. [CrossRef] [PubMed]
132. Tabe, L.; Wirtz, M.; Molvig, L.; Droux, M.; Hell, R. Overexpression of serine acetyltransferase produced large increases in O-acetylserine and free cysteine in developing seeds of a grain legume. *J. Exp. Bot.* **2010**, *61*, 721–733. [CrossRef] [PubMed]
133. Aragão, F.J.L.; Vianna, G.R.; Albino, M.M.C.; Rech, E.L. Transgenic dry bean tolerant to the herbicide glufosinate ammonium. *Crop Sci.* **2002**, *42*, 1298–1302. [CrossRef]
134. Rech, E.L.; Vianna, G.R.; Aragao, F.J. High-efficiency transformation by biolistics of soybean, common bean and cotton transgenic plants. *Nat. Protoc.* **2008**, *3*, 410–418. [CrossRef] [PubMed]
135. Collado, R.; Bermúdez-Caraballoso, I.; García, L.R.; Veitía, N.; Torres, D.; Romero, C.; Angenon, G. Epicotyl sections as targets for plant regeneration and transient transformation of common bean using *Agrobacterium tumefaciens*. *In Vitro Cell. Dev. Biol. Plant* **2016**, *52*, 500–511. [CrossRef]
136. Tiwari, S.; Mishra, D.K.; Singh, A.; Singh, P.K.; Tuli, R. Expression of a synthetic *cry1EC* gene for resistance against *Spodoptera litura* in transgenic peanut (*Arachis hypogaea* L.). *Plant Cell Rep.* **2008**, *27*, 1017–1025. [CrossRef] [PubMed]
137. Tiwari, S.; Mishra, D.K.; Chandrasekhar, K.; Singh, P.K.; Tuli, R. Expression of delta-endotoxin *Cry1EC* from an inducible promoter confers insect protection in peanut (*Arachis hypogaea* L.) plants. *Pest Manag. Sci.* **2011**, *67*, 137–145. [CrossRef] [PubMed]
138. Chu, Y.; Deng, X.Y.; Faustinelli, P.; Ozias-Akins, P. *Bcl-xL* transformed peanut (*Arachis hypogaea* L.) exhibits paraquat tolerance. *Plant Cell Rep.* **2008**, *27*, 85–92. [CrossRef]
139. Krishna, G.; Singh, B.K.; Kim, E.K.; Morya, V.K.; Ramteke, P.W. Progress in genetic engineering of peanut (*Arachis hypogaea* L.)—A review. *Plant Biotechnol. J.* **2015**, *13*, 147–162. [CrossRef] [PubMed]
140. Das, S.K.; Shethi, K.J.; Hoque, M.I.; Sarker, R.H. *Agrobacterium*-mediated genetic transformation in lentil (*Lens culinaris* Medik.) followed by in vitro flowering and seed formation. *Plant Tissue Cult. Biotechnol.* **2012**, *22*, 13–26. [CrossRef]
141. O'Sullivan, D.M.; Angra, D. Advances in faba bean genetics and genomics. *Front. Genet.* **2016**, *7*, 150. [CrossRef]
142. Jaganathan, D.; Ramasamy, K.; Sellamuthu, G.; Jayabalan, S.; Venkataraman, G. CRISPR for crop improvement: An update review. *Front. Plant Sci.* **2018**, *9*. [CrossRef]
143. Burglin, T.R. Analysis of TALE superclass homeobox genes (MEIS, PBC, KNOX, Iroquois, TGIF) reveals a novel domain conserved between plants and animals. *Nucleic Acids Res.* **1997**, *25*, 4173–4180. [CrossRef]

144. Magnani, E.; Hake, S. KNOX lost the OX: The Arabidopsis *KNATM* gene defines a novel class of KNOX transcriptional regulators missing the homeodomain. *Plant Cell* **2008**, *20*, 875–887. [CrossRef] [PubMed]
145. Endrizzi, K.; Moussian, B.; Haecker, A.; Levin, J.Z.; Laux, T. The *SHOOT MERISTEMLESS* gene is required for maintenance of undifferentiated cells in Arabidopsis shoot and floral meristems and acts at a different regulatory level than the meristem genes *WUSCHEL* and *ZWILLE*. *Plant J.* **1996**, *10*, 967–979. [CrossRef] [PubMed]
146. Chuck, G.; Lincoln, C.; Hake, S. *KNAT1* induces lobed leaves with ectopic meristems when overexpressed in Arabidopsis. *Plant Cell* **1996**, *8*, 1277–1289. [CrossRef] [PubMed]
147. Vollbrecht, E.; Veit, B.; Sinha, N.; Hake, S. The developmental gene *Knotted-1* is a member of a maize homeobox gene family. *Nature* **1991**, *350*, 241–243. [CrossRef] [PubMed]
148. Tamaoki, M.; Kusaba, S.; Kano-Murakami, Y.; Matsuoka, M. Ectopic expression of a tobacco homeobox gene, *NTH15*, dramatically alters leaf morphology and hormone levels in transgenic tobacco. *Plant Cell Physiol.* **1997**, *38*, 917–927. [CrossRef] [PubMed]
149. Frugis, G.; Giannino, D.; Mele, G.; Nicolodi, C.; Innocenti, A.M.; Chiappetta, A.; Bitonti, M.B.; Dewitte, W.; Van Onckelen, H.; Mariotti, D. Are homeobox knotted-like genes and cytokinins the leaf architects? *Plant Physiol.* **1999**, *119*, 371–374. [CrossRef] [PubMed]
150. Frugis, G.; Giannino, D.; Mele, G.; Nicolodi, C.; Chiappetta, A.; Bitonti, M.B.; Innocenti, A.M.; Dewitte, W.; Van Onckelen, H.; Mariotti, D. Overexpression of *KNAT1* in lettuce shifts leaf determinate growth to a shoot-like indeterminate growth associated with an accumulation of isopentenyl-type cytokinins. *Plant Physiol.* **2001**, *126*, 1370–1380. [CrossRef]
151. Kakimoto, T. Identification of plant cytokinin biosynthetic enzymes as dimethylallyl diphosphate: ATP/ADP isopentenyltransferases. *Plant Cell Physiol.* **2001**, *42*, 677–685. [CrossRef]
152. Takei, K.; Sakakibara, H.; Sugiyama, T. Identification of genes encoding adenylate isopentenyltransferase, a cytokinin biosynthesis enzyme, in *Arabidopsis thaliana*. *J. Biol. Chem.* **2001**, *276*, 26405–26410. [CrossRef]
153. Yanai, O.; Shani, E.; Dolezal, K.; Tarkowski, P.; Sablowski, R.; Sandberg, G.; Samach, A.; Ori, N. Arabidopsis KNOXI proteins activate cytokinin biosynthesis. *Curr. Biol.* **2005**, *15*, 1566–1571. [CrossRef]
154. Jasinski, S.; Piazza, P.; Craft, J.; Hay, A.; Woolley, L.; Rieu, I.; Phillips, A.; Hedden, P.; Tsiantis, M. KNOX action in *Arabidopsis* is mediated by coordinate regulation of cytokinin and gibberellin activities. *Curr. Biol.* **2005**, *15*, 1560–1565. [CrossRef] [PubMed]
155. Sakamoto, T.; Sakakibara, H.; Kojima, M.; Yamamoto, Y.; Nagasaki, H.; Inukai, Y.; Sato, Y.; Matsuoka, M. Ectopic expression of KNOTTED1-like homeobox protein induces expression of cytokinin biosynthesis genes in rice. *Plant Physiol.* **2006**, *142*, 54–62. [CrossRef]
156. Shani, E.; Burko, Y.; Ben-Yaakov, L.; Berger, Y.; Amsellem, Z.; Goldshmidt, A.; Sharon, E.; Ori, N. Stage-specific regulation of *Solanum lycopersicum* leaf maturation by class 1 KNOTTED1-LIKE HOMEOBOX proteins. *Plant Cell* **2009**, *21*, 3078–3092. [CrossRef] [PubMed]
157. Shani, E.; Ben-Gera, H.; Shleizer-Burko, S.; Burko, Y.; Weiss, D.; Ori, N. Cytokinin regulates compound leaf development in tomato. *Plant Cell* **2010**, *22*, 3206–3217. [CrossRef] [PubMed]
158. Sakamoto, T.; Kamiya, N.; Ueguchi-Tanaka, M.; Iwahori, S.; Matsuoka, M. KNOX homeodomain protein directly suppresses the expression of a gibberellin biosynthetic gene in the tobacco shoot apical meristem. *Genes Dev.* **2001**, *15*, 581–590. [CrossRef] [PubMed]
159. Tanaka-Ueguchi, M.; Itoh, H.; Oyama, N.; Koshioka, M.; Matsuoka, M. Over-expression of a tobacco homeobox gene, *NTH15*, decreases the expression of a gibberellin biosynthetic gene encoding GA 20-oxidase. *Plant J.* **1998**, *15*, 391–400. [CrossRef]
160. Bolduc, N.; Hake, S. The maize transcription factor KNOTTED1 directly regulates the gibberellin catabolism gene *ga2ox1*. *Plant Cell* **2009**, *21*, 1647–1658. [CrossRef]
161. Di Giacomo, E.; Iannelli, M.A.; Frugis, G. TALE and Shape: How to Make a Leaf Different. *Plants* **2013**, *2*, 317–342. [CrossRef]
162. Bolduc, N.; Yilmaz, A.; Mejia-Guerra, M.K.; Morohashi, K.; O'Connor, D.; Grotewold, E.; Hake, S. Unraveling the KNOTTED1 regulatory network in maize meristems. *Genes Dev.* **2012**, *26*, 1685–1690. [CrossRef]
163. Scofield, S.; Murison, A.; Jones, A.; Fozard, J.; Aida, M.; Band, L.R.; Bennett, M.; Murray, J.A.H. Coordination of meristem and boundary functions by transcription factors in the *SHOOT MERISTEMLESS* regulatory network. *Development* **2018**, *145*, 157081. [CrossRef]

164. Clouse, S.D.; Sasse, J.M. BRASSINOSTEROIDS: Essential regulators of plant growth and development. *Annu. Rev. Plant Physiol. Plant Mol. Biol.* **1998**, *49*, 427–451. [CrossRef] [PubMed]
165. Sun, Y.; Fan, X.Y.; Cao, D.M.; Tang, W.; He, K.; Zhu, J.Y.; He, J.X.; Bai, M.Y.; Zhu, S.; Oh, E.; et al. Integration of brassinosteroid signal transduction with the transcription network for plant growth regulation in Arabidopsis. *Dev. Cell* **2010**, *19*, 765–777. [CrossRef] [PubMed]
166. Tsuda, K.; Kurata, N.; Ohyanagi, H.; Hake, S. Genome-wide study of *KNOX* regulatory network reveals brassinosteroid catabolic genes important for shoot meristem function in rice. *Plant Cell* **2014**, *26*, 3488–3500. [CrossRef] [PubMed]
167. Gendron, J.M.; Liu, J.S.; Fan, M.; Bai, M.Y.; Wenkel, S.; Springer, P.S.; Barton, M.K.; Wang, Z.Y. Brassinosteroids regulate organ boundary formation in the shoot apical meristem of Arabidopsis. *Proc. Natl. Acad. Sci. USA* **2012**, *109*, 21152–21157. [CrossRef] [PubMed]
168. Cano-Delgado, A.; Yin, Y.; Yu, C.; Vafeados, D.; Mora-Garcia, S.; Cheng, J.C.; Nam, K.H.; Li, J.; Chory, J. BRL1 and BRL3 are novel brassinosteroid receptors that function in vascular differentiation in Arabidopsis. *Development* **2004**, *131*, 5341–5351. [CrossRef] [PubMed]
169. Mele, G.; Ori, N.; Sato, Y.; Hake, S. The *knotted1*-like homeobox gene *BREVIPEDICELLUS* regulates cell differentiation by modulating metabolic pathways. *Genes Dev.* **2003**, *17*, 2088–2093. [CrossRef] [PubMed]
170. Testone, G.; Condello, E.; Verde, I.; Nicolodi, C.; Caboni, E.; Dettori, M.T.; Vendramin, E.; Bruno, L.; Bitonti, M.B.; Mele, G.; et al. The peach (*Prunus persica* L. Batsch) genome harbours 10 *KNOX* genes, which are differentially expressed in stem development, and the class 1 *KNOPE1* regulates elongation and lignification during primary growth. *J. Exp. Bot.* **2012**, *63*, 5417–5435. [CrossRef]
171. Hake, S.; Smith, H.M.; Holtan, H.; Magnani, E.; Mele, G.; Ramirez, J. The role of *knox* genes in plant development. *Annu. Rev. Cell Dev. Biol.* **2004**, *20*, 125–151. [CrossRef]
172. Kerstetter, R.; Vollbrecht, E.; Lowe, B.; Veit, B.; Yamaguchi, J.; Hake, S. Sequence analysis and expression patterns divide the maize *knotted1*-like homeobox genes into two classes. *Plant Cell* **1994**, *6*, 1877–1887. [CrossRef]
173. Furumizu, C.; Alvarez, J.P.; Sakakibara, K.; Bowman, J.L. Antagonistic roles for *KNOX1* and *KNOX2* genes in patterning the land plant body plan following an ancient gene duplication. *PLoS Genet.* **2015**, *11*, e1004980. [CrossRef]
174. Di Giacomo, E.; Sestili, F.; Iannelli, M.A.; Testone, G.; Mariotti, D.; Frugis, G. Characterization of *KNOX* genes in *Medicago truncatula*. *Plant Mol. Biol.* **2008**, *67*, 135–150. [CrossRef] [PubMed]
175. Di Giacomo, E.; Laffont, C.; Sciarra, F.; Iannelli, M.A.; Frugier, F.; Frugis, G. KNAT3/4/5-like class 2 KNOX transcription factors are involved in *Medicago truncatula* symbiotic nodule organ development. *New Phytol.* **2017**, *213*, 822–837. [CrossRef]
176. Poethig, R.S. Vegetative phase change and shoot maturation in plants. *Curr. Top. Dev. Biol.* **2013**, *105*, 125–152. [PubMed]
177. Hanke, M.; Flachowsky, H.; Peil, A.; Hättasch, C. No Flower no Fruit—Genetic Potentials to Trigger Flowering in Fruit Trees. In *Genes, Genomes and Genomics*; Books, G.S., Ed.; Global Science Books Ltd.: Ikenobe, Japan, 2007; Volume 1, pp. 1–20.
178. Layne, D.; Bassi, D. *The Peach: Botany, Production and Uses*; CABI: Oxfordshire, UK, 2008; pp. 1–615.
179. Bitonti, M.B.; Cozza, R.; Chiappetta, A.; Giannino, D.; Ruffini Castiglione, M.; Dewitte, W.; Mariotti, D.; Van Onckelen, H.; Innocenti, A.M. Distinct nuclear organization, DNA methylation pattern and cytokinin distribution mark juvenile, juvenile-like and adult vegetative apical meristems in peach (*Prunus persica* (L.) Batsch). *J. Exp. Bot.* **2002**, *53*, 1047–1054. [CrossRef] [PubMed]
180. Giannino, D.; Mele, G.; Cozza, R.; Bruno, L.; Testone, G.; Ticconi, C.; Frugis, G.; Bitonti, M.B.; Innocenti, A.M.; Mariotti, D. Isolation and characterization of a maintenance DNA-methyltransferase gene from peach (*Prunus persica* [L.] Batsch): Transcript localization in vegetative and reproductive meristems of triple buds. *J. Exp. Bot.* **2003**, *54*, 2623–2633. [CrossRef] [PubMed]
181. Giannino, D.; Frugis, G.; Ticconi, C.; Florio, S.; Mele, G.; Santini, L.; Cozza, R.; Bitonti, M.B.; Innocenti, A.; Mariotti, D. Isolation and molecular characterisation of the gene encoding the cytoplasmic ribosomal protein S28 in *Prunus persica* [L.] Batsch. *Mol. Gen. Genet.* **2000**, *263*, 201–212. [CrossRef] [PubMed]
182. Van Nocker, S.; Gardiner, S.E. Breeding better cultivars, faster: Applications of new technologies for the rapid deployment of superior horticultural tree crops. *Hortic. Res.* **2014**, *1*, 14022. [CrossRef]

183. Gordon, D.; Damiano, C.; DeJong, T.M. Preformation in vegetative buds of *Prunus persica*: Factors influencing number of leaf primordia in overwintering buds. *Tree Physiol.* **2006**, *26*, 537–544. [CrossRef]
184. Reinoso, H.; Luna, V.; Pharis, R.; Bottini, R. Dormancy in peach (*Prunus persica*) flower buds. V. Anatomy of bud development in relation to phenological stage. *Can. J. Bot.* **2002**, *80*, 656–663. [CrossRef]
185. Yamane, H. Regulation of bud dormancy and bud break in japanese apricot (*Prunus mume* Siebold & Zucc.) and peach [*Prunus persica* (L.) Batsch]: A summary of recent studies. *Hortic. J.* **2014**, *83*, 187–202.
186. Hyun, Y.; Richter, R.; Coupland, G. Competence to flower: Age-controlled sensitivity to environmental cues. *Plant Physiol.* **2017**, *173*, 36–46. [CrossRef] [PubMed]
187. Zhang, L.; Hu, Y.; Wang, H.; Feng, S.; Zhang, Y. Involvement of miR156 in the regulation of vegetative phase change in plants. *J. Am. Soc. Hortic. Sci.* **2015**, *140*, 387–395.
188. Wang, J.W.; Park, M.Y.; Wang, L.J.; Koo, Y.; Chen, X.Y.; Weigel, D.; Poethig, R.S. miRNA control of vegetative phase change in trees. *PLoS Genet.* **2011**, *7*, e1002012. [CrossRef] [PubMed]
189. Xu, X.; Li, X.; Hu, X.; Wu, T.; Wang, Y.; Xu, X.; Zhang, X.; Han, Z. High miR156 expression is required for auxin-induced adventitious root formation via *MxSPL26* independent of PINs and ARFs in *Malus xiaojinensis*. *Front. Plant Sci.* **2017**, *8*, 1059. [CrossRef] [PubMed]
190. Jia, X.L.; Chen, Y.K.; Xu, X.Z.; Shen, F.; Zheng, Q.B.; Du, Z.; Wang, Y.; Wu, T.; Xu, X.F.; Han, Z.H.; et al. miR156 switches on vegetative phase change under the regulation of redox signals in apple seedlings. *Sci. Rep.* **2017**, *7*, 14223. [CrossRef] [PubMed]
191. Bastías, A.; Almada, R.; Rojas, P.; Donoso, J.M.; Hinrichsen, P.; Sagredo, B. Aging gene pathway of microRNAs 156/157 and 172 is altered in juvenile and adult plants from in vitro propagated *Prunus* sp. *Cienc. Investig. Agrar.* **2016**, *43*, 429–441. [CrossRef]
192. Sgamma, T.; Cirilli, M.; Caboni, E.; Maurizio, M.; Thomas, B.; Muleo, R. In vitro plant culture system induces phase transition in fruit-bearing plants. *Acta Hortic.* **2016**, *1110*, 13–20. [CrossRef]
193. Albani, M.C.; Coupland, G. Comparative analysis of flowering in annual and perennial plants. *Curr. Top. Dev. Biol.* **2010**, *91*, 323–348.
194. Wells, C.E.; Vendramin, E.; Jimenez Tarodo, S.; Verde, I.; Bielenberg, D.G. A genome-wide analysis of MADS-box genes in peach [*Prunus persica* (L.) Batsch]. *BMC Plant Biol.* **2015**, *15*, 41. [CrossRef]
195. Hong, Y.; Jackson, S. Floral induction and flower formation—The role and potential applications of miRNAs. *Plant Biotechnol. J.* **2015**, *13*, 282–292. [CrossRef]
196. Li, S.; Shao, Z.; Fu, X.; Xiao, W.; Li, L.; Chen, M.; Sun, M.; Li, D.; Gao, D. Identification and characterization of *Prunus persica* miRNAs in response to UVB radiation in greenhouse through high-throughput sequencing. *BMC Genom.* **2017**, *18*, 938. [CrossRef] [PubMed]
197. Sun, L.; Wang, Y.; Yan, X.; Cheng, T.; Ma, K.; Yang, W.; Pan, H.; Zheng, C.; Zhu, X.; Wang, J.; et al. Genetic control of juvenile growth and botanical architecture in an ornamental woody plant, *Prunus mume* Sieb. et Zucc. as revealed by a high-density linkage map. *BMC Genet.* **2014**, *15*, S1. [CrossRef] [PubMed]
198. Romeu, J.F.; Monforte, A.J.; Sanchez, G.; Granell, A.; Garcia-Brunton, J.; Badenes, M.L.; Rios, G. Quantitative trait loci affecting reproductive phenology in peach. *BMC Plant Biol.* **2014**, *14*, 52. [CrossRef] [PubMed]
199. Fan, S.; Bielenberg, D.G.; Zhebentyayeva, T.N.; Reighard, G.L.; Okie, W.R.; Holland, D.; Abbott, A.G. Mapping quantitative trait loci associated with chilling requirement, heat requirement and bloom date in peach (*Prunus persica*). *New Phytol.* **2010**, *185*, 917–930. [CrossRef] [PubMed]
200. Hernandez Mora, J.R.; Micheletti, D.; Bink, M.; Van de Weg, E.; Cantin, C.; Nazzicari, N.; Caprera, A.; Dettori, M.T.; Micali, S.; Banchi, E.; et al. Integrated QTL detection for key breeding traits in multiple peach progenies. *BMC Genom.* **2017**, *18*, 404. [CrossRef] [PubMed]
201. Sabbadini, S.; Pandolfini, T.; Girolomini, L.; Molesini, B.; Navacchi, O. Peach (*Prunus persica* L.). In *Agrobacterium Protocols*; Wang, K., Ed.; Springer: New York, NY, USA, 2015; Volume 2, pp. 205–215.
202. Liu, H.; Qian, M.; Song, C.; Li, J.; Zhao, C.; Li, G.; Wang, A.; Han, M. Down-regulation of *PpBGAL10* and *PpBGAL16* delays fruit softening in peach by reducing polygalacturonase and pectin methylesterase activity. *Front Plant Sci.* **2018**, *9*, 1015. [CrossRef] [PubMed]
203. Cui, H.; Wang, A. An efficient viral vector for functional genomic studies of *Prunus* fruit trees and its induced resistance to Plum pox virus via silencing of a host factor gene. *Plant Biotechnol. J.* **2017**, *15*, 344–356. [CrossRef]
204. Nagle, M.; Dejardin, A.; Pilate, G.; Strauss, S.H. Opportunities for innovation in genetic transformation of forest trees. *Front. Plant Sci.* **2018**, *9*, 1443. [CrossRef]

205. Srinivasan, C.; Dardick, C.; Callahan, A.; Scorza, R. Plum (*Prunus domestica*) trees transformed with poplar *FT1* result in altered architecture, dormancy requirement, and continuous flowering. *PLoS ONE* **2012**, *7*, e40715. [CrossRef]
206. Petri, C.; Alburquerque, N.; Faize, M.; Scorza, R.; Dardick, C. Current achievements and future directions in genetic engineering of European plum (*Prunus domestica* L.). *Transgenic Res.* **2018**, *27*, 225–240. [CrossRef]
207. Yamagishi, N.; Kishigami, R.; Yoshikawa, N. Reduced generation time of apple seedlings to within a year by means of a plant virus vector: A new plant-breeding technique with no transmission of genetic modification to the next generation. *Plant Biotechnol. J.* **2014**, *12*, 60–68. [CrossRef] [PubMed]
208. Malnoy, M.; Viola, R.; Jung, M.H.; Koo, O.J.; Kim, S.; Kim, J.S.; Velasco, R.; Nagamangala Kanchiswamy, C. DNA-Free Genetically Edited Grapevine and Apple Protoplast Using CRISPR/Cas9 Ribonucleoproteins. *Front. Plant Sci.* **2016**, *7*, 1904. [CrossRef] [PubMed]
209. Us-Camas, R.Y.; Rivera-Solís, G.; Duarte-Aké, F.; De-la-Peña, C. In vitro culture: An epigenetic challenge for plants. *Plant Cell Tissues Organ Cult.* **2014**, *118*, 187–201. [CrossRef]
210. Wang, J.; Zhang, X.; Yan, G.; Zhou, Y.; Zhang, K. Over-expression of the *PaAP1* gene from sweet cherry (*Prunus avium* L.) causes early flowering in *Arabidopsis thaliana*. *J. Plant Physiol.* **2013**, *170*, 315–320. [CrossRef] [PubMed]
211. Zhang, X.; An, L.; Nguyen, T.H.; Liang, H.; Wang, R.; Liu, X.; Li, T.; Qi, Y.; Yu, F. The Cloning and Functional Characterization of Peach *CONSTANS* and *FLOWERING LOCUS T* Homologous Genes *PpCO* and *PpFT*. *PLoS ONE* **2015**, *10*, e0124108. [CrossRef] [PubMed]
212. Yarur, A.; Soto, E.; Leon, G.; Almeida, A.M. The sweet cherry (*Prunus avium*) *FLOWERING LOCUS T* gene is expressed during floral bud determination and can promote flowering in a winter-annual Arabidopsis accession. *Plant Reprod.* **2016**, *29*, 311–322. [CrossRef] [PubMed]
213. Xu, Y.; Zhang, L.; Xie, H.; Zhang, Y.-Q.; Oliveira, M.M.; Ma, R.-C. Expression analysis and genetic mapping of three *SEPALLATA*-like genes from peach (*Prunus persica* (L.) Batsch). *Tree Genet. Genom.* **2008**, *4*, 693–703. [CrossRef]
214. Li, Y.; Xu, Z.; Yang, W.; Cheng, T.; Wang, J.; Zhang, Q. Isolation and functional characterization of *SOC1*-like genes in *Prunus mume*. *J. Am. Soc. Hortic. Sci.* **2016**, *141*, 315–326.
215. Wisniewski, M.; Norelli, J.; Bassett, C.; Artlip, T.; Macarisin, D. Ectopic expression of a novel peach (*Prunus persica*) CBF transcription factor in apple (*Malus* × *domestica*) results in short-day induced dormancy and increased cold hardiness. *Planta* **2011**, *233*, 971–983. [CrossRef]
216. Sasaki, R.; Yamane, H.; Ooka, T.; Jotatsu, H.; Kitamura, Y.; Akagi, T.; Tao, R. Functional and expressional analyses of *PmDAM* genes associated with endodormancy in Japanese apricot. *Plant Physiol.* **2011**, *157*, 485–497. [CrossRef]
217. Chen, Y.; Jiang, P.; Thammannagowd, S.; Liang, H. Characterization of Peach *TFL1* and comparison with FT/TFL1 gene families of the *Rosaceae*. *J. Am. Soc. Hortic. Sci.* **2013**, *138*, 12–17.

© 2019 by the authors. Licensee MDPI, Basel, Switzerland. This article is an open access article distributed under the terms and conditions of the Creative Commons Attribution (CC BY) license (http://creativecommons.org/licenses/by/4.0/).

Review

Patterning the Axes: A Lesson from the Root

Riccardo Di Mambro [1], Sabrina Sabatini [2] and Raffaele Dello Ioio [2,*]

[1] Department of Biology, University of Pisa, via L. Ghini, 13-56126 Pisa, Italy; riccardo.dimambro@unipi.it
[2] Dipartimento di Biologia e Biotecnologie, Laboratory of Functional Genomics and Proteomics of Model Systems, Università di Roma "Sapienza", via dei Sardi, 70-00185 Rome, Italy; sabrina.sabatini@uniroma1.it
* Correspondence: raffaele.delloioio@uniroma1.it

Received: 22 November 2018; Accepted: 24 December 2018; Published: 31 December 2018

Abstract: How the body plan is established and maintained in multicellular organisms is a central question in developmental biology. Thanks to its simple and symmetric structure, the root represents a powerful tool to study the molecular mechanisms underlying the establishment and maintenance of developmental axes. Plant roots show two main axes along which cells pass through different developmental stages and acquire different fates: the root proximodistal axis spans longitudinally from the hypocotyl junction (proximal) to the root tip (distal), whereas the radial axis spans transversely from the vasculature tissue (centre) to the epidermis (outer). Both axes are generated by stereotypical divisions occurring during embryogenesis and are maintained post-embryonically. Here, we review the latest scientific advances on how the correct formation of root proximodistal and radial axes is achieved.

Keywords: *Arabidopsis*; root; stem cells; root development; differentiation; ground tissue; radial patterning; proximodistal patterning

---

1. Introduction

One of the most intriguing questions in developmental biology is how the body plan is established. To answer this question, for decades scientists have focused on the formation of developmental axes, utilizing different model systems. Most of our knowledge on axes formation derives from studies on vertebrate limb development [1,2]. However, these systems present several limitations due to their complex structure that limits analysis at a single cell resolution. On the contrary, plant roots display a simple and organized structure, where cell lineages are easily distinguishable by shape and position [3,4]. Furthermore, due to the presence of the cell wall, plant cells do not migrate; hence, cell fate and identity can be easily followed during different stages of organ development [3–5]. For these reasons, roots represent a powerful tool to study the molecular mechanisms on how developmental axes are established and maintained. Roots can be represented as a series of concentric cylinders, where epidermis is the outermost tissue while the vasculature bundles lie in the centre [4] (Figure 1). Roots display two main developmental axes: the proximodistal axis, extending longitudinally from the root–shoot junction (proximal) to the root apex (distal); the radial axis, spreading transversally from the vasculature bundles to the epidermis [4] (Figure 1). Like other animal model systems, root axes are established during embryogenesis and maintained post-embryonically by the activity of meristems [3–6]. Meristems are localized structures that sustain post embryonic indeterminate plant organ growth due to the activity of stem cell niches (SCNs) [3–5]. In the *Arabidopsis* root meristem, there are five sets of stem cells (initials) that give rise to all root tissues: epidermis and lateral root cap initials (EPI LRC STEM CELLS), cortex and endodermis initial (CEI), pericycle initials, vasculature initials and distally columella initials (Figure 1) [3,4]. These sets of stem cells surround the QC (Quiescent Center) which maintains, by contacting them, their stem cell identity (Figure 1) [7,8]. The stem cells divide asymmetrically and anticlinally generating daughter cells (Figure 1) that generates both the proximodistal and radial axes through stereotypical cell divisions. Along the proximodistal axis, the stem cell daughters divide anticlinally a fixed number of times, generating the division zone of the meristem. In the proximal area of the meristem, those cells cease to divide when they reach a boundary called the transition zone (TZ). Here they start to elongate and differentiate, generating the elongation/differentiation zone [3,5,9,10] (Figure 1). In this zone, cells acquire characteristic differentiation features such as root hairs for the epidermis or tracheids for the vascular cells [11,12]. The position of the TZ is fundamental for proximodistal axis specification, as it marks the boundary between undifferentiated and differentiated cells [9,13].

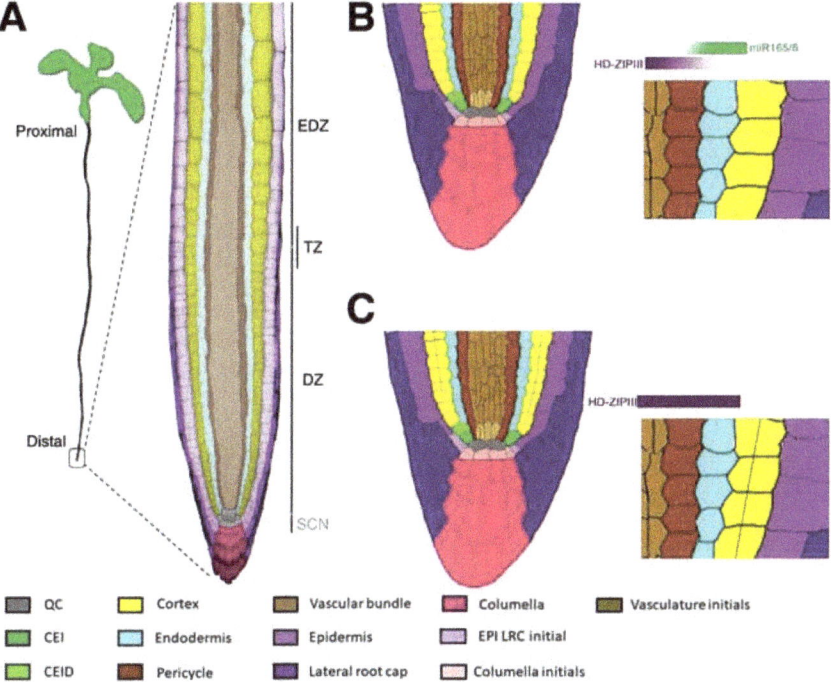

Figure 1. *Arabidopsis* root structure. (**A**) Representation of an *Arabidopsis* seedling where the proximodistal axis is indicated. In the blow up, a representation of the *Arabidopsis* root apex is shown where false colours highlight the different tissues. Root zonation: stem cell niche, SCN; division zone, DZ; elongation/differentiation zone, EDZ; transition zone, TZ. (**B**) Cartoon reporting the longitudinal section of a wild type (Wt) *Arabidopsis* root stem cell niche. Different colours represent root tissues and initials, as indicated in the legend. The blow up highlights the typical ground tissue (GT) architecture (one layer of endodermis and one layer of cortex) resulting from the opposite graded distribution of miR165/6 and Class III Homeodomain Leucine Zipper (HD-ZIPIII) (triangle shapes above blow up). In particular, miR165/6 (green) presents low expression in the vascular bundle and high expression in the endodermis, constraining HD-ZIPIII expression. As a result, HD-ZIPIII (red) present high expression in the vascular bundle and low expression in the endodermis. (**C**) Cartoon reporting the longitudinal section of an *Arabidopsis* stem cell niche lacking miR165/6 expression. The blow up highlights the HD-ZIPIII expanded expression in the whole ground tissue (GT). This results in the formation of an extra layer of the cortical tissue (dashed line). QC, quiescent centre; CEI, cortex and endodermis initial; CEID, cortex and endodermis initial daughter cell; EPI, epidermis; LRC, lateral root cap.

Radially, most of the stem cells daughters divide periclinally, giving rise to two tissues with different identities. For example, cortex and endodermis are derived from the periclinal division of the daughter of the cortex and endodermis initial (CEI), whereas epidermis and lateral root cap originate from the EPI LRC initial [14–16]. The control of the asymmetric divisions occurring in the stem cell daughters is key for the correct patterning of the radial axis. Indeed, alteration of the position and timing of those divisions causes the formation of aberrant body plan and shape (Figure 1).

Thanks to the generation of new tools and the improvement of molecular methodologies, several molecular mechanisms underlying the establishment and maintenance of the proximodistal and radial are in the process of being discovered and fully comprehended. In this review, we report the current view on how these two axes are patterned.

2. Root Radial Axis

The root radial axis organization depends on the coordinated activity of periclinal divisions of the stem cell daughters. One of the most studied mechanisms patterning the radial axis is the one controlling the formation of the cortex and the endodermis root tissues. These tissues originate from a single stem cell (CEI) that firstly divides anticlinally, thereby generating a daughter cell (CEID). This cell divides periclinally, generating the cortex and the endodermis that together are called Ground Tissue (GT). GT specification starts in the embryo when a periclinal division at early globular embryonic stage separates the pro-vasculature tissues from the GT precursor cell. Only later, at the heart embryonic stage, a pro-GT division leads to the specification of the cortex and the endodermis [6]. It was recently shown that the establishment of the pro-GT at early embryonic stages depends on the plant hormone auxin. A maximum level of auxin activity driven by the auxin responsive factor MONOPTEROS/AUXIN RESPONSIVE FACTOR 5 (MP/ARF5) in the GT precursor cells is required for GT formation [17]. Indeed, *mp* null mutants display impaired GT establishment [17].

Two GRAS family transcription factors, SHORTROOT (SHR) and SCARECROW (SCR), are involved in the formation of the cortex and endodermis layers, as they are necessary and sufficient to promote the CEID periclinal division [18–21]. SHR is a mobile transcription factor expressed in the vasculature. SHR moves toward the CEID, CEI and endodermis via plasmodesmata, where it is sequestered into the nucleus [22,23] (Figures 2 and 3). SHR movements restriction is fundamental for GT patterning, as overexpression of SHR results in additional GT layer formation [24,25]. In the vasculature, SHR is maintained mostly in the cytoplasm by the activity of SCARECROW-LIKE23 (SCL23) [26]. In the CEID, SHR forms a molecular complex with SCR and it is sequestered in the nucleus by the activity of SCR. In the nucleus, SHR/SCR complex sustains the expression of *SCR* itself and induces the expression of *INDETERMINATE DOMAIN C2H2 zinc finger* (BIRD) transcription factors such as *JACKDAW* (*JKD*), *NUTCRACKER* (*NUC*) and *MAGPIE* (*MGP*) [21,27–31]. BIRD proteins physically interact with the SHR/SCR complex, restricting SHR movements to the stele [27,29,32]. SHR/SCR complex promotes the expression of the cell cycle regulator *CYCLIND6* (*CYCD6;1*) in the CEID, inducing here a periclinal division [33,34]. Via a combination of mathematical modelling and wet biology, it has been proposed that the SHR/SCR/CYCD6;1 module, together with the cell cycle inhibitor RETINOBLASTOMA-RELATED (RBR) protein, acts via a bistable circuit to regulate the CEID asymmetric division [33–35]. In the CEI, the CYCD6;1 together with the CDKB1;1 (CYCLIN DEPENDENT KINASE 1;1) or CDKB1;2 induces the phosphorylation of RBR, reducing its activity in the CEID, thus promoting the periclinal division [34]. Auxin is a key factor for the promotion of this periclinal division. Indeed, an auxin maximum in the CEI promotes *CYCD6;1* expression [34]. On the contrary, RBR was shown to directly interact with SCR, reducing its transcriptional activator activity in the endodermis [36]. The RBR and SCR interaction, together with the activity of the RBR regulator CYCD6;1, limits the asymmetric cell division in the SCN, thus allowing the formation of the endodermal and cortical layers (Figure 2). Recently, a sophisticated molecular mechanism was proposed for a SHR/SCR-dependent switching on of the CYCD6;1 involving the RNA POLYMERASE II cofactor Mediator. Depending on the SHR concentration, SCR interacts with the subunit 31 of the Mediator to promote *CYCD6;1* expression [37].

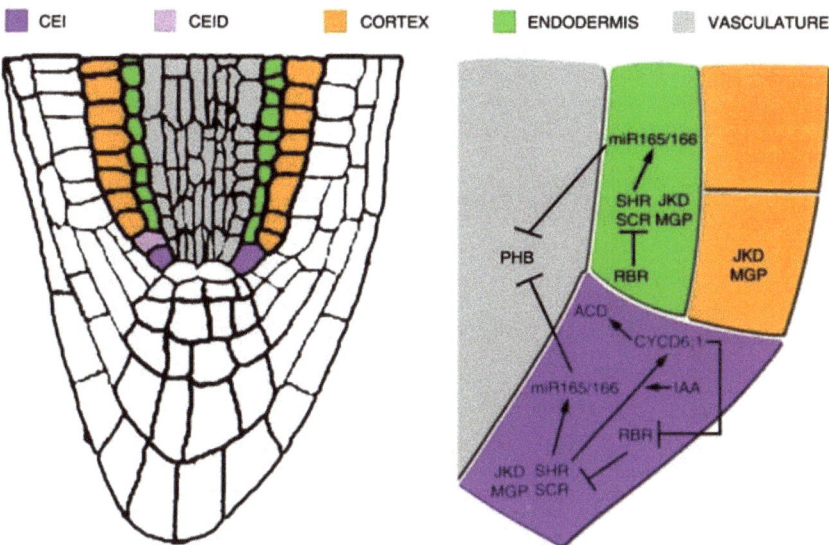

Figure 2. Schematic representation of the gene regulatory network acting for the cortex and endodermis initial periclinal division. On the left, representation is provided of the *Arabidopsis* root tip, where the cortex and endodermis initial (CEI) and its daughter cell (CEID), cortex, endodermis and vascular tissues are depicted in colour. In the blow up, the gene regulatory network supporting the CEI asymmetric cell division (ACD) is shown. In the CEI, the SHR/SCR complex sustains the expression of SCR and promotes the expressions of CYCD1;6 and of JKD and MGP. CYCD6;1 expression is also sustained by high levels of auxin (IAA) in the CEI. CYCD6;1 represses RBR activity, which in turn regulates negatively the ACD by a direct repression of SCR activity. SHR/SCR complex also promotes the expression of miR165/6, thus restricting PHB expression in the vascular tissue.

Once CEID divides, several factors coordinate the formation of the cortical and endodermal layers. SHR also promotes endodermal fate, as suggested by the loss of endodermis identity in *shr* mutants [18,28,38,39]. It was shown that BIRD proteins, other than regulating SHR movements, play a key role in determining cortical identity, as multiple mutant combinations of BIRD members show GT with no cortical identity [32]. Therefore, the combined activity of SHR, SCR and BIRD proteins is necessary to pattern the GT. Interestingly SHR and SCR are involved only in the maintenance of GT and not in its establishment. Once MP initiates the ground tissue lineage, it acts upstream of the SHR/SCR module, controlling ground tissue patterning and maintenance.

SCHIZORIZA (SCZ), a member of the Heat Shock Transcription Factor family, is also involved in GT patterning and its activity depends on SHR and SCR [40–42]. Interestingly, SCZ is expressed in all root tissues except for the lateral root cap. It was shown that SCZ, together with JKD, MGP, and NUC proteins, promotes cortical identity (Figures 2 and 3) [32]. It must be pointed out that *scz* mutants present additional tissue layers with mixed cortical, endodermal and epidermal identities, suggesting a role for this gene in tissue fate separation [40–42]. The analysis of SCZ target genes will help to establish an understanding of how SCZ patterns the radial axis.

Besides organizing the GT, SHR/SCR complex is involved in vasculature patterning. *Arabidopsis* root vasculature consists of an inner xylem bundle (metaxylem in the centre, protoxylem aside) with two juxtaposed phloematic bundles [4] (Figure 3). The formation and development of the metaxylem depends on the redundant activity of the Class III Homeodomain Leucine Zipper (HD-ZIPIII) members, a family of five transcription factors targeted by microRNA 165/6 (miR165/6). SHR/SCR promotes in the endodermis the expression of miR165/6 that, moving toward the stele via plasmodesmata, generates an opposite gradient of the HD-ZIPIII proteins, with a maximum in the metaxylem and a minimum in the endodermis [22,43,44] (Figures 1 and 3). The formation of a radial gradient of HD-ZIPIIIs is sufficient to pattern the xylem fate specification, as high HD-ZPIIIs levels promote metaxylem formation, whereas low ones promote protoxylem [43,44] (Figure 3). In the stele, HD-ZIPIIIs control the biosynthesis and activity of the phytohormone cytokinin, which in turn regulates auxin distribution and signalling [44,45]. This finely regulated mechanism is sufficient to pattern the stele.

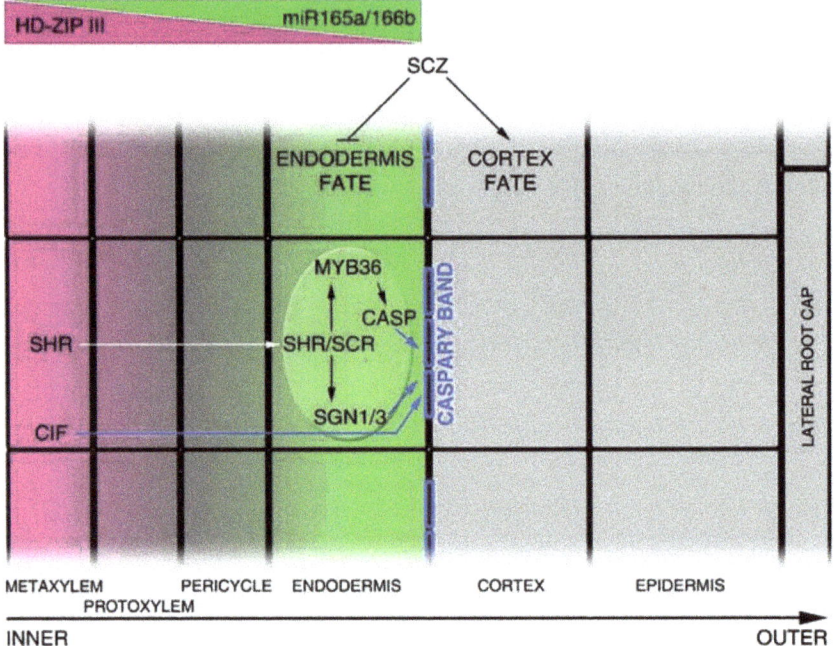

Figure 3. Image showing the molecular mechanisms controlling radial axis patterning. The figure shows a half radial section of the *Arabidopsis* root. Each square file corresponds to a different tissue layer, where the central file (inner) corresponds to the metaxylem and the outer file to the lateral root cap, as indicated in the scheme. Class III Homeodomain Leucine Zipper III (HD-ZIPIII) and miR165a/6b (opposite gradients) are indicated in purple and green, respectively. Blue squares on endodermal cells represent Casparian strips. White arrow indicates SHR protein movement from the vascular tissue into endodermal cell nucleus. Blue arrows indicate the CIF-, CASP- and SGN1-dependent regulation of Caspary band formation. SCZ promotes the cortical identity supporting tissue fate separation.

It was recently shown that miR165/6 distribution is not only crucial for vasculature development but also for GT patterning [46,47] (Figures 1 and 2). A mir165/6-dependent minimum of HD-ZIPIIIs in the CEI/CEID and endodermis is required to restrict the number of cortical layers, as miR165/6-insensitive HD-ZIPIII mutants show additional cortical layers [47]. HD-ZIPIIIs expression in the GT results in ectopic *CYCD6;1* activation, prompting additional GT divisions. Intriguingly, the HD-ZIPIII member PHABULOSA (PHB) indirectly sustains *CYCD6;1* expression in a SHR-independent manner, but how PHB triggers periclinal divisions is still not known [46–48]. It was recently shown that PHB directly targets MP to pattern the vasculature tissue [49]. Nevertheless, whether PHB/MP circuit is important for GT establishment is not known.

It was recently shown that SHR, together with SCR, also specifies endodermis differentiation. Functional endodermis is characterized by Casparian strips, lignified structures deposited on the radial and transverse side of the endodermal cell wall [24]. SHR directs the formation of the Casparian strips by inducing the MYB DOMAIN transcription factor *MYB36* and the receptor-like kinase *SGN1* and

SGN3 [50,51] (Figure 3). In endodermal cells, MYB36 induces the expression of the transmembrane proteins CASPARIAN STRIP MEMBRANE DOMAIN PROTEIN (CASP) [52,53], which are involved in the recruitment of lignin synthesis enzymes on the plasma-membrane. SGN1/3 position CASP proteins on the plasma membrane (Figure 3). Nonetheless, SHR promotes the formation of a non-functional Caspary band and it requires the activity of vasculature-deriving small peptides, *CASPARIAN STRIP INTEGRITY FACTOR* (*CIF*), to generate a functional strip [50,51] (Figure 3). Hence, SHR and SCR constitute an important module to control endodermis differentiation.

3. Root Proximodistal Axis

Different from the radial axis, where most of the cells show different identities but similar developmental stages, along the proximodistal axis cells display different stages of development. Positioning of the TZ plays a key role for patterning the proximodistal axis, as the TZ separates proliferating meristematic cells from the elongated ones [54] (Figures 1 and 4). The position of the TZ depends on the dynamic equilibrium between cell division and cell differentiation; alterations of this equilibrium cause the TZ position to shift toward the distal or the proximal area of the root, thus varying the proximodistal zonation.

Auxin plays a pivotal role in establishing the root proximodistal axis, acting as a local morphogen [55,56]. Already at the globular stage of embryogenesis, a maximum of auxin in the basal pole of the embryo determines the position of the SCN [57]. This auxin maximum is controlled by the activity of the auxin polar transport efflux facilitators PIN FORMED (PINs) that distribute this hormone [58–60]. Auxin signalling is necessary for the formation of the SCN [55,61]. Interestingly, *mp* loss of function mutants or gain of function mutants of its repressor, the AUX/IAA auxin signalling repressor BODENLOS (BDL), display no root formation [61–63]. Together with auxin, four AP2 transcription factors, PLETHORA 1,2,3 and 4 (PLT1,2,3 and 4), control stem cell activity and root growth from embryogenesis onwards [64,65]. Multiple combinations of the loss of function mutants *plt1,2,3,4* show no root SCN formation, whereas constitutive expressions of *PLT* genes induces shoot homeotic transformation into root [65]. The GATA transcription factor HANABA TARANU/MONOPOLE (HAN) forms the boundary between embryonic apical and basal pole, confining PLT expression and the auxin maximum to the root precursors domain [66]. PLTs also play also an active role in the repression of the apical pole identity. Indeed, PLT, together with miR165/6, represses the apical embryonic SCN formation by restricting HD-ZIPIIIs expression [67,68]. Lack of this repression leads to the homeotic transformation of the root into shoot, suggesting a master role for PLT in determining the embryonic apical-basal axis [68]. It has been recently demonstrated that PLT regulates the expression of *HAN* and the synthesis of auxin via direct control of *YUCCA3*, a gene involved in auxin biosynthesis [69]. One possibility is that PLTs regulate the expression of genes involved in apical fate determination, such as HD-ZIPIII directly acting on *HAN* or on auxin synthesis. Future studies will clarify this point.

Post-embryonically, PLTs and auxin are required to maintain SCN activity in the root, forming a gradient with a maximum in this zone (Figure 4) [55,64,65,70]. Ectopic inductions of auxin or PLTs maximum in the meristem convert other cell types in stem cells, underlying the importance of these maxima for stem cell specification. Post-embryonically, PLTs mRNAs and auxin are distributed in a gradient along the meristematic proximodistal axis [70]. PLTs and auxin gradients are strictly interconnected. Indeed, auxin promotes PLTs expression, whereas PLTs regulate auxin distribution, controlling PINs expression and auxin biosynthesis [58,64,69]. Different concentrations of auxin or PLTs result in different developmental outputs, i.e., high PLTs and auxin levels are necessary for SCN specification, whereas minimum auxin and PLT levels are necessary to induce cell differentiation at the TZ [55,64,65,70,71] (Figure 4). Recent studies have shown that the PLTs gradient along the proximodistal axis is partially independent from auxin, while the capacity of these proteins to diffuse along this axis plays an important role [65,70].

PIN-dependent polar auxin transport is necessary to position the auxin maximum at the root distal part [55,59] and the manner in which an auxin minimum is positioned at the proximal TZ has recently been elucidated. Indeed, the role of the plant hormone cytokinin in shaping the auxin gradient has been revealed. To position this minimum cytokinin triggers a module that involves the cytokinin receptor AHK3 (*ARABIDOPSIS* HISTIDINE KINASE 3), the cytokinin-dependent transcription factor ARR1 (*ARABIDOPSIS* RESPONSE REGULATOR1), the auxin signalling repressor SHY2/IAA3 (SHORT HYPOCOTYL2/INDOLE-3-ACETIC ACID INDUCIBLE 3) and the auxin catabolic enzyme GH3.17 (GRETCHEN HAGEN 3.17) [54,71–73] (Figure 4). Cytokinin via AHK3 activates ARR1 directly inducing the expression of *SHY2* in the vasculature at the TZ. Here, SHY2 downregulates *PINs* expression, thus reducing the shoot to root auxin efflux and, hence, cell division activity; auxin instead induces proteasome-dependent SHY2 degradation, supporting PINs expression [54,72–74] (Figure 4). At the same time, cytokinin via ARR1 induces the *GH3.17* gene in the lateral root cap and epidermis [71], where it mediates auxin degradation (Figure 4). The coordinated regulation of both auxin signalling and catabolism localizes a developmental instructive auxin minimum that positions the TZ [71].

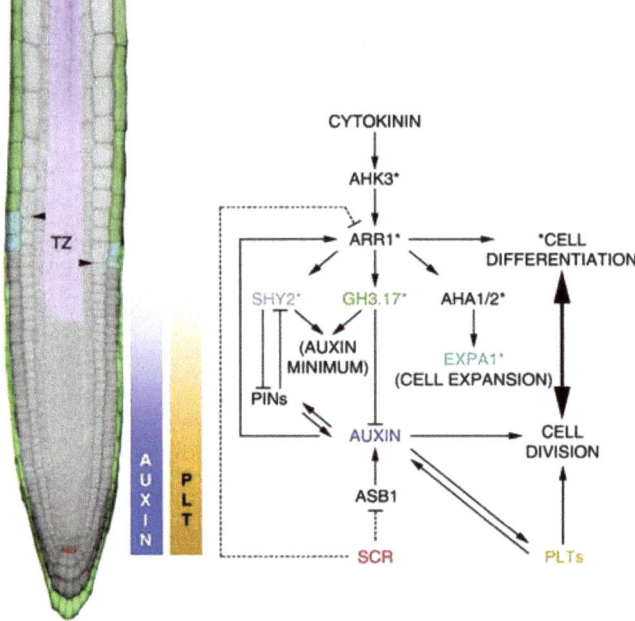

Figure 4. Molecular mechanisms leading to the root proximodistal axis patterning. On the left, representation of the *Arabidopsis* root. False colours indicate the activity domain of genes involved in the regulation of the equilibrium between cell division and cell differentiation processes established in the root. Arrowheads indicate the position of the TZ. Auxin and PLT-graded distributions are indicated in the piecewise colour bar, where the maximum is in the SCN and the minimum at the TZ. The molecular mechanisms acting in the control of meristem activity are indicated in the diagram. The QC SCR domain is represented by red; in green, the root cap and differentiated epidermis GH3.17 domain; in purple, the vascular bundles SHY2 domain; and in light blue, the first elongating epidermal cells EXPA1 domain. Asterisks represent genes involved in the regulation of the cell differentiation process mediated by cytokinin activity. Dashed lines indicate SCR-dependent indirect regulation of ASB1 and ARR1.

The coordination of SCN activity with cell differentiation process at the TZ is fundamental for proper proximodistal axis patterning. This spatial coordination is controlled by SCR and SHR, which repress cytokinin activity in the SCN, thus controlling auxin production [34,75,76]. In particular, SCR represses ARR1 expression in the QC, which in turn positively regulates the expression of the auxin biosynthesis gene *ASB1* (*ANTHRANILATE SYNTHASE BETA SUBUNIT 1*). Since ARR1 is induced by auxin at the TZ SCR, by regulating auxin biosynthesis in the QC, this controls stem cell division in the SCN and cell differentiation at the TZ [34].

Considering the key role of cytokinin in positioning the TZ, the regulation of cytokinin synthesis is fundamental for proper root patterning [13,24]. ISOPENTENYL TRANSFERASE (IPT) enzymes are key regulators of cytokinin synthesis [24]. In the meristem, the transcription factor PHB promotes cytokinin synthesis via direct induction of *IPT1* and *IPT7* [77]. The PHB-dependent cytokinin production is sufficient to activate the ARR1/SHY2 module, positioning the TZ. Intriguingly, ARR1 represses both PHB and its repressors miR165/6 expression, triggering a negative incoherent feedforward loop that finely tunes cytokinin levels and prevents meristem from differentiating [77].

Recently, it was demonstrated that cytokinin also promotes the transition of cells from the meristematic zone to the elongation zone, regulating apoplastic acidification and cell expansion. ARR1 directly regulates enzymes involved in cellular expansion such as the α-expansin EXPANSIN1 (*EXPA1*), controlling cell wall loosening and the plasma membrane H^+-ATPases (HA) 1 and 2 (AHA1 and AHA2) that transport protons (H^+) out of the cell. Interestingly, *expa1* mutants show a shift of the TZ toward the root proximal zone without interfering with the final cell size, suggesting that EXPA1-dependent cell expansion is mostly controlling the timing of cell exit from the division zone more than the final cell size itself [78] (Figure 4).

4. Future Perspectives

In recent years we have increased our understanding of the mechanisms controlling the formation of both radial and proximodistal axes of the root apical meristem. It is interesting to notice that the molecular mechanisms patterning both the radial and the proximodistal axes involve the same main players (i.e., auxin, PHB and SHR/SCR). Future studies will elucidate how the mechanisms controlling the development of these two axes coordinate in order to generate a structured body plan. Moreover, the effectors of the master genes governing the zonation of both those axes are starting to be understood, but several players are still missing and will be discovered in the future. In this optic, with respect to the proximodistal axis, the lists of targets of ARR1 and PLTs were published. This knowledge will allow us to better understand how those genes are interconnected and how they coordinate to ensure continuous growth.

In multicellular organisms, cell elongation is accompanied by endoreduplication, genome duplication in absence of mitosis [79,80]. Similarly, it was shown that cells at the TZ show enhanced the number of genome copies compared to their meristematic progenitors. Moreover, cytokinin is known to promote endoreduplication [81]. It will be interesting in the future to investigate the role of endoreduplication in patterning the root proximodistal axis.

Most of the molecular mechanisms patterning the root axes were discovered in *Arabidopsis*. In recent years, several variations in those mechanisms were found to be the basis for interspecific variability in plants.

Indeed, root axis structure are largely variable among species. For example, the root radial axis is extremely variable, as number and features of tissue layers is strictly dependent on the species [24,82,83]. One of the tissues along the radial axis that is most variable between species is the cortex [24]. For example, *Cardamine hirsuta*, a close relative of *Arabidopsis*, displays two cortex layers. It was recently shown that the second cortical layer of *Cardamine* emerges from the activity of a developmental domain absent in *Arabidopsis*, the cortex endodermis mixed identity tissue (CEM). Activity of HD-ZIPIII in this domain is crucial for the formation of the second cortical, as knockdown of these genes results in the loss of this additional layer [47]. How PHB controls the CEM periclinal division is still not known. In *Arabidopsis*, ectopic expression of PHB indirectly promotes *CYCD6;1* expression independently from SHR, therefore PHB might control CEM division acting on this gene. Moreover, whether this mechanism is conserved in other distant relatives with multiple cortical layers or whether this mechanism adds to the one controlled by SHR is still not known. Future research will allow us to uncover the answers to this interesting question. It was shown that SHR and SCR also play a key role in patterning the differences in radial axis anatomy among species. The multi cortical layered species *Oryza sativa* (rice), indeed, maintains the SHR/SCR interaction, but SHR movements are subject to lower restriction, for promoting multiple cortical layers formation [38,84].

The mechanisms governing root axis formation and maintenance in *Arabidopsis* might be valid for most of the species but may not be universal. It was shown that in the root of most of the species, auxin controls cell division, whereas cytokinin controls differentiation. However, in the fern *Azolla filiculoides*, cytokinin promotes cell division, whereas auxin promotes cell differentiation [85]. Analysis of the molecular mechanisms controlling root axes in species other than *Arabidopsis* will permit us to understand how and when these mechanisms arose and diverged. In this optic, utilization of close relatives of *Arabidopsis* might allow us to understand this crucial point.

Author Contributions: R.D.M., S.S. and R.D.I. wrote the manuscript and prepared the figures.

Funding: This work was supported by a FIRB2013 (FUTURO IN RICERCA) to R.D.I. and by The European Research Council (grant number 260368) to S.S.

Acknowledgments: We extend our sincere thanks to P. Costantino, N. Svolacchia, E. Salvi, G. Bertolotti and A. Rocchetti for critical comments on the manuscript.

Conflicts of Interest: The authors claim no conflict of interest.

References

1. Tabin, C.; Wolpert, L. Rethinking the proximodistal axis of the vertebrate limb in the molecular era. *Genes Dev.* **2007**, *21*, 1433–1442. [CrossRef] [PubMed]
2. Zuniga, A. Next generation limb development and evolution: Old questions, new perspectives. *Development* **2015**, *142*, 3810–3820. [CrossRef]
3. Dolan, L.; Janmaat, K.; Willemsen, V.; Linstead, P.; Poethig, S.; Roberts, K.; Scheres, B. Cellular organisation of the *Arabidopsis* thaliana root. *Development* **1993**, *119*, 71–84. [PubMed]
4. Scheres, B.; Benfey, P.; Dolan, L. Root development. *Arabidopsis Book* **2002**, *1*, e0101. [CrossRef] [PubMed]
5. Heidstra, R.; Sabatini, S. Plant and animal stem cells: Similar yet different. *Nat. Rev. Mol. Cell Biol.* **2014**, *15*, 301–312. [CrossRef]
6. Ten Hove, C.A.; Lu, K.J.; Weijers, D. Building a plant: Cell fate specification in the early *Arabidopsis* embryo. *Development* **2015**, *142*, 420–430. [CrossRef] [PubMed]
7. Van den Berg, C.; Willemsen, V.; Hendriks, G.; Weisbeek, P.; Scheres, B. Short-range control of cell differentiation in the *Arabidopsis* root meristem. *Nature* **1997**, *390*, 287–289.
8. Sabatini, S.; Heidstra, R.; Wildwater, M.; Scheres, B. Scarecrow is involved in positioning the stem cell niche in the *Arabidopsis* root meristem. *Genes Dev.* **2003**, *17*, 354–358. [CrossRef]
9. Perilli, S.; Di Mambro, R.; Sabatini, S. Growth and development of the root apical meristem. *Curr. Opin. Plant Biol.* **2012**, *15*, 17–23. [CrossRef] [PubMed]
10. Di Mambro, R.; Sabatini, S. Developmental analysis of *Arabidopsis* root meristem. *Methods Mol. Biol.* **2018**, *1761*, 33–45.
11. Drapek, C.; Sparks, E.E.; Benfey, P.N. Uncovering gene regulatory networks controlling plant cell differentiation. *Trends Genet.* **2017**, *33*, 529–539. [CrossRef]
12. Benfey, P.N. Defining the path from stem cells to differentiated tissue. *Curr. Top. Dev. Biol.* **2016**, *116*, 35–43. [PubMed]
13. Dello Ioio, R.; Linhares, F.S.; Sabatini, S. Emerging role of cytokinin as a regulator of cellular differentiation. *Curr. Opin. Plant Biol.* **2008**, *11*, 23–27. [CrossRef] [PubMed]
14. Kumpf, R.P.; Nowack, M.K. The root cap: A short story of life and death. *J. Exp. Bot.* **2015**, *66*, 5651–5662. [CrossRef] [PubMed]
15. Pauluzzi, G.; Divol, F.; Puig, J.; Guiderdoni, E.; Dievart, A.; Perin, C. Surfing along the root ground tissue gene network. *Dev. Biol.* **2012**, *365*, 14–22. [CrossRef] [PubMed]

16. Choi, J.W.; Lim, J. Control of asymmetric cell divisions during root ground tissue maturation. *Mol. Cells* **2016**, *39*, 524–529. [CrossRef] [PubMed]
17. Moller, B.K.; Ten Hove, C.A.; Xiang, D.; Williams, N.; Lopez, L.G.; Yoshida, S.; Smit, M.; Datla, R.; Weijers, D. Auxin response cell-autonomously controls ground tissue initiation in the early *Arabidopsis* embryo. *Proc. Natl. Acad. Sci. USA* **2017**, *114*, E2533–E2539. [CrossRef] [PubMed]
18. Heidstra, R.; Welch, D.; Scheres, B. Mosaic analyses using marked activation and deletion clones dissect *Arabidopsis* scarecrow action in asymmetric cell division. *Genes Dev.* **2004**, *18*, 1964–1969. [CrossRef]
19. Helariutta, Y.; Fukaki, H.; Wysocka-Diller, J.; Nakajima, K.; Jung, J.; Sena, G.; Hauser, M.T.; Benfey, P.N. The short-root gene controls radial patterning of the *Arabidopsis* root through radial signaling. *Cell* **2000**, *101*, 555–567. [CrossRef]
20. Nakajima, K.; Sena, G.; Nawy, T.; Benfey, P.N. Intercellular movement of the putative transcription factor shr in root patterning. *Nature* **2001**, *413*, 307–311. [CrossRef]
21. Clark, N.M.; Hinde, E.; Winter, C.M.; Fisher, A.P.; Crosti, G.; Blilou, I.; Gratton, E.; Benfey, P.N.; Sozzani, R. Tracking transcription factor mobility and interaction in *Arabidopsis* roots with fluorescence correlation spectroscopy. *Elife* **2016**, *5*. [CrossRef] [PubMed]
22. Vaten, A.; Dettmer, J.; Wu, S.; Stierhof, Y.D.; Miyashima, S.; Yadav, S.R.; Roberts, C.J.; Campilho, A.; Bulone, V.; Lichtenberger, R.; et al. Callose biosynthesis regulates symplastic trafficking during root development. *Dev. Cell* **2011**, *21*, 1144–1155. [CrossRef] [PubMed]
23. Hofhuis, H.F.; Heidstra, R. Transcription factor dosage: More or less sufficient for growth. *Curr. Opin. Plant Biol.* **2018**, *45*, 50–58. [CrossRef] [PubMed]
24. Di Ruocco, G.; Di Mambro, R.; Dello Ioio, R. Building the differences: A case for the ground tissue patterning in plants. *Proc. Biol. Sci.* **2018**, *285*, 20181746. [CrossRef] [PubMed]
25. Sena, G.; Jung, J.W.; Benfey, P.N. A broad competence to respond to short root revealed by tissue-specific ectopic expression. *Development* **2004**, *131*, 2817–2826. [CrossRef] [PubMed]
26. Long, Y.; Goedhart, J.; Schneijderberg, M.; Terpstra, I.; Shimotohno, A.; Bouchet, B.P.; Akhmanova, A.; Gadella, T.W., Jr.; Heidstra, R.; Scheres, B.; et al. Scarecrow-like23 and scarecrow jointly specify endodermal cell fate but distinctly control short-root movement. *Plant J.* **2015**, *84*, 773–784. [CrossRef] [PubMed]
27. Long, Y.; Smet, W.; Cruz-Ramirez, A.; Castelijns, B.; de Jonge, W.; Mahonen, A.P.; Bouchet, B.P.; Perez, G.S.; Akhmanova, A.; Scheres, B.; et al. *Arabidopsis* bird zinc finger proteins jointly stabilize tissue boundaries by confining the cell fate regulator short-root and contributing to fate specification. *Plant Cell* **2015**, *27*, 1185–1199. [CrossRef] [PubMed]
28. Cui, H.; Levesque, M.P.; Vernoux, T.; Jung, J.W.; Paquette, A.J.; Gallagher, K.L.; Wang, J.Y.; Blilou, I.; Scheres, B.; Benfey, P.N. An evolutionarily conserved mechanism delimiting shr movement defines a single layer of endodermis in plants. *Science* **2007**, *316*, 421–425. [CrossRef]
29. Welch, D.; Hassan, H.; Blilou, I.; Immink, R.; Heidstra, R.; Scheres, B. *Arabidopsis* jackdaw and magpie zinc finger proteins delimit asymmetric cell division and stabilize tissue boundaries by restricting short-root action. *Genes Dev.* **2007**, *21*, 2196–2204. [CrossRef]
30. Long, Y.; Stahl, Y.; Weidtkamp-Peters, S.; Postma, M.; Zhou, W.; Goedhart, J.; Sanchez-Perez, M.I.; Gadella, T.W.J.; Simon, R.; Scheres, B.; et al. In vivo fret-flim reveals cell-type-specific protein interactions in *Arabidopsis* roots. *Nature* **2017**, *548*, 97–102. [CrossRef]
31. Long, Y.; Stahl, Y.; Weidtkamp-Peters, S.; Smet, W.; Du, Y.; Gadella, T.W.J., Jr.; Goedhart, J.; Scheres, B.; Blilou, I. Optimizing fret-flim labeling conditions to detect nuclear protein interactions at native expression levels in living *Arabidopsis* roots. *Front. Plant Sci.* **2018**, *9*, 639. [CrossRef] [PubMed]
32. Moreno-Risueno, M.A.; Sozzani, R.; Yardimci, G.G.; Petricka, J.J.; Vernoux, T.; Blilou, I.; Alonso, J.; Winter, C.M.; Ohler, U.; Scheres, B.; et al. Transcriptional control of tissue formation throughout root development. *Science* **2015**, *350*, 426–430. [CrossRef] [PubMed]
33. Sozzani, R.; Cui, H.; Moreno-Risueno, M.A.; Busch, W.; Van Norman, J.M.; Vernoux, T.; Brady, S.M.; Dewitte, W.; Murray, J.A.; Benfey, P.N. Spatiotemporal regulation of cell-cycle genes by shortroot links patterning and growth. *Nature* **2010**, *466*, 128–132. [CrossRef] [PubMed]

34. Perilli, S.; Perez-Perez, J.M.; Di Mambro, R.; Peris, C.L.; Diaz-Trivino, S.; Del Bianco, M.; Pierdonati, E.; Moubayidin, L.; Cruz-Ramirez, A.; Costantino, P.; et al. Retinoblastoma-related protein stimulates cell differentiation in the *Arabidopsis* root meristem by interacting with cytokinin signaling. *Plant Cell* **2013**, *25*, 4469–4478. [CrossRef] [PubMed]
35. Wildwater, M.; Campilho, A.; Perez-Perez, J.M.; Heidstra, R.; Blilou, I.; Korthout, H.; Chatterjee, J.; Mariconti, L.; Gruissem, W.; Scheres, B. The retinoblastoma-related gene regulates stem cell maintenance in *Arabidopsis* roots. *Cell* **2005**, *123*, 1337–1349. [CrossRef] [PubMed]
36. Cruz-Ramirez, A.; Diaz-Trivino, S.; Blilou, I.; Grieneisen, V.A.; Sozzani, R.; Zamioudis, C.; Miskolczi, P.; Nieuwland, J.; Benjamins, R.; Dhonukshe, P.; et al. A bistable circuit involving scarecrow-retinoblastoma integrates cues to inform asymmetric stem cell division. *Cell* **2012**, *150*, 1002–1015. [CrossRef] [PubMed]
37. Zhang, X.; Zhou, W.; Chen, Q.; Fang, M.; Zheng, S.; Scheres, B.; Li, C. Mediator subunit med31 is required for radial patterning of *Arabidopsis* roots. *Proc. Natl. Acad. Sci. USA* **2018**, *115*, E5624–E5633. [CrossRef]
38. Wu, S.; Lee, C.M.; Hayashi, T.; Price, S.; Divol, F.; Henry, S.; Pauluzzi, G.; Perin, C.; Gallagher, K.L. A plausible mechanism, based upon short-root movement, for regulating the number of cortex cell layers in roots. *Proc. Natl. Acad. Sci. USA* **2014**, *111*, 16184–16189. [CrossRef]
39. Koizumi, K.; Hayashi, T.; Wu, S.; Gallagher, K.L. The short-root protein acts as a mobile, dose-dependent signal in patterning the ground tissue. *Proc. Natl. Acad. Sci. USA* **2012**, *109*, 13010–13015. [CrossRef]
40. Mylona, P.; Linstead, P.; Martienssen, R.; Dolan, L. Schizoriza controls an asymmetric cell division and restricts epidermal identity in the *Arabidopsis* root. *Development* **2002**, *129*, 4327–4334.
41. Ten Hove, C.A.; Willemsen, V.; de Vries, W.J.; van Dijken, A.; Scheres, B.; Heidstra, R. Schizoriza encodes a nuclear factor regulating asymmetry of stem cell divisions in the *Arabidopsis* root. *Curr. Biol.* **2010**, *20*, 452–457. [CrossRef] [PubMed]
42. Pernas, M.; Ryan, E.; Dolan, L. Schizoriza controls tissue system complexity in plants. *Curr. Biol.* **2010**, *20*, 818–823. [CrossRef] [PubMed]
43. Carlsbecker, A.; Lee, J.Y.; Roberts, C.J.; Dettmer, J.; Lehesranta, S.; Zhou, J.; Lindgren, O.; Moreno-Risueno, M.A.; Vaten, A.; Thitamadee, S.; et al. Cell signalling by microrna165/6 directs gene dose-dependent root cell fate. *Nature* **2010**, *465*, 316–321. [CrossRef] [PubMed]
44. Muraro, D.; Mellor, N.; Pound, M.P.; Help, H.; Lucas, M.; Chopard, J.; Byrne, H.M.; Godin, C.; Hodgman, T.C.; King, J.R.; et al. Integration of hormonal signaling networks and mobile micrornas is required for vascular patterning in *Arabidopsis* roots. *Proc. Natl. Acad. Sci. USA* **2014**, *111*, 857–862. [CrossRef] [PubMed]
45. Bishopp, A.; Help, H.; El-Showk, S.; Weijers, D.; Scheres, B.; Friml, J.; Benkova, E.; Mahonen, A.P.; Helariutta, Y. A mutually inhibitory interaction between auxin and cytokinin specifies vascular pattern in roots. *Curr. Biol.* **2011**, *21*, 917–926. [CrossRef] [PubMed]
46. Miyashima, S.; Koi, S.; Hashimoto, T.; Nakajima, K. Non-cell-autonomous microrna165 acts in a dose-dependent manner to regulate multiple differentiation status in the *Arabidopsis* root. *Development* **2011**, *138*, 2303–2313. [CrossRef] [PubMed]
47. Di Ruocco, G.; Bertolotti, G.; Pacifici, E.; Polverari, L.; Tsiantis, M.; Sabatini, S.; Costantino, P.; Dello Ioio, R. Differential spatial distribution of mir165/6 determines variability in plant root anatomy. *Development* **2018**, *145*. [CrossRef]
48. Miyashima, S.; Hashimoto, T.; Nakajima, K. Argonaute1 acts in *Arabidopsis* root radial pattern formation independently of the shr/scr pathway. *Plant Cell Physiol.* **2009**, *50*, 626–634. [CrossRef]
49. Muller, C.J.; Valdes, A.E.; Wang, G.; Ramachandran, P.; Beste, L.; Uddenberg, D.; Carlsbecker, A. Phabulosa mediates an auxin signaling loop to regulate vascular patterning in *Arabidopsis*. *Plant Physiol.* **2016**, *170*, 956–970. [CrossRef]
50. Li, P.; Yu, Q.; Gu, X.; Xu, C.; Qi, S.; Wang, H.; Zhong, F.; Baskin, T.I.; Rahman, A.; Wu, S. Construction of a functional casparian strip in non-endodermal lineages is orchestrated by two parallel signaling systems in *Arabidopsis thaliana*. *Curr. Biol.* **2018**, *28*, 2777.e2.–2786.e2. [CrossRef]
51. Drapek, C.; Sparks, E.E.; Marhavy, P.; Taylor, I.; Andersen, T.G.; Hennacy, J.H.; Geldner, N.; Benfey, P.N. Minimum requirements for changing and maintaining endodermis cell identity in the *Arabidopsis* root. *Nat. Plants* **2018**, *4*, 586–595. [CrossRef] [PubMed]
52. Kamiya, T.; Borghi, M.; Wang, P.; Danku, J.M.; Kalmbach, L.; Hosmani, P.S.; Naseer, S.; Fujiwara, T.; Geldner, N.; Salt, D.E. The myb36 transcription factor orchestrates casparian strip formation. *Proc. Natl. Acad. Sci. USA* **2015**, *112*, 10533–10538. [CrossRef] [PubMed]

53. Roppolo, D.; De Rybel, B.; Denervaud Tendon, V.; Pfister, A.; Alassimone, J.; Vermeer, J.E.; Yamazaki, M.; Stierhof, Y.D.; Beeckman, T.; Geldner, N. A novel protein family mediates casparian strip formation in the endodermis. *Nature* **2011**, *473*, 380–383. [CrossRef]
54. Dello Ioio, R.; Linhares, F.S.; Scacchi, E.; Casamitjana-Martinez, E.; Heidstra, R.; Costantino, P.; Sabatini, S. Cytokinins determine *Arabidopsis* root-meristem size by controlling cell differentiation. *Curr. Biol.* **2007**, *17*, 678–682. [CrossRef] [PubMed]
55. Sabatini, S.; Beis, D.; Wolkenfelt, H.; Murfett, J.; Guilfoyle, T.; Malamy, J.; Benfey, P.; Leyser, O.; Bechtold, N.; Weisbeek, P.; et al. An auxin-dependent distal organizer of pattern and polarity in the *Arabidopsis* root. *Cell* **1999**, *99*, 463–472. [CrossRef]
56. Leyser, O. Auxin distribution and plant pattern formation: How many angels can dance on the point of pin? *Cell* **2005**, *121*, 819–822. [CrossRef]
57. Yoshida, S.; Saiga, S.; Weijers, D. Auxin regulation of embryonic root formation. *Plant Cell Physiol.* **2013**, *54*, 325–332. [CrossRef]
58. Blilou, I.; Xu, J.; Wildwater, M.; Willemsen, V.; Paponov, I.; Friml, J.; Heidstra, R.; Aida, M.; Palme, K.; Scheres, B. The pin auxin efflux facilitator network controls growth and patterning in *Arabidopsis* roots. *Nature* **2005**, *433*, 39–44. [CrossRef]
59. Wisniewska, J.; Xu, J.; Seifertova, D.; Brewer, P.B.; Ruzicka, K.; Blilou, I.; Rouquie, D.; Benkova, E.; Scheres, B.; Friml, J. Polar pin localization directs auxin flow in plants. *Science* **2006**, *312*, 883. [CrossRef]
60. Friml, J.; Vieten, A.; Sauer, M.; Weijers, D.; Schwarz, H.; Hamann, T.; Offringa, R.; Jurgens, G. Efflux-dependent auxin gradients establish the apical-basal axis of *Arabidopsis*. *Nature* **2003**, *426*, 147–153. [CrossRef]
61. Weijers, D.; Schlereth, A.; Ehrismann, J.S.; Schwank, G.; Kientz, M.; Jurgens, G. Auxin triggers transient local signaling for cell specification in *Arabidopsis* embryogenesis. *Dev. Cell* **2006**, *10*, 265–270. [CrossRef] [PubMed]
62. Hamann, T.; Mayer, U.; Jurgens, G. The auxin-insensitive bodenlos mutation affects primary root formation and apical-basal patterning in the *Arabidopsis* embryo. *Development* **1999**, *126*, 1387–1395. [PubMed]
63. Hamann, T.; Benkova, E.; Baurle, I.; Kientz, M.; Jurgens, G. The *Arabidopsis* bodenlos gene encodes an auxin response protein inhibiting monopteros-mediated embryo patterning. *Genes Dev.* **2002**, *16*, 1610–1615. [CrossRef] [PubMed]
64. Aida, M.; Beis, D.; Heidstra, R.; Willemsen, V.; Blilou, I.; Galinha, C.; Nussaume, L.; Noh, Y.S.; Amasino, R.; Scheres, B. The plethora genes mediate patterning of the *Arabidopsis* root stem cell niche. *Cell* **2004**, *119*, 109–120. [CrossRef] [PubMed]
65. Galinha, C.; Hofhuis, H.; Luijten, M.; Willemsen, V.; Blilou, I.; Heidstra, R.; Scheres, B. Plethora proteins as dose-dependent master regulators of *Arabidopsis* root development. *Nature* **2007**, *449*, 1053–1057. [CrossRef] [PubMed]
66. Nawy, T.; Bayer, M.; Mravec, J.; Friml, J.; Birnbaum, K.D.; Lukowitz, W. The gata factor hanaba taranu is required to position the proembryo boundary in the early *Arabidopsis* embryo. *Dev. Cell* **2010**, *19*, 103–113. [CrossRef] [PubMed]
67. Grigg, S.P.; Galinha, C.; Kornet, N.; Canales, C.; Scheres, B.; Tsiantis, M. Repression of apical homeobox genes is required for embryonic root development in *Arabidopsis*. *Curr. Biol.* **2009**, *19*, 1485–1490. [CrossRef]
68. Smith, Z.R.; Long, J.A. Control of *Arabidopsis* apical-basal embryo polarity by antagonistic transcription factors. *Nature* **2010**, *464*, 423–426. [CrossRef]
69. Santuari, L.; Sanchez-Perez, G.F.; Luijten, M.; Rutjens, B.; Terpstra, I.; Berke, L.; Gorte, M.; Prasad, K.; Bao, D.; Timmermans-Hereijgers, J.L.; et al. The plethora gene regulatory network guides growth and cell differentiation in *Arabidopsis* roots. *Plant Cell* **2016**, *28*, 2937–2951. [CrossRef]
70. Mahonen, A.P.; Ten Tusscher, K.; Siligato, R.; Smetana, O.; Diaz-Trivino, S.; Salojarvi, J.; Wachsman, G.; Prasad, K.; Heidstra, R.; Scheres, B. Plethora gradient formation mechanism separates auxin responses. *Nature* **2014**, *515*, 125–129. [CrossRef]
71. Di Mambro, R.; De Ruvo, M.; Pacifici, E.; Salvi, E.; Sozzani, R.; Benfey, P.N.; Busch, W.; Novak, O.; Ljung, K.; Di Paola, L.; et al. Auxin minimum triggers the developmental switch from cell division to cell differentiation in the *Arabidopsis* root. *Proc. Natl. Acad. Sci. USA* **2017**, *114*, E7641–E7649. [CrossRef] [PubMed]

72. Dello Ioio, R.; Nakamura, K.; Moubayidin, L.; Perilli, S.; Taniguchi, M.; Morita, M.T.; Aoyama, T.; Costantino, P.; Sabatini, S. A genetic framework for the control of cell division and differentiation in the root meristem. *Science* **2008**, *322*, 1380–1384. [CrossRef] [PubMed]
73. Moubayidin, L.; Perilli, S.; Dello Ioio, R.; Di Mambro, R.; Costantino, P.; Sabatini, S. The rate of cell differentiation controls the *Arabidopsis* root meristem growth phase. *Curr. Biol.* **2010**, *20*, 1138–1143. [CrossRef] [PubMed]
74. Tiwari, S.B.; Wang, X.J.; Hagen, G.; Guilfoyle, T.J. Aux/IAA proteins are active repressors, and their stability and activity are modulated by auxin. *Plant Cell* **2001**, *13*, 2809–2822. [CrossRef] [PubMed]
75. Moubayidin, L.; Di Mambro, R.; Sozzani, R.; Pacifici, E.; Salvi, E.; Terpstra, I.; Bao, D.; van Dijken, A.; Dello Ioio, R.; Perilli, S.; et al. Spatial coordination between stem cell activity and cell differentiation in the root meristem. *Dev. Cell* **2013**, *26*, 405–415. [CrossRef]
76. Salvi, E.; Di Mambro, R.; Pacifici, E.; Dello Ioio, R.; Costantino, P.; Moubayidin, L.; Sabatini, S. Scarecrow and shortroot control the auxin/cytokinin balance necessary for embryonic stem cell niche specification. *Plant Signal Behav.* **2018**, *13*, e1507402.
77. Dello Ioio, R.; Galinha, C.; Fletcher, A.G.; Grigg, S.P.; Molnar, A.; Willemsen, V.; Scheres, B.; Sabatini, S.; Baulcombe, D.; Maini, P.K.; et al. A phabulosa/cytokinin feedback loop controls root growth in *Arabidopsis*. *Curr. Biol.* **2012**, *22*, 1699–1704. [CrossRef]
78. Pacifici, E.; Di Mambro, R.; Dello Ioio, R.; Costantino, P.; Sabatini, S. Acidic cell elongation drives cell differentiation in the *Arabidopsis* root. *EMBO J.* **2018**, *37*, e99134. [CrossRef]
79. Vuolo, F.; Kierzkowski, D.; Runions, A.; Hajheidari, M.; Mentink, R.A.; Gupta, M.D.; Zhang, Z.; Vlad, D.; Wang, Y.; Pecinka, A.; et al. Lmi1 homeodomain protein regulates organ proportions by spatial modulation of endoreduplication. *Genes Dev.* **2018**, *32*, 1361–1366. [CrossRef]
80. Hayashi, K.; Hasegawa, J.; Matsunaga, S. The boundary of the meristematic and elongation zones in roots: Endoreduplication precedes rapid cell expansion. *Sci. Rep.* **2013**, *3*, 2723. [CrossRef]
81. Takahashi, N.; Kajihara, T.; Okamura, C.; Kim, Y.; Katagiri, Y.; Okushima, Y.; Matsunaga, S.; Hwang, I.; Umeda, M. Cytokinins control endocycle onset by promoting the expression of an apc/c activator in *Arabidopsis* roots. *Curr. Biol.* **2013**, *23*, 1812–1817. [CrossRef] [PubMed]
82. *Esau's Plant Anatomy: Meristems, Cells, and Tissues of the Plant Body: Their Structure, Function, and Development*, 3rd ed.; Wiley Interscience, John Wiley & Sons: Hoboken, NJ, USA, 2006; p. 601. ISBN 13: 978-0-471-73843-5.
83. Ron, M.; Dorrity, M.W.; de Lucas, M.; Toal, T.; Hernandez, R.I.; Little, S.A.; Maloof, J.N.; Kliebenstein, D.J.; Brady, S.M. Identification of novel loci regulating interspecific variation in root morphology and cellular development in tomato. *Plant Physiol.* **2013**, *162*, 755–768. [CrossRef] [PubMed]
84. Henry, S.; Dievart, A.; Divol, F.; Pauluzzi, G.; Meynard, D.; Swarup, R.; Wu, S.; Gallagher, K.L.; Perin, C. Shr overexpression induces the formation of supernumerary cell layers with cortex cell identity in rice. *Dev. Biol.* **2017**, *425*, 1–7. [CrossRef] [PubMed]
85. De Vries, J.; Fischer, A.M.; Roettger, M.; Rommel, S.; Schluepmann, H.; Brautigam, A.; Carlsbecker, A.; Gould, S.B. Cytokinin-induced promotion of root meristem size in the fern azolla supports a shoot-like origin of euphyllophyte roots. *New Phytol.* **2016**, *209*, 705–720. [CrossRef] [PubMed]

© 2018 by the authors. Licensee MDPI, Basel, Switzerland. This article is an open access article distributed under the terms and conditions of the Creative Commons Attribution (CC BY) license (http://creativecommons.org/licenses/by/4.0/).

Article

Cytokinin-Dependent Control of *GH3* Group II Family Genes in the *Arabidopsis* Root

Emanuela Pierdonati [1,†], Simon Josef Unterholzner [1,†,‡], Elena Salvi [1,†], Noemi Svolacchia [1,†], Gaia Bertolotti [1,†], Raffaele Dello Ioio [1], Sabrina Sabatini [1] and Riccardo Di Mambro [2,*]

1. Dipartimento di Biologia e Biotecnologie, Laboratory of Functional Genomics and Proteomics of Model Systems, Università di Roma, Sapienza-via dei Sardi, 70–00185 Rome, Italy; emanuela.pier@gmail.com (E.P.); unterholzner.simonjosef@gmail.com (S.J.U.); elena.salvi2@gmail.com (E.S.); svolacchia.noemi@gmail.com (N.S.); gaia.bertolotti@uniroma1.it (G.B.); raffaele.delloioio@uniroma1.it (R.D.I.); sabrina.sabatini@uniroma1.it (S.S.)
2. Department of Biology, University of Pisa-via L. Ghini, 13–56126 Pisa, Italy
* Correspondence: riccardo.dimambro@unipi.it
† These authors contributed equally to this work.
‡ Present address: Faculty of Science and Technology, Free University of Bolzano–Piazza Università, 5-I-39100 Bolzano, Italy.

Received: 22 March 2019; Accepted: 6 April 2019; Published: 8 April 2019

Abstract: The *Arabidopsis* root is a dynamic system where the interaction between different plant hormones controls root meristem activity and, thus, organ growth. In the root, a characteristic graded distribution of the hormone auxin provides positional information, coordinating the proliferating and differentiating cell status. The hormone cytokinin shapes this gradient by positioning an auxin minimum in the last meristematic cells. This auxin minimum triggers a cell developmental switch necessary to start the differentiation program, thus, regulating the root meristem size. To position the auxin minimum, cytokinin promotes the expression of the *IAA-amido synthase group II* gene *GH3.17*, which conjugates auxin with amino acids, in the most external layer of the root, the lateral root cap tissue. Since additional *GH3* genes are expressed in the root, we questioned whether cytokinin to position the auxin minimum also operates via different *GH3* genes. Here, we show that cytokinin regulates meristem size by activating the expression of *GH3.5* and *GH3.6* genes, in addition to *GH3.17*. Thus, cytokinin activity provides a robust control of auxin activity in the entire organ necessary to regulate root growth.

Keywords: GRETCHEN HAGEN 3 (GH3) IAA-amido synthase group II; root apical meristem; auxin; cytokinin; lateral root cap; auxin minimum; auxin conjugation

1. Introduction

Organ growth in plants is supported by the meristems, regions providing a reservoir of undifferentiated cells whose activity depends on the stem cell niche [1]. In the root, the stem cells daughters proliferate establishing the division zone of the meristem and, more distally from the root tip along the longitudinal axis, those cells differentiate generating the differentiation zone [2–5]. The boundary between proliferating and differentiating cells is called transition zone (TZ). The position of this cell boundary depends on the coordinated activity of the stem cell niche, the division zone, and the differentiation zone. The activities of these zones are controlled by a dynamic equilibrium between cell division and cell differentiation. The regulation of this equilibrium results in a shoot-ward or a root-ward shift of the TZ position along the root longitudinal axis [2–4,6]. The position of the TZ depends on the antagonistic interaction between cytokinin and auxin hormones [7,8]. It has been demonstrated that cytokinin controls TZ localization by positioning an auxin minimum specifically in the last meristematic cells of each root tissue [9]. In particular, cytokinin through the primary cytokinin response transcription factor *ARABIDOPSIS* RESPONSE REGULATOR 1 (ARR1), positively regulates the expression of the *Aux/IAA SHORT HYPOCOTYL 2* (*SHY2*) gene, which in turn negatively controls the polar auxin efflux carriers *PIN1*, *PIN3* and *PIN7* genes at the vascular tissue TZ. At the same time, ARR1 positively regulates the expression of the *IAA-amino synthase* of the *GH3* Group II gene family *GRETCHEN HAGEN 3.17* (*GH3.17*) [8,9].

The roots of *Arabidopsis thaliana* can be represented as a series of concentric cylinders where the vascular bundles lie in the center [1–4]. In the radial axis of the root, the lateral root cap (LRC) represents the most external tissue that surrounds all tissues of the root meristem [1–4]. The LRC serves to facilitate root penetration in the soil, it acts as a physical protective barrier of the root meristem, and it plays an important role in meristem maintenance [10–17]. It was previously demonstrated that a molecular mechanism acting specifically in the LRC controls root meristem size and, thus, root growth, by positioning the TZ [17]. In particular, ARR1, besides *GH3.17*, regulates auxin levels by promoting the transcription of the auxin intracellular transporter *PIN-FORMED 5* (*PIN5*) gene [17]. GH3.17 irreversibly conjugates free auxin with amino acids specifically in the LRC cells, thus, promoting hormone inactivation, whereas PIN5 operates on auxin intracellular homeostasis mediating auxin compartmentalization in the endoplasmic reticulum. As a result, the LRC acts as an auxin sink where the regulation of auxin levels, controlled by the cytokinin activity, influences auxin distribution within the entire meristem regulating root meristem size and, thus, root growth [17].

Due to the importance of the tissue-specific activity of cytokinin in the LRC, we question whether cytokinin controls meristem size from this tissue by acting on additional genes. It has been already reported that the induction of cytokinin activity in the LRC regulates the expression of *GH3.5*, *GH3.6* and *GH3.9* genes [17], members of the *GH3* Group II gene family [18–20].

Here we show that *GH3.5* and *GH3.6* genes are expressed in the LRC and that their expression is cytokinin-dependent. We also show that those genes, similarly to *GH3.17*, are involved in meristem size regulation. These findings highlight the pivotal role of cytokinin in localizing a strong auxin inactivation process in the LRC to regulate meristem activity.

2. Results

In order to unveil cytokinin-dependent mechanisms acting in the LRC to control meristem size, we took advantage of the already published microarray data reporting genes differentially regulated in the LRC in response to ARR1 induction. These data, resulting from the transcriptional profiling of LRC cells upon induction of a constitutive active form of ARR1 (ARR1ΔDDK) in the LRC, revealed that genes belonging to "auxin homeostasis regulation" gene ontology category are highly represented [17]. Interestingly, among these genes, *GH3.5*, *GH3.6*, and *GH3.9* were positively regulated by ARR1 in the LRC [17]. It was demonstrated that GH3.5, GH3.6, and GH3.9 IAA-amido synthases participate in maintaining auxin homeostasis by conjugating amino acids to the hormone [18,19,21–24] and thereby affect the levels of free auxin molecules that are biologically active and suited for binding to their receptors.

Considering the LRC specific ARR1-dependent positive regulation of *GH3.5*, *GH3.6*, and *GH3.9*, and the LRC specific domain of activity of GH3.17, we thus questioned whether the expression domain of *GH3.5*, *GH3.6*, and *GH3.9* localizes in the LRC. To this end, we generated GREEN FLUORESCENT PROTEIN (GFP) translational fusions of these three *GH3s* (*pGH3.5::GH3.5-GFP*, *pGH3.6::GH3.6-GFP*, and *pGH3.9::GH3.9-GFP* lines, respectively). The GFP signal was undetectable for all those lines (data not shown), most likely because of the low expression of those genes in the root as also previously reported [25]. Therefore, we developed transcriptional fluorescent reporters for each of the *GH3s* using a three-time YELLOW FLUORESCENT PROTEIN (3xYFP) fusion (*pGH3.9–3xYFP*, *pGH3.5–3xYFP*, and *pGH3.6–3xYFP* lines, respectively). Additionally, we also generated a *GH3.17* transcriptional fusion line with the same reporter (3xYFP) (*pGH3.17–3xYFP* line) to verify the overlap of the expression domains of the translational and the transcriptional fusions of *GH3.17*. The *pGH3.17–3xYFP* line revealed a localized YFP expression in the more external layer of the LRC and in the differentiated epidermal cells (Figure 1A), resembling that of the *pGH3.17:GH3.17-GFP* translational fusion [9], and upon cytokinin treatment *pGH3.17–3xYFP* expression was significantly increased (Figure 1A,B). Although YFP expression was not detectable in *pGH3.9–3xYFP* line (data not shown), the analysis of *pGH3.5–3xYFP* and *pGH3.6–3xYFP* lines revealed, similarly to *pGH3.17–3xYFP* line, a fluorescent signal in the LRC tissue (Figure 1C,E). Moreover, based on the microarray data [17] and given that cytokinin promotes *GH3.17* expression (Figure 1) [9], we verified if the expression of *GH3.5* and *GH3.6* are responsive to cytokinin analyzing *pGH3.5–3xYFP* and *pGH3.6–3xYFP* lines upon cytokinin treatment. The fluorescence signal of *pGH3.5–3xYFP* and *pGH3.6–3xYFP* was detected in the youngest cells of the outermost LRC layer, in the columella and in the vascular tissues (Figure 1C,E). After four hours of cytokinin treatment, the fluorescence signal of *pGH3.5–3xYFP* and *pGH3.6–3xYFP* lines was significantly increased compared to untreated lines (Figure 1C–F). Furthermore, while *GH3.17* expression was localized in the more external tissues of the root (the LRC and the differentiated epidermal cells), *GH3.5* and *GH3.6* expression was induced also in the vascular tissue (Figure 1C,E). This hinted at the possibility that *GH3.5* and *GH3.6* are involved in the regulation of auxin levels not only in the LRC, where the regulation of auxin levels affects meristem size but also in the vascular tissue, possibly in coordination with the robust auxin flux active in this tissue.

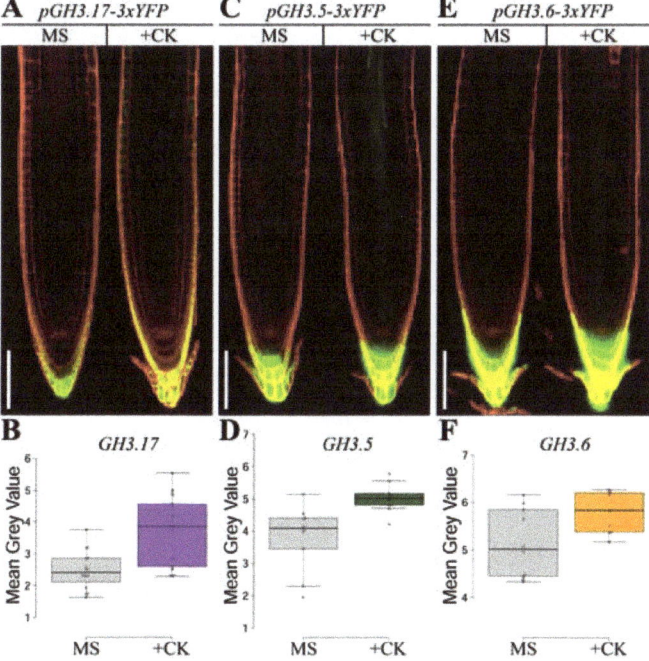

Figure 1. Cytokinin induces *GH3.5* and *GH3.6* expressions. (**A,C,E**) Confocal images of five days after germination (dag) roots expressing *pGH3.17–3xYFP*, *pGH3.5–3xYFP* and *pGH3.6–3xYFP* constructs untreated (MS) and treated for four hours with 5 μM of cytokinin (+CK) (see Materials and Methods). Scale bar, 100 μm. (**B,D,F**) *Mean Grey Value* quantification of *pGH3.17–3xYFP*, *pGH3.5–3xYFP* and *pGH3.6–3xYFP* lines untreated (grey) and treated with cytokinin 5μM for four hours (+CK) (purple-*pGH3.17–3xYFP*; green-*pGH3.5–3xYFP*; orange-*pGH3.6–3xYFP* lines, respectively) at 5 dag where center lines show the medians. Box limits indicate the 25th and 75th percentiles as determined by R software. Whiskers extend 1.5 times the interquartile range from the 25th and 75th percentiles, data points are plotted as open circles. Statistical significance: (**B**) Two biological replicates. *p*-value < 0.005, Student's t-test, $n = 16, 17$ sample points, (**D**) two biological replicates. *p*-value < 0.05, Student's t-test, $n = 10, 9$ sample points, (**F**) *p*-value < 0.005, Student's t-test, $n = 10, 10$ sample points.

Taken together, these results corroborate the idea that a cytokinin-dependent mechanism regulates the auxin inactivation process by controlling the expression of several members of the *GH3* Group II gene family.

It has been shown that GH3.17 activity in the LRC is necessary and sufficient for the regulation of the meristem size [9,17]. To understand if GH3.5 and GH3.6 are involved in the control of the meristem activity, we analyzed the meristem size of *gh3.5–1* and *gh3.6–1* loss of function mutants. The meristem size is measured taking into account the number of meristematic cells of the cortex tissue [3]. In a similar way to *gh3.17–1* mutant, *gh3.5–1* and *gh3.6–1* mutants showed increased meristem size when compared to wild type plants (Figure 2A,B). These data indicate that GH3.5 and GH3.6 activities together with GH3.17 are required for meristem size regulation. To unveil if GH3.5 and GH3.6 regulate root meristem size by acting downstream of cytokinin we analyzed the root meristem size of these mutants upon cytokinin treatment. As previously reported, wild type plants treated with cytokinin, show a reduction of the meristem size [7,8], while *gh3.17–1* meristem is not affected [9]. Differently from *gh3.17–1*, *gh3.5–1* and *gh3.6–1* mutants showed a slight decrease in the number of meristematic cells when comparing the untreated and the cytokinin treated mutant plants (Figure 2C). These data indicate that the root meristem size of those mutants is only partially affected by cytokinin activity, suggesting that cytokinin regulates root meristem size also via GH3.5 and GH3.6. To further investigate the relation between GH3.17, GH3.5, and GH3.6 activities in controlling meristem size, we generated the *gh3.17–1;gh3.5–1* and *gh3.17–1;gh3.6–1* double mutants. Both the double mutants *gh3.17–1;gh3.5–1* and *gh3.17–1;gh3.6–1* showed a meristem size similar to that of the parental single mutants *gh3.17–1*, *gh3.5–1* and *gh3.6–1* (Figure 3), corroborating the idea that these GH3s act in the same pathway in regulating auxin levels to control meristem size. We, thus, inferred that cytokinin globally promotes the auxin inactivation process triggering *GH3.17*, *GH3.5*, and *GH3.6* expressions, and as a consequence, determining root meristem size.

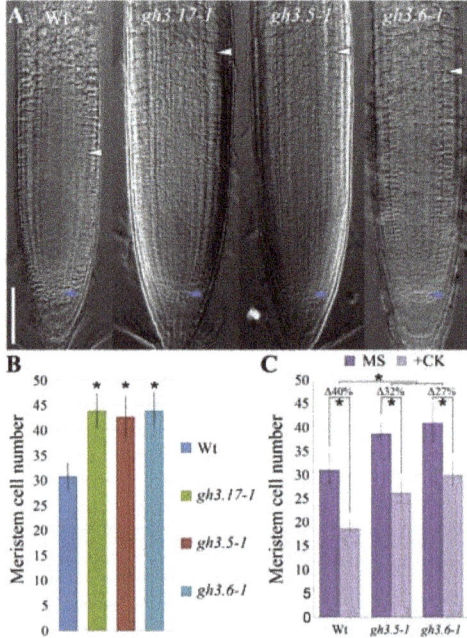

Figure 2. *GH3.5* and *GH3.6* are involved in the control of root meristem size. (**A**) Root meristems at 5 dag of Wt, *gh3.17–1*, *gh3.5–1* and *gh3.6–1* plants. Blue and white arrowheads indicate the quiescent

center (QC) and the cortex transition zone (i.e., meristem size), respectively. Scale bar, 100 µm. (**B**) Analysis of meristematic cortical cell number of Wt, *gh3.17–1*, *gh3.5–1*, and *gh3.6–1* plants. Error bars indicate standard deviation (SD). Two biological replicates were performed. Asterisk (*) indicates a significance with a *p*-value < 0.005, Student's *t*-test, *n* = 18, 15, 16, 17. (**C**) Analysis of meristematic cortical cell number of Wt, *gh3.5–1* and *gh3.6–1* untreated (MS) plants and after 22 h of cytokinin treatment (+CK) (see *Materials and Methods*). "∆" indicates the relative decrease percentage in the number of meristematic cells of the cortex after cytokinin treatment. Error bars indicate SD. Two biological replicates were performed. * indicates a significance with a *p*-value < 0.005, Student's *t*-test, $n_{(MS)}$ = 14, 14, 18 and $n_{(+CK)}$ = 15, 16, 22.

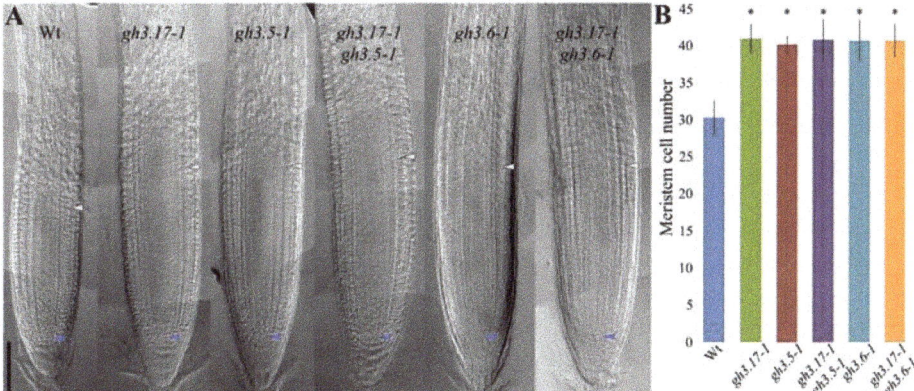

Figure 3. GH3.5, GH3.6, and GH3.17 synergistically act in the control of meristem size. (**A**) Bright field microscopy images of root apical meristems at 5 dag of Wt, *gh3.17–1*, *gh3.5–1*, *gh3.17–1;gh3.5–1*, *gh3.6–1* and *gh3.17–1;gh3.6–1* plants. Blue and white arrowheads indicate the QC and the cortex transition zone (i.e., meristem size), respectively. Scale bar, 100 µm. (**B**) Analysis of the number of meristematic cells of the cortex of Wt, *gh3.17–1*, *gh3.5–1*, *gh3.17–1;gh3.5–1*, *gh3.6–1* and *gh3.17–1;gh3.6–1* plants. Error bars indicate SD. Two biological replicates were performed. * indicates a significance with a *p*-value < 0.001, Student's t-test, *n* = 26, 22, 20, 32, 30, 30.

3. Discussion

In plants, the hormone auxin is distributed as a gradient with morphogenetic properties, similarly to retinoic acid in animals [26–29]. Indeed, variations in auxin distribution profoundly change cell developmental programs [30]. In the root, an auxin maximum controls stem cell activities [30–32] while an auxin minimum establishes the position of the TZ, a cell boundary where stem cell daughters stop to divide and start to differentiate [9]. Indeed, differences in auxin contents between cells of the same tissue are translated into a developmental switch from proliferation to differentiation. The position of the auxin minimum in the root depends on the activity of the GH3.17 enzyme that specifically acts in the LRC tissue [9,17].

Here, we demonstrated that cytokinin supports TZ positioning and, hence, cell differentiation by controlling in the LRC the expression of multiple genes belonging to the *GH3* Group II gene family, such as *GH3.17*, *GH3.5*, and *GH3.6*. These GH3s conjugate auxin to different amino acids, thus, adjusting the levels of active auxin within each cell [9,18,19,21–24].

Cytokinin-dependent control of *GH3.17* expression [9] and the simultaneous activation of *GH3.5* and *GH3.6* gene expression (this work) highlights that auxin inactivation process strongly depends on cytokinin activity in the LRC. Interestingly, it has been already reported that a coordinated *GH3.5*, *GH3.6*, and *GH3.17* activity is necessary during hypocotyl elongation [33].

Although auxin negatively regulates its own levels by promoting *GH3.5* and *GH3.6* expression [25,34,35], *GH3.17* is not controlled by auxin itself [9]. Thus, *GH3.17* cytokinin-dependent control determines a change of auxin levels without suffering from any auxin feedback.

The data collected here show that the specific localized expression of three *GH3* Group II genes, regulating auxin inactivation in the LRC tissue, is crucial for meristem activity. Moreover, from these results, the LRC emerges as an important tissue where GH3-dependent auxin conjugation takes place and, hence, the site where the control of auxin levels is finely imposed in the root. Intriguingly, cytokinin-dependent *GH3.5* and *GH3.6* regulation happens in both LRC and vascular bundle. It will be interesting to know whether GH3.5 and GH3.6 are required in both of those tissues to control root meristem size. Further studies are required to address this crucial point. Nonetheless, the expression domain of the *GH3s* genes prompts the hypothesis that the control of auxin inactivation has to be confined to specific tissues rather than to the whole root to control root meristem size and, therefore, organ growth.

4. Materials and Methods

4.1. Plant Material and Growth Conditions

The *Arabidopsis thaliana* ecotypes Columbia-0 (Col-0) was used as a control because the *gh3.17–1* [9], *gh3.5–1* and *gh3.6–1* mutants are in *Col-0* background. *gh3.5–1* and *gh3.6–1* lines were obtained from the NASC collection (SALK_033434C and SALK_082530). Homozygous mutants from the Salk T-DNA were identified by PCR as described (http://signal.salk.edu/tdnaprimers.html). For growth conditions, *Arabidopsis* seeds were surface sterilized, and seedlings were grown on one-half strength Murashige and Skoog (MS) medium containing 0.8% agar at 22 °C in long-day conditions (16-h-light/8-h-dark cycle) as previously described [3].

Arabidopsis locus IDs from this article: *GH3.17* (AT1G28130), *GH3.5* (AT4G27260), *GH3.6* (AT5G54510) and *GH3.9* (AT2G47750).

4.2. Generation of GH3s Transgenic Plants

Standard molecular biology techniques and the Gateway system (Invitrogen) were used for the cloning procedures. For the *pGH3.5::GH3.5-GFP*, *pGH3.6::GH3.6-GFP*, and *pGH3.9::GH3.9-GFP* transgenic plants, the promoter sequences of *GH3.5* (2959 bp), *GH3.6* (1993 bp), *GH3.9* (2312 bp), and *GH3.17* (2128 bp) and genomic sequences of *GH3.5* (2189 bp), *GH3.6* (2244 bp), and *GH3.9* (2668 bp) were amplified from genomic DNA of *Arabidopsis thaliana* Columbia ecotype using specific primers (pGH3.5 FW 5′-TTTTTCATTGGATGTGAGGAA-3′, pGH3.5 REV 5′-GGTTTAAGAGAAAGAGAGA AGTCTGAG-3′, pGH3.6 FW 5′-AAAACCCATTAACAGCAGACG-3′, pGH3.6 REV 5′-CGTTTAGGT TTTGTGTTTAAAATTC-3′, pGH3.9 FW 5′-TGTCCTTGCAAGTGCAAAAT-3′, pGH3.9 REV 5′-TTCTC AGCTAACCCAAAGAAAG-3′, pGH3.17 FW 5′-GGGCGTTACGTATCAGGAAA-3′, pGH3.17 REV 5′-TGTCTGAAAGCAGACACAAACA-3′, gGH3.5 FW 5′-ATGCCTGAGGCACCAAAGAA-3′, gGH3.5 REV 5′-GTTACTCCCCCACTGTTTGTG-3′, gGH3.6 FW 5′-ATGCCTGAGGCACCAAAG-3′, gGH3.6 REV 5′-GTTACTCCCCCATTGCTTGT-3′, gGH3.9 FW 5′-ATGGATGTAATGAAGCTTGATCA-3′, gGH3.9 REV 5′-TGGAACCCAAGTCGGGTC-3′) and cloned in a *pDONOR-P4P1* and *pDONOR-221* vectors: *pDONOR-P4P1-pGH3.5*, *pDONOR-P4P1-pGH3.6*, *pDONOR-P4P1-pGH3.9*, and *pDONOR-P4P1-pGH3.17*, promoter sequences, respectively, *pDONOR-221-pGH3.5*, *pDONOR-221-pGH3.6*, and *pDONOR-221-pGH3.9*, genomic sequences, respectively. The LR reactions were then conducted by using the *pDONOR-P4P1-pGH3.5/pGH3.6/pGH3.9*, the *pDONOR221-gGH3.5/gGH3.6/gGH3.9* and a *pDONORP2P3-GFP* vectors.

For *pGH3.5–3xYFP*, *pGH3.6–3xYFP*, *pGH3.9–3xYFP*, and *pGH3.17–3xYFP* transgenic plants, the promoter sequences of *GH3.5*, *GH3.6*, *GH3.9*, and *GH3.17* cloned in the *pDONOR-P4P1* vectors, as described above, were used. The LR reactions were then conducted by using the *pDONOR-P4P1-pGH3.5/pGH3.6/pGH3.9/pGH3.17*, a *pDONOR221–3xYFP*, and *pDONORP2P3-NOST2* vectors [36].

The obtained LR products were then sub-cloned in the Gateway *pBm43GW* destination vector. Plasmids were transformed into *Col-0* plants by floral dipping [37]. Each expression domain of the T2 generation of the *3xYFP* transcriptional fusion lines was analyzed to verify homogeneous expression. Each transgenic line revealed the same YFP expression pattern.

4.3. Hormonal Treatments

Five days after germination (dag) seedlings were transferred onto solid one-half MS medium containing 0.025% DMSO solvent (mock condition) or onto solid medium containing a final concentration of 5 µM trans-Zeatin (tZ, Duchefa) dissolved in DMSO (0.025% final concentration). A twenty-two-hour hormone treatment was used for meristem size analysis in response to cytokinin and a four-hour hormone treatment was used for *GH3s* transcriptional reporter lines expression analysis.

4.4. Bright Field and Confocal Microscopy Analysis

Differential interference contrast (DIC) with Nomarski technology microscopy (Zeiss Axio Imager A2 microscope) was used to count meristem cell number with bright field microscopy. Root meristem size of each plant was measured based on the number of cortex cells in a file extending from the quiescent center to the first elongated cortex cell excluded [3]. Plants were mounted in a chloral hydrate solution [3]. Confocal images were obtained using a Zeiss LSM 780 confocal laser scanning microscope. For confocal laser scanning analysis, a propidium iodide 10 µM staining was used. For each experiment, two biological replicates were performed, and the number of samples analyzed were reported in the relative figure legend. Results were comparable in all experiments. The statistical significance was determined by Student's t-test (http://graphpad.com/quickcalcs/ttest2.cfm), data were reported in the relative figure legend.

4.5. GH3s Reporter Lines Fluorescence Quantification

The fluorescence intensity of *pGH3.5–3xYFP*, *pGH3.6–3xYFP* and *pGH3.17–3xYFP* lines untreated and treated with cytokinin 5 µM for four hours (Figure 1) was quantified as reported in [3]. Mean Grey Value of YFP channel of confocal laser scanning microscope images was measured with the software *ImageJ* (https://imagej.nih.gov/ij/). Fluorescence signal was measured taking into consideration the same area for untreated and treated lines (length 550 µm × width 187 µm) starting from the tip of the root. Student's t-test was used to determine the statistical significance (http://graphpad.com/quickcalcs/ttest2.cfm) as reported in the relative figure legend.

4.6. Statistical Analysis Criteria

All the experiments were performed with a number of samples large enough to ensure the statistical significance of the analysis, as reported in corresponding figure legends. Representative sample pictures of the experiments were chosen in all figures.

Author Contributions: Conceptualization, R.D.M; methodology, E.P., S.J.U., E.S., N.S., R.D.I., G.B, and R.D.M.; formal analysis, E.P., S.J.U., E.S., N.S., R.D.I., G.B, and R.D.M.; investigation, E.P., S.J.U., E.S., N.S., R.D.I., G.B, and R.D.M.; data curation, E.P., S.J.U., E.S., N.S., R.D.I., and G.B.; writing – original draft preparation, R.D.M.; writing – review and editing, S.S., R.D.I and R.D.M.; supervision, R.D.M.; project administration, R.D.M.; funding acquisition, S.S. and R.D.M.

Funding: Postdoc fellowship of the German Academic Exchange Service (DAAD) (to S.J.U.), European Research Council grant 260368 (to S.S.) and MIUR (to S.S. and R.D.M.).

Conflicts of Interest: The authors declare no conflicts of interest.

References

1. Dolan, L.; Janmaat, K.; Willemsen, V.; Linstead, P.; Poethig, S.; Roberts, K.; Scheres, B. Cellular organisation of the arabidopsis thaliana root. *Development* **1993**, *119*, 71–84.

2. Di Ruocco, G.; Di Mambro, R.; Dello Ioio, R. Building the differences: A case for the ground tissue patterning in plants. *Proc. Biol. Sci.* **2018**, *285*. [CrossRef]
3. Di Mambro, R.; Sabatini, S. Developmental analysis of arabidopsis root meristem. *Methods. Mol. Biol.* **2018**, *1761*, 33–45.
4. Di Mambro, R.; Sabatini, S.; Dello Ioio, R. Patterning the axes: A lesson from the root. *Plants (Basel)* **2018**, *8*, 8. [CrossRef]
5. Salvi, E.; Di Mambro, R.; Pacifici, E.; Dello Ioio, R.; Costantino, P.; Moubayidin, L.; Sabatini, S. Scarecrow and shortroot control the auxin/cytokinin balance necessary for embryonic stem cell niche specification. *Plant Signal Behav.* **2018**, *13*, e1507402.
6. Pacifici, E.; Di Mambro, R.; Dello Ioio, R.; Costantino, P.; Sabatini, S. Acidic cell elongation drives cell differentiation in the arabidopsis root. *EMBO J.* **2018**, *37*, e99134. [CrossRef]
7. Dello Ioio, R.; Linhares, F.S.; Scacchi, E.; Casamitjana-Martinez, E.; Heidstra, R.; Costantino, P.; Sabatini, S. Cytokinins determine arabidopsis root-meristem size by controlling cell differentiation. *Curr. Biol.* **2007**, *17*, 678–682. [CrossRef]
8. Dello Ioio, R.; Nakamura, K.; Moubayidin, L.; Perilli, S.; Taniguchi, M.; Morita, M.T.; Aoyama, T.; Costantino, P.; Sabatini, S. A genetic framework for the control of cell division and differentiation in the root meristem. *Science* **2008**, *322*, 1380–1384. [CrossRef]
9. Di Mambro, R.; De Ruvo, M.; Pacifici, E.; Salvi, E.; Sozzani, R.; Benfey, P.N.; Busch, W.; Novak, O.; Ljung, K.; Di Paola, L.; et al. Auxin minimum triggers the developmental switch from cell division to cell differentiation in the arabidopsis root. *Proc. Natl. Acad. Sci. USA* **2017**, *114*, E7641–E7649. [CrossRef]
10. Swarup, R.; Kramer, E.M.; Perry, P.; Knox, K.; Leyser, H.M.; Haseloff, J.; Beemster, G.T.; Bhalerao, R.; Bennett, M.J. Root gravitropism requires lateral root cap and epidermal cells for transport and response to a mobile auxin signal. *Nat. Cell. Biol.* **2005**, *7*, 1057–1065. [CrossRef]
11. Bennett, T.; van den Toorn, A.; Sanchez-Perez, G.F.; Campilho, A.; Willemsen, V.; Snel, B.; Scheres, B. Sombrero, bearskin1, and bearskin2 regulate root cap maturation in arabidopsis. *Plant Cell* **2010**, *22*, 640–654. [CrossRef]
12. Xuan, W.; Audenaert, D.; Parizot, B.; Moller, B.K.; Njo, M.F.; De Rybel, B.; De Rop, G.; Van Isterdael, G.; Mahonen, A.P.; Vanneste, S.; et al. Root cap-derived auxin pre-patterns the longitudinal axis of the arabidopsis root. *Curr. Biol.* **2015**, *25*, 1381–1388. [CrossRef]
13. Kanno, S.; Arrighi, J.F.; Chiarenza, S.; Bayle, V.; Berthome, R.; Peret, B.; Javot, H.; Delannoy, E.; Marin, E.; Nakanishi, T.M.; et al. A novel role for the root cap in phosphate uptake and homeostasis. *Elife* **2016**, *5*, e14577. [CrossRef]
14. Xuan, W.; Band, L.R.; Kumpf, R.P.; Van Damme, D.; Parizot, B.; De Rop, G.; Opdenacker, D.; Moller, B.K.; Skorzinski, N.; Njo, M.F.; et al. Cyclic programmed cell death stimulates hormone signaling and root development in arabidopsis. *Science* **2016**, *351*, 384–387. [CrossRef]
15. Blancaflor, E.B.; Fasano, J.M.; Gilroy, S. Laser ablation of root cap cells: Implications for models of graviperception. *Adv. Space Res.* **1999**, *24*, 731–738. [CrossRef]
16. Tsugeki, R.; Fedoroff, N.V. Genetic ablation of root cap cells in arabidopsis. *Proc. Natl. Acad. Sci. USA* **1999**, *96*, 12941–12946. [CrossRef]
17. Di Mambro, R.; Svolacchia, N.; Dello Ioio, R.; Pierdonati, E.; Salvi, E.; Pedrazzini, E.; Vitale, A.; Perilli, S.; Sozzani, R.; Benfey, P.N.; et al. The lateral root cap acts as an auxin sink that controls meristem size. *Curr. Biol.* **2019**, *29*, 1199–1205. [CrossRef]
18. Nakazawa, M.; Yabe, N.; Ichikawa, T.; Yamamoto, Y.Y.; Yoshizumi, T.; Hasunuma, K.; Matsui, M. Dfl1, an auxin-responsive gh3 gene homologue, negatively regulates shoot cell elongation and lateral root formation, and positively regulates the light response of hypocotyl length. *Plant J.* **2001**, *25*, 213–221. [CrossRef]
19. Staswick, P.E.; Serban, B.; Rowe, M.; Tiryaki, I.; Maldonado, M.T.; Maldonado, M.C.; Suza, W. Characterization of an arabidopsis enzyme family that conjugates amino acids to indole-3-acetic acid. *Plant Cell* **2005**, *17*, 616–627. [CrossRef]
20. Hagen, G.; Guilfoyle, T. Auxin-responsive gene expression: Genes, promoters and regulatory factors. *Plant Mol. Biol.* **2002**, *49*, 373–385. [CrossRef]
21. Khan, S.; Stone, J.M. Arabidopsis thaliana gh3.9 influences primary root growth. *Planta* **2007**, *226*, 21–34. [CrossRef]

22. Westfall, C.S.; Sherp, A.M.; Zubieta, C.; Alvarez, S.; Schraft, E.; Marcellin, R.; Ramirez, L.; Jez, J.M. Arabidopsis thaliana gh3.5 acyl acid amido synthetase mediates metabolic crosstalk in auxin and salicylic acid homeostasis. *Proc. Natl. Acad. Sci. USA* **2016**, *113*, 13917–13922. [CrossRef]
23. LeClere, S.; Tellez, R.; Rampey, R.A.; Matsuda, S.P.; Bartel, B. Characterization of a family of iaa-amino acid conjugate hydrolases from arabidopsis. *J. Biol. Chem.* **2002**, *277*, 20446–20452. [CrossRef]
24. Staswick, P.E.; Tiryaki, I.; Rowe, M.L. Jasmonate response locus jar1 and several related arabidopsis genes encode enzymes of the firefly luciferase superfamily that show activity on jasmonic, salicylic, and indole-3-acetic acids in an assay for adenylation. *Plant Cell* **2002**, *14*, 1405–1415. [CrossRef]
25. Bargmann, B.O.; Vanneste, S.; Krouk, G.; Nawy, T.; Efroni, I.; Shani, E.; Choe, G.; Friml, J.; Bergmann, D.C.; Estelle, M.; et al. A map of cell type-specific auxin responses. *Mol. Syst. Biol.* **2013**, *9*, 688. [CrossRef]
26. Shimozono, S.; Iimura, T.; Kitaguchi, T.; Higashijima, S.; Miyawaki, A. Visualization of an endogenous retinoic acid gradient across embryonic development. *Nature* **2013**, *496*, 363–366. [CrossRef]
27. Begemann, G.; Schilling, T.F.; Rauch, G.J.; Geisler, R.; Ingham, P.W. The zebrafish neckless mutation reveals a requirement for raldh2 in mesodermal signals that pattern the hindbrain. *Development* **2001**, *128*, 3081–3094.
28. Ozaki, R.; Kuroda, K.; Ikemoto, Y.; Ochiai, A.; Matsumoto, A.; Kumakiri, J.; Kitade, M.; Itakura, A.; Muter, J.; Brosens, J.J.; et al. Reprogramming of the retinoic acid pathway in decidualizing human endometrial stromal cells. *PLoS One* **2017**, *12*, e0173035. [CrossRef]
29. Sosnik, J.; Zheng, L.; Rackauckas, C.V.; Digman, M.; Gratton, E.; Nie, Q.; Schilling, T.F. Noise modulation in retinoic acid signaling sharpens segmental boundaries of gene expression in the embryonic zebrafish hindbrain. *Elife* **2016**, *5*, e14034. [CrossRef]
30. Sabatini, S.; Beis, D.; Wolkenfelt, H.; Murfett, J.; Guilfoyle, T.; Malamy, J.; Benfey, P.; Leyser, O.; Bechtold, N.; Weisbeek, P.; et al. An auxin-dependent distal organizer of pattern and polarity in the arabidopsis root. *Cell* **1999**, *99*, 463–472. [CrossRef]
31. Grieneisen, V.A.; Xu, J.; Maree, A.F.; Hogeweg, P.; Scheres, B. Auxin transport is sufficient to generate a maximum and gradient guiding root growth. *Nature* **2007**, *449*, 1008–1013. [CrossRef] [PubMed]
32. Petersson, S.V.; Johansson, A.I.; Kowalczyk, M.; Makoveychuk, A.; Wang, J.Y.; Moritz, T.; Grebe, M.; Benfey, P.N.; Sandberg, G.; Ljung, K. An auxin gradient and maximum in the arabidopsis root apex shown by high-resolution cell-specific analysis of iaa distribution and synthesis. *Plant Cell* **2009**, *21*, 1659–1668. [CrossRef] [PubMed]
33. Tian, C.E.; Muto, H.; Higuchi, K.; Matamura, T.; Tatematsu, K.; Koshiba, T.; Yamamoto, K.T. Disruption and overexpression of auxin response factor 8 gene of arabidopsis affect hypocotyl elongation and root growth habit, indicating its possible involvement in auxin homeostasis in light condition. *Plant J.* **2004**, *40*, 333–343. [CrossRef]
34. Chaiwanon, J.; Wang, Z.Y. Spatiotemporal brassinosteroid signaling and antagonism with auxin pattern stem cell dynamics in arabidopsis roots. *Curr. Biol.* **2015**, *25*, 1031–1042. [CrossRef] [PubMed]
35. Paponov, I.A.; Paponov, M.; Teale, W.; Menges, M.; Chakrabortee, S.; Murray, J.A.; Palme, K. Comprehensive transcriptome analysis of auxin responses in arabidopsis. *Mol. Plant* **2008**, *1*, 321–337. [CrossRef] [PubMed]
36. Di Ruocco, G.; Bertolotti, G.; Pacifici, E.; Polverari, L.; Tsiantis, M.; Sabatini, S.; Costantino, P.; Dello Ioio, R. Differential spatial distribution of mir165/6 determines variability in plant root anatomy. *Development* **2018**, *145*. [CrossRef] [PubMed]
37. Clough, S.J.; Bent, A.F. Floral dip: A simplified method for agrobacterium-mediated transformation of arabidopsis thaliana. *Plant J.* **1998**, *16*, 735–743. [CrossRef] [PubMed]

 © 2019 by the authors. Licensee MDPI, Basel, Switzerland. This article is an open access article distributed under the terms and conditions of the Creative Commons Attribution (CC BY) license (http://creativecommons.org/licenses/by/4.0/).

Communication

Stress-Triggered Long-Distance Communication Leads to Phenotypic Plasticity: The Case of the Early Root Protoxylem Maturation Induced by Leaf Wounding in Arabidopsis

Ilaria Fraudentali [1], Renato Alberto Rodrigues-Pousada [2], Alessandro Volpini [1], Paraskevi Tavladoraki [1], Riccardo Angelini [1] and Alessandra Cona [1,*]

1. Department of Science, University "Roma Tre", 00146 Rome, Italy; ilaria.fraudentali@uniroma3.it (I.F.); ale.volpini@stud.uniroma3.it (A.V.); paraskevi.tavladoraki@uniroma3.it (P.T.); riccardo.angelini@uniroma3.it (R.A.)
2. Department of Life, Health and Environmental Sciences, University of L'Aquila, 67100 L'Aquila, Italy; pousada@univaq.it
* Correspondence: alessandra.cona@uniroma3.it; Tel.: +39-06-5733-6360

Received: 26 October 2018; Accepted: 30 November 2018; Published: 4 December 2018

Abstract: Root architecture and xylem phenotypic plasticity influence crop productivity by affecting water and nutrient uptake, especially under those environmental stress, which limit water supply or imply excessive water losses. Xylem maturation depends on coordinated events of cell wall lignification and developmental programmed cell death (PCD), which could both be triggered by developmental- and/or stress-driven hydrogen peroxide (H_2O_2) production. Here, the effect of wounding of the cotyledonary leaf on root protoxylem maturation was explored in *Arabidopsis thaliana* by analysis under Laser Scanning Confocal Microscope (LSCM). Leaf wounding induced early root protoxylem maturation within 3 days from the injury, as after this time protoxylem position was found closer to the tip. The effect of leaf wounding on protoxylem maturation was independent from root growth or meristem size, that did not change after wounding. A strong H_2O_2 accumulation was detected in root protoxylem 6 h after leaf wounding. Furthermore, the H_2O_2 trap N,N^1-dimethylthiourea (DMTU) reversed wound-induced early protoxylem maturation, confirming the need for H_2O_2 production in this signaling pathway.

Keywords: wounding; root plasticity; hydrogen peroxide; protoxylem

1. Introduction

Plant adaptive capacity and acclimatization resources play a pivotal role in increasing plant fitness and survival, especially in fast-changing environmental conditions. Thus, the unravelling of variation in phenotypic plasticity in traits of agronomic interest could provide us with beneficial tools for the development of crops more efficiently adaptable to a changing environment. Phenotypic plasticity integrates genetically determined developmental processes and environmental influences [1], and because of this, identifying phenotypic traits showing favourable adaptive plasticity will provide the basis for further studies focused on assessing the underlying genetic basis.

Root systems play a prominent role in crop health and productivity, especially under resource-limited environmental conditions, and plasticity of root traits, such as root growth and architecture, confers functional adaptivity to soils that are poor in water and nutrients [2]. In this regard, root development and differentiation follow different dynamics and may respond to different signalling pathways under physiological or stress conditions, allowing adaptive plasticity in sub-optimal growth conditions. Under physiological conditions, the boundaries defining the division, elongation and maturation zones of the root are developmentally regulated by the cytokinin/auxin [3–5] and/or reactive oxygen species (ROS) pathways [6], and changes in their positions are coordinated with each other [7]. Vascular patterning is finely integrated in the root developmental program by the cytokinin/auxin/thermospermine pathway responsible for the specification of the identity of the protoxylem [8,9], which is going to mature later in the proximal region beyond the zone of maximum elongation growth, where it undertakes the deposition of secondary walls [10].

However, the correlation among root length, meristem size and protoxylem element position may be disrupted under stress or phytotoxic conditions [7], and both ROS [11,12] and stress signalling hormones, such as the wound signal jasmonic acid (JA) [13,14], may assume a role in root length and meristem size specification independently from or interfering with the cytokinin/auxin pathway. In this regard, root xylem phenotypic plasticity has been shown to occur in response to drought stress [15,16], as well as to various stress-simulating conditions [17]. During acclimation to drought, plasticity of root xylem tissues may enhance water absorption from the soil improving plant performance and protecting yield [18]. Moreover, an early xylem differentiation was observed in maize (*Zea mays*), tobacco (*Nicotiana tabacum*), and Arabidopsis (*Arabidopsis thaliana*) roots under stress-simulated conditions, such as those induced by polyamine (PA)-treatment or amine oxidase (AO)-overexpression [19,20], as well as those signalled by methyl jasmonate (MeJA) treatment [20] or by a compromised status of cell-wall pectin integrity [21]. In these conditions a hydrogen peroxide (H_2O_2)-triggered early root xylem maturation, measured as the distance of the first xylem elements with fully developed secondary wall thickenings from the apical meristem, repositions xylem precursors closer to the tip. Furthermore, a higher number of xylem elements in tobacco plants over-expressing a fungal endo-polygalacturonase (PG plants) [21] and in water-stressed soybean has been reported [18]. In the latter system, it has been suggested that this xylem adaptive plasticity enhances water uptake by improving root hydraulic conductivity under drought [18].

Noteworthy, plant dehydration may occur not only under drought, but also as a consequence of those stresses that may lead to excessive water losses, such as leaf mechanical damage caused by herbivore feeding or atmospheric agents. Indeed, it has been reported that several wound-inducible genes were likewise induced by dehydration, implying that water stress is an important component in the plant responses to mechanical wounding [22]. Consistently, other evidence supports the occurrence of cross-tolerance mechanisms between JA-signalled wounding or insect feeding and those stresses that involve perturbation of water potential [23,24]. In this regard, it has been reported that wounding increases salt tolerance in tomato plants [23] and that whitefly infestation promotes drought resistance in maize plants [24], in both cases by a mechanism involving JA biosynthesis [23,24]. In this regard, the phenotypic plasticity of the root xylem system elicited by leaf wounding has never been explored. Here, we provide evidence that wounding of the cotyledonary leaf triggers leaf to root long-distance communication resulting in early root protoxylem differentiation in Arabidopsis. The proposed approach may represent a model for future investigations focused on unravelling the occurrence of phenotypic plasticity induced by long-distance communication triggered by biotic/abiotic stresses imposed at a specific distal site.

2. Results

2.1. Leaf Wounding Promotes Alteration of Protoxylem Maturation in Root without Affecting Root Length and Meristem Size

To explore the effect of leaf wounding on root xylem phenotypic plasticity, 7-day-old Arabidopsis seedlings were injured by cutting a cotyledonary leaf, and then roots were observed under Laser Scanning Confocal Microscope (LSCM) 3 days after the injury, for the investigation of the distance from the root apical meristem of the first protoxylem cell with fully developed secondary wall thickenings (whose location is here referred as "protoxylem position") and meristem size. Figure 1 shows images acquired under LSCM after PI staining and relative bright-field images of root apexes from unwounded control and leaf-wounded seedlings. Plantlets in which a cotyledonary leaf was cut present an anticipation of the maturation of protoxylem, as shown by the earlier presence of cells with fully developed secondary wall thickenings that appear closer to the apical meristem as compared to unwounded control plants. Figure 2 demonstrates that these qualitative data were confirmed by statistically significant quantitative analysis. In fact, the mean distance of the first protoxylem cell with fully developed secondary wall thickenings from the root apical meristem was approximately 1620 µm in leaf-wounded plants as compared to unwounded control plants, showing a distance of approximately 2060 µm. The effect of leaf wounding on protoxylem position was specific and not dependent upon variation in root growth or meristem size, which were unchanged in leaf-wounded plants compared to unwounded control plants (Figure 3; Table 1).

Figure 1. Analysis under LSCM after PI staining of root apexes and respective bright-field images from 10-day-old unwounded control (**A–E**) and leaf-wounded (**F–J**) seedlings. (**A–E**) bright-field of the root from unwounded control seedlings (**A**), PI staining (**B**) bright-field (**C**) and overlay image (**D**) of the magnified zone of the root from unwounded control seedlings, in which protoxylem position (defined by the position of the first protoxylem cell with fully developed secondary cell wall thickenings) is located, PI staining of the root shown in A (**E**); (**F–J**) PI staining of the root from leaf-wounded seedling injured at the age of 7 days by cutting the cotyledonary leaf, analysed 3 days after injury (**F**), PI staining (**G**) bright-field (**H**) and overlay image (**I**) of the magnified zone of the root from leaf-wounded seedlings, in which protoxylem position is located, bright-field of the root shown in F (**J**). The images presented are representative of experiments repeated at least five times with ten seedlings analysed each time. Shown images were obtained aligning serial overlapping micrographs of the same root by Photoshop Software (Adobe). Bars: 100 µm (**A,E,F,J**) and 10 µm (**B–D,G–I**).

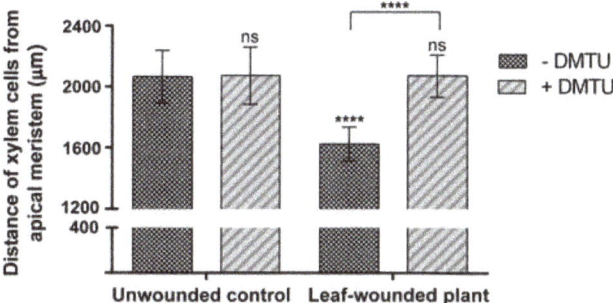

Figure 2. Analysis of differences in protoxylem maturation in leaf-wounded seedlings grown in medium with or without the H_2O_2-scavenger DMTU. Distances from the apical meristem of the protoxylem position (defined by the position of the first protoxylem cell with fully developed secondary cell wall thickenings) are reported. These experiments were repeated at least five times with ten seedlings analysed each time (mean values ± SD; $n = 50$). The statistical significance levels between unwounded control DMTU-untreated plants and DMTU-treated and/or wounded plants were evaluated by p levels as follows: ****, $p \leq 0.0001$; ns, not significant. The significance levels between wounded DMTU-untreated and DMTU-treated plants are reported above the horizontal square bracket.

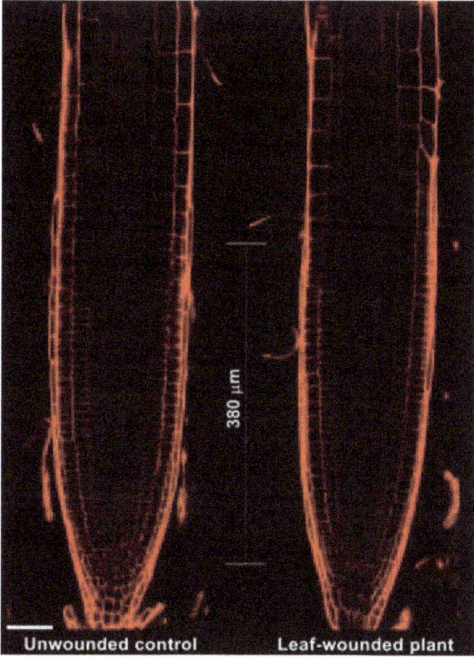

Figure 3. Analysis under LSCM after PI staining of the leaf wounding effect on the length of the meristematic zone, determined by measuring the distance between the quiescent centre and the first elongating cell in the cortex cell file. The images presented show roots from 10-day-old unwounded control and leaf-wounded seedlings, injured at the age of 7 days by cutting the cotyledonary leaf with scissors and analysed 3 days after the injury; roots presented are representative of experiments repeated at least five times with ten seedlings analysed each time. Shown images were obtained aligning serial overlapping micrographs of the same root by Photoshop Software (Adobe). Bar: 50 μm.

Table 1. Analysis of differences in root growth and meristem size in leaf-wounded seedlings grown in medium with or without the H_2O_2-scavenger DMTU. The effect of leaf wounding on root growth was evaluated as the difference between the length measured at the onset of the wounding and that measured after 3 days. The length of the meristematic zone was determined by measuring the distance between the quiescent centre and the first elongating cell in the cortex cell file. These experiments were repeated at least five times with ten seedlings analysed each time (mean values ± SD; $n = 50$). The statistical significance levels between unwounded control and wounded plants were evaluated by p levels as follows: ns, not significant.

	Root Growth (cm)		Meristem Size (μm)	
	Unwounded Control	Leaf-Wounded Plant	Unwounded Control	Leaf-Wounded Plant
−DMTU	2.55 ± 0.20	2.39 ± 0.15 ns	374.0 ± 28.4	372.3 ± 34.6 ns
+DMTU	2.54 ± 0.25	2.39 ± 0.24 ns	373.3 ± 15.9	370.1 ± 17.3 ns

2.2. Early Xylem Maturation in Arabidopsis Roots upon Leaf Wounding Requires H_2O_2

Figure 2 also shows that the H_2O_2-scavenger N,N^1-dimethylthiourea (DMTU), provided at the working concentration of 100 μM, according to a previous report [25], opposes the effect of leaf wounding on early protoxylem maturation consistently with what was previously demonstrated for the MeJA-mediated induction of protoxylem differentiation [20]. To confirm that the effect of the wound-induced signalling on the early maturation of protoxylem cells require H_2O_2, this compound was detected in situ in Arabidopsis roots following leaf wounding by exploiting the fluorogenic peroxidase substrate Amplex Ultra Red (AUR). Figure 4 shows that 6 h after leaf wounding, a strong AUR signal was revealed in the root zone where the first protoxylem cell with fully developed secondary cell wall thickenings is found, which was not detectable in unwounded control roots, which is suggestive of a tissue-specific H_2O_2 production triggered by a long-distance leaf-to-root communication and leading to early protoxylem differentiation.

3. Discussion

Leaf-to-root long-distance communication is crucial in coordinating biochemical and physiological events between aerial and underground organs, especially in response to changes in environmental conditions [26–29]. Leaf damage is a frequent injury during the plant lifespan, and may be caused by both herbivores, such as chewing insects, and atmospheric conditions. The wound site is an easy passage for both pathogen entry and water loss, and the presence of leaf mechanical damage triggers several local responses devoted to healing the wound [21,30–32]. Furthermore, complex signalling networks propagate information from the wound site through the whole plant body, allowing systemic responses [29], among which, xylem root remodelling could represent a strategy for enhancing water uptake and counteracting the excessive water loss caused by the wound.

The analysis of root growth, protoxylem position and meristem size in plants in which the cotyledonary leaf has been cut shows a DMTU-reversible early protoxylem differentiation occurring 3 days after injury (Figures 1 and 2), which is independent from variation in meristem size and root growth, which were unchanged (Table 1; Figure 3). A root protoxylem-specific accumulation of H_2O_2 was detectable 6 h after the injury, supporting its involvement in the variation of protoxylem position (Figure 4). This response is consistent with previous data, where roots of MeJA-treated plants showed a H_2O_2-dependent remodelling of the protoxylem, which appeared to be closer to the root tip, independent of root growth or meristem size [20]. Based on the effects of MeJA treatment on protoxylem differentiation, it has been hypothesized that under stress conditions, extracellular H_2O_2 production may drive early xylem differentiation independently from the auxin/cytokinin/T-Spm loop [17]. In particular, in differentiating protoxylem elements, the H_2O_2 production driven by cell wall-localized oxidation of PAs was suggested to be involved in both developmental programmed cell death (PCD) and peroxidase-mediated lignin polymerization [17,19,20], which represent key steps in the terminal phase of the xylem differentiation process. PAs are oxidized to aminoaldehydes by

AOs, which include copper-containing amine oxidases (CuAOs) and flavin adenine dinucleotide (FAD)-dependent polyamine oxidases (PAOs), with the production of a corresponding amine moiety and the biologically active compound H_2O_2 [33]. Among the cell-wall sources of ROS, it has been known for a long time that AOs are involved in wound-healing responses [30,33] and root xylem differentiation [19–21]. Our results suggest the occurrence of a systemic signalling linking an abiotic stress such as leaf wounding with distal root phenotypic plasticity such as variation in protoxylem position, and open the question of unravelling the responsible ROS source.

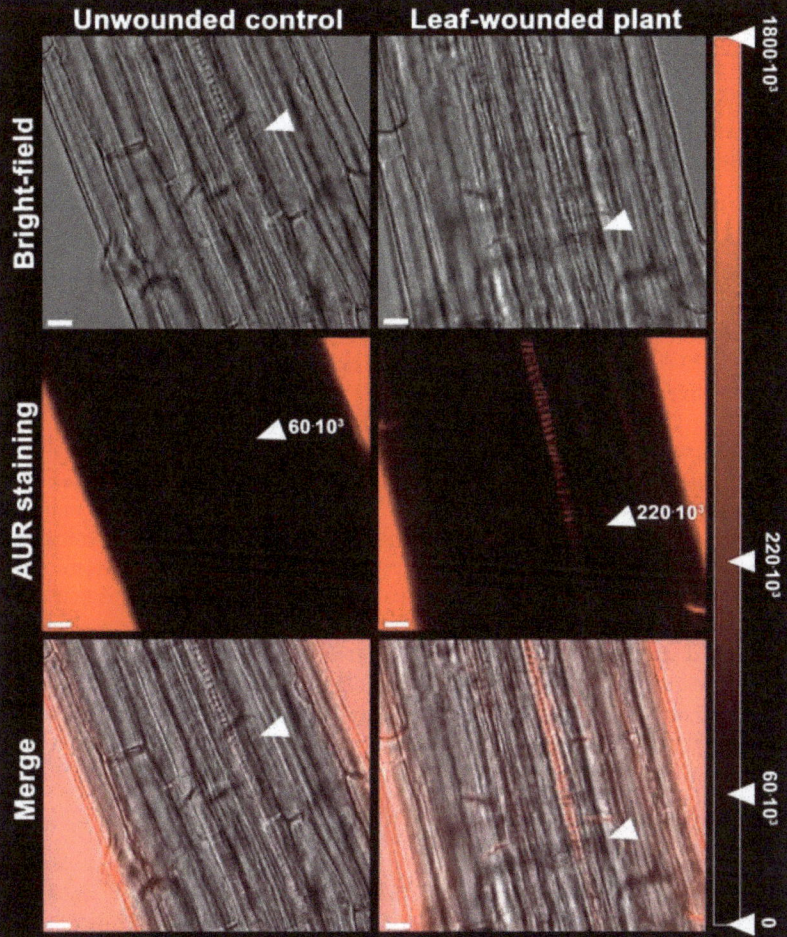

Figure 4. In situ H_2O_2 detection by analysis under LSCM after AUR staining of roots from 7-day-old unwounded control and leaf-wounded seedlings 6 h after injury. The corresponding bright-field and overlay images are shown. Micrographs show the root zone corresponding to the site of appearance of the first protoxylem cell with fully developed secondary cell wall thickenings (arrows) and have been taken at the level of the central root section. Images are representative of those obtained from ten seedlings from five independent experiments. In the red degrading scale, the average values of fluorescence intensity, measured as the sum of the pixels of each 65 µm^2 rectangle, are indicated for unwounded control and leaf-wounded plants, and these were $60 \times 10^3 \pm 19 \times 10^3$ and $220 \times 10^3 \pm 38 \times 10^3$, respectively (mean values ± SD; n = 25). The maximum pixel sum for a completely saturated square was approximately 1800×10^3. Bar: 10 µm.

4. Materials and Methods

4.1. Plant Materials, Treatments and Root Growth Analysis

Arabidopsis seedlings (Columbia-0 ecotype) were grown in vitro in a growth chamber at 23 °C and 55% relative humidity under a photoperiod of 16 h light and 8 h dark. Sterilization of seeds was carried out according to Valvekens et al. [34]. After cold stratification at 4 °C, seeds were grown in one-half-strength Murashige and Skoog salt mixture added with 0.5% (w/v) sucrose and 0.8% (w/v) agar. Plates were kept in vertical position to allow root growth on the solid medium surface. For analysis under LSCM of root protoxylem position and meristem size, 7-day-old seedlings were selected for homogeneity in root length and then transferred onto fresh medium with or without 100 µM DMTU. After the transfer, seedlings were injured by cutting the cotyledonary leaf with scissors, and after 6 h (AUR staining) or 3 days (PI staining), they were collected for analysis under LSCM. The effect of leaf wounding on root growth was evaluated as the difference between the length measured at the onset of the wounding and that measured after 3 days.

4.2. Protoxylem Position and Meristem Size Analysis under LSCM by Cell Wall PI Staining and Bright-Field Examination of Root Tissues

Root apices from 10-day-old unwounded control and leaf-wounded seedlings treated or not with 100 µM DMTU for the last 3 days, were incubated for 5/10 min in PI (10 µg mL^{-1}) to highlight cell wall and protoxylem [35] and then observed under LSCM using a 488 nm argon laser, with a 600–680 nm band-pass filter and a 40× oil immersion objective. The PI staining was allowed to proceed until protoxylem was completely highlighted. Roots were concurrently analysed by bright-field microscopy, using the same laser beam as described above. To analyse protoxylem maturation, the distance from the root apical meristem of the first protoxylem cell with fully developed secondary wall thickenings was measured following the method described by Ghuge et al. [20] considering the point where a sharp intensification of protoxylem PI staining was detectable as indicative of fully differentiated secondary cell wall thickenings (this point is referred to here as the protoxylem position). Analysis of protoxylem position was validated by the correspondence between the site where the sharp increase in the PI-induced fluorescence was revealed under LSCM and that of protoxylem appearance under bright-field microscope [20]. The length of the meristematic zone was determined by measuring the distance between the quiescent centre and the first elongating cell in the cortex cell file [13,36,37]. The images shown were obtained by aligning serial overlapping micrographs of the same root using Photoshop Software (Adobe, San Jose, CA, USA). Protoxylem position and meristem size were estimated exploiting the Leica Application Suite Advanced Fluorescence software, and then used for statistical analysis.

4.3. Hydrogen Peroxide In Situ Detection

To reveal the in situ extracellular H_2O_2 accumulation, the fluorogenic peroxidase substrate AUR (Molecular Probes, Invitrogen, Carlsbad, CA, USA) was exploited [38], and the fluorescence of the peroxidase reaction product was detected under LSCM in root apices from 7-day-old unwounded control and leaf-wounded seedlings 6 h after injury, as hereafter described. Root apices were stained by incubation in 100 µM AUR for 5/10 min and then observed under LSCM using a 543 nm helium-neon laser with a 550–700 nm band-pass filter. For the measurement of the AUR fluorescence intensity in roots of unwounded control and leaf-wounded plants, five rectangles of approximate 65 µm^2 for each analysed root were drawn over the protoxylem maturation zone and the sum of the pixels corresponding to the fluorescence present in each rectangle was measured exploiting the quantitative analysis of the LAS-AF software used to acquire the confocal images.

4.4. Statistics

The analyses under LSCM of protoxylem position, meristem size and H_2O_2 accumulation after PI and AUR staining, as well as root growth analysis, were performed on five independent experiments on a minimum of ten plants per treatment, yielding reproducible results. Images from single representative experiments are shown. Statistical tests of protoxylem position, meristem size and root growth were performed using GraphPad Prism (GraphPad Software, San Diego, CA, USA) with one-way ANOVA. The statistical significance of differences was evaluated by *p* levels as follows: ns, not significant; *, $p \leq 0.05$; **, $p \leq 0.01$; ***, $p \leq 0.001$; and ****, $p \leq 0.0001$. The average values of fluorescence intensity for unwounded control and leaf-wounded plants were obtained by analysing five roots for treatment, and five rectangles of approximately 65 μm^2 for each analysed root.

Author Contributions: I.F., A.C. and R.A.R.-P. conceived the project. I.F., R.A.R.-P. and A.C. designed the study, I.F. and A.V. performed most of the experiments. I.F., R.A.R.-P. and A.C. wrote the manuscript with contributions of R.A and P.T.

Funding: Grant to Department of Science, Roma Tre University (MIUR-Italy Dipartimenti di Eccellenza, ARTICOLO 1, COMMI 314–337 LEGGE 232/2016); RIA2017, RIA2018 from the University of L'Aquila—Department of Life, Health, and Environmental Sciences.

Acknowledgments: The Grant to Department of Science, Roma Tre University (MIUR-Italy Dipartimenti di Eccellenza, ARTICOLO 1, COMMI 314–337 LEGGE 232/2016) is gratefully acknowledged. RIA2017, RIA2018 from the University of L'Aquila—Department of Life, Health, and Environmental Sciences, is gratefully acknowledge (R.A.R.-P.).

Conflicts of Interest: The authors declare no conflict of interest.

References

1. Nicotra, A.B.; Atkin, O.K.; Bonser, S.P.; Davidson, A.M.; Finnegan, E.J.; Mathesius, U.; Poot, P.; Purugganan, M.D.; Richards, C.L.; Valladares, F.; et al. Plant phenotypic plasticity in a changing climate. *Trends Plant Sci.* **2010**, *15*, 684–692. [CrossRef] [PubMed]
2. Topp, C.N. Hope in Change: The Role of Root Plasticity in Crop Yield Stability. *Plant Physiol.* **2016**, *172*, 5–6. [CrossRef] [PubMed]
3. Verbelen, J.P.; de Cnodder, T.; Le, J.; Vissenberg, K.; Baluska, F. The root apex of Arabidopsis thaliana consists of four distinct zones of growth activities: Meristematic zone, transition zone, fast elongation zone and growth terminating zone. *Plant Signal. Behav.* **2006**, *1*, 296–304. [CrossRef] [PubMed]
4. Petricka, J.J.; Winter, C.M.; Benfey, P.N. Control of Arabidopsis root development. *Annu. Rev. Plant Biol.* **2012**, *63*, 563–590. [CrossRef] [PubMed]
5. Di Mambro, R.; De Ruvo, M.; Pacifici, E.; Salvi, E.; Sozzani, R.; Benfey, P.N.; Busch, W.; Novak, O.; Ljung, K.; Di Paola, L.; et al. Auxin minimum triggers the developmental switch from cell division to cell differentiation in the Arabidopsis root. *Proc. Natl. Acad. Sci. USA* **2017**, *114*, 7641–7649. [CrossRef] [PubMed]
6. Tsukagoshi, H.; Busch, W.; Benfey, P.N. Transcriptional regulation of ROS controls transition from proliferation to differentiation in the root. *Cell* **2010**, *143*, 606–616. [CrossRef] [PubMed]
7. Rost, T.L.; Baum, S. On the correlation of primary root length, meristem size and protoxylem tracheary element position in pea seedlings. *Am. J. Bot.* **1988**, *75*, 414–424. [CrossRef]
8. Muñiz, L.; Minguet, E.G.; Singh, S.K.; Pesquet, E.; Vera-Sirera, F.; Moreau-Courtois, C.L.; Carbonell, J.; Blázquez, M.A.; Tuominen, H. ACAULIS5 controls Arabidopsis xylem specification through the prevention of premature cell death. *Development* **2008**, *135*, 2573–2582. [CrossRef] [PubMed]
9. Bishopp, A.; Help, H.; El-Showk, S.; Weijers, D.; Scheres, B.; Friml, J.; Benková, E.; Mähönen, A.P.; Helariutta, Y. A mutually inhibitory interaction between auxin and cytokinin specifies vascular pattern in roots. *Curr. Biol.* **2011**, *21*, 917–926. [CrossRef] [PubMed]
10. Kobayashi, K.; Takahashi, F.; Suzuki, M.; Suzuki, H. Examination of morphological changes in the first formed protoxylem in Arabidopsis seedlings. *J. Plant Res.* **2002**, *115*, 107–112. [CrossRef]
11. Garrido, I.; García-Sánchez, M.; Casimiro, I.; Casero, P.J.; García-Romera, I.; Ocampo, J.A.; Espinosa, F. Oxidative stress induced in sunflower seedling roots by aqueous dry olive-mill residues. *PLoS ONE* **2012**, *7*, e46137. [CrossRef] [PubMed]

12. Lv, B.; Tian, H.; Zhang, F.; Liu, J.; Lu, S.; Bai, M.; Li, C.; Ding, Z. Brassinosteroids regulate root growth by controlling reactive oxygen species homeostasis and dual effect on ethylene synthesis in Arabidopsis. *PLoS Genet.* **2018**, *14*, e1007144. [CrossRef] [PubMed]
13. Chen, Q.; Sun, J.; Zhai, Q.; Zhou, W.; Qi, L.; Xu, L.; Wang, B.; Chen, R.; Jiang, H.; Qi, J.; et al. The basic helix-loop-helix transcription factor MYC2 directly represses PLETHORA expression during jasmonate-mediated modulation of the root stem cell niche in Arabidopsis. *Plant Cell* **2011**, *23*, 3335–3352. [CrossRef] [PubMed]
14. Jang, G.; Chang, S.H.; Um, T.Y.; Lee, S.; Kim, J.K.; Choi, Y.D. Antagonistic interaction between jasmonic acid and cytokinin in xylem development. *Sci Rep.* **2017**, *7*, 10212. [CrossRef] [PubMed]
15. Jang, G.; Choi, Y.D. Drought stress promotes xylem differentiation by modulating the interaction between cytokinin and jasmonic acid. *Plant Signal. Behav.* **2018**, *13*, e1451707. [CrossRef]
16. Ramachandran, P.; Wang, G.; Augstein, F.; de Vries, J.; Carlsbecker, A. Continuous root xylem formation and vascular acclimation to water deficit involves endodermal ABA signalling via miR165. *Development* **2018**, *145*, dev159202. [CrossRef]
17. Ghuge, S.A.; Tisi, A.; Carucci, A.; Rodrigues-Pousada, R.A.; Franchi, S.; Tavladoraki, P.; Angelini, R.; Cona, A. Cell wall amine oxidases: New players in root xylem differentiation under stress conditions. *Plants* **2015**, *4*, 489–504. [CrossRef]
18. Prince, S.; Murphy, M.; Mutava, R.N.; Durnell, L.A.; Valliyodan, B.; Shannon, J.G.; Nguyen, H.T. Root xylem plasticity to improve water use and yield in water-stressed soybean. *J. Exp. Bot.* **2017**, *68*, 2027–2036. [CrossRef]
19. Tisi, A.; Federico, R.; Moreno, S.; Lucretti, S.; Moschou, P.N.; Roubelakis-Angelakis, K.A.; Angelini, R.; Cona, A. Perturbation of polyamine catabolism can strongly affect root development and xylem differentiation. *Plant Physiol.* **2011**, *157*, 200–215. [CrossRef]
20. Ghuge, S.A.; Carucci, A.; Rodrigues Pousada, R.A.; Tisi, A.; Franchi, S.; Tavladoraki, P.; Angelini, R.; Cona, A. The apoplastic copper AMINE OXIDASE1 mediates jasmonic acid-induced protoxylem differentiation in Arabidopsis roots. *Plant Physiol.* **2015**, *168*, 690–707. [CrossRef]
21. Cona, A.; Tisi, A.; Ghuge, S.A.; Franchi, S.; de Lorenzo, G.; Angelini, R. Wound healing response and xylem differentiation in tobacco plants over-expressing a fungal endopolygalacturonase is mediated by copper amine oxidase activity. *Plant Physiol. Biochem.* **2014**, *82*, 54–65. [CrossRef]
22. Reymond, P.; Weber, H.; Damond, M.; Farmer, E.E. Differential gene expression in response to mechanical wounding and insect feeding in Arabidopsis. *Plant Cell* **2000**, *12*, 707–720. [CrossRef] [PubMed]
23. Capiati, D.A.; País, S.M.; Téllez-Iñón, M.T. Wounding increases salt tolerance in tomato plants: Evidence on the participation of calmodulin-like activities in cross-tolerance signalling. *J. Exp. Bot.* **2006**, *57*, 2391–2400. [CrossRef] [PubMed]
24. Park, Y.S.; Ryu, C.M. Insect stings to change gear for healthy plant: Improving maize drought tolerance by whitefly infestation. *Plant Signal. Behav.* **2016**, *11*, e1179420. [CrossRef] [PubMed]
25. Murali Achary, V.M.; Panda, B.B. Aluminium-induced DNA damage and adaptive response to genotoxic stress in plant cells are mediated through reactive oxygen intermediates. *Mutagenesis* **2010**, *25*, 201–209. [CrossRef] [PubMed]
26. Huber, A.E.; Bauerle, T.L. Long-distance plant signaling pathways in response to multiple stressors: The gap in knowledge. *J. Exp. Bot.* **2016**, *67*, 2063–2079. [CrossRef] [PubMed]
27. Katz, E.; Chamovitz, D.A. Wounding of Arabidopsis leaves induces indole-3-carbinol-dependent autophagy in roots of Arabidopsis thaliana. *Plant J.* **2017**, *91*, 779–787. [CrossRef]
28. Chen, L.; Wang, G.; Chen, P.; Zhu, H.; Wang, S.; Ding, Y. Shoot-Root Communication Plays a Key Role in Physiological Alterations of Rice (*Oryza sativa*) Under Iron Deficiency. *Front. Plant Sci.* **2018**, *9*, 757. [CrossRef]
29. Hilleary, R.; Gilroy, S. Systemic signaling in response to wounding and pathogens. *Curr. Opin. Plant Biol.* **2018**, *43*, 57–62. [CrossRef]
30. Angelini, R.; Tisi, A.; Rea, G.; Chen, M.M.; Botta, M.; Federico, R.; Cona, A. Involvement of polyamine oxidase in wound healing. *Plant Physiol.* **2008**, *146*, 162–177. [CrossRef]
31. Savatin, D.V.; Gramegna, G.; Modesti, V.; Cervone, F. Wounding in the plant tissue: The defense of a dangerous passage. *Front. Plant Sci.* **2014**, *5*, 470. [CrossRef] [PubMed]

32. Heyman, J.; Canher, B.; Bisht, A.; Christiaens, F.; De Veylder, L. Emerging role of the plant ERF transcription factors in coordinating wound defense responses and repair. *J. Cell Sci.* **2018**, *131*, jcs208215. [CrossRef] [PubMed]
33. Tavladoraki, P.; Cona, A.; Angelini, R. Copper-containing amine oxidases and FAD-dependent polyamine oxidases are key players in plant tissue differentiation and organ development. *Front. Plant Sci.* **2016**, *7*, 824. [CrossRef] [PubMed]
34. Valvekens, D.; Montagu, M.V.; Van Lijsebettens, M. Agrobacterium tumefaciens-mediated transformation of Arabidopsis thaliana root explants by using kanamycin selection. *Proc. Natl. Acad. Sci. USA* **1988**, *85*, 5536–5540. [CrossRef] [PubMed]
35. Mähönen, A.P.; ten Tusscher, K.; Siligato, R.; Smetana, O.; Díaz-Triviño, S.; Salojärvi, J.; Wachsman, G.; Prasad, K.; Heidstra, R.; Scheres, B. PLETHORA gradient formation mechanism separates auxin responses. *Nature* **2014**, *515*, 125–129. [CrossRef] [PubMed]
36. Casamitjana-Martínez, E.; Hofhuis, H.F.; Xu, J.; Liu, C.M.; Heidstra, R.; Scheres, B. Root-specific CLE19 overexpression and the sol1/2 suppressors implicate a CLV-like pathway in the control of Arabidopsis root meristem maintenance. *Curr. Biol.* **2003**, *13*, 1435–1441. [CrossRef]
37. Dello Ioio, R.; Nakamura, K.; Moubayidin, L.; Perilli, S.; Taniguchi, M.; Morita, M.T.; Aoyama, T.; Costantino, P.; Sabatini, S. A genetic framework for the control of cell division and differentiation in the root meristem. *Science* **2008**, *322*, 1380–1384. [CrossRef] [PubMed]
38. Ashtamker, C.; Kiss, V.; Sagi, M.; Davydov, O.; Fluhr, R. Diverse subcellular locations of cryptogein-induced reactive oxygen species production in tobacco Bright Yellow-2 cells. *Plant Physiol.* **2007**, *143*, 1817–1826. [CrossRef]

© 2018 by the authors. Licensee MDPI, Basel, Switzerland. This article is an open access article distributed under the terms and conditions of the Creative Commons Attribution (CC BY) license (http://creativecommons.org/licenses/by/4.0/).

Review

From *A. rhizogenes* RolD to Plant P5CS: Exploiting Proline to Control Plant Development

Maurizio Trovato *, Roberto Mattioli and Paolo Costantino

Department of Biology and Biotechnology, Sapienza University of Rome, 00185 Rome, Italy; roberto.mattioli@uniroma1.it (R.M.); paolo.costantino@uniroma1.it (P.C.)
* Correspondence: maurizio.trovato@uniroma1.it; Tel.: +39-06-4991-2922

Received: 12 October 2018; Accepted: 1 December 2018; Published: 6 December 2018

Abstract: The capability of the soil bacterium *Agrobacterium rhizogenes* to reprogram plant development and induce adventitious hairy roots relies on the expression of a few root-inducing genes (*rol A, B, C* and *D*), which can be transferred from large virulence plasmids into the genome of susceptible plant cells. Contrary to *rolA, B* and *C*, which are present in all the virulent strains of *A. rhizogenes* and control hairy root formation by affecting auxin and cytokinin signalling, *rolD* appeared non-essential and not associated with plant hormones. Its role remained elusive until it was discovered that it codes for a proline synthesis enzyme. The finding that, in addition to its role in protein synthesis and stress adaptation, proline is also involved in hairy roots induction, disclosed a novel role for this amino acid in plant development. Indeed, from this initial finding, proline was shown to be critically involved in a number of developmental processes, such as floral transition, embryo development, pollen fertility and root elongation. In this review, we present a historical survey on the *rol* genes focusing on the role of *rolD* and proline in plant development.

Keywords: plant development and organogenesis; proline biosynthesis; RolD; *rol* genes

---

1. Hairy Roots and *rol* Genes

Rhizobium rhizogenes, formerly known as *Agrobacterium rhizogenes* [1–5] is the etiological agent of the hairy root disease, consisting of abundant root proliferation at the site of bacterial infection. The capability of *Rhizobium rhizogenes* to induce hairy roots on susceptible dicotyledonous plants relies on its extraordinary ability to transfer a DNA fragment, called T-DNA, from a large Ri (root-inducing) plasmid to the genome of a plant cell [6–8]. The mechanism of T-DNA transfer [9] represents a natural form of genetic engineering, whose comprehension and exploitation has paved the way to the development of plant genetic transformation [10–13].

Hairy roots can be easily cultivated in vitro on hormone free medium [14] (Figure 1) and, in most plant species, can also be regenerated into whole fertile plants [15]. In addition, hairy roots produce unusual amino acid-sugar conjugates, called opines (Figure 2) which are not present in normal plant tissues. Depending on the specific Ri plasmid the transforming T-DNA comes from, one of four possible opines, that is agropine, cucumopine, mannopine and mikimopine, is synthesized by enzymes encoded by genes borne on the T-DNA and catabolized by enzymes encoded by genes located on the non-transferred plasmid portion. Because of the tight correlation between the synthesis of a given opine in hairy roots and the utilization of the same opine by the bacterium [16], a further opine-based classification of *Agrobacterium* strains has been proposed and will be adopted in this review. The T-DNA of all the Ri-plasmids have been characterized and sequenced [17–20]. The T-DNA of cucumopine-, mannopine- and mikimopine-type Ri plasmids turned out to consist in a continuous stretch of DNA, while the T-DNA of the agropine-type Ri plasmid is split in two T-DNA, called TR- and TL-DNA, which are independently transferred and integrated into the plant cell. Subsequent genetic work has clearly shown that the TL-DNA is uniquely responsible for hairy root induction, while the TR-DNA plays an accessory role to facilitate hairy root induction in some recalcitrant plant species. In a seminal work by White et al. [21] an extensive mutagenesis analysis was carried out, by transposon tagging, on the agropine-type pA4 plasmid. The genetic analysis led to the identification of four classes of mutations capable to affect the rooting phenotype and denominated, accordingly, *rol* (root loci) *A, B, C* and *D*. To further identify their functions, different *rol* combinations were cloned into binary vectors and transferred to *Agrobacterium* [22] to be used either for infection experiments on different plant hosts or for generating transgenic plants. The first analyses confirmed that the rol genes were the only Ri T-DNA segments responsible for hairy root induction and showed that a DNA fragment encompassing *rolA, B* and *C* was almost as effective in inducing hairy roots as the whole Ri T-DNA [23]. Accordingly, because of the functional importance of *rolA, B* and *C* and because these genes are present in all virulent strains of *Agrobacterium rhizogenes*, most of the studies initially focused on these oncogenes, particularly on *rolB*, while little attention was paid to *rolD*. Most of the aspects related to *Agrobacterium*, hairy roots and *rolA, B* and *C*, have been covered by excellent reviews [24–26] and will not be further expanded.

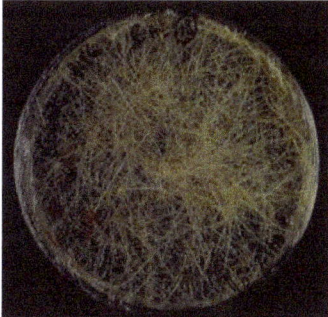

Figure 1. In vitro culture of roots induced on carrot discs by co-inoculation with an *Agrobacterium* strain containing a mannopine-type pRi8196. Once a hairy root culture is established, it can be maintained in vitro without the need of plant hormone supplementation. Fully fertile transgenic plants can be regenerated by these hairy roots.

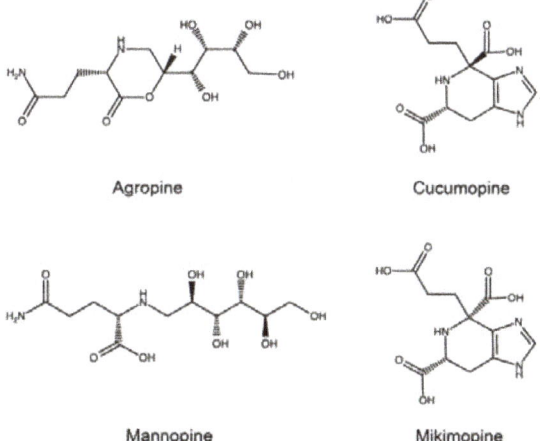

Figure 2. Chemical structure of agropine, cucumopine, mannopine and mikimopine, the four opines found in *A. rhizogenes* strains. The genes responsible for the synthesis of these unusual amino acid-sugar conjugates are borne on the T-DNA, while the genes coding for the catabolic enzymes are found on the non-transferred plasmid portion.

2. RolD

As already noted, *rolD* is not present in all virulent *A. rhizogenes* strains and therefore plays a marginal role in hairy root induction. However, transformation experiments [27–29] showed that expression of *rolD* is developmentally regulated in and can deeply affect the development of plant cells. Tobacco plants expressing *rolD* driven by its own promoter have been reported to reach anthesis in average 60 days (in some cases as many as 75) before untransformed plants [28]. The inflorescence was richer and long-lasting, compared to controls plants and the overall morphology of the plants was deeply altered, with a strong reduction in height and with tiny and bract-like leaves. Furthermore, organogenesis experiments on thin cell layers (TCL) from *rolD* and control plants cultured on different synthetic media confirmed and extended in vitro the notion that *rolD* has the potentiality to enhance and anticipate flower formation [28]. Similar results were obtained in tomato [30] and *Arabidopsis* [31]. The small size typical of all *rolD*-expressing transgenic plants, may be accounted for by the early and abundant proliferation of axillary buds, leading to highly branched shoots. Down-regulation of *CYP79F1/SUPERSHOOT/BUSHY* (*SPS*), a gene involved in glucosinolate biosynthesis [32], was reported in *Arabidopsis* transformed with *rolD* [31]. Since *SPS* normally inhibits the formation of lateral shoots by altering cytokinin balance, the proliferation of axillary branches of Arabidopis transgenic for *rolD* may be accounted for by a (secondary) effect of *rolD* on *SPS* expression.

It is not clear how *SPS* downregulation can affect the cytokinin/auxin ratio, since the synthesis of indole glucosinolates in *Arabidopsis* proceeds from the transformation of tryptophane to indole-3-acetaldoxime catalysed by CYP79B2 and CYP79B3 [33], while CYP79F1 is involved in the biosynthesis of aliphatic glucosinolates [32,34]. However, CYP79F1/SPS has the potentiality to affect cytokinin/ auxin balance through the synthesis of a common aldoxime precursor. Consistently, a null *CYP79F1* mutant (*bus1-1f*), totally devoid of short-chain methionine-derived glucosinolates, was also found enriched in indole-3-methyl-glucosinolate, indole-3-acetic acid and indole-3-acetonitrile [34].

Histochemical analysis of tobacco plants expressing the GUS reporter gene driven by the *rolD* promoter, revealed that this gene has a complex pattern of expression under strict developmental control [29]. Unlike other *rol* genes, which are always expressed in meristematic tissues, the promoter of *rolD* is not active in plant meristems but rather works in all growing and differentiating tissues throughout development, from the embryo to the adult plant. In particular, the expression of *rolD* characterizes the region of elongation and expansion of every tissue and organ. Intriguingly, as already mentioned, mutations in *rolD* prevent the T-DNA-induced hairy roots from elongating after initiation [21]. This suggests the possibility that *rolD* may be functionally involved in the process of elongation and/or maturation of roots and, possibly, of other organs.

A similarity search, based on a combination of iterative and noniterative methods, detected a highly significant sequence similarity between *rolD* and the gene coding for ornithine cyclodeaminase (OCD), an unusual enzyme of bacterial origin that catalyses the direct conversion of ornithine and NAD+ into proline and NH_4^+ [35] (Figure 3). This bioinformatic prediction was experimentally confirmed by enzymatic assays on RolD expressed and purified in *E. coli* and on soluble extracts from plants overexpressing the oncogene under the control of a CaMV35S promoter. The enzymatic assays revealed a specific ornithine-dependent proline production, associated to NAD+ reduction, that could only be accounted for by OCD activity. No functional OCDs have been detected so far in plants [36], where ornithine is converted to proline only via pyridoxal phosphate-dependent reactions. OCD seems to be a specialized enzyme that has been found only in a limited number of prokaryotic species, such as *Agrobacterium*, *Sinorhizobium*, *Rhodobacterium* and *Brucella* as well as in some extremophile archaea, such as *Archeoglobus* and *Methanobacterium*, where it is involved in the catabolism of unusual carbon and nitrogen sources like opines or methane. Interestingly, in *A. tumefaciens* OCD is encoded by genes localized in the non-transferred part of the Ti plasmid [7,37] to be used for opine catabolism, while in *A. rhizogenes*, OCD has become part of the T-DNA and it is expressed only in the plant cells. Intriguingly, in animals the mu-crystallins family of proteins shares significant similarities with OCD. This is not surprising, because in the mammalian eye often lens proteins derive from metabolic enzymes or stress proteins, which acquire reflective properties while, in some cases, maintaining their original metabolic activity [38].

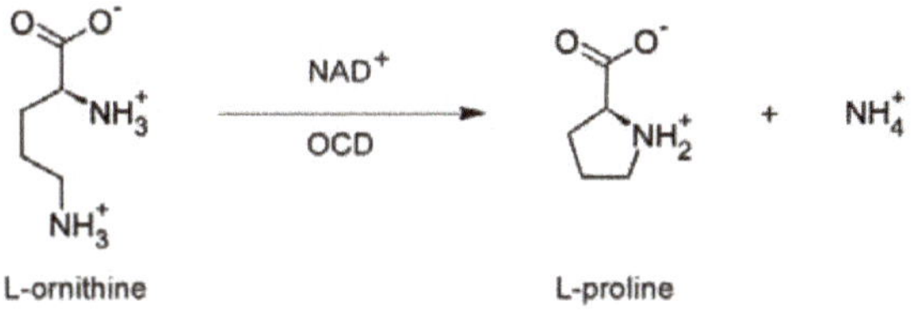

Figure 3. Proline synthesis from ornithine. The enzyme ornithine cyclo deaminase (OCD), an enzyme frequently found in bacteria but uncommon in plants, catalyses the NAD$^+$-dependent conversion of ornithine into proline and NH$_4^+$.

Since *rolD* is only present in the TL-DNA of the agropine-type Ri plasmids, its expression seems not strictly required for hairy root elongation. Although not experimentally demonstrated, it is tempting to speculate that *rolD*, similarly to the ancillary role played by the T$_R$-DNA-borne *iaaH* and *iaaM* genes in the process of hairy root induction, might play an auxiliary role in hairy root elongation by providing more proline in hosts with low levels of endogenous proline or during environmental stresses requiring higher proline demand. In support of this hypothesis, proline has been shown to accumulate during the elongation of the maize primary roots at low water potential [39].

As alternative explanation, other genes, either belonging to the Ri T-DNA or to the plant genome itself, could functionally substitute for *rolD* expression. This hypothesis is based on the work of Levesque et al (1988) [40] who observed that the Ri TL-DNA genes are functionally redundant and may derive from a common ancestral T-DNA. Redundancy, according to authors, would serve as an adaptive strategy to ensure function in a variety of host species and environmental conditions [40]. In the case of *rolD*, a recent duplication has apparently occurred between ORF 15 (*rolD*) and the ORFs 18 and 17, which, assembled together, restore a direct repetition of *rolD* [40]. Furthermore, portions of the Ri TL-DNA plasmid, including *rolD*, have been detected in the genome of some plant species (ct-TDNA), probably as a result of ancient *Agrobacterium* transformations [41,42]. It must be noted that, since OCD activity, in addition to proline, also produces NH4, a major nitrogen source which behaves as a signalling molecule capable of triggering multiple physiological and morphological responses in plants [43], we cannot rule out the possibility that some of the developmental alterations attributed to OCD may be accounted for, or contributed to, a perturbed ammonium homeostasis.

3. The Role of Proline in Plant Development

The discovery that RolD is a proline-synthesizing enzyme involved in root elongation [21,29] but also in flowering time [21,28,30], implied the possibility that this cyclic amino acid may have a role in plant development. It was already well-established that proline, in addition to its role in protein synthesis, is involved in the plant cell response to many types of stresses, essentially because a strong proline accumulation is observed soon after stress occurrence in many plant species [44].

However, proline accumulation was also described, in non-stressed conditions, in the tissues and organs of different plant species, particularly during the reproductive phase [45–50], supporting the idea that proline may play a role in plant reproductive development in normo-osmotic conditions. In the total amino acid pool of *Arabidopsis*, the percentage of proline raises from 1–3% in vegetative tissues before floral transition, to 26% in reproductive tissues after floral transition [49]. Similarly, Schwacke et al. (1999) [50] observed that the content of free proline in tomato flowers was 60-fold higher than in any other organ analysed. Although proline is a relatively common amino acid in plants, because of the frequent occurrence of long stretches of proline or hydroxyproline residues in cell wall proteins, particularly extensins [51], it is unlikely that, in non-stressed conditions, such large amount of proline can be accumulated for the needs of protein synthesis.

Differently from OCD, which catalyses the direct conversion of ornithine to proline, in higher plants proline is mainly synthesized in the cytosol from glutamate in a two-step reaction involving the enzyme δ-pyrroline-5-carboxylate synthetase (P5CS) and δ-pyrroline-5-carboxylate reductase (P5CR). Subsequently, proline is exported to the mitochondrion where it is catabolized back to glutamate by the enzymes proline dehydrogenase (ProDH) and δ-pyrroline-5-carboxylate synthetase P5CDH [44]. An alternative route starting from ornithine and mediated by ornithine δ-aminotransferase (δOAT) has also been reported [52], at least in some physiological conditions but its functional significance in maintaining proline homeostasis is strongly controversial [53,54].

The genes coding for the anabolic and catabolic enzymes of proline synthesis are highly conserved among plant species, although *P5CS* and *ProDH*, the genes coding for the rate-limiting steps of the anabolic and, respectively, catabolic pathways, may be present in multiple variants [55]. In *Arabidopsis*, P5CS is encoded by two paralog genes *P5CS1* and *P5CS2* [56], whose deduced amino acid sequences share 98% amino acid identity. In spite of the high similarity of these isoforms, *P5CS1* and *P5CS2* have a different tissue specificity and play non-redundant but partially overlapping functions, as inferred by the analysis of transgenic *Arabidopsis* carrying mutations in either *P5CS1* or *P5CS2* [57,58]. *P5CS1* is responsive to stress induction, while *P5CS2* is constitutively expressed at low levels in all tissues and organs and at high level in meristematic tissues, floral organs and in embryos [57,58].

3.1. Floral Transition

Consistent with the strong anticipation and stimulation of flowering induced by the ectopic expression of *rolD* [28,30,31], a number of authors reported, in absence of stressing stimuli, upregulation of both proline biosynthesis (*P5CS, P5CR*) and transport genes (*ProT*) in reproductive tissues [50,59,60], such as flowers, inflorescences and anthers, suggesting a possible role of proline in flowering. Intriguingly, the expression of the proline catabolic genes (*ProDH, P5CDH*) was also reported to increase in reproductive tissues under normo-osmotic conditions [61–63], in striking contrast with the strong downregulation of these genes observed under stressed conditions [64,65]. In agreement with these data, Kavi Kishor et al. (1995) [66] reported that constitutive overexpression of *P5CS1* in tobacco plants enhances flower development under drought conditions, while Nanjo et al. (1999) [67] reported that antisense expression of *P5CS1* inhibits bolting in *Arabidopsis*. *Arabidopsis p5cs1* mutants and to a greater extent, *p5cs1 p5cs2/P5CS2* sesquimutants, exhibited a strong delay in floral transition [58,68,69] (Figure 4A), while transgenic *Arabidopsis* overexpressing *P5CS1* under the control of the strong and constitutive CaMV35S promoter, showed a striking anticipation of flowering time and a proliferation of coflorescences, particularly in short day conditions [68]. In transgenic *35S::P5CS1* plants, the expression of the recombinant *P5CS1* was downregulated after flower transition, along with the endogenous allele of *P5CS1* and *P5CS2*, and, accordingly, *P5CS1* was overexpressed only for a short time, up to floral transition [68]. Altogether, these data suggest that proline plays a key role in flower transition, bolting and coflorescence architecture.

Presently, the molecular mechanism through which proline affects flowering time is not clear but it seems quite different from the mechanism through which proline protects plant cells from stress injuries [70]. One major difference between these mechanisms is the concentration of proline involved: during floral transition proline reaches only a localized and transient increase in the shoot apical meristem (SAM), while under stress conditions, it accumulates at high levels in all the tissues of the plant [68]. The accumulation of proline measured in *35S-P5CS1* plants (up to 3-fold the level of the wildtype), seems modest compared to that achieved under stress, where proline levels are 10 to 20-fold higher than in unstressed plants [68], suggesting that this amino acid may behave as a floral signal able to interact with flower regulators. It is well known that floral transition, i.e., the transition from a vegetative shoot apical meristem (SAM) to a floral SAM, involves a profound change in the identity of the apical meristem that starts producing flowers rather than leaves [71]. This switch to vegetative to reproductive development is regulated by a number of environmental and endogenous inputs, which converge to regulate master flowering regulators, which, in turn, activate floral identity genes. By genetic analysis four major pathways have been identified, photoperiodic, autonomous, vernalisation and gibberellin pathway [72–74], which are controlled by the master regulators *CONSTANS (CO)* and *FLOWERING LOCUS C (FLC)*, which, in turn, control the floral integrators *LEAFY (LFY)*, *SUPPRESSOR OF CONSTANS 1 (SOC1)* and *FLOWERING TIME (FT)* to eventually activate the floral identity genes *APETALA 1 (AP1)*, *APETALA 2 (AP2)*, *FRUITFULL (FUL)* and *CAULIFLOWER (CAL)*. In agreement with the hypothesis that proline may behave as a floral signalling molecule, *P5CS2* has been identified as an early regulatory target of *CONSTANS (CO)*, a master transcriptional regulator of the photoperiodic pathway [75]. However, because of the importance of proline as redox buffer [76] and ROS scavenger [77], we cannot rule out the possibility that proline may act as an active metabolite involved in metabolic signalling [78]. Overall, the body of accumulated evidence points to proline as a modulator of floral transition although a full comprehension of its mechanism of action and of the floral pathway it interacts with is still to be gained.

Figure 4. Effects of proline on plant development. (**A**) *Arabidopsis p5cs1* mutants (middle) and to a greater extent, *p5cs1 p5cs2/P5CS2* sesquimutants (right), defective in proline synthesis, are late flowering, compared to a wildtype (left) [68]. (**B**) Aberrant orientations of cellular division planes observed in an octant embryo from a segregating population of heterozygous *p5cs2/+* [58]. (**C**) *Arabidopsis* anthers from *p5cs1 p5cs2/P5CS2* sesquimutants stained with the vital Alexander's staining show a population of stained and viable pollen grains mixed with a population of unstained and unviable aberrant pollen grains (indicated by arrows) [69,79]. (**D**) GUS staining of *CYCB:GUS* roots in a wildtype (leftmost side) and a *p5cs1 p5cs2/P5CS2* (rightmost side) background reveals the effects of proline on cell division [80]. Arrows show the aberrant division planes in an octant embryo in (B) and the unstained, unviable pollen grains in a *p5cs1 p5cs2/P5cs2* anther in C. Bars = 10 µm (B), 50 µm (C) and 20 µm (D).

3.2. Embryo Development

Proline seems to play an important role also in plant embryogenesis. Quite surprisingly, despite the high sequence similarity shared by *P5CS1* and *P5CS2* and although both genes share the same pattern of expression in shoot apical meristems and embryos [57,58], *p5cs2* but not *p5cs1* mutants, are embryo lethal suggesting a specific role of *P5CS2* in embryogenesis. The embryo defects (Figure 4B) can be partially [57,58] or totally [69] complemented by treatment with L-proline suggesting that mutations in *P5CS2* specifically affect proline accumulation in developing seeds. The reason for such striking differences between P5CS1 and P5CS2 are not fully understood but it may be related to different subcellular localization of these two proteins in the embryo, as proposed by Szekely et al (2008) [57] who detected a P5CS1:GFP fusion protein outside the cytoplasm within subcellular bodies, while P5CS2:GFP had a cytoplasmic localization. A microscopic analysis of the mutant embryos revealed a number of aberrations typically associated with defects in cell cycle progression, such as anomalous orientations of the cellular division planes (Figure 4B), multi-nucleate suspensor cells and adventitious embryos [58], suggesting a possible relation between proline and cell cycle.

3.3. Pollen Fertility

One the best-known and less-explained fact on pollen composition is the exceedingly large amount of proline found in different plant species [49,50,81–83] suggesting a special role for proline in pollen development and function. At present, it is not known how proline can accumulate in pollen in such large quantities. In principle, proline could be synthesized inside gametophytic pollen grains, in surrounding sporophytic tissues, such as the tapetum or the intermediate layer, or be transported through phloem or xylem vessels from far away tissues. Because single, double and triple knockout mutants for all the genes belonging to the AtProT family—the best known group of proline transporter in plants—showed no differences, compared to wildtype, [84] and because microarray data detect strong expression of proline biosynthesis genes in pollen and anthers [85], proline accumulation in pollen grains likely derive from endogenous synthesis either in sporophytic or in gametophytic tissues of the anther.

Since pollen grains are subjected to a process of natural dehydration during their maturation, a role for proline as a compatible osmolyte has been proposed by some authors [49], while others [86] postulated that this amino acid may act as source of energy or metabolic precursor to fuel the rapid elongation of the pollen tube. A sound scientific base to settle this contrasting views was independently given by two research groups [69,79] who demonstrated, by a combination of genetic and physiological experiments, that a *p5cs1 p5cs2/P5CS2* sesquimutant, homozygous for *p5cs1* and heterozygous for *p5cs2*, was strongly impaired in pollen fertility [87].

The fertility defects of the sesquimutants were accounted for by defects in pollen grains, a number of which-presumably those with a *p5cs1*, *p5cs2* haploid genotype-were degenerated and unviable (Figure 4C). The proline content of the sesquimutant pollen population was measured and found to be less than a third compared to wildtype pollen. Moreover, exogenous proline supplied from the beginning of anther development was shown to partially complement both morphological and functional defects of the aberrant pollen grains. All in all, these data indicate that proline is required for pollen development and fertility and further corroborate the notion of the crucial importance of proline in reproductive development.

3.4. Root Elongation

In addition to its role in plant reproductive development, a novel role as modulator of root growth has been recently ascribed to proline [80]. In plants, postembryonic root growth is driven by the activity of the root meristem, which continuously regenerates itself in the staminal niche, while generating transit-amplifying cells, which undergo additional division in the proximal meristem and eventually, differentiate in the meristem transition zone. The balance between cell proliferation and cell differentiation determines root meristem size and, in turn, root growth and is largely controlled by plant hormones, particularly cytokinin and auxin [88].

The size of the root meristem, expressed as the number of cortex cells spanning from the quiescent centre (QC) to the first elongated cell in the transition zone (TZ) [89], was analysed in *p5cs1 p5cs2/P5CS2* sesquimutants relative to wildtype. Proline-deficient mutants were found to have root meristems remarkably smaller than the wildtype and the addition of micromolar concentrations of exogenous proline fully rescued the sesquimutant root meristem to wildtype size. Importantly, the effect of exogenous proline was also tested on wildtype roots and shown to have a specific and dose-dependent stimulatory effect at low concentrations and an inhibitory effect at high concentrations [80].

Considering the role played by *rolD* in the hairy roots syndrome, it not surprising that proline can modulate root elongation. Indeed, in the genesis of hairy roots RolD/OCD is involved in the elongation of roots generated by the combined action of RolA, B and C [21]. In addition, exogenous proline at micromolar concentration was shown to induce elongation of both primary and secondary roots in Arabidopsis [58].

The action of proline on root meristem seems independent of the plant hormones auxin, cytokinin and gibberellin as shown by a combination of pharmacological, molecular and genetic experiments [80]. Proline was found to regulate cell division in early stages of root development modulating the expression of *CYCB1;1*, the gene coding for the G2/M-specific CYCLINB1;1 (Figure 4D).

Other hormone-independent mechanisms are known to modulate root growth, such as the superoxide/hydrogen peroxide ratio reported by Tsukagoshi et al. (2010) [90] but the case of proline is quite surprising because the accumulation of this amino acid in the root is under strict abscisic acid (ABA) control under stress conditions [91]. However, proline has also been shown to be regulated by non-ABA-dependent factors [60,92] and it is possible that two parallel signalling pathways can independently control proline-dependent root regulation under stressed and, respectively, non-stressed conditions.

4. Conclusions and Perspectives

Much like the hairy root syndrome, which was originally thought as a simpler variant of the crown gall disease but eventually turned out to be a highly sophisticated and, as yet, not fully understood biological mechanism, the role of proline in plant development is unveiling unexpected complexities in plant development. Thanks to the study of *rolD*, we now know that proline can modulate the size of the root meristem independently of plant hormones and finely tune development in reproductive organs, although we are still far from a full comprehension of the underlying mechanisms of action. In a way, much like plant hormones, proline may behave as a second messenger. Because of the remarkable chemical-physical properties of this cyclic amino acid, however, a mechanism mediated by or dependent on metabolic regulations cannot be ruled out.

The long scientific journey from hairy roots to RolD to plant P5CS has produced more open questions than definitive answers. Our understanding of the genetic and molecular mechanisms through which proline exerts its effects on plant development is still rudimentary. We do not know whether the effects of proline on different developmental processes are mediated by different mechanisms or share a common molecular machinery. Clearly, further work is needed to fully understand the complex molecular mechanism/s by which proline can finely tune developmental processes as diverse as hairy root elongation, floral transition or pollen fertility.

Author Contributions: M.T. and P.C. wrote the review. R.M. collected the bibliography and prepared the figures. All the authors checked the draft and discussed improvements.

Funding: This work was partially supported by research grants from Sapienza Università to MT, from the Italian Ministry of Education, University and Research (Progetti di Ricerca di Interesse Nazionale) to PC.

Acknowledgments: We wish to dedicate this review to the memory of Prof. Domenico Mariotti, one of the first researchers in Italy to pioneer experiments on *Agrobacterium*-mediated plant genetic transformation and regeneration, paving the way to the development of the modern plant genetic engineering and leading to exciting new discoveries on plant development.

Conflicts of Interest: The authors declare no conflicts of interest.

References

1. Holmes, B.; Roberts, P. The Classification, Identification and Nomenclature of Agrobacteria. *J. Appl. Microbiol.* **1981**, *50*, 443–467. [CrossRef]
2. Young, J.M.; Kuykendall, L.D.; Martínez-Romero, E.; Kerr, A.; Sawada, H. A revision of Rhizobium Frank 1889, with an emended description of the genus, and the inclusion of all species of Agrobacterium Conn 1942 and Allorhizobium undicola de Lajudie et al. 1998 as new combinations: Rhizobium radiobacter, R. rhizogenes, R. rubi, R. undicola and R. vitis. *Int. J. Syst. Evol. Microbiol.* **2001**, *51*, 89–103. [CrossRef] [PubMed]
3. Young, J.B. Implications of alternative classifications and horizontal gene transfer for bacterial taxonomy. *Int. J. Syst. Evol. Microbiol.* **2001**, *51*, 945–953. [CrossRef] [PubMed]
4. Young, J.M.; Kuykendall, L.D.; Martínez-Romero, E.; Kerr, A.; Sawada, H. Classification and nomenclature of Agrobacterium and Rhizobium—A reply to Farrand et al. (2003). *Int. J. Syst. Evol. Microbiol.* **2003**, *53*, 1689–1695. [CrossRef] [PubMed]
5. Kerr, A.; Brisbane, P.G. Agrobacterium. In *Plant Bacterial Diseases, A Diagnostic Guide*; Fahy, P.C., Persley, G.J., Eds.; Academic Press: Sydney, Australia, 1983; pp. 27–43.
6. Chilton, M.-D.; Drummond, M.H.; Merio, D.J.; Sciaky, D.; Montoya, A.L.; Gordon, M.P.; Nester, E.W. Stable incorporation of plasmid DNA into higher plant cells: The molecular basis of crown gall tumorigenesis. *Cell* **1977**, *11*, 263–271. [CrossRef]
7. Chilton, M.-D.; Tepfer, D.A.; Petit, A.; David, C.; Casse-Delbaurt, F.; Tempé, J. *Agrobacterium rhizogenes* inserts T-DNA into the genomes of the host plant root cells. *Nature* **1982**, *295*, 432–434. [CrossRef]
8. Lemmers, M.; De Beuckeleer, M.; Holster, P.; Zambryski, P.; De Picker, A.; Hernalsteens, J.P.; Van Montagu, M.; Schell, J. Internal organization, boundaries and integration of Ti-plasmid DNA in nopaline crown gall tumours. *J. Mol. Biol.* **1980**, *144*, 353–376. [CrossRef]
9. Gelvin, S.B. *Agrobacterium* Mediated Plant Transformation: The Biology behind the "Gene-Jockeying" Tool. *Microbiol. Mol. Biol. Rev.* **2003**, *67*, 16–37. [CrossRef]
10. Hwang, H.-H.; Yu, M.; Lai, E.-M. Agrobacterium-Mediated Plant Transformation: Biology and Applications. *Arabidopsis Book* **2017**, *15*, e0186. [CrossRef]
11. Rugini, E.; Pellegrineschi, A.; Mencuccini, M.; Mariotti, D. Increase of rooting ability in the woody species kiwi (*Actinidia deliciosa* A. Chev.) by transformation with Agrobacterium rhizogenes rol genes. *Plant Cell Rep.* **1991**, *10*, 291–295. [CrossRef]
12. Fontana, G.S.; Santini, L.; Caretto, S.; Frugis, G.; Mariotti, D. Genetic transformation in the grain legume Cicer arietinum L. (chickpea). *Plant Cell Rep.* **1993**, *12*, 194–198. [CrossRef] [PubMed]
13. Frugis, G.; Caretto, S.; Santini, L.; Mariotti, D. Agrobacterium rhizogenes rol genes induce productivity-related phenotypical modifications in "creeping-rooted" alfalfa types. *Plant Cell Rep.* **1995**, *14*, 488–492. [CrossRef] [PubMed]
14. White, F.F.; Braun, A.C. A cancerous neoplasm of plants. *Cancer Res.* **1942**, *2*, 597–617.
15. Spanò, L.; Costantino, P. Regeneration of plants from callus cultures of roots induced by Agrobacterium rhizogenes on tobacco. *J. Phytopathol.* **1982**, *106*, 87–92. [CrossRef]
16. Petit, A.; Delhaye, S.; Tempé, J.; Morel, G. Recherches sur les guanidines des tissus de crown gall. Mise en évidence d'une relation biochimique spécifique entre les souches d'*Agrobacterium tumefaciens* et les tumeurs qu'elles induisent. *Physiol. Vég.* **1970**, *8*, 205–213.
17. Slightom, J.L.; Durand-Tardiff, M.; Jouanin, L.; Tepfer, D. Nucleotide sequence analysis of TL-DNA of *Agrobacterium rhizogenes* agropine type plasmid. Identification of open reading frames. *J. Biol. Chem.* **1986**, *261*, 108–121. [PubMed]
18. Hansen, G.; Larribe, M.; Vaubert, D.; Tempé, J.; Biermann, B.J.; Montoya, A.L.; Chilton, M.D.; Brevet, J. Agrobacterium rhizogenes pRi8196 T-DNA: Mapping and DNA sequence of functions involved in mannopine synthesis and hairy root differentiation. *Proc. Natl. Acad. Sci. USA* **1991**, *88*, 7763–7767. [CrossRef]
19. Tanaka, N.; Ikeda, T.; Oka, A. Nucleotide Sequence of the rol Region of the Mikimopine-type Root-inducing Plasmid pRi1724. *Biosci. Biotechnol. Biochem.* **1994**, *58*, 548–551. [CrossRef]

20. Mankin, S.L.; Hill, D.S.; Olhoft, P.M.; Toren, E.; Wenck, A.R.; Nea, L.; Xing, L.; Brown, J.A.; Fu, H.; Ireland, L.; et al. Disarming and sequencing of Agrobacterium rhizogenes strain K599 (NCPPB2659) plasmid pRi2659. *In Vitro Cell. Dev. Biol. Plant* **2007**, *43*, 521–535. [CrossRef]
21. White, F.F.; Taylor, B.H.; Huffman, G.A.; Nester, E.W. Molecular and genetic analysis of the transferred DNA regions of the root-inducing plasmid of *Agrobacterium rhizogenes*. *J. Bacteriol.* **1985**, *164*, 33–44.
22. Cardarelli, M.; Mariotti, D.; Pomponi, M.; Spanò, L.; Capone, I.; Costantino, P. *Agrobacterium rhizogenes* T-DNA genes capable of inducing hairy root phenotype. *Mol. Gen. Genet.* **1987**, *209*, 475–480. [CrossRef] [PubMed]
23. Capone, I.; Spanò, L.; Cardarelli, M.; Bellincampi, D.; Petit, A.; Costantino, P. Induction and growth properties of carrot roots with different complements of Agrobacterium rhizogenes T-DNA. *Plant Mol. Biol.* **1989**, *13*, 43–52.
24. Kado, C.I. Historical account on gaining insights on the mechanism of crown gall tumorigenesis induced by Agrobacterium tumefaciens. *Front. Microbiol.* **2014**, *5*, 1–15. [CrossRef] [PubMed]
25. Binns, A.N.; Costantino, P. The Agrobacterium Oncogenes. In *The Rhizobiaceae*; Spaink, H.P., Kondorosi, A., Hooykaas, P.J.J., Eds.; Springer: Dordrecht, The Netherlands, 1998. [CrossRef]
26. Costantino, P.; Capone, I.; Cardarelli, M.; De Paolis, A.; Mauro, M.L.; Trovato, M. Bacterial plant oncogenes: The *rol* genes' saga. *Genetica* **1994**, *94*, 203–211. [CrossRef] [PubMed]
27. Leach, F.; Aoyagi, K. Promoter analysis of the highly expressed rolC and roID root-inducing genes of *Agrobacterium rhizogenes*: Enhancer and tissue-specific DNA determinants are dissociated. *Plant Sci.* **1991**, *79*, 69–76. [CrossRef]
28. Mauro, M.L.; Trovato, M.; De Paolis, A.; Gallelli, A.; Costantino, P.; Altamura, M.M. The plant oncogene *rolD* stimulates flowering in transgenic tobacco plants. *Dev. Biol.* **1996**, *180*, 693–700. [CrossRef] [PubMed]
29. Trovato, M.; Mauro, M.L.; Costantino, P.; Altamura, M.M. The *rolD* gene from *Agrobacterium rhizogenes* is developmentally regulated in transgenic tobacco. *Protoplasma* **1997**, *197*, 111–120. [CrossRef]
30. Bettini, P.; Michelotti, S.; Bindi, D.; Giannini, R.; Capuana, M.; Buaiatti, M. Pleiotropic effect of the insertion of *Agrobacterium rhizogenes rolD* gene in tomato (*Lycopersicon esculentum* Mill). *Theor. Appl. Genet.* **2003**, *107*, 831–836. [CrossRef]
31. Falasca, G.; Altamura, M.M.; D'Angeli, S.; Zaghi, D.; Costantino, P.; Mauro, M.L. The rolD oncogene promotes axillary bud and adventitious root meristems in Arabidopsis. *Plant Physiol. Biochem.* **2010**, *48*, 797–804. [CrossRef]
32. Tantikanjana, T.; Mikkelsen, M.D.; Hussain, M.; Halkier, B.A.; Sundaresan, V. Functional analysis of the tandem duplicated P450 genes SPS/BUS/CYP79F1 and CYP79F2 in glucosinolate biosynthesis and plant development by Ds transposition generated double mutants. *Plant Physiol.* **2004**, *135*, 840–848. [CrossRef]
33. Hull, A.K.; Vij, R.; Celenza, J.L. Arabidopsis cytochrome P450s that catalyze the first step of tryptophan-dependent indole-3-acetic acid biosynthesis. *Proc. Natl. Acad. Sci. USA* **2000**, *97*, 2379–2384. [CrossRef] [PubMed]
34. Reintanz, B.; Lehnen, M.; Reichelt, M.; Gershenzon, J.; Kowalczyk, M.; Sandberg, G.; Godde, M.; Uhl, R.; Palme, K. Bus, a bushy Arabidopsis CYP79F1 knockout mutant with abolished synthesis of short-chain aliphatic glucosinolates. *Plant Cell* **2001**, *13*, 351–367. [CrossRef] [PubMed]
35. Trovato, M.; Maras, B.; Linhares, F.; Costantino, P. The plant oncogene *rolD* encodes a functional ornithine cyclodeaminase. *Proc. Natl. Acad. Sci. USA* **2001**, *98*, 13449–13453. [CrossRef] [PubMed]
36. Sharma, S.; Shinde, S.; Verslues, P.E. Functional characterization of an ornithine cyclodeaminase-like protein of Arabidopsis thaliana. *BMC Plant Biol.* **2013**, *13*, 182. [CrossRef] [PubMed]
37. Thomashow, M.F.; Nutter, A.L.; Montoya, A.L.; Gordon, M.P.; Nester, E.W. Integration and organization of Ti plasmid sequences in crown gall tumors. *Cell* **1980**, *19*, 729–739. [CrossRef]
38. Piatigorsky, J.; Kantorow, M.; Gopal-Srivastava, R.; Tomarev, S.I. Recruitment of enzymes and stress proteins as lens crystallins. *EXS* **1994**, *71*, 241–250. [PubMed]
39. Verslues, P.E.; Sharp, R.E. Proline accumulation in maize (*Zea mays* L.) primary roots at low water potentials. II. Metabolic source of increased proline deposition in the elongation zone. *Plant Physiol.* **1999**, *119*, 1349–1360. [CrossRef]
40. Levesque, H.; Delepelaire, P.; Rouzé, P.; Slightom, J.L.; Tepfer, D. Common evolutionary origin of the central portions of the Ri TL-DNA of *Agrobacterium rhizogenes* and the Ti T-DNAs of *Agrobacterium tumefaciens*. *Plant Mol. Biol.* **1988**, *11*, 731–744. [CrossRef]

41. Furner, I.J.; Huffman, G.A.; Amasino, R.M.; Garfinkel, D.J.; Gordon, M.P.; Nester, E.W. An Agrobacterium transformation in the evolution of the genus Nicotiana. *Nature* **1986**, *319*, 422. [CrossRef]
42. White, F.F.; Garfinkel, D.J.; Huffman, G.A.; Gordon, M.P.; Nester, E.W. Sequences homologous to Agrobacterium rhizogenes T-DNA in the genomes of uninfected plants. *Nature* **1983**, *301*, 348. [CrossRef]
43. Liu, Y.; von Wirén, N. Ammonium as a signal for physiological and morphological responses in plants. *J. Exp. Bot.* **2017**, *68*, 2581–2592. [CrossRef] [PubMed]
44. Trovato, M.; Mattioli, R.; Costantino, P. Multiple Roles of Proline in Plant Stress Tolerance and Development. *Rend. Lincei Sci. Fis. Nat.* **2008**, *19*, 325–346. [CrossRef]
45. Vansuyt, G.; Vallee, J.-C.; Prevost, J. La pyrroline-5-carboxylate réductase et la proline déhydrogénase chez *Nicotiana tabacum* var. Xanthi n.c. en fonction de son développement. *Physiol. Vég.* **1979**, *19*, 95–105.
46. Venekamp, J.H.; Koot, J.T.M. The sources of free proline and asparagine in field bean plants, *Vicia faba* L., during and after a short period of water withholding. *J. Plant Physiol.* **1988**, *32*, 102–109. [CrossRef]
47. Mutters, R.G.; Ferreira, L.G.R.; Hall, A.E. Proline content of the anthers and pollen of heat-tolerant and heat-sensitive cowpea subjected to different temperatures. *Crop Sci.* **1989**, *29*, 1497–1500. [CrossRef]
48. Walton, E.F.; Clark, C.J.; Boldingh, H.L. Effect of hydrogen cyanamide on amino acid profiles in kiwifruit buds during bud-break. *Plant Physiol.* **1991**, *97*, 1256–1259. [CrossRef] [PubMed]
49. Chiang, H.H.; Dandekar, A.M. Regulation of proline accumulation in *Arabidopsis thaliana* (L) Heynh during development and in response to desiccation. *Plant Cell Environ.* **1995**, *18*, 1280–1290. [CrossRef]
50. Schwacke, R.; Grallath, S.; Breitkreuz, K.E.; Stransky, E.; Stransky, H.; Frommer, W.B.; Rentsch, D. LeProT1, a transporter for proline, glycine betaine, and gamma-amino butyric acid in tomato pollen. *Plant Cell* **1999**, *11*, 377–392. [CrossRef]
51. Snowalter, A.M. Structure and function of plant cell wall proteins. *Plant Cell* **1993**, *5*, 9–23. [CrossRef]
52. Roosens, N.H.; Thu, T.T.; Iskandar, H.M.; Jacobs, M. Isolation of the ornithine-delta-aminotransferase cDNA and effect of salt stress on its expression in *Arabidopsis thaliana*. *Plant Physiol.* **1998**, *117*, 263–271. [CrossRef]
53. Funck, D.; Stadelhofer, B.; Koch, W. Ornithine-δ-aminotransferase is essential for arginine catabolism but not for proline biosynthesis. *BMC Plant Biol.* **2008**, *8*, 40. [CrossRef] [PubMed]
54. Winter, G.; Todd, D.C.; Trovato, M.; Forlani, G.; Funck, D. Physiological implications of arginine metabolism in plants. *Front. Plant Sci.* **2015**, *6*, 1–14. [CrossRef] [PubMed]
55. Fichman, Y.; Gerdes, S.Y.; Kovács, H.; Szabados, L.; Zilberstein, A.; Csonka, L. Evolution of proline biosynthesis: Enzymology, bioinformatics, genetics, and transcriptional regulation. *Biol. Rev. Camb. Philos. Soc.* **2015**, *90*, 1065–1099. [CrossRef] [PubMed]
56. Strizhov, N.; Ábrahám, E.; Ökrész, L.; Blickling, S.; Zilberstein, A.; Schell, J.; Koncz, C.; Szabados, L. Differential expression of two *P5CS* genes controlling proline accumulation during salt-stress requires ABA and is regulated by ABA1, ABI1 and AXR2 in Arabidopsis. *Plant J.* **1997**, *12*, 557–569. [CrossRef] [PubMed]
57. Székely, G.; Ábrahám, E.; Cséplő, A.; Rigó, G.; Zsigmond, L.; Csiszár, J.; Ayaydin, F.; Strizhov, N.; Jásik, J.; Schmelzer, E.; et al. Duplicated *P5CS* genes of Arabidopsis play distinct roles in stress regulation and developmental control of proline biosynthesis. *Plant J.* **2008**, *53*, 11–28. [CrossRef] [PubMed]
58. Mattioli, R.; Falasca, G.; Sabatini, S.; Altamura, M.M.; Costantino, P.; Trovato, M. The proline biosynthetic genes *P5CS1* and *P5CS2* play overlapping roles in Arabidopsis flower transition but not in embryo development. *Physiol. Plant.* **2009**, *137*, 72–85. [CrossRef]
59. Verbruggen, N.; Villarroroel, R.; Van Montagu, M. Osmoregulation of a Pyrroline-5-Carboxylate Reductase Gene in *Arabidopsis thaliana*. *Plant Physiol.* **1993**, *103*, 771–781. [CrossRef] [PubMed]
60. Savouré, A.; Jaoua, S.; Hua, X.J.; Ardiles, W.; Van Montagu, M.; Verbruggen, N. Abscisic acid-independent and abscisic acid-dependent regulation of proline biosynthesis following cold and osmotic stresses in *Arabidopsis thaliana*. *Mol. Gen. Genet.* **1997**, *254*, 104–109. [CrossRef]
61. Verbruggen, N.; Hua, X.J.; May, M.; Van Montagu, M. Environmental and developmental signals modulate proline homeostasis: Evidence for a negative transcriptional regulator. *Proc. Natl. Acad. Sci. USA* **1996**, *93*, 8787–8791. [CrossRef]
62. Deuschle, K.; Funck, D.; Hellmann, H.; Däschner, K.; Binder, S.; Frommer, W.B. A nuclear gene encoding mitochondrial Δ1-pyrroline-5-carboxylate dehydrogenase and its potential role in protection from proline toxicity. *Plant J.* **2001**, *27*, 345–356. [CrossRef]
63. Funck, D.; Eckard, S.; Müller, G. Non-redundant functions of two proline dehydrogenase isoforms in Arabidopsis. *BMC Plant Biol.* **2010**, *10*, 70. [CrossRef]

64. Kiyosue, T.; Yoshiba, Y.; Yamaguchi-Shinozaki, K.; Shinozaki, K. A nuclear gene encoding mitochondrial proline dehydrogenase, an enzyme involved in proline metabolism, is upregulated by proline but downregulated by dehydration in Arabidopsis. *Plant Cell* **1996**, *8*, 1323–1335. [CrossRef] [PubMed]
65. Peng, Z.; Lu, Q.; Verma, D.P.S. Reciprocal regulation of 1-pyrroline-5-carboxylate synthetase and proline dehydrogenase genes controls proline levels during and after osmotic stress in plants. *Mol. Gen. Genet.* **1996**, *253*, 334–341. [CrossRef] [PubMed]
66. Kavi Kishor, P.B.; Hong, Z.; Miao, G.-H.; Hu, C.-A.A.; Verma, D.P.S. Overexpression of δ-Pyrroline-5-Carboxylate Synthetase Increases Proline Production and Confers Osmotolerance in Transgenic Plants. *Curr. Sci.* **1995**, *88*, 1387–1394. [CrossRef]
67. Nanjo, T.; Kobayashi, M.; Yoshiba, Y.; Sanada, Y.; Wada, K.; Tukaya, H.; Kakubari, Y.; Yamaguchi-Shinozaki, K.; Shinozaki, K. Biological functions of proline in morphogenesis and osmotolerance revealed in antisense transgenic *Arabidopsis thaliana*. *Plant J.* **1999**, *18*, 185–193. [CrossRef] [PubMed]
68. Mattioli, R.; Marchese, D.; D'Angeli, S.; Altamura, M.M.; Costantino, P.; Trovato, M. Modulation of intracellular proline levels affects flowering time and inflorescence architecture in Arabidopsis. *Plant Mol. Biol.* **2008**, *66*, 277–288. [CrossRef] [PubMed]
69. Funck, D.; Winter, G.; Baumgarten, L.; Forlani, G. Requirement of proline synthesis during Arabidopsis reproductive development. *BMC Plant Biol.* **2012**, *12*, 191. [CrossRef]
70. Mattioli, R.; Costantino, P.; Trovato, M. Proline accumulation in plants. *Plant Signal. Behav.* **2009**, *4*, 1016–1018. [CrossRef]
71. Simpson, G.G.; Dean, C. *Arabidopsis*, the Rosetta Stone of Flowering Time? *Science* **2002**, *296*, 285–289. [CrossRef]
72. Amasino, R.M.; Michaels, S.D. The timing of flowering. *Plant Physiol.* **2010**, *154*, 516–520. [CrossRef]
73. Andrés, F.; Coupland, G. The genetic basis of flowering responses to seasonal cues. *Nat. Rev. Genet.* **2012**, *13*, 627. [CrossRef] [PubMed]
74. Conti, L. Hormonal control of the floral transition: Can one catch them all? *Dev. Biol.* **2017**, *430*, 288–301. [CrossRef] [PubMed]
75. Samach, A.; Onouchi, H.; Gold, S.E.; Ditta, G.S.; Schwarz-Sommer, Z.S.; Yanofsky, M.F.; Coupland, G. Distinct Roles of CONSTANS Target Genes in Reproductive Development of Arabidopsis. *Science* **2000**, *288*, 1613–1616. [CrossRef] [PubMed]
76. Hare, P.; Cress, W. Metabolic implications of stress-induced proline accumulation in plants. *Plant Growth Regul.* **1997**, *21*, 79–102. [CrossRef]
77. Smirnoff, N.; Cumbes, Q.J. Hydroxyl radical scavenging activity of compatible solutes. *Phytochemistry* **1989**, *28*, 1057–1060. [CrossRef]
78. Nunes-Nesi, A.; Fernie, A.R.; Stitt, M. Metabolic and Signaling Aspects Underpinning the Regulation of Plant Carbon Nitrogen Interactions. *Mol. Plant* **2010**, *3*, 973–996. [CrossRef]
79. Mattioli, R.; Biancucci, M.; Lonoce, C.; Costantino, P.; Trovato, M. Proline is required for male gametophyte development in Arabidopsis. *BMC Plant Biol.* **2012**, *12*, 236. [CrossRef]
80. Biancucci, M.; Mattioli, R.; Moubayidin, L.; Sabatini, S.; Costantino, P.; Trovato, M. Proline affects the size of the root meristematic zone in Arabidopsis. *BMC Plant Biol.* **2015**, *15*, 263. [CrossRef]
81. Krogaard, H.; Andersen, A.S. Free Amino-Acids of Nicotiana-Alata Anthers during Development Invivo. *Physiol. Plant.* **1983**, *57*, 527–531. [CrossRef]
82. Khoo, U.; Stinson, H.T. Free amino acid differences between cytoplasmic male sterile and normal fertile anthers. *Proc. Natl. Acad. Sci. USA* **1957**, *43*, 603–607. [CrossRef]
83. Lansac, A.R.; Sullivan, C.Y.; Johnson, B.E. Accumulation of free proline in sorghum (*Sorghum bicolor*) pollen. *Can. J. Bot./Rev. Can. Bot.* **1996**, *74*, 40–45. [CrossRef]
84. Lehmann, S.; Gumy, C.; Blatter, E.; Boeffel, S.; Fricke, W.; Rentsch, D. In planta function of compatible solute transporters of the AtProT family. *J. Exp. Bot.* **2011**, *62*, 787–796. [CrossRef] [PubMed]
85. Honys, D.; Twell, D. Comparative analysis of the Arabidopsis pollen transcriptome. *Plant Physiol.* **2003**, *132*, 640–652. [CrossRef] [PubMed]
86. Zhang, H.Q.; Croes, A.F. Proline metabolism in pollen - degradation of proline during germination and early tube growth. *Planta* **1983**, *159*, 46–49.
87. Biancucci, M.; Mattioli, R.; Forlani, G.; Funck, D.; Costantino, P.; Trovato, M. Role of proline and GABA in sexual reproduction of angiosperms. *Front. Plant Sci.* **2015**, *6*, 680. [CrossRef]

88. Moubayidin, L.; Perilli, S.; Dello Ioio, R.; Di Mambro, R.; Costantino, P.; Sabatini, S. The Rate of Cell Differentiation Controls the Arabidopsis Root Meristem Growth Phase. *Curr. Biol.* **2010**, *20*, 1138–1143. [CrossRef]
89. Dello Ioio, R.; Linhares, F.S.; Scacchi, E.; Casamitjana-Martinez, E.; Heidstra, R.; Costantino, P.; Sabatini, S. Cytokinins Determine *Arabidopsis* Root-Meristem Size by Controlling Cell Differentiation. *Curr. Biol.* **2007**, *17*, 678–682. [CrossRef]
90. Tsukagoshi, H.; Busch, W.; Benfey, P.N. Transcriptional Regulation of ROS Controls Transition from Proliferation to Differentiation in the Root. *Cell* **2010**, *143*, 606–616. [CrossRef]
91. Stewart, C.R. The Mechanism of Abscisic Acid-induced Proline Accumulation in Barley Leaves. *Plant Physiol.* **1980**, *66*, 230–233. [CrossRef]
92. Ábrahám, E.; Rigó, G.; Székely, G.; Nagy, R.; Koncz, C.; Szabados, L. Light-dependent induction of proline biosynthesis by abscisic acid and salt stress is inhibited by brassinosteroid in Arabidopsis. *Plant Mol. Biol.* **2003**, *51*, 363–372. [CrossRef]

© 2018 by the authors. Licensee MDPI, Basel, Switzerland. This article is an open access article distributed under the terms and conditions of the Creative Commons Attribution (CC BY) license (http://creativecommons.org/licenses/by/4.0/).

Review

The CLV-WUS Stem Cell Signaling Pathway: A Roadmap to Crop Yield Optimization

Jennifer C. Fletcher [1,2]

[1] Plant Gene Expression Center, United States Department of Agriculture-Agricultural Research Service, Albany, CA 94710, USA; jfletcher@berkeley.edu; Tel.: +1-510-559-5917; Fax: +1-510-559-5678
[2] Department of Plant and Microbial Biology, University of California, Berkeley, CA 94720, USA

Received: 12 September 2018; Accepted: 10 October 2018; Published: 19 October 2018

Abstract: The shoot apical meristem at the growing shoot tip acts a stem cell reservoir that provides cells to generate the entire above-ground architecture of higher plants. Many agronomic plant yield traits such as tiller number, flower number, fruit number, and kernel row number are therefore defined by the activity of the shoot apical meristem and its derivatives, the floral meristems. Studies in the model plant *Arabidopsis thaliana* demonstrated that a molecular negative feedback loop called the CLAVATA (CLV)-WUSCHEL (WUS) pathway regulates stem cell maintenance in shoot and floral meristems. CLV-WUS pathway components are associated with quantitative trait loci (QTL) for yield traits in crop plants such as oilseed, tomato, rice, and maize, and may have played a role in crop domestication. The conservation of these pathway components across the plant kingdom provides an opportunity to use cutting edge techniques such as genome editing to enhance yield traits in a wide variety of agricultural plant species.

Keywords: CLE; CLV; WUS; stem cells; meristem; SAM; signaling; locule

---

1. Introduction

Plants are unique among living organisms in their ability to continuously grow and develop new organs throughout their life cycles. This continuous growth strategy produces leaves, stems, and flowers in architectures that can vary widely between species, from squat yellow dandelions to tall, leafy trees. The sources of cells for continuous organ formation are the apical meristems at the growing shoot and root tips. The shoot apical meristem (SAM) forms in the embryo and consists of a small reservoir of stem cells whose descendants generate all of the above-ground structures of the plant [1]. Following germination, the vegetative SAM produces a series of leaves from its flanks. At the transition to flowering the vegetative meristem becomes a reproductive inflorescence meristem (IFM) that produces axillary meristems followed by floral meristems that generate the flowers and seeds. Thus, SAM activity is the ultimate source of many yield traits in agronomic crop plants, because the direct outcome of plant organogenesis is the production of leaves, fruits, pods, seeds, and other structures that humans harvest and eat.

The SAM has the dual function of maintaining an active stem cell population while concurrently generating new organs. The organs form as primordia on the meristem flanks, while the self-renewing stem cell reservoir at the apex replenishes the cells that depart from the meristem into the primordia (Figure 1A). The stem cell pool is sustained by the activity of an underlying group of cells in the core of the SAM called the organizing center (OC). The maintenance of SAM homeostasis via a balance between stem cell loss and renewal is critical for plant development, because plants with reduced SAM activity prematurely cease growth before forming their full complement of organs [2,3] whereas those with over-active meristems have enlarged stems and can produce many extra branches, flowers, fruits, and seeds [4,5].

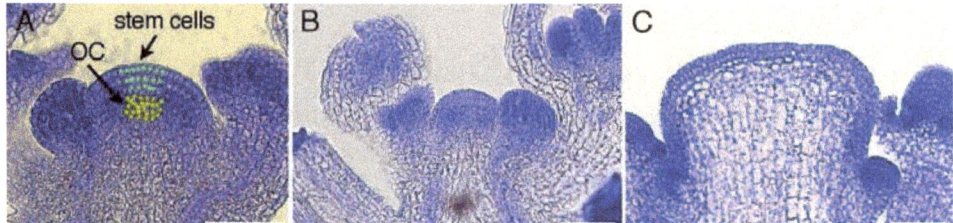

Figure 1. Shoot apical meristems of *Arabidopsis* wild-type and *clv3* mutant plants during the inflorescence phase. (**A**) Key domains within the shoot apical meristem. The apical stem cells are colored in green and the underlying organizing center (OC) cells in yellow. Primordia arise as dome-shaped structures on the meristem flanks. (**B**) Wild-type Columbia-0 inflorescence meristem (IFM) and flanking floral meristem primordia. (**C**) Enlarged *clv3* null mutant IFM and flanking floral meristem primordia. Scale bars, 50 μm.

Communication between individual cells is crucial to coordinate the various aspects of SAM function. Classical experiments demonstrated that the fate of each SAM cell is determined by positional information rather than by its lineage-specific heritage [6–8], and that the distinct functional domains within the SAM exchange cell fate information cues [9]. The SAM is further stratified into clonally distinct cell layers [10–12] that participate in both SAM maintenance and organ formation [13,14], requiring that these activities be orchestrated between all cell layers. Therefore, signaling between SAM cells is necessary for the cells to assess their relative positions in the meristem and behave coordinately with their neighbors. As described below, a molecular network called the CLAVATA (CLV)-WUSCHEL (WUS) pathway conveys intercellular signals that are critical for shoot and floral meristem maintenance in higher plants.

Crop plants have undergone vigorous selection by humans during the past 10,000 years [15,16], especially for yield traits such as larger and more numerous inflorescence meristems, fruits, and seeds. The CLV-WUS pathway in particular has been a target of selection during crop domestication to enhance agricultural yields [17]. Here, I review our understanding of the CLV-WUS signaling system in *Arabidopsis* shoot meristems and discuss studies demonstrating that components of the pathway are associated with variation in yield traits in agronomic crops such as mustard, tomato, rice, and maize.

2. CLV-WUS Shoot Apical Meristem Maintenance Pathway

The CLV-WUS signaling pathway plays a central role in maintaining shoot and floral stem cell homeostasis in *Arabidopsis* (Figure 2A). The *WUS* gene is dispensable for establishing the embryo stem cell reservoir [18], but is required to sustain stem cell fate during vegetative and reproductive development [3]. *WUS* is expressed exclusively in the SAM organizing center and encodes a homeodomain transcription factor of the WUSCHEL-LIKE HOMEOBOX (WOX) family [19]. WUS is a bi-functional protein that can both repress and activate gene transcription in the SAM [20]. Among the key targets of direct WUS repression in the OC are negative regulators of cytokinin activity, a hormone that promotes cell proliferation across the SAM [21]. WUS also directly represses the transcription of cell differentiation-inducing transcription factor genes that are normally expressed in organ primordia, to prevent premature stem cell differentiation at the apex of the SAM [22]. In addition, WUS protein moves between cells through plasmodesmata into the apical stem cell domain [23] where it maintains stem cell fate and induces the expression of the *CLV3* gene in a dosage-dependent fashion [24,25]. WUS functions together with members of the HAIRY MERISTEM (HAM) family of GRAS domain transcriptional regulators to regulate stem cell production [26] and to ensure that *CLV3* transcription is activated exclusively in the outermost apical layers of the SAM [27].

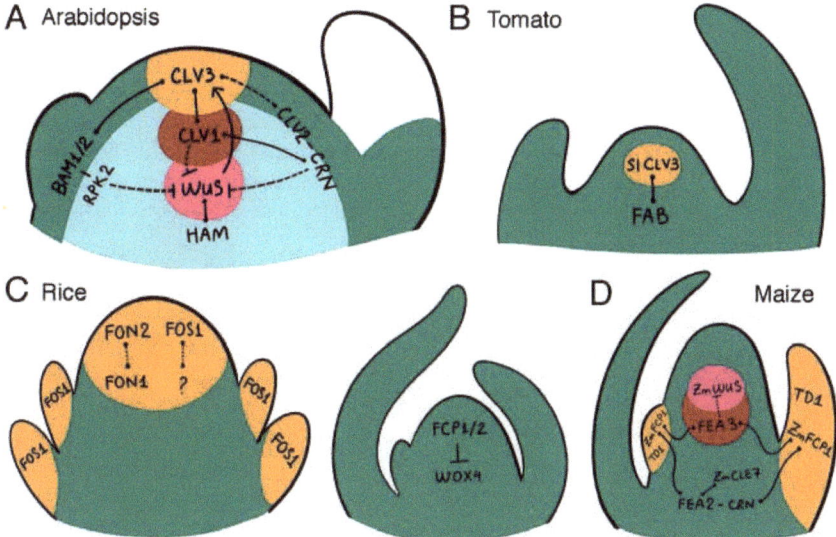

Figure 2. CLV-WUS signaling pathways in model and crop plant meristems. (**A**) *Arabidopsis* SAM. (**B**) Tomato SAM. (**C**) Rice FM and SAM. (**D**) Maize SAM. Genes with characterized genetic and/or biochemical interactions are shown. Arrows depict positive regulation and bars depict negative regulation. Solid lines represent direct interactions and dashed lines represent indirect interactions. Solid lines with rounded ends depict direct peptide–receptor interactions. Unidentified receptors for peptides are denoted by question marks.

The CLV signal transduction pathway negatively regulates stem cell accumulation in above-ground meristems. Mutations in *Arabidopsis CLV* genes cause progressive enlargement of the shoot and floral stem cell pools (Figure 1B,C), resulting in plants with enlarged stems and excess flowers, as well as flowers with extra sepals, petals and stamens, and siliques with more than two locules [4,28]. *CLV3* encodes a founding member of the CLAVATA3/EMBRYO SURROUNDING REGION (CLE) family of polypeptides [29], which are present throughout the plant kingdom [30,31]. *CLV3* is expressed within the shoot and floral stem cell domain [32] and encodes a pre-propeptide that is processed into a 12–13 amino acid arabinosylated glycoprotein [33,34]. This glycoprotein moves through the extracellular space to communicate stem cell fate information with neighboring cells [35].

The CLV3 signal is perceived and transduced at the plasma membrane by several distinct sets of receptors (Figures 2A and 3). CLV3 peptides are bound by the CLV1 leucine-rich repeat receptor-like kinase (LRR-RLK) that is produced in cells beneath the stem cell reservoir [36,37]. A second distinct receptor complex consists of heterodimers of the CLV2 LRR receptor-like protein [38] and the CORYNE (CRN) protein, a presumptive pseudokinase that functions as a CLV2 co-receptor [39,40]. CRN mediates localization of CLV2/CRN complexes to the plasma membrane [41], where they can directly interact with CLV1 heterodimers [41–43]. Yet in contrast to *CLV1*, *CLV2* and *CRN* are expressed throughout the entire SAM, and the CLV2-CRN complex functions largely independently of CLV1 in CLV3 signal transduction [39,41,43]. Reports differ as to whether the CLV2 receptor itself directly binds the CLV3 ligand or if an additional co-receptor is required [42,44]. Other receptors appear to mediate CLV3 signaling predominantly on the flanks of the meristem. Three LRR-RLK genes that form a monophyletic group with *CLV1*, termed *BARELY ANY MERISTEM1, 2* and *3 (BAM1–3)*, act redundantly to promote stem cell maintenance on the meristem periphery [45], and both BAM1 and BAM2 directly bind CLV3 peptides [42,44]. The BAM1 protein physically associates with the LRR receptor-like kinase RECEPTOR-LIKE PROTEIN KINASE2 (RPK2) [46], which itself does not bind CLV3 peptides and thus is proposed to regulate meristem maintenance by transmitting the CLV3 signal through the BAM1 pathway [44]. An additional group of four LRR-RLKs termed the CLAVATA3 INSENSITIVE RECEPTOR KINASES (CIKs) undergo rapid phosphorylation in response to CLV3 signaling, and appear to function as co-receptors for the CLV1, CLV2-CRN, and BAM-RPK2 receptor pathways [47]. CLV3-mediated signaling through these receptor complexes limits stem cell accumulation by restricting the *WUS* expression domain to the OC [48,49]. Thus, the CLV-WUS pathway functions as a dynamic negative feedback loop that allows the stem cell domain and the underlying OC to continually adjust their size relative to one another to maintain SAM homeostasis.

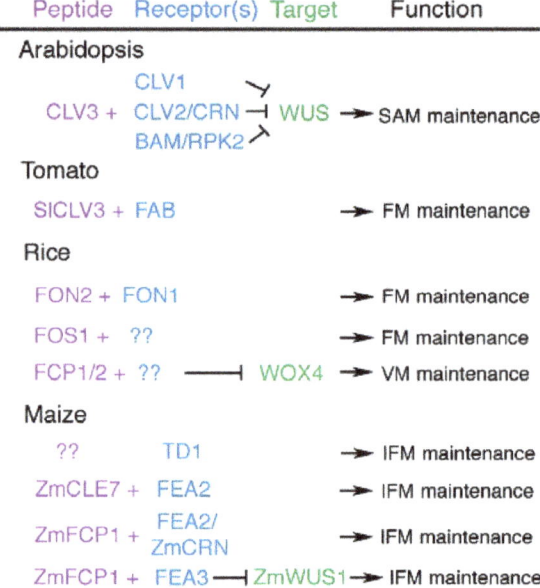

Figure 3. Components of CLV-WUS signaling pathways and their functions in model and crop plants. Proteins with characterized genetic and/or biochemical interactions are listed. Unidentified peptides and receptors are denoted by question marks. Arrows depict positive regulation and bars depict negative regulation. SAM, shoot apical meristem; FM, floral meristem; VM, vegetative meristem; IFM, inflorescence meristem.

3. CLV-WUS Pathway in Dicotyledonous Crop Plants

Arabidopsis thaliana is related to cultivated mustard varieties—such as *Brassica rapa*, *Brassica juncea*, and *Brassica napus*—which are agriculturally important oil crops that provide edible oils for human diets as well as raw material for animal feed and industrial processes such as biodiesel production [50]. Like *Arabidopsis*, oilseed floral meristems produce sepals, petals, stamens, and two carpels, the latter of which develop into the two locules of the siliques. The oil and protein products of Brassica plants are contained inside the seeds that develop within the siliques, and thus enhancing silique yield traits has long been a major goal of oilseed production and genetic improvement [51,52].

Several multilocular Brassica lines with more than two locules have been identified in natural populations [53], and recent studies have implicated CLV-WUS pathway components in the appearance of this trait (Table 1). The *B. rapa* var. *yellow sarson ml4* mutant exhibits a multilocular phenotype caused by a single nucleotide mutation in a *CLV3* gene homolog that produces an amino acid substitution in the CLE domain [54]. Similarly, a multilocular phenotype found in the *B. juncea* Duoshi cultivar results from mutations in a *CLV1* gene homolog, *BjLn1* [55,56], while a trilocular phenotype in *B. juncea* J163-4 plants is caused by the insertion of a copia-LTR retrotransposable element into the coding region of a second *CLV1* homolog, *BjMc1*, interrupting its transcription [52]. These multilocular Brassica plants have significantly higher yield than the corresponding bilocular plants without affecting viability [54,56,57], suggesting that selectively targeting *CLV* genes can be a powerful method of obtaining high-yield oilseed cultivars. This has been tested by the use of CRISPR-Cas9 genome editing to target *CLV* pathway components in allotetraploid *B. napus* plants, which contain two copies each of the *CLV1*, *CLV2*, and *CLV3* genes [50]. Simultaneous mutation of both copies of any of the three *BnCLV* genes resulted in plants with enlarged IFMs, multilocular siliques, and higher seed yield, with mutations in the *BnCLV3* genes producing the most severe effects [50].

Table 1. *CLV* peptide and receptor gene orthologs in crop plants

Gene Ortholog					
Arabidopsis	Brassica	Tomato	Rice	Maize	References
CLV3	BrCLV3	SlCLV3	FON2	ZmCLE7	[32,54,58–61]
CLV1	BjMc1; BjLn1	FAB	FON1	TD1	[36,52,56,62,63]
CLV2	BnA02CLV2; BnC02CLV2	?	?	FEA2	[5,38,50]
CRN	?	?	?	ZmCRN	[39,61]

Members of the CLV-WUS pathway also play key roles in regulating locule number in tomato (Table 1, Figure 2B). The wild ancestor of tomato had a small, bilocular fruit, whereas modern tomato varieties contain eight or more locules [64]. The *fasciated (fas)* and *locule number (lc)* genes are the major quantitative trait loci (QTL) controlling the number of tomato fruit locules, and most cultivated tomato varieties contain mutations in either the *fas* or the *fas* and *lc* genes [65]. The multilocular *fas* phenotype results from a mutation in the regulatory region of a *CLV3*-related gene, *SlCLV3* [59], whereas the *lc* trait is caused by two single nucleotide polymorphisms (SNPs) in a repressor element downstream of a *WUS* gene homolog [66,67]. Evidence suggests that selection at both loci took place during tomato domestication to produce plants with increased fruit locule number [59,66]. Generation of a suite of novel *SlCLV3* promoter alleles using genome editing produced plants with a continuum of variation in fruit locule number [67], providing a blueprint for engineering quantitative variation in yield traits for breeding purposes.

In addition, a forward genetic screen for tomato mutants with increased inflorescence branching and fruit locule number identified both known and novel *CLV* pathway members [59]. Plants with the *fasciated and branched (fab)* multilocular phenotype contain a missense mutation in the closest tomato homolog of *CLV1*, which affects the kinase domain. Interestingly, the *fasciated inflorescence (fin)* and *fab2* phenotypes are caused by mutations in arabinosyltransferase genes. Arabinosyltransferases catalyze the transfer of L-arabinose to the hydroxyl group of hydroxyproline (Hyp) residues in their target proteins [68]. In *Arabidopsis*, the Hyp^7 residue of the active CLV3 polypeptide is post-translationally modified with three L-arabinose residues [34,69], and the addition of arabinosylated SlCLV3 peptides can rescue the tomato *fin* phenotype [59]. Therefore, arabinosyltransferase genes are critical components of the CLV-WUS stem cell signaling pathway that can impact crop productivity traits.

4. CLV-WUS Pathway in Grasses

The broad function of the CLV-WUS pathway in mediating shoot meristem maintenance is conserved in agronomically important grass species (Table 1, Figure 2C), with some important modifications. In rice (*Oryza sativa* L.), stem cell maintenance appears to be regulated by several distinct pathways, the relative contributions of which depend on the type of meristem. The *FLORAL ORGAN NUMBER (FON1)* and *FON2* genes restrict stem cell accumulation specifically in floral meristems, without affecting vegetative or inflorescence meristem activity [70]. *FON1* encodes the rice ortholog of the CLV1 receptor kinase [62]. It is expressed within the floral meristems but also within the shoot meristem throughout development, suggesting that related receptor kinase genes share functional redundancy with *FON1* in vegetative and inflorescence tissues [62]. Such genes, however, remain to be characterized. The *FON2* gene, also referred to as *FON4*, functions in the same genetic pathway as *FON1* and encodes a CLV3-related protein [58,60]. Like *CLV3*, *FON2* is expressed at the apex of both shoot and floral meristems [58,60]. Thus, in rice floral meristems, the FON1-FON2 system corresponds to the CLV1-CLV3 peptide-receptor kinase signaling system in *Arabidopsis* (Figure 3).

Several other *CLE* genes also play roles in orchestrating rice meristem maintenance (Figures 2C and 3). QTL analysis identified the *FON2 SPARE1 (FOS1)* gene in indica varieties as a suppressor of the *fon2* floral organ number phenotype in japonica, indicating that *FOS1* can substitute for *FON2* activity in rice floral meristems [71]. Constitutive expression of *FOS1* leads to termination of the vegetative SAM, suggesting a potential function for FOS1 in vegetative SAM cell maintenance [71]. The FOS1 CLE domain is more similar to *Arabidopsis* CLE8 and CLE13 than to CLV3, and because FOS1 activity does not require FON1, FOS1 signaling is thought to occur largely in parallel with the FON1-FON2 pathway [71]. Two other *CLE* genes, *FON2-LIKE CLE PROTEIN1 (FCP1)* and *FCP2*, encode proteins that differ in the CLE domain by one amino acid [72] and act redundantly to negatively regulate vegetative stem cell activity and promote leaf initiation [73]. FCP1 represses the expression of rice *WOX4*, an ortholog of *Arabidopsis WOX4* [74], which promotes the undifferentiated state of the vegetative SAM [73]. Thus the rice *WOX4* gene functions similarly to the *Arabidopsis WUS* gene [19], whereas the *WUS* ortholog in rice [74], called *TILLERS ABSENT1 (TAB1)*, is required for axillary meristem initiation but not for shoot or floral meristem maintenance [75]. These studies identify additional *CLE* signaling peptide genes besides *CLV3* as potential targets for genome editing to enhance yield traits in crop plants, particularly grasses.

Maize is a monoecious plant that develops two distinct inflorescence meristem structures: the terminal IFM, called the tassel, that bears male flowers; and the axillary IFMs, called the ears, that bear female flowers. The ear inflorescence meristems produce multiple rows of secondary meristems called spikelet pair meristems, which branch to form spikelet meristems. The spikelet meristems then branch to form two floral meristems, one of which develops into a flower (and after fertilization, a seed kernel) while the other aborts. Modern cultivated corn varieties contain between 8 and 20 rows of kernels within their ears [76], compared to the two rows of kernels found in teosinte, the ancestor of maize, and the ability of the ear IFM to produce additional rows of spikelet meristems appears to have been a major factor in the maize domestication process [5,15]. Molecular evidence indicates that CLV-WUS pathway components underlie much of the variation in this key yield trait.

Mutations at multiple maize loci generate fasciated phenotypes in which the male and/or female inflorescences are enlarged and display increased numbers of spikelet pair and/or spikelet meristems [77]. One of the first such mutants cloned was *thick tassel dwarf1 (td1)*, which displays increased tassel and ear IFM size and results from a mutation in the maize ortholog of the *CLV1* gene [63] (Table 1). The *TD1* locus maps near QTL for tassel spikelet density and for kernel row number [63], whereas the *FASCIATED EAR2 (FEA2)* gene encodes the maize ortholog of *CLV2* [5] and corresponds to a distinct QTL for kernel row number [78]. Thus multiple CLV receptors are likely to have been targets of selection during maize domestication (Figures 2D and 3).

The FEA2 receptor-like protein is proposed to regulate meristem maintenance by transmitting signals from two different CLE peptides through two distinct downstream pathways. FEA2 physically associates in vivo with COMPACT PLANT2 (CT2), the alpha subunit of the heterotrimeric GTP binding protein [79] that along with other Gα domain-containing eXtra Large GTP-binding proteins (XLGs) contribute to restricting IFM size [80]. In CLE peptide response assays both *fea2* and *ct2* plants are resistant to application of ZmCLE7, the maize CLV3 ortholog, suggesting that ZmCLE7 peptide signaling is transmitted across the plasma membrane by a FEA2-CT2 receptor-G protein complex [61]. FEA2 also heterodimerizes with ZmCRN, which acts in separate pathway from CT2. *Zmcrn* plants are sensitive to ZmCLE7 application, but both *fea2* and *Zmcrn* plants are resistant to the application of a related CLE peptide ZmFCP1. In contrast to *ZmCLE7*, *ZmFCP1* is not expressed in the SAM but is detected in incipient and initiating leaf primordia [81]. FEA2, therefore, also appears capable of transmitting a ZmFCP1 signal from organ primordia to regulate IFM activity through a ZmCRN-mediated pathway. Interestingly, the *ZmCRN* locus has significant association with kernel row number variability [82], suggesting that it too contributes to quantitative variation in this trait.

Finally, the CLE peptide ZmFCP1 signals through the LRR receptor-like protein FASCIATED EAR3 (FEA3) to suppress the expression of *ZmWUS1* in the region below the organizing center [81] (Figure 2D). Computational models suggest that ZmFCP1 signaling from developing organ primordia is sufficient to restrict stem cell accumulation in the neighboring SAM by limiting the size of the *ZmWUS1* expression domain [81]. Whether the other maize *WUS* ortholog, *ZmWUS2*, is also a target of ZmFCP1-FEA3 signaling is unknown. FEA3 acts in a separate pathway than FEA2 and weak alleles of *FEA3* and *FEA2* independently enhance kernel row number, although weak *fea2* alleles do not increase overall yield due to a compensatory reduction in kernel size [78,81]. Nonetheless, in maize as in other crop plants, the reduction of stem cell regulatory gene activity can lead to improvement of agronomic traits.

5. Perspectives

Gene homologies between *Arabidopsis* and agronomic plants continue to be robust tools for technology transfer, facilitating the translation of basic genetic and genomic information into direct crop improvements. A recent study of the moss *Physcomitrella patens* reveals that the core components of the CLV signaling pathway, namely a CLE peptide and a CLV1/BAM-like RLK, originated with land plants, and that their ability to regulate stem cell proliferation and cell fate is likely to be an ancestral feature of land plants that enabled three-dimensional growth [31]. To date, *CLE* genes have been identified in over 50 plant species, including *Medicago truncatula*, *Lotus japonicas*, wheat, potato, soybean, common bean, banana, and poplar [30]. Additionally, members of the *WUS* clade of *WOX* genes with stem cell-related functions appeared after the divergence of vascular plants from bryophytes [83]. Thus, the potential for modulating the *CLV-WUS* pathway and related *CLE* genes to enhance yield traits exists in a very large number of agricultural plant species.

To date, a major challenge to manipulating yield trait genes in agronomic plants has been the presence of multiple genes within the genome that encode redundant or overlapping stem cell maintenance functions. For example, several homologous copies of the *CLV1*, *CLV2*, and *CLV3* genes exist within polyploid genomes such as *Brassica napa* [50] and wheat (*Triticum aestivum* L.) [84]. In addition, genetic evidence indicates that multiple *CLE* genes as well as multiple *CLV1/BAM* LRR-RLK gene paralogs are involved in the regulation of stem cell maintenance. The advent of multiplex genome editing, which directs the simultaneous targeting of multiple members of a gene family as well as multiple components of a molecular pathway [85], offers great potential to produce beneficial architecture modifications in both dicot and monocot crop species. In this respect, it is worth noting that hypomorphic mutations that reduce *CLV-WUS* gene function, such as mutations in tomato *CLV3* or *WUS* regulatory regions [59,66,67] or missense mutations in maize *CLV1* or *CLV2* receptor kinase genes [78,81], can be sufficient to achieve significant yield increases without the need to completely eliminate gene function. Thus, novel approaches such as genome editing of stem cell maintenance gene promoters [67] may also be a fruitful approach to fine-tune CLV-WUS signaling and thus tailor yield trait optimization within individual crop species.

Author Contributions: The author is responsible for all aspects of the manuscript.

Funding: This research was funded by the United States Department of Agriculture CRIS grant number 2030-21000-048-00D.

Acknowledgments: The author thanks Thai Q. Dao for the illustrations used in Figure 2, and Dezhi Du, Thai Q. Dao and Wassim Hage for helpful comments.

Conflicts of Interest: The author declares no conflict of interest. The funding sponsors had no role in the design of the study; in the collection, analyses, or interpretation of data; in the writing of the manuscript, or in the decision to publish the results.

References

1. Steeves, T.A.; Sussex, I.M. *Patterns in Plant Development*; Cambridge University Press: New York, NY, USA, 1989.
2. Kieffer, M.; Stern, Y.; Cook, H.; Clerici, E.; Maulbetsch, C.; Laux, T.; Davies, B. Analysis of the transcription factor WUSCHEL and its functional homologue in *Antirrhinum* reveals a potential mechanism for their roles in meristem maintenance. *Plant Cell* **2006**, *18*, 560–573. [CrossRef] [PubMed]
3. Laux, T.; Mayer, K.F.X.; Berger, J.; Jurgens, G. The *WUSCHEL* gene is required for shoot and floral meristem integrity in *Arabidopsis*. *Development* **1996**, *122*, 87–96. [PubMed]
4. Clark, S.E.; Running, M.P.; Meyerowitz, E.M. *CLAVATA1*, a regulator of meristem and flower development in *Arabidopsis*. *Development* **1993**, *119*, 397–418. [PubMed]
5. Taguchi-Shiobara, F.; Yuan, Z.; Hake, S.; Jackson, D. The *FASCIATED EAR2* gene encodes a leucine-rich repeat receptor-like protein that regulates shoot meristem proliferation in maize. *Genes Dev.* **2001**, *15*, 2755–2766. [CrossRef] [PubMed]
6. Furner, I.J.; Pumfrey, J.E. Cell fate in the shoot apical meristem of *Arabidopsis thaliana*. *Development* **1992**, *115*, 755–764.
7. Irish, V.F.; Sussex, I.M. A fate map of the *Arabidopsis* embryonic shoot apical meristem. *Development* **1992**, *115*, 745–753.
8. Poethig, R.S.; Coe, E.H.J.; Johri, M.M. Cell lineage patterns in maize *Zea mays* embryogenesis: A clonal analysis. *Dev. Biol.* **1986**, *117*, 392–404. [CrossRef]
9. Sussex, I.M. Experiments on the cause of dorsiventrality in leaves. *Nature* **1954**, *174*, 351–352. [CrossRef]
10. Poethig, R.S. Clonal analysis of cell lineage patterns in plant development. *Am. J. Bot.* **1987**, *74*, 581–594. [CrossRef]
11. Satina, S.; Blakeslee, A.F.; Avery, A.G. Demonstration of the three germ layers in the shoot apex of *Datura* by means of induced polyploidy in periclinal chimeras. *Am. J. Bot.* **1940**, *27*, 895–905. [CrossRef]
12. Tilney-Bassett, R.A.E. *Plant Chimeras*; E. Arnold: London, UK, 1986.
13. Poethig, R.S.; Sussex, I.M. The cellular parameters of leaf development in tobacco: A clonal analysis. *Planta* **1985**, *165*, 170–184. [CrossRef] [PubMed]
14. Poethig, R.S.; Sussex, I.M. The developmental morphology and growth dynamics of the tobacco leaf. *Planta* **1985**, *165*, 158–169. [CrossRef] [PubMed]
15. Doebley, J.F.; Gaut, B.A.; Smith, B.D. The molecular genetics of crop domestication. *Cell* **2006**, *127*, 1309–1321. [CrossRef] [PubMed]
16. Kuittinen, H.; Aguade, M. Nucleotide variation at the CHALCONE ISOMERASE locus in *Arabidopsis thaliana*. *Genetics* **2000**, *155*, 863–872. [PubMed]
17. Somssich, M.; Je, B.I.; Simon, R.; Jackson, D. CLAVATA-WUSCHEL signalling in the shoot meristem. *Development* **2016**, *143*, 3238–3248. [CrossRef] [PubMed]
18. Zhang, Z.; Tucker, E.; Hermann, M.; Laux, T. A molecular framework for the embryonic initiation of shoot meristem stem cells. *Dev. Cell* **2017**, *40*, 264–277. [CrossRef] [PubMed]
19. Mayer, K.F.X.; Schoof, H.; Haecker, A.; Lenhard, M.; Jurgens, G.; Laux, T. Role of WUSCHEL in regulating stem cell fate in the *Arabidopsis* shoot meristem. *Cell* **1998**, *95*, 805–815. [CrossRef]
20. Ikeda, M.; Mitsuda, N.; Ohme-Takagi, M. *Arabidopsis* WUSCHEL is a bifunctional transcription factor that acts as a repressor in stem cell regulation and as an activator in floral patterning. *Plant Cell* **2009**, *21*, 3493–3505. [CrossRef] [PubMed]
21. Leibfried, A.; To, J.P.C.; Busch, W.; Stehling, S.; Kehle, A.; Demar, M.; Kieber, J.J.; Lohmann, J.U. WUSCHEL controls meristem function by direct regulation of cytokinin-inducible response regulators. *Nature* **2005**, *438*, 1172–1175. [CrossRef] [PubMed]
22. Yadav, R.K.; Perales, M.; Gruel, J.; Ohno, C.; Heisler, M.; Girke, T.; Jonsson, H.; Reddy, G.V. Plant stem cell maintenance involves direct transcriptional repression of differentiation program. *Mol. Syst. Biol.* **2013**, *9*, 654. [CrossRef] [PubMed]
23. Daum, G.; Medzihradszky, A.; Suzaki, T.; Lohmann, J.U. A mechanistic framework for noncell autonomous stem cell induction in *Arabidopsis*. *Proc. Natl. Acad. Sci. USA* **2014**, *111*, 14619–14624. [CrossRef] [PubMed]
24. Perales, M.; Rodriguez, K.; Snipes, S.; Yadav, R.K.; Diaz-Mendoza, M.; Reddy, G.V. Threshold-dependent transcriptional discrimination underlies stem cell homeostasis. *Proc. Natl. Acad. Sci. USA* **2016**, *113*, E6298–E6306. [CrossRef] [PubMed]

25. Yadav, R.K.; Perales, M.; Gruel, J.; Girke, T.; Jonsson, H.; Reddy, G.V. WUSCHEL protein movement mediates stem cell homeostasis in the *Arabidopsis* shoot apex. *Genes Dev.* **2011**, *25*, 2025–2030. [CrossRef] [PubMed]
26. Zhou, Y.; Liu, X.; Engstrom, E.M.; Nimchuk, Z.L.; Pruneda-Paz, J.L.; Tarr, P.T.; Yan, A.; Kay, S.A.; Meyerowitz, E.M. Control of plant stem cell function by conserved interacting transcriptional regulators. *Nature* **2015**, *517*, 377–380. [CrossRef] [PubMed]
27. Zhou, Y.; Yan, A.; Han, H.; Li, T.; Geng, Y.; Liu, X.; Meyerowitz, E.M. HAIRY MERISTEM with WUSCHEL confines CLAVATA3 expression to the outer apical meristem layers. *Science* **2018**, *361*, 502–506. [CrossRef] [PubMed]
28. Clark, S.E.; Running, M.P.; Meyerowitz, E.M. *CLAVATA3* is a specific regulator of shoot and floral meristem development affecting the same processes as *CLAVATA1*. *Development* **1995**, *121*, 2057–2067.
29. Cock, J.M.; McCormick, S. A large family of genes that share homology with *CLAVATA3*. *Plant Physiol.* **2001**, *126*, 939–942. [CrossRef] [PubMed]
30. Goad, D.M.; Zhu, C.; Kellogg, E.A. Comprehensive identification and clustering of CLV3/ESR-related (CLE) genes in plants finds groups with potentially shared function. *New Phytol.* **2016**, *16*, 605–616. [CrossRef] [PubMed]
31. Whitewoods, C.D.; Cammarata, J.; Venza, Z.N.; Sang, S.; Crook, A.D.; Aoyama, T.; Wang, X.Y.; Waller, M.; Kamisugi, Y.; Cuming, A.C.; et al. *CLAVATA* was a genetic novelty for the morphological innovation of 3D growth in land plants. *Curr. Biol.* **2018**, *28*, 2365–2376. [CrossRef] [PubMed]
32. Fletcher, J.C.; Brand, U.; Running, M.P.; Simon, R.; Meyerowitz, E.M. Signaling of cell fate decisions by CLAVATA3 in *Arabidopsis* shoot meristems. *Science* **1999**, *283*, 1911–1914. [CrossRef] [PubMed]
33. Kondo, T.; Sawa, S.; Kinoshita, A.; Mizuno, S.; Kakimoto, T.; Fukuda, H.; Sakagami, Y. A plant peptide encoded by *CLV3* identified by in situ MALDI-TOF MS analysis. *Science* **2006**, *313*, 845–848. [CrossRef] [PubMed]
34. Ohyama, K.; Shinohara, H.; Ogawa-Ohnishi, M.; Matsubayashi, Y. A glycopeptide regulating stem cell fate in *Arabidopsis thaliana*. *Nat. Chem. Biol.* **2009**, *5*, 578–580. [CrossRef] [PubMed]
35. Rojo, E.; Sharma, V.K.; Kovaleva, V.; Raikhel, N.V.; Fletcher, J.C. CLV3 is localized to the extracellular space, where it activates the *Arabidopsis* CLAVATA stem cell signaling pathway. *Plant Cell* **2002**, *14*, 969–977. [CrossRef] [PubMed]
36. Clark, S.E.; Williams, R.W.; Meyerowitz, E.M. The *CLAVATA1* gene encodes a putative receptor kinase that controls shoot and floral meristem size in *Arabidopsis*. *Cell* **1997**, *89*, 575–585. [CrossRef]
37. Ogawa, M.; Shinohara, H.; Sakagami, Y.; Matsubayashi, Y. *Arabidopsis* CLV3 peptide directly binds the CLV1 ectodomain. *Science* **2008**, *319*, 294. [CrossRef] [PubMed]
38. Jeong, S.; Trotochaud, A.E.; Clark, S.E. The *Arabidopsis CLAVATA2* gene encodes a receptor-like protein required for the stability of the CLAVATA1 receptor-like kinase. *Plant Cell* **1999**, *11*, 1925–1933. [CrossRef] [PubMed]
39. Muller, R.; Bleckmann, A.; Simon, R. The receptor kinase CORYNE of *Arabidopsis* transmits the stem cell-limiting signal CLAVATA3 independently of CLAVATA1. *Plant Cell* **2008**, *20*, 934–946. [CrossRef] [PubMed]
40. Nimchuk, Z.L.; Tarr, P.T.; Meyerowitz, E.M. An evolutionarily conserved pseudokinase mediates stem cell production in plants. *Plant Cell* **2011**, *23*, 851–854. [CrossRef] [PubMed]
41. Bleckmann, A.; Weidtkamp-Peters, S.; Seidel, C.A.M.; Simon, R. Stem cell signaling in *Arabidopsis* requires CRN to localize CLV2 to the plasma membrane. *Plant Physiol.* **2010**, *152*, 166–176. [CrossRef] [PubMed]
42. Guo, Y.; Han, L.; Hymes, M.; Denver, R.; Clark, S.E. CLAVATA2 forms a distinct CLE-binding receptor complex regulating *Arabidopsis* stem cell specification. *Plant J.* **2010**, *63*, 889–900. [CrossRef] [PubMed]
43. Zhu, Y.; Wang, Y.; Li, R.; Song, X.; Wang, Q.; Huang, S.; Jin, J.; Liu, C.; Lin, J. Analysis of interactions among the CLAVATA3 receptors reveals a direct interaction between CLAVATA2 and CORNYE in *Arabidopsis*. *Plant J.* **2010**, *61*, 223–233. [CrossRef] [PubMed]
44. Shinohara, H.; Matsubayashi, Y. Reevaluation of the CLV3-receptor interaction in the shoot apical meristem: Dissection of the CLV3 signaling pathway from a direct ligand-binding point of view. *Plant J.* **2015**, *82*, 328–336. [CrossRef] [PubMed]
45. DeYoung, B.; Bickle, K.L.; Schrage, K.J.; Muskett, P.; Patel, K.; Clark, S.E. The CLAVATA1-related BAM1, BAM2 and BAM3 receptor kinase-like proteins are required for meristem function in *Arabidopsis*. *Plant J.* **2006**, *45*, 1–16. [CrossRef] [PubMed]

46. Kinoshita, A.; Betsuyaku, S.; Osakabe, Y.; Mizuno, S.; Nagawa, S.; Stahl, Y.; Simon, R.; Yamaguchi-Shinozaki, K.; Fukuda, H.; Sawa, S. RPK2 is an essential receptor-like kinase that transmits the CLV3 signal in *Arabidopsis*. *Development* **2010**, *137*, 3911–3920. [CrossRef] [PubMed]
47. Hu, C.; Zhu, Y.; Cui, Y.; Cheng, K.; Liang, W.; Wei, Z.; Zhu, M.; Yin, H.; Zeng, L.; Xiao, Y.; et al. A group of receptor kinases are essential for CLAVATA signalling to maintain stem cell homeostasis. *Nat. Plants* **2018**, *4*, 205–211. [CrossRef] [PubMed]
48. Brand, U.; Fletcher, J.C.; Hobe, M.; Meyerowitz, E.M.; Simon, R. Dependence of stem cell fate in *Arabidopsis* on a feedback loop regulated by CLV3 activity. *Science* **2000**, *289*, 617–619. [CrossRef] [PubMed]
49. Schoof, H.; Lenhard, M.; Haecker, A.; Mayer, K.F.X.; Jurgens, G.; Laux, T. The stem cell population of *Arabidopsis* shoot meristems is maintained by a regulatory loop between the *CLAVATA* and *WUSCHEL* genes. *Cell* **2000**, *100*, 635–644. [CrossRef]
50. Yang, Y.; Zhu, K.; Li, H.; Han, S.; Meng, Q.; Khan, S.U.; Fan, C.; Xie, K.; Zhou, Y. Precise editing of *CLAVATA* genes in *Brassica napus* L. regulates multilocular silique development. *Plant Biotechnol. J.* **2017**, *16*, 1322–1335. [CrossRef] [PubMed]
51. Zhang, L.W.; Li, S.P.; Chen, L.; Yang, G.S. Identification and mapping of a major dominant quantitative trait locus controlling seeds per silique as a single mendelian factor in *Brassica napus* L. *Theor. Appl. Genet.* **2012**, *125*, 695–705. [CrossRef] [PubMed]
52. Xu, P.; Cao, S.; Hu, K.; Wang, X.; Huang, W.; Wang, G.; Lv, Z.; Liu, Z.; Wen, J.; Yi, B.; et al. Trilocular phenotype in *Brassica juncea* L. resulted from interruption of *CLAVATA1* gene homologue *(BjMc1)* transcription. *Sci. Rep.* **2017**, *7*, 3498. [CrossRef] [PubMed]
53. Lui, H. *Brassica napus Genetics and Breeding*; China Agricultural University Press: Beijing, China, 2000.
54. Fan, C.; Wu, Y.; Yang, Q.; Meng, Q.; Zhang, K.; Li, J.; Wang, J.; Zhou, Y. A novel single-nucleotide mutation in a *CLAVATA3* gene homolog controls a multilocular silique trait in *Brassica rapa* L. *Mol. Plant* **2014**, *7*, 1788–1792. [CrossRef] [PubMed]
55. Zhao, H.C.; Du, D.Z.; Lui, Q.Y.; Yu, Q.L.; Wang, R.S. Study on multilocular heredity of *B. juncea*. *J. Northwest Si-Tech Univ. Agric. For. (Nat. Sci. Ed.)* **2003**, *31*, 90–92.
56. Xiao, L.; Li, X.; Liu, F.; Zhao, Z.; Xu, L.; Chen, C.; Wang, Y.; Shang, G.; Du, D. Mutations in the CDS and promoter of *BjuA07.CLV1* cause a multilocular trait in *Brassica juncea*. *Sci. Rep.* **2018**, *8*, 5339. [CrossRef] [PubMed]
57. Lv, Z.W.; Xu, P.; Zhang, X.; Wen, J.; Yi, B.; Ma, C.; Tu, J.; Fu, T.; Shen, J. Primary study on anatomic and genetic analyses of multi-loculus in *Brassica juncea*. *Chin. J. Oil Crop Sci.* **2012**, *34*, 461–466.
58. Suzaki, T.; Toriba, T.; Fujimoto, M.; Tsutsumi, Y.; Kitano, H.; Hirano, H. Conservation and diversification of meristem mechanism in *Oryza sativa*: Function of the *FLORAL ORGAN NUMBER2* gene. *Plant Cell Physiol.* **2006**, *47*, 1591–1602. [CrossRef] [PubMed]
59. Xu, C.; Liberatore, K.L.; MacAlister, C.A.; Huang, Z.; Chu, Y.-H.; Jiang, K.; Brooks, C.; Ogawa-Ohnishi, M.; Xiong, G.; Pauly, M.; et al. A cascade of arabinosyltransferases controls shoot meristem size in tomato. *Nat. Genet.* **2015**, *47*, 784–792. [CrossRef] [PubMed]
60. Chu, H.; Qian, Q.; Liang, W.; Yin, C.; Tan, H.; Yao, X.; Yuan, Z.; Yang, J.; Huang, H.; Luo, D.; et al. The *FLORAL ORGAN NUMBER4* gene encoding a putative ortholog of *Arabidopsis* CLAVATA3 regulates apical meristem size in rice. *Plant Physiol.* **2006**, *142*, 1039–1052. [CrossRef] [PubMed]
61. Je, B.I.; Xu, F.; Wu, Q.; Liu, L.; Meeley, R.; Gallagher, J.P.; Corcilius, L.; Payne, R.J.; Bartlett, M.E.; Jackson, D. The CLAVATA receptor FASCIATED EAR2 responds to distinct CLE peptides by signaling through two downstream effectors. *eLife* **2018**, *7*, e35673. [CrossRef] [PubMed]
62. Suzaki, T.; Sato, M.; Ashikari, M.; Miyoshi, M.; Nagato, Y.; Hirano, H.-Y. The gene *FLORAL ORGAN NUMBER1* regulates floral meristem size in rice and encodes a leucine-rich repeat receptor kinase orthologous to *Arabidopsis* CLAVATA1. *Development* **2004**, *131*, 5649–5657. [CrossRef] [PubMed]
63. Bommert, P.; Lunde, C.; Nardmann, J.; Vollbrecht, E.; Running, M.P.; Jackson, D.; Hake, S.; Werr, W. Thick tassel dwarf1 encodes a putative maize ortholog of the *Arabidopsis* CLAVATA1 leucine-rich repeat receptor-like kinase. *Development* **2005**, *132*, 1235–1245. [CrossRef] [PubMed]
64. Tanksley, S.D. The genetic, developmental, and molecular bases of fruit size and shape variation in tomato. *Plant Cell* **2004**, *16*, S181–S189. [CrossRef] [PubMed]

65. Lippman, Z.; Tanksley, S.D. Dissecting the genetic pathway to extreme fruit size in tomato using a cross between the small-fruited wild species *L. pimpinellifolium* and *L. esculentum* var. Giant Heirloom. *Genetics* **2001**, *158*, 413–422. [PubMed]
66. Munos, S.; Ranc, N.; Botton, E.; Berard, A.; Rolland, S.; Duffe, P.; Carretero, Y.; Le Paslier, M.-C.; Delalande, C.; Bouzayen, M.; et al. Increase in tomato locule number is controlled by two single-nucleotide polymorphisms located near *WUSCHEL*. *Plant Physiol.* **2011**, *156*, 2244–2254. [CrossRef] [PubMed]
67. Rodriguez-Leal, D.; Lemmon, Z.H.; Man, J.; Bartlett, M.; Lippman, Z. Engineering quantitative trait variation for crop improvement by genome editing. *Cell* **2017**, *171*, 470–480. [CrossRef] [PubMed]
68. Ogawa-Ohnishi, M.; Matsushita, W.; Matsubayashi, Y. Identification of three hydroxyproline o-arabinosyltransferases in *Arabidopsis*. *Nat. Chem. Biol.* **2013**, *9*, 726–730. [CrossRef] [PubMed]
69. Shinohara, H.; Matsubayashi, Y. Chemical synthesis of *Arabidopsis* CLV3 glycopeptide reveals the impact of hydroxyproline arabinosylation on peptide confirmation and activity. *Plant Cell Physiol.* **2013**, *54*, 369–374. [CrossRef] [PubMed]
70. Nagasawa, N.; Miyoshi, M.; Kitano, H.; Satoh, H.; Nagato, Y. Mutations associated with floral organ number in rice. *Planta* **1996**, *198*, 627–633. [CrossRef] [PubMed]
71. Suzaki, T.; Ohneda, M.; Toriba, T.; Yoshida, A.; Hirano, H. *FON2 SPARE1* redundantly regulates floral meristem maintenance with *FLORAL ORGAN NUMBER2* in rice. *PLoS Genet.* **2009**, *5*, e1000693. [CrossRef] [PubMed]
72. Suzaki, T.; Yoshida, A.; Hirano, H.Y. Functional diversification of CLAVATA3-related CLE proteins in meristem maintenance in rice. *Plant Cell* **2008**, *20*, 2049–2058. [CrossRef] [PubMed]
73. Ohmori, Y.; Tanaka, W.; Kojima, M.; Sakakibara, H.; Hirano, H. *WUSCHEL-RELATED HOMEOBOX4* is involved in meristem maintenance and is negatively regulated by the CLE gene *FCP1* in rice. *Plant Cell* **2013**, *25*, 229–241. [CrossRef] [PubMed]
74. Nardmann, J.; Werr, W. The shoot stem cell niche in angiosperms: Expression patterns of *WUS* orthologues in rice and maize imply major modifications in the course of mono- and dicot evolution. *Mol. Biol. Evol.* **2006**, *23*, 2492–2504. [CrossRef] [PubMed]
75. Tanaka, W.; Ohmori, Y.; Ushijima, T.; Matsusaka, H.; Matsushita, T.; Kumamaru, T.; Kawano, S.; Hirano, H. Axillary meristem formation in rice requires the *WUSCHEL* ortholog *TILLERS ABSENT1*. *Plant Cell* **2013**, *27*, 1173–1184. [CrossRef] [PubMed]
76. Doebley, J. The genetics of maize evolution. *Annu. Rev. Genet.* **2004**, *38*, 37–59. [CrossRef] [PubMed]
77. Jackson, D.; Hake, S. The genetics of ear fasciation in maize. *Maize Genet. Coop. Newsl.* **1999**, *73*, 2.
78. Bommert, P.; Nagasawa, N.S.; Jackson, D. Quantitative variation in maize kernel row number is controlled by the FASCIATED EAR2 locus. *Nat. Genet.* **2013**, *45*, 334–337. [CrossRef] [PubMed]
79. Bommert, P.; Je, B.I.; Goldschmidt, A.; Jackson, D. The maize Gα gene *COMPACT PLANT2* functions in *CLAVATA* signaling to control shoot meristem size. *Nature* **2013**, *502*, 555–558. [CrossRef] [PubMed]
80. Wu, Q.; Regan, M.; Furukawa, H.; Jackson, D. Role of heterotrimeric Gα proteins in maize development and enhancement of agronomic traits. *PLoS Genet.* **2018**, *14*, e1007374. [CrossRef] [PubMed]
81. Je, B.I.; Gruel, J.; Lee, Y.K.; Bommert, P.; Arevalo, E.D.; Eveland, A.L.; Wu, Q.; Goldschmidt, A.; Meeley, R.; Bartlett, M.; et al. Signaling from maize organ primordia via FASCIATED EAR3 regulates stem cell proliferation and yield traits. *Nat. Genet.* **2016**, *48*, 785–791. [CrossRef] [PubMed]
82. Liu, L.; Du, Y.; Huo, D.; Wang, M.; Shen, X.; Yue, B.; Qiu, F.; Zheng, Y.; Yan, J.; Zhang, Z. Genetic architecture of maize kernel row number and whole genome prediction. *Theor. Appl. Genet.* **2015**, *128*, 2243–2254. [CrossRef] [PubMed]
83. Nardmann, J.; Werr, W. The invention of WUS-like stem cell-promoting functions in plants predates leptosporangiate ferns. *Plant Mol. Biol.* **2012**, *78*, 123–134. [CrossRef] [PubMed]
84. International Wheat Genome Sequencing Consortium. Shifting the limits in wheat research and breeding using a fully annotated reference genome. *Science* **2018**, *361*, eaar7191. [CrossRef] [PubMed]
85. Ma, X.; Zhang, Q.; Zhu, Q.; Liu, W.; Chen, Y.; Qiu, R.; Wang, B.; Yang, Z.; Li, H.; Lin, Y.; et al. A robust CRISPR/Cas9 system for convenient, high-efficiency multiplex genome editing in monocot and dicot plants. *Mol. Plant* **2015**, *8*, 1274–1284. [CrossRef] [PubMed]

© 2018 by the author. Licensee MDPI, Basel, Switzerland. This article is an open access article distributed under the terms and conditions of the Creative Commons Attribution (CC BY) license (http://creativecommons.org/licenses/by/4.0/).

Review
Organogenesis at the Shoot Apical Meristem

Jan Traas

Laboratoire de Reproduction et Développement des Plantes, Université de Lyon, ENS de Lyon, UCBL, INRA, CNRS, 46 Allée d'Italie, 69364 Lyon CEDEX O7, France; Jan.Traas@ens-lyon.fr

Received: 1 November 2018; Accepted: 21 December 2018; Published: 28 December 2018

Abstract: Lateral organ initiation at the shoot apical meristem involves complex changes in growth rates and directions, ultimately leading to the formation of leaves, stems and flowers. Extensive molecular analysis identifies auxin and downstream transcriptional regulation as major elements in this process. This molecular regulatory network must somehow interfere with the structural elements of the cell, in particular the cell wall, to induce specific morphogenetic events. The cell wall is composed of a network of rigid cellulose microfibrils embedded in a matrix composed of water, polysaccharides such as pectins and hemicelluloses, proteins, and ions. I will discuss here current views on how auxin dependent pathways modulate wall structure to set particular growth rates and growth directions. This involves complex feedbacks with both the cytoskeleton and the cell wall.

Keywords: shoot meristem; morphogenesis; molecular regulation; cell wall; cytoskeleton

1. Introduction

Plants continuously make organs and tissues, thanks to the activity of meristems. Thus, the shoot meristems—at the tip of the stems and branches—initiate all the aerial parts, while the root meristems are responsible for the underground organs. The secondary meristems maintain the secondary growth of stems. I will focus here on lateral organ formation at the shoot apical meristem. Approaching the problem from a multi-scale perspective, we will discuss current evidence showing how molecular activity is translated into changes in geometry, while organs and tissues grow.

2. The Shoot Meristem: Molecular Regulation

The shoot meristem is a complex structure, divided into domains with specific functions [1]. At the tip of the meristematic dome is the so-called central zone, which contains the true stem cells. An intricate regulatory network determines the size and position of this population. At its heart is a signalling loop, which involves the transcription factor WUSCHEL (WUS), the receptor kinase (CLAVATA 1) CLV1, the receptor like protein CLV2 and the ligand CLV3 [2]. WUS is expressed in the so-called organizing centre at the base of the central zone, two or three cell layers deep. It activates CLV3 in the cells above and the ligand subsequently diffuses into the surrounding cells. Here, it interacts with the receptor complex CLV1/CLV2 to inhibit WUS. Many additional regulators have been identified, including partners of CLV1, components of diverse hormone signalling pathways, in particular cytokinin, as well as additional transcription factors active in other parts of the meristem. The meristem centre contains auxin, but there is evidence that proves it is not sensitive to the hormone [3]. Other non-elucidated interactions with meristem regulators such as SHOOTMERISTEMLESS also play a role. I will not discuss central zone regulation in further detail, but rather concentrate on what is happening at the periphery of the meristem when cells produced by the central zone enter differentiation.

Cell growth and division push certain daughters of central zone cells to the periphery. These cells are in principle pluripotent and their daughters will be incorporated in organs or stem tissues. A major molecular signalling network involved in cell differentiation at the periphery is auxin (see e.g., [3,4]).

The hormone is transported from cell-to-cell by membrane-localised transporters of the PIN family and accumulates at certain spots where it will launch the initiation of organ primordia. The importance of auxin transport in organ formation is illustrated by the phenotype of the *pin1* mutant in Arabidopsis [4]. This mutant is no longer able to transport auxin along its surface, and as a result forms naked inflorescence stems, unable to form flowers.

Auxin feeds into a complex regulatory molecular network. At the meristem, a range of transcriptional regulators is implicated in the early transduction cascade [3] that subsequently initiates further downstream events. In addition, cross talk with other signalling pathways, in particular that of cytokinin, is essential for correct organ initiation ([5–8] and references therein). Interestingly, many of the auxin-activated regulators are highly expressed at the periphery and only weakly in the meristem centre, although the auxin concentrations are high there [3]. This would suggest that auxin mainly acts in the peripheral zone. One of the main transcription factors activated directly by auxin is MONOPTEROS (MP) [9]. When MP is mutated, auxin can still accumulate, but organ formation is affected (see e.g., [4]). This is particularly striking at the inflorescence meristems, as the full knock-out *mp* mutant forms a naked, pin-like stem with very few or no flowers forming. An extensive analysis identified three other transcription factors as direct downstream targets of MP: AINTEGUMENTA (ANT), AINTEGUMENTA LIKE 6 (AIL6) and LEAFY (LFY) [10]. The triple *ant ail6 lfy* barely forms any organs, suggesting that all three genes are involved in organ outgrowth. Although this general model of auxin induced MP directly activating ANT/AIL6/LFY still stands, the triple mutant still produces some outgrowths that are still sensitive to auxin transport inhibitors, suggesting that other factors are involved [10].

Recent studies have revealed a more complex role of auxin in the more global coordination of meristem function. This involves transcription factors of the so-called APETALA 2 (AP2) family, DORNRÖSCHEN (DRN) and DORNRÖSCHEN-LIKE (DRNL) [11–15]. Both transcription factors are expressed in complementary domains at the SAM: DRN mainly at the central zone, and DRNL in the organ founder cells. Although this would suggest complementary roles, there is good evidence that both factors act synergistically in controlling CLV3 expression. Hereby, DRN directly binds the CLV3 promoter to positively regulate its expression. How DRNL affects CLV3 expression at a distance is not known at this stage [14]. Interestingly, DRN and DRNL, together with PUCHI, another transcription factor of the AP2 family, act synergistically in the control of floral organ number and even flower identity [12]. MP directly inhibits DRN at the peripheral zone. MP expression itself occurs along a gradient, with low expression at the meristem centre, thus allowing DRN to participate in the activation of CLV3 there [14]. In this manner, MP is also important in controlling the balance between meristem maintenance and organ formation at the periphery.

The regulators described above, only represent a very partial view of the molecular network. Other factors have been identified, and transcriptomic analysis has revealed that many genes are differentially regulated between the meristem centre and the periphery (e.g., [16]). The challenge for the future will be to produce a more complete, integrated model of the molecular network coordinating meristem function.

3. Translating Molecular Regulation into Changes in Geometry

So far, I have only considered the molecular regulation of meristem function. The next question is how this network of transcription factors and signalling molecules leads to the actual changes in shape we observe during organ outgrowth at the SAM. Growth is a physical process and the deformation of living tissues requires mechanical forces, which cause cells to grow at a certain rate and into a certain direction. We should therefore, not only look at morphogenesis from a geometrical point of view, but also consider the physical, structural components of the growing cells, in particular the extracellular matrix, called the cell wall. In the rapidly growing meristematic cells, these walls can be described as dense networks of cellulose fibres (microfibrils) cross-linked to a matrix that is largely composed of pectins and hemicelluloses (for reviews see: e.g., [17–20]. The matrix components can occur in different forms with different properties, defining their mechanical characteristics and capacity to bind to other wall elements.

The regulation of plant cell growth is closely linked to this cell wall structure ([18–22] and references therein). The cell walls are constantly under tension because of the internal turgor pressure. In addition, since the walls form a continuum linking the cells together, differences in growth rates between neighbouring cells can also influence the tensile forces acting on the individual walls. Together these forces form a tissue-wide stress field, causing the elastic deformation of the walls. According to widely accepted hypotheses, growth occurs when the cell walls yield to these forces and start to deform plastically. The yielding threshold depends on the degree of cross-linking between the wall components and can be modified, for instance, through the activity of wall-modifying enzymes. In the meristem, the major targets of wall-modifying enzymes are pectins and hemicelluloses [23]. The plastic deformation of the wall causes it to become thinner, which is compensated by synthesis and the insertion of new polymers. Whereas the overall growth rate largely depends on parameters like wall stiffness (the degree of cross-linking between the polymers) or wall synthesis, growth directions depend mostly on the orientation of the cellulose microfibrils, which restrict growth along their length. This orientation depends on the trajectories of the membrane bound cellulose synthases, which are guided by the microtubule cytoskeleton at the cell cortex [24,25].

In order to control morphogenesis, the molecular regulatory networks have to interfere with the local composition and texture of the cell wall. This process is conceptually simple, but in fact extremely complex and involves hundreds of wall-synthesizing and wall-modifying enzymes, often with redundant functions [26]. In principle, turgor pressure can also vary, but since little or nothing is known about its regulation at the shoot apex, it will not be further discussed here. In the following paragraphs, I will briefly summarize some of the current knowledge regarding the regulation of wall properties during growth at the shoot apical meristem.

4. Controlling Growth Rates at the Meristem

As indicated above, it is thought that growth rates are determined at the level of individual cells, largely by controlling wall stiffness and synthesis. Although we are only at the beginning of our understanding, there is strong evidence to suggest that local wall properties are very actively regulated during organ formation.

In an extensive analysis of over 150 enzymes involved in the synthesis of wall polymers, Yang and colleagues (2016) [27] found that most of them showed distinct patterns at the shoot meristem with a striking difference between the meristem proper and the young outgrowing organs. Armezzani et al. (2018) [23] also described strong differences in the expression of wall-modifying enzymes, in particular Expansins and XTHs, which in principle target hemicellulose and have the capacity to change wall stiffness.

How the expression of these genes is controlled is not precisely known, although a range of cell wall modifying enzymes have been identified as putative targets of meristem expressed transcription factors ([28,29]). Peaucelle and colleagues also identified potential roles of pectin modifications in organ outgrowth [30–33]. Pectin gels can be stiffer or looser depending on the degree of cross-linking of the individual polymers by Ca^{2+}. Transgenic plants showing modified levels of specific forms of pectin show a dramatic reduction or increase in organ formation. In contrast to what these results might suggest, the intense activity of wall modifying genes does not lead to dramatic changes in wall mechanics. Measurements using atomic force microscopy have shown that wall stiffness is reduced during organ formation, but this remains within a limit of 20–50% at most [34].

5. Controlling Growth Directions at the Meristem

What about growth directions? Although differences in stiffness between individual walls can be involved [35] there is a general consensus that growth directions are mostly determined by the anisotropic properties of the cellulose network. If most microfibrils are aligned in one particular direction, they will restrict growth in that direction. As said above, microfibril orientation is regulated by the microtubule network, which guides the cellulose synthase complexes in the membrane. Accordingly, microtubule arrays are often (but not always) very precisely aligned perpendicular to the main growth direction [25]. How are these arrangements controlled? Since microtubule dynamics is not the main topic here, I will only give a very short overview, and highlight two general non-exclusive hypotheses—linked to the capacity of microtubules to self-organize into bundles. This depends in principle on a limited set of basic properties, such as polymerization/depolymerization, alignment ('zippering') and severing (cutting) [36], which involves an extensive set of associated proteins. The first hypothesis proposes cell geometry as an important organising factor [36–39]. Since microtubules and especially microtubule bundles are relatively stiff, they do not easily bend around the sharp cell corners in the small meristematic cells. In addition, the obstacles formed by these corners can affect microtubule stability and cause rapid depolymerisation. Therefore, cell geometry might play a significant role in microtubule organisation. This does not explain, however, why microtubules can show coherent alignment in neighbouring cells with sometimes very different shapes. We will consider the second hypothesis, which proposes that microtubules align along mechanical stresses [40] in somewhat more detail. The general idea here is that tissue-wide stress patterns generated by turgor pressure and differential growth (rapidly growing tissues 'pulling' on the more slowly growing ones) provide directional cues to the cytoskeleton. This generates a negative feedback loop, where the microtubules align the cellulose microfibrils along the main stress direction, thus causing the cells to resist the forces in that direction. Mechanical models show that in principle this should be sufficient to generate basic shapes such as cylindrical stems or dome shaped structures [40]. Evidence comes from work on the shoot apical meristem, where strong correlations between predicted stress patterns and microtubule alignments are found. Evidence also comes from hypocotyls and experiments where the stress patterns are perturbed, for instance using ablation or by applying external constraints [40,41]. This stress-based hypothesis for microtubule alignment provides a straightforward explanation for the coordinated behaviour of the structural elements in neighbouring cells. Although a mechanism involved in translating stress patterns into microtubule alignments has remained elusive, there are a number of interesting indications of how this could work. First of all, the direction of microtubule movements driven by motor proteins on artificial substrates in vitro is sensitive to stress, although the effects of this property in the living cell remains to be established [42]. In the context of morphogenesis at the shoot meristem, KATANIN (KTN), a protein involved in microtubule dynamics, stands out [43]. KTN is a so-called microtubule severing protein that destabilises local interactions between tubulin molecules. This supposedly promotes partial microtubule disassembly, efficient movement and, in rapidly growing plant cells, favours microtubule alignment. Interestingly, in mutants where KTN is impaired, the microtubule arrays are less organised and show a decreased capacity to align along predicted force patterns, even during mechanical perturbation [43]. Importantly, KTN directly interacts with RHO INTERACTING CRIB CONTAINING PROTEIN 1 (RIC1), which in turn interacts with RHO in PLANTS 6 (ROP6), thus potentially linking KTN function to cellular signalling [44]. Activation of the ROP pathway itself has been associated with auxin signal transduction, but how auxin is precisely perceived in this context remains a matter of debate [45].

How does the cytoskeleton behave at the shoot apical meristem? At the very tip of the meristem, microtubules mostly occur in isotropic (disorganised), dynamic networks. Towards the periphery they become highly anisotropic (organised, aligned) and the cells form tissue-wide microtubule arrangements surrounding the meristematic dome. This is particularly evident in organ boundaries [40,43]. As mentioned above, these supracellular arrangements correspond also qualitatively to the predicted stress pattern at the meristem surface. Important changes in these concentric patterns occur during organ formation. Soon after auxin accumulates, the microtubule arrays disorganise to become fully isotropic. In the context of the mechanical feedback hypothesis, this can be interpreted as the local inactivation of this feedback. The effect of auxin on the microtubules is thought to be a relatively direct effect, potentially involving ROP signalling [34].

Importantly, it seems to be sufficient to disorganise the microtubule arrays at the periphery to cause outgrowth, as drug treatments or mutations affecting microtubule alignments also lead to the formation of ectopic outgrowths or bulges on the meristem [34]. Mutations in KTN and treatment with the microtubule depolymerising drug, Oryzalin, even induce the formation of organs in the absence of auxin accumulation in the *pin1* mutant. Mechanical models have shown, that this shift to isotropic microfibril deposition could act in synergy with the relatively limited reduction in wall stiffness described above to induce rapid primordium outgrowth [23,34].

6. Not that Simple: Some Open Questions

A scenario emerges, where auxin accumulation through transport activates downstream transcriptional regulation, leading to the activation of certain wall-modifying or synthesizing enzymes and a slight reduction in wall stiffness. In parallel, auxin—potentially via a KTN based signalling cascade—causes an inactivation of the mechanical feedback on microtubules (Figure 1). This leads to the disorganisation of the microtubule arrays and a switch to the isotropic deposition of cellulose microfibrils. Together these two effects of auxin act in synergy to cause the organ to bulge out, driven by turgor pressure.

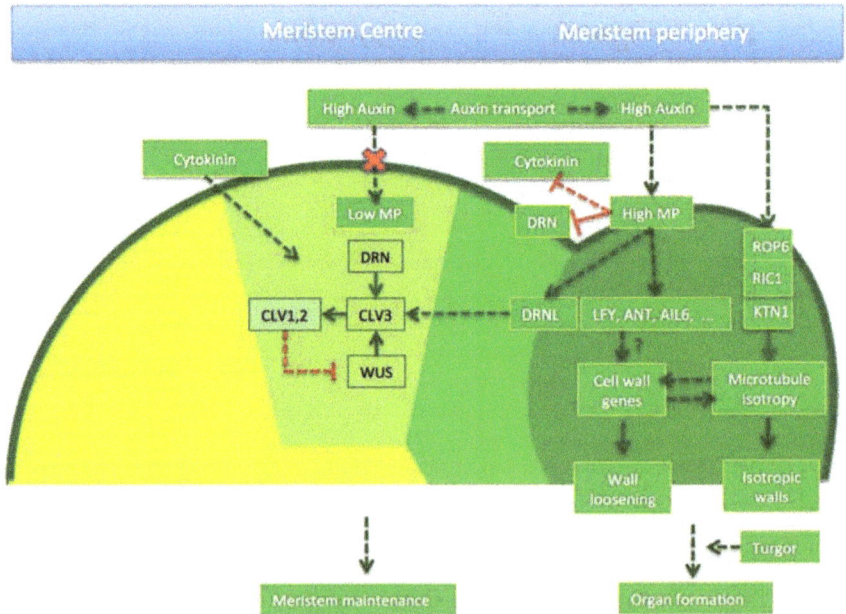

Figure 1. Molecular and cellular regulation of organ initiation at the periphery. Auxin transport generates auxin maxima at the meristem centre (light green area) and periphery (darker green), but since the centre is relatively insensitive to auxin (red cross), its effects seem to be limited and cytokinin driven meristem maintenance dominates. Auxin at the periphery causes wall loosening and cell isotropy. This involves both transcriptional and cellular responses. Depending on their wall properties, cells will then grow at particular rates and in particular directions, driven by turgor pressure. Dotted arrows represent indirect effects, solid lines direct, molecular relationships. Green arrows stand for positive control and red lines for inhibitions.

This scenario leaves many questions open regarding the molecular players or the cellular signalling cascades involved. The precise changes in composition and mechanics of the cell wall during organ formation also remain almost a complete unknown. For the sake of simplicity, I have mainly discussed auxin here as an upstream regulator. However, there is strong evidence to support the idea that the localisation of auxin transporters is influenced by cell wall properties [46,47], pointing to the existence of some type of feedback towards signalling, which remains not understood at all. Here, I would like to highlight the following two points that are of particular interest (see Figure 2 for overview).

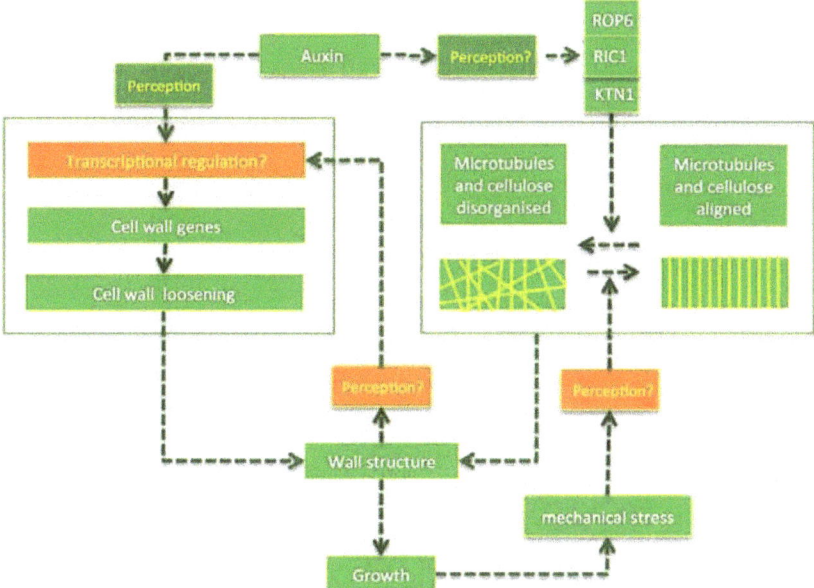

Figure 2. Some open questions. High auxin concentrations caused by auxin transport affects the cell wall structure in two manners during organ outgrowth: wall loosening (box at the left) and microtubule/ microfibril organisation (pictured in the box on the right). Wall loosening involves transcriptional regulation. High auxin concentrations also promote a disorganisation of the microtubules, probably via a ROP/KTN based pathway (see also Figure 1), although this remains to be established. There is strong evidence that the cells perceive wall properties and mechanical stress and feed this information back to transcription and the cytoskeleton. Mechanical stress, for example, promotes microtubule alignment, while changes in wall anisotropy induce transcriptional responses. It is not known how wall structure and mechanical stress are perceived and transduced. Green arrows indicate positive regulation; orange boxes refer to the poorly understood processes that are discussed in the text.

The first point concerns transcriptional regulation. As mentioned above, the presence of isotropic microtubule arrays at the meristem periphery is sufficient to cause organ outgrowth [34]. This outgrowth can even lead to the formation of flower-like structures in the absence of auxin transport as in *pin1 ktn* mutants. Importantly, this involves for example the transcriptional activation of cell wall modifying enzymes [23]. Therefore, a local switch to the isotropic deposition of cellulose fibres can also have effects on transcription and activate certain transcription factors required for flower formation and wall modification, even in the absence of high concentrations of auxin. In other words, there seems to be a feedback from the cytoskeleton to transcriptional regulation. How this works is completely unknown. In this context, it is worth noting that a range of membrane bound receptors have been associated with wall related signalling [48–50]. These receptors could potentially sense the mechanical status of the cell wall. This could even involve the direct binding of particular wall components such as pectins.

The second point of interest worth highlighting, concerns the mechanical feedback itself. As discussed, a number of components potentially involved in directional mechano-sensing have been identified. In addition, there is a strong correlation between microtubule alignment and predicted force patterns. To date, it is the only possible directional signal that coincides at least qualitatively with MT alignments at the meristem. Nevertheless, a negative feedback where microtubules align along the main force direction and cause the cells to resist to this direction leaves us with a fundamental contradiction. In principle, movement (strain) must be at the basis of force sensing. By reinforcing the wall along the main force direction, the microtubules also cause the cell to grow (i.e., to move) in a different direction. In other words, the main movement is no longer in the direction of the main force. Why is this movement not sensed by the microtubule arrays? How can they sense the main stress direction and not react to strain? The answer to this question is not yet known, but the evidence indicates that the effect of stress on microtubules must be indirect.

Funding: This research received no external funding.

Conflicts of Interest: The author declares no conflict of interest.

References

1. Gaillochet, C.; Daum, G.; Lohmann, J.U. O cell, where art thou? The mechanisms of shoot meristem patterning. *Curr. Opin. Plant. Biol.* **2015**, *23*, 91–97. [CrossRef] [PubMed]
2. Schoof, H.; Lenhard, M.; Haecker, A.; Mayer, K.F.; Jurgens, G.; Laux, T. The stem cell population of Arabidopsis shoot meristems in maintained by a regulatory loop between the CLAVATA and WUSCHEL genes. *Cell* **2000**, *100*, 635–644. [CrossRef]
3. Vernoux, T.; Brunoud, G.; Farcot, E.; Morin, V.; Van den Daele, H.; Legrand, J.; Oliva, M.; Das, P.; Larrieu, A.; Wells, D.; et al. The auxin signalling network translates dynamic input into robust patterning at the shoot apex. *Mol. Syst. Biol.* **2011**, *7*, 508. [CrossRef] [PubMed]
4. Reinhardt, D.; Pesce, E.R.; Stieger, P.; Mandel, T.; Baltensperger, K.; Bennett, M.; Traas, J.; Friml, J.; Kuhlemeier, C. Regulation of phyllotaxis by polar auxin transport. *Nature* **2003**, *426*, 255–260. [CrossRef] [PubMed]
5. Leibfried, A.; To, J.P.C.; Busch, W.; Stehling, S.; Kehle, A.; Demar, M.; Kieber, J.J.; Lohmann, J.U. WUSCHEL controls meristem function by direct regulation of cytokinin-inducible response regulators. *Nature* **2005**, *438*, 1172–1175. [CrossRef] [PubMed]
6. Zhao, Z.; Andersen, S.U.; Ljung, K.; Dolezal, K.; Miotk, A.; Schultheiss, S.J.; Lohmann, J.U. Hormonal control of the shoot stem-cell niche. *Nature* **2010**, *465*, 1089-U1154. [CrossRef] [PubMed]
7. Gordon, S.P.; Chickarmane, V.S.; Ohno, C.; Meyerowitz, E.M. Multiple feedback loops through cytokinin signaling control stem cell number within the Arabidopsis shoot meristem. *Proc. Natl. Acad. Sci. USA* **2009**, *106*, 16529–16534. [CrossRef]
8. Besnard, F.; Refahi, Y.; Morin, V.; Marteaux, B.; Brunoud, G.; Chambrier, P.; Rozier, F.; Mirabet, V.; Legrand, J.; Laine, S.; et al. Cytokinin signalling inhibitory fields provide robustness to phyllotaxis. *Nature* **2014**, *505*, 417. [CrossRef]
9. Hardtke, C.S.; Berleth, T. The Arabidopsis gene MONOPTEROS encodes a transcription factor mediating embryo axis formation and vascular development. *EMBO J.* **1998**, *17*, 1405–1411. [CrossRef]
10. Yamaguchi, N.; Wu, M.F.; Winter, C.M.; Berns, M.C.; Nole-Wilson, S.; Yamaguchi, A.; Coupland, G.; Krizek, B.A.; Wagner, D. A Molecular Framework for Auxin-Mediated Initiation of Flower Primordia. *Dev. Cell* **2013**, *24*, 271–282. [CrossRef]
11. Chandler, J.W.; Cole, M.; Flier, A.; Grewe, B.; Werr, W. The AP2 transcription factors DORNROSCHEN and DORNROSCHEN-LIKE redundantly control *Arabidopsis* embryo patterning via interaction with PHAVOLUTA. *Development* **2007**, *134*, 1653–1662. [CrossRef] [PubMed]
12. Chandler, J.W.; Werr, W. DORNROSCHEN, DORNROSCHEN-LIKE, and PUCHI redundantly control floral meristem identity and organ initiation in *Arabidopsis*. *J. Exp. Bot.* **2017**, *68*, 3457–3472. [CrossRef] [PubMed]
13. Cole, M.; Chandler, J.; Weijers, D.; Jacobs, B.; Comelli, P.; Werr, W. DORNROSCHEN is a direct target of the auxin response factor MONOPTEROS in the *Arabidopsis* embryo. *Development* **2009**, *136*, 1643–1651. [CrossRef] [PubMed]

14. Luo, L.; Zeng, J.; Wu, H.; Tian, Z.; Zhao, Z. A Molecular Framework for Auxin-Controlled Homeostasis of Shoot Stem Cells in *Arabidopsis*. *Mol. Plant* **2018**. [CrossRef] [PubMed]
15. Traas, J. Molecular Networks Regulating Meristem Homeostasis. *Mol. Plant* **2018**, *11*, 883–885. [CrossRef] [PubMed]
16. Yadav, R.K.; Girke, T.; Pasala, S.; Xie, M.T.; Reddy, V. Gene expression map of the Arabidopsis shoot apical meristem stem cell niche. *Proc. Natl. Acad. Sci. USA* **2009**, *106*, 4941–4946. [CrossRef] [PubMed]
17. Cosgrove, D.J. Plant expansins: Diversity and interactions with plant cell walls. *Curr. Opin. Plant. Biol.* **2015**, *25*, 162–172. [CrossRef]
18. Cosgrove, D.J. Plant cell wall extensibility: Connecting plant cell growth with cell wall structure, mechanics, and the action of wall-modifying enzymes. *J. Exp. Bot.* **2016**, *67*, 463–476. [CrossRef]
19. Cosgrove, D.J. Catalysts of plant cell wall loosening. *F1000Research* **2016**, *5*. [CrossRef]
20. Ali, O.; Traas, J. Force-Driven Polymerization and Turgor-Induced Wall Expansion. *Trends Plant Sci.* **2016**, *21*, 398–409. [CrossRef]
21. Boudaoud, A. An introduction to the mechanics of morphogenesis for plant biologists. *Trends Plant Sci.* **2010**, *15*, 353–360. [CrossRef] [PubMed]
22. Mirabet, V.; Das, P.; Boudaoud, A.; Hamant, O. The role of mechanical forces in plant morphogenesis. *Annu. Rev. Plant Biol.* **2011**, *62*, 365–385. [CrossRef]
23. Armezzani, A.; Abad, U.; Ali, O.; Andres Robin, A.; Vachez, L.; Larrieu, A.; Mellerowicz, E.J.; Taconnat, L.; Battu, V.; Stanislas, T.; et al. Transcriptional induction of cell wall remodelling genes is coupled to microtubule-driven growth isotropy at the shoot apex in *Arabidopsis*. *Development* **2018**, *145*. [CrossRef] [PubMed]
24. Paredez, A.R.; Somerville, C.R.; Ehrhardt, D.W. Visualization of cellulose synthase demonstrates functional association with microtubules. *Science* **2006**, *312*, 1491–1495. [CrossRef] [PubMed]
25. Chan, J. Microtubule and cellulose microfibril orientation during plant cell and organ growth. *J. Microsc.* **2012**, *247*, 23–32. [CrossRef] [PubMed]
26. Tucker, M.R.; Lou, H.; Aubert, M.K.; Wilkinson, L.G.; Little, A.; Houston, K.; Pinto, S.C.; Shirley, N.J. Exploring the Role of Cell Wall-Related Genes and Polysaccharides during Plant Development. *Plants (Basel)* **2018**, *7*, 42. [CrossRef] [PubMed]
27. Yang, W.; Schuster, C.; Beahan, C.T.; Charoensawan, V.; Peaucelle, A.; Bacic, A.; Doblin, M.S.; Wightman, R.; Meyerowitz, E.M. Regulation of Meristem Morphogenesis by Cell Wall Synthases in *Arabidopsis*. *Curr. Biol.* **2016**, *26*, 1404–1415. [CrossRef]
28. Yant, L.; Mathieu, J.; Dinh, T.T.; Ott, F.; Lanz, C.; Wollmann, H.; Chen, X.; Schmid, M. Orchestration of the floral transition and floral development in *Arabidopsis* by the bifunctional transcription factor APETALA2. *Plant Cell* **2010**, *22*, 2156–2170. [CrossRef]
29. Schlereth, A.; Moller, B.; Liu, W.; Kientz, M.; Flipse, J.; Rademacher, E.H.; Schmid, M.; Jurgens, G.; Weijers, D. MONOPTEROS controls embryonic root initiation by regulating a mobile transcription factor. *Nature* **2010**, *464*, 913–916. [CrossRef]
30. Braybrook, S.A.; Hofte, H.; Peaucelle, A. Probing the mechanical contributions of the pectin matrix: Insights for cell growth. *Plant Signal Behav.* **2012**, *7*, 1037–1041. [CrossRef]
31. Peaucelle, A.; Braybrook, S.A.; Le Guillou, L.; Bron, E.; Kuhlemeier, C.; Hofte, H. Pectin-induced changes in cell wall mechanics underlie organ initiation in *Arabidopsis*. *Curr. Biol.* **2011**, *21*, 1720–1726. [CrossRef] [PubMed]
32. Peaucelle, A.; Louvet, R.; Johansen, J.N.; Hofte, H.; Laufs, P.; Pelloux, J.; Mouille, G. Arabidopsis phyllotaxis is controlled by the methyl-esterification status of cell-wall pectins. *Curr. Biol.* **2008**, *18*, 1943–1948. [CrossRef] [PubMed]
33. Peaucelle, A.; Louvet, R.; Johansen, J.N.; Salsac, F.; Morin, H.; Fournet, F.; Belcram, K.; Gillet, F.; Hofte, H.; Laufs, P.; et al. The transcription factor BELLRINGER modulates phyllotaxis by regulating the expression of a pectin methylesterase in *Arabidopsis*. *Development* **2011**, *138*, 4733–4741. [CrossRef] [PubMed]
34. Sassi, M.; Ali, O.; Boudon, F.; Cloarec, G.; Abad, U.; Cellier, C.; Chen, X.; Gilles, B.; Milani, P.; Friml, J.; et al. An auxin-mediated shift toward growth isotropy promotes organ formation at the shoot meristem in *Arabidopsis*. *Curr. Biol.* **2014**, *24*, 2335–2342. [CrossRef] [PubMed]
35. Peaucelle, A.; Wightman, R.; Hofte, H. The Control of Growth Symmetry Breaking in the *Arabidopsis* Hypocotyl. *Curr. Biol.* **2015**, *25*, 1746–1752. [CrossRef] [PubMed]

36. Wasteneys, G.O.; Ambrose, J.C. Spatial organization of plant cortical microtubules: Close encounters of the 2D kind. *Trends Cell Biol* **2009**, *19*, 62–71. [CrossRef] [PubMed]
37. Ambrose, C.; Wasteneys, G.O. Nanoscale and geometric influences on the microtubule cytoskeleton in plants: Thinking inside and outside the box. *Protoplasma* **2012**, *249* (Suppl. S1), S69–S76. [CrossRef]
38. Chakrabortty, B.; Blilou, I.; Scheres, B.; Mulder, B.M. A computational framework for cortical microtubule dynamics in realistically shaped plant cells. *PLoS Comput. Biol.* **2018**, *14*, e1005959. [CrossRef]
39. Chakrabortty, B.; Willemsen, V.; de Zeeuw, T.; Liao, C.Y.; Weijers, D.; Mulder, B.; Scheres, B. A Plausible Microtubule-Based Mechanism for Cell Division Orientation in Plant Embryogenesis. *Curr. Biol.* **2018**, *28*, 3031–3043. [CrossRef]
40. Hamant, O.; Heisler, M.G.; Jonsson, H.; Krupinski, P.; Uyttewaal, M.; Bokov, P.; Corson, F.; Sahlin, P.; Boudaoud, A.; Meyerowitz, E.M.; et al. Developmental patterning by mechanical signals in *Arabidopsis*. *Science* **2008**, *322*, 1650–1655. [CrossRef]
41. Robinson, S.; Kuhlemeier, C. Global Compression Reorients Cortical Microtubules in Arabidopsis Hypocotyl Epidermis and Promotes Growth. *Curr. Biol.* **2018**, *28*, 1794–1802. [CrossRef] [PubMed]
42. Inoue, D.; Nitta, T.; Kabir, A.M.; Sada, K.; Gong, J.P.; Konagaya, A.; Kakugo, A. Sensing surface mechanical deformation using active probes driven by motor proteins. *Nat. Commun.* **2016**, *7*, 12557. [CrossRef] [PubMed]
43. Uyttewaal, M.; Burian, A.; Alim, K.; Landrein, B.T.; Borowska-Wykret, D.; Dedieu, A.; Peaucelle, A.; Ludynia, M.; Traas, J.; Boudaoud, A.; et al. Mechanical Stress Acts via Katanin to Amplify Differences in Growth Rate between Adjacent Cells in *Arabidopsis*. *Cell* **2012**, *149*, 439–451. [CrossRef]
44. Lin, D.S.; Cao, L.Y.; Zhou, Z.Z.; Zhu, L.; Ehrhardt, D.; Yang, Z.B.; Fu, Y. Rho GTPase Signaling Activates Microtubule Severing to Promote Microtubule Ordering in *Arabidopsis*. *Curr. Biol.* **2013**, *23*, 290–297. [CrossRef]
45. Xu, T.; Wen, M.; Nagawa, S.; Fu, Y.; Chen, J.G.; Wu, M.J.; Perrot-Rechenmann, C.; Friml, J.; Jones, A.M.; Yang, Z. Cell surface- and rho GTPase-based auxin signaling controls cellular interdigitation in *Arabidopsis*. *Cell* **2010**, *143*, 99–110. [CrossRef] [PubMed]
46. Braybrook, S.A.; Peaucelle, A. Mechano-chemical aspects of organ formation in *Arabidopsis thaliana*: The relationship between auxin and pectin. *PLoS ONE* **2013**, *8*, e57813. [CrossRef] [PubMed]
47. Tameshige, T.; Hirakawa, Y.; Torii, K.U.; Uchida, N. Cell walls as a stage for intercellular communication regulating shoot meristem development. *Front. Plant Sci.* **2015**, *6*, 324. [CrossRef] [PubMed]
48. Feng, W.; Kita, D.; Peaucelle, A.; Cartwright, H.N.; Doan, V.; Duan, Q.; Liu, M.C.; Maman, J.; Steinhorst, L.; Schmitz-Thom, I.; et al. The FERONIA Receptor Kinase Maintains Cell-Wall Integrity during Salt Stress through Ca(2+) Signaling. *Curr. Biol.* **2018**, *28*, 666–675. [CrossRef] [PubMed]
49. Hematy, K.; Sado, P.E.; Van Tuinen, A.; Rochange, S.; Desnos, T.; Balzergue, S.; Pelletier, S.; Renou, J.P.; Hofte, H. A receptor-like kinase mediates the response of Arabidopsis cells to the inhibition of cellulose synthesis. *Curr. Biol.* **2007**, *17*, 922–931. [CrossRef]
50. Wolf, S. Plant cell wall signalling and receptor-like kinases. *Biochem. J.* **2017**, *474*, 471–492. [CrossRef]

© 2018 by the author. Licensee MDPI, Basel, Switzerland. This article is an open access article distributed under the terms and conditions of the Creative Commons Attribution (CC BY) license (http://creativecommons.org/licenses/by/4.0/).

Review

Drawing a Line: Grasses and Boundaries

Annis E Richardson [1] and Sarah Hake [1,2,*]

1. Plant and Microbial Biology, University of California, Berkeley, CA 94720, USA; annisrichardson@berkeley.edu
2. USDA Plant Gene Expression Center, 800 Buchanan Street, Albany, CA 94710, USA
* Correspondence: hake@berkeley.edu; Tel.: +01-510-559-5907

Received: 3 November 2018; Accepted: 18 December 2018; Published: 25 December 2018

Abstract: Delineation between distinct populations of cells is essential for organ development. Boundary formation is necessary for the maintenance of pluripotent meristematic cells in the shoot apical meristem (SAM) and differentiation of developing organs. Boundaries form between the meristem and organs, as well as between organs and within organs. Much of the research into the boundary gene regulatory network (GRN) has been carried out in the eudicot model *Arabidopsis thaliana*. This work has identified a dynamic network of hormone and gene interactions. Comparisons with other eudicot models, like tomato and pea, have shown key conserved nodes in the GRN and species-specific alterations, including the recruitment of the boundary GRN in leaf margin development. How boundaries are defined in monocots, and in particular the grass family which contains many of the world's staple food crops, is not clear. In this study, we review knowledge of the grass boundary GRN during vegetative development. We particularly focus on the development of a grass-specific within-organ boundary, the ligule, which directly impacts leaf architecture. We also consider how genome engineering and the use of natural diversity could be leveraged to influence key agronomic traits relative to leaf and plant architecture in the future, which is guided by knowledge of boundary GRNs.

Keywords: grass; ligule; organogenesis; boundaries

1. Organogenesis

Organogenesis is the self-organizing process in which complex tissues arise from pluripotent progenitors and is common to all multicellular organisms. In plants, the process of organogenesis extends beyond embryogenesis, which enables them to continually produce organs. All aerial organs arise as relatively simple-shaped primordium on the periphery of the shoot apical meristem (SAM), which contains the pluripotent stem cells. The first molecular marker of organogenesis is the downregulation of class 1 *KNOTTED-LIKE HOMEOBOX* (*KNOX*) genes in the peripheral zone of the SAM [1–3]. This earliest stage of primordium growth is referred to as the P0, with the plastochron stage (P) as the time between successive primordium initiations.

The spacing of organ primordia around a SAM (the phyllotaxy) is self-organizing and highly robust. Phyllotaxy is determined by the distribution of the phytohormone auxin, which is influenced by the directional export of auxin by the PIN-FORMED transporters (PIN). This process is a self-organizing feedback loop, and the spacing between each primordium is predicted to be influenced by the size of the region of auxin depletion around the older primordium [4–10]. The formation of PIN1 convergence points in the SAM of the model eudicot plant *Arabidopsis thaliana* is essential for organ initiation [11–15]. This PIN1 convergence point leads to the formation of an auxin maximum and the subsequent downregulation of *KNOX* genes, which allows differentiation and outgrowth of organ primordia.

2. Boundaries and Plant Development

A fundamental step in organogenesis of multicellular organisms is the delineation of distinct populations of cells by forming boundaries. Boundary formation is essential for the function of the mature organ since it allows correct patterning and the segregation of different activities. In the case of vegetative development in plants, the formation of a boundary between the SAM and the incipient primordia is essential for both maintenance of the stem cell population and the correct shape of the mature organ [16]. This meristem/organ boundary allows for the separation of the cells that will become determinate and form the organ, while those that retain an indeterminate state maintain the meristem.

Meristem/organ boundaries are characterized by low division and expansion rates, parallel oriented microtubules, and relatively stiff cell walls [17]. These features contrast with the high cell division and cell expansion rates, low cell wall stiffness, and perpendicular oriented microtubules in the primordium tissue. The difference between the tissue properties of boundaries and the primordium generates conflict within the tissue, which allows the physical bulging of the primordium from the surface of the meristem [18–20]. The distribution of differentially growing regions can then generate distinct shapes [21]. Therefore, in addition to roles in separation of functionally different tissues, boundaries also contribute to organ shape through differential growth patterning [22].

Boundaries also form within the organ itself, delineating different tissues. These within-organ boundaries can have central roles in the final organ shape. For example, the juxtaposition of the abaxial and adaxial tissues in the leaf are essential for lamina outgrowth [23–25]. Within-organ boundaries can also be elaborated, contributing to morphological diversity. For example, stipules form at the base of the petiole in eudicot leaves such as peas [26]. Boundary regions can also be elaborated in mutants in response to ectopic gene expression. For example, ectopic *KNOTTED1* expression in the lemma/awn boundary in the barley *Hooded* mutant, results in the formation of a "hood" structure consisting of an ectopic floral meristem and triangular lateral outgrowths [27–31]. Similarly, ectopic *KNAT1* expression leads to meristems forming in the boundary regions of the lobed leaf [32].

2.1. The Boundary Gene Regulatory Network

Most of our understanding in how meristem/organ boundaries are defined has come from genetic studies in *Arabidopsis*. Of particular importance are mutants that have a fused organ phenotype, including *cup-shaped-cotyledon1/2/3 (cuc)*, *growth regulating factor (grf)*, and *lateral organ boundary (lob)* mutants [16,33–36], which highlighted key boundary genes. This body of work has shown that boundary specification requires a complex network of transcription factors, miRNAs, and hormone interactions summarized in Figure 1. Central players include the NAC domain transcription factors, *NO APICAL MERISTEM (NAM*, or *AtCUC1*, and *AtCUC3)*, which are regulated by miR164, and are part of a feedback network with the *KNOX* gene, *SHOOTMERISTEMLESS (STM)* [33–35,37–46]. The CUC transcription factors also directly regulate the expression of other boundary genes, such as *LIGHT-DEPENDENT SHORT HYPOCOTYLS 3* and *4 (LSH3* and *4)*, which are proposed to suppress organ differentiation [37]. Downstream of the *CUC* genes, *GRFs* are also expressed in the boundary, which play a role in the suppression of cell division and expansion [34].

Low growth rates in the boundary are also influenced by the spatial distribution of growth promoting hormones like auxin and brassinosteroids (BR) [47]. Both auxin and BR maintain higher levels in the meristem and developing primordia, and low levels in the boundary. Low auxin levels in the boundary are influenced by JAGGED LATERAL ORGANS (JLO) [39]. High BR in the primordium feeds back to regulate the spatial expression of the *CUC* genes, which limits them to the boundary domain. This inhibition is through BR promotion of *BRASSINAZOLE-RESISTANT 1 (BZR1)* expression, which inhibits *CUC* expression. Low BR levels in the boundary are influenced by the expression of *PHYB ACTIVATION TAGGED SUPPRESSOR1 (BAS1)*, which is a BR inactivating enzyme [35]. The expression of *BAS1* is regulated directly by the boundary gene *LOB1*, and BR can influence *LOB* expression forming a reinforcing feedback loop [35], restricting low BR to the boundary domain.

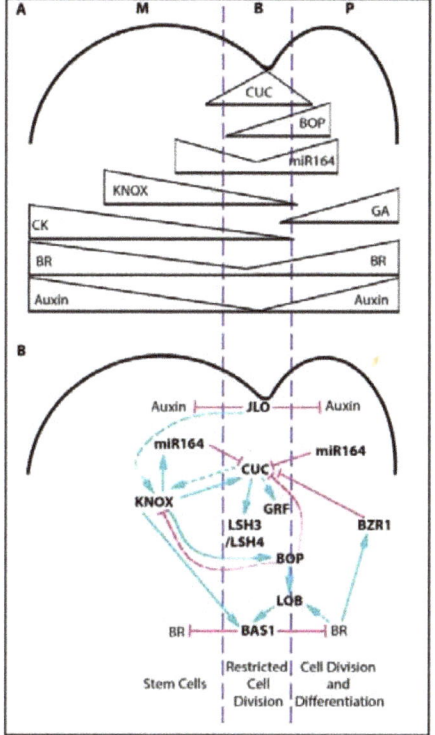

Figure 1. Regulation of meristem-organ boundaries in *Arabidopsis*. (**A**) Gradients of a selection of gene expression patterns and hormones across the meristem/organ boundary. (**B**) A summary of the gene regulatory network involved in meristem/organ boundary specification. Blue arrows indicate positive regulation while magenta lines indicate negative regulation. Solid lines represent direct regulation, dashed lines, indirect regulation. The dark blue lines delineate the meristem (M), boundary (B), and primordium (P) regions.

The boundary region is further refined through the activity of the ankyrin repeat proteins BLADE-ON-PETIOLE 1 and 2 (BOP1 and 2), which are localized to the base of the developing primordium. BOP activity results in the repression of *CUC* gene expression in the base of the primordium, and promotes the expression of *LOB* transcription factors [48,49].

Overall, these feedback networks maintain a clear boundary domain delineating the two distinct populations of cells (the meristem and the differentiating primordium) and spatially pattern distinct growth patterns.

2.2. The Boundary Gene Regulatory Network and Leaf Margin Development

The boundary gene regulatory network also has a role in the elaboration of leaf margin development, in particular, influencing serration and compound leaf development in eudicot systems including *Arabidopsis*, *Cardamine hirsuta*, tomato, and pea [50–55].

Despite the fact that many of the boundary components are shared between species, work in diverse eudicots has highlighted key differences in the network when it has been co-opted for margin development. For example, in tomato the *KNOX* gene *TKn1* is sufficient to initiate compound leaf formation. However, in peas *KNOX* genes do not have a role in compound leaf development and the pea ortholog of LEAFY (UNI) is sufficient to initiate compound leaf development [56,57]. LFY/UNI in peas allow formation of compound leaves by promoting indeterminacy in the margin, while LFY in inflorescences cause determinate growth in flowers. In tomatoes, gibberellic acid (GA) inhibits leaf complexity but, in peas, GA promotes leaf complexity [50]. This co-option of the boundary network in margin elaboration and the variations between eudicot species illustrates that different plant species can use the same regulators to induce opposite effects. Different eudicots also use specific factors to modulate leaf margin development. For example, the homeodomain protein RCO functions to inhibit growth in the boundary of developing leaflets in *Cardamine hirsuta*. RCO is specific to the core *Brassicacae* but was lost in *Arabidopsis* [58].

Given the profound effect that boundary specification can have on leaf shape and plant productivity, translating this research into crop species is vital. This translation is especially important when considering future aims of developing accurate computational models of crop growth and development to help predict the effects of a changing climate on crop productivity. In this paper, we review the current understanding of boundary specification during vegetative development, and the effects on leaf morphology in grass crops in comparison with eudicot models (Table 1).

Table 1. Glossary of studied related genes in *Arabidopsis*, maize, and barley mentioned in the review.

Arabidopsis	Maize	Barley	Description
SHOOTMERISTEMLESS (STM)	KNOTTED1 (KN1)	BARLEY KNOTTED 3 (BKn3)	KNOX transcription factor
CUP-SHAPED COTYLEDON 1,2,3 (CUC1, 2, 3)	NO APICAL MERISTEM 1 and 2 (NAM 1,2), CUC3		NAC domain transcription factor
BLADE ON PETIOLE 1 (BOP)	TASSELS REPLACE UPPER EARS 1 (TRU1) and TRU1-like	UNICULME4 (CUL4)	Ankyrin repeat domain protein
miR164 a/b/c	miR164 a/b/c/d/e/f/g/h		microRNA
PIN-FORMED 1 (PIN1)	PINFORMED 1a and 1b (PIN1a, PIN1b)		Auxin transporter
not present in *Arabidopsis*	SISTER OF PIN1 (SoPIN1)		Auxin transporter

3. Vegetative Organogenesis in Grass Crops

Most of the major food crops, including wheat, rice and maize, are members of the grass family (*Poaceae*) and are part of the monophyletic clade called the monocots. The monocots diverged from eudicot species 150 mya [59]. Monocots have distinct leaf shapes, generally sharing an ensheathing leaf base and parallel venation. These shape differences between monocots and eudicots are clear from the earliest stages of organogenesis (Figure 2A–H).

Unlike eudicot models in which the P0 is a point on the SAM that grows out to form a peg-like outgrowth (Figure 2F–G), in the grasses, the leaf P0 encircles the SAM (Figure 2B–C), and is referred to as the disc of insertion [60,61]. This disc of insertion forms the ensheathing leaf base. Each successive leaf base encircles both the meristem and all younger leaves, forming whorls containing a single leaf (Figure 2C). Like eudicot models, auxin accumulation followed by the downregulation of class 1 *KNOX* genes [1] is central to organ initiation in the grasses (Figure 2C). When auxin signaling is disrupted, as is the case when maize SAMs are treated with the auxin inhibitor NPA (N-1-naphthylphthalamic acid), organ initiation and *KNOX* downregulation is halted [62]. Auxin signaling is, therefore, central to recruitment of cells into the primordia in grasses, which is similar to eudicots.

Auxin maxima are formed by convergence points of PIN1 in *Arabidopsis* meristems. In contrast to *Arabidopsis*, the grass model, *Brachypodium distachyon*, has two *PIN1* orthologues known as *PIN1a* and *PIN1b*, and a sister clade to *PIN1*, *SISTER OF PIN1* (*SoPIN1*), with each showing sub-functionalization. This is independent of transcriptional control. PIN1a and PIN1b accumulate in the vasculature and the *pin1a/pin1b* double mutant has short internodes. SoPIN1 forms convergence points in the inflorescence meristem and the mutant has organ initiation defects similar to the *Arabidopsis pin1* mutant [63,64]. The *SoPIN1* clade is not unique to the grasses and is found in eudicots, including *Medicago truncatula* and tomato, but was lost in the *Brassicaceae* family. Mutants in the *SoPIN1* clade in

Medicago and tomato (*entire2*) show pleiotropic effects including defects in leaf development [65,66], which suggests that *SoPIN1* could have a role in grass leaf development even though it was not reported for *Brachypodium* [63].

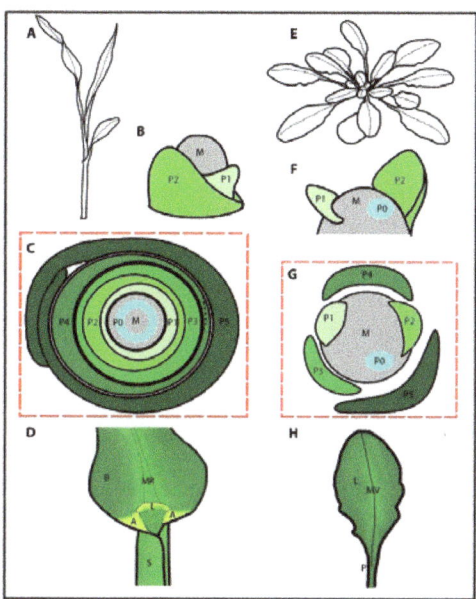

Figure 2. Grasses have distinct leaf and primordium shapes. A comparison of the morphology of the grasses versus *Arabidopsis* during vegetative development. (**A**) A cartoon of a grass seedling. (**B**) A cartoon representation of a grass vegetative meristem (M) with the first and second leaf primordia shown (P1 and P2) encircling the meristem. (**C**) A cartoon of a transverse cross section through a grass seedling, showing how each successive leaf (P1–P5) encircles the meristem and the younger leaves. The P0 is the region of *KNOX* gene expression down-regulation, which forms a ring. The sheath margin boundaries are not defined until P4 (dotted line) after which the margins are separate (P5). (**D**) A cartoon of the blade/sheath boundary in a mature grass leaf, depicting the blade (B), midrib (MR), ligule (L), auricles (A), and the sheath (S). (**E**) A cartoon of an *Arabidopsis* plant during vegetative growth. (**F**) A cartoon representation of an *Arabidopsis* vegetative meristem (M) with the first and second leaf primordia shown (P1 and P2), which do not encircle the meristem. (**G**) A cartoon of a transverse cross section through an *Arabidopsis* seedling, showing each successive leaf (P1-P5). (**H**) A cartoon of a mature *Arabidopsis* leaf, depicting the lamina (L), midvein (MV), and the petiole (P).

Once the disc of insertion has been specified, the ring-shaped P1 primordium (Figure 2B), goes on to develop into a grass leaf with a modular structure (Figure 2A). The wrapped lower leaf region known as the sheath provides structural support. The middle hinge region regulates a leaf angle and develops two distinct structures including the ligule, which is a fringe of tissue proposed to act as a sliding gasket, and two triangular auricle regions at the leaf margin, which influence the leaf angle (Figure 2D). The upper region known as the blade bends away from the plant and intercepts light. The interaction between these three regions influences plant height and the leaf angle, which has significant impacts on plant productivity [67–75]. These traits are of high agronomic importance since they can directly affect the yield of an individual plant, and the yield of an entire field when planting density is taken into consideration.

Clonal sector analyses of grass leaf development have shown that during the earliest stages of leaf primordium development from P0 to P3, only the blade forms. Then from P3–P4, the sheath margins arise from an overlapping region in the disc of insertion [61,76] and the ligule/auricle region begins to differentiate [77]. Therefore, during the earliest stages of grass leaf development, three different boundaries need to be specified for correct leaf shape; the ring-shaped meristem/organ boundaries (which go on to become the boundaries between the leaf whorls), the intra-whorl boundaries (the boundary between the sheath margins) and a within-organ boundary (the boundary between the sheath and blade where the ligule and auricle form) (Figure 2C–D).

4. Boundary Specification in Grass Crops

4.1. Meristem/Organ Boundaries in the Grasses

Meristem/organ boundaries in the grasses form an encircling ring (Figure 2B,C). Without correct specification of this boundary, delineation between the stem cells (the SAM) and the differentiating primordium cells fails to occur. A lack of separation between differentiating and pluripotent cells results in the termination of the meristem, as observed in mutants such as the *cupuliformis* in *Antirrhinum*, *cuc1/2/3* in *Arabidopsis*, and *nam* in Petunia, which all have mutations in the *NAC* domain transcription factor family *NAM/CUC3* [16,36,78,79]. These classic meristem-organ boundary mutants also develop fused leaves and floral organs due to the lack of *KNOX* gene downregulation in the organ boundaries. Thus, highlighting the role of boundaries in maintaining the separation between successive whorls of organs as well as maintaining the meristem. Conversely, meristem-like activity can spread into the leaf if the boundary is not maintained. In the case of the *blade-on-petiole* (*bop1*) mutant in *Arabidopsis*, *KNOX* gene activity, which is indicative of a meristem-like identity, spreads into the leaf base resulting in the formation of ectopic leaf tissue [80].

Grass genomes have representatives of the core genetic elements in the *Arabidopsis* meristem-organ boundary regulatory network (Table 2). For example, the *NAM* and *CUC3* genes, and the core miRNA164-NAM module likely predates the monocot/eudicot split [81–83]. Rice has one representative of NAM (Os06g0267500) and CUC3 (Os08g0511200), but maize, like *Arabidopsis*, has two NAM genes (GRMZM2G139700 and GRMZM2G393433) and one CUC3 (GRMZM2G430522), which illustrates gene duplication of the NAM family outside of the Brassicas [51]. The expression pattern of ZmCUC3 in maize lateral organs and the SAM mirrors that in eudicots, although the patterns of ZmCUC3 and the ZmNAM1/2 genes during embryo development differ [81]. Similarly, maize has a recent duplication of the *AtBOP1/2* genes called *TASSELS REPLACE UPPER EARS 1* (*ZmTRU1*, GRMZM2G039867) and *TRU1-like* (*ZmTRL1*, GRMZM2G060723), whereas rice has only a single gene, *OsBOP* (Os01g72020) [84].

Unlike the *NAM/CUC3* and *BOP* genes, some of the gene families implicated in meristem-organ boundary specification are enlarged in the grasses. For example, where there are three members of the miR164 family in *Arabidopsis* (miR164a,b,c), there are six reported in rice (miR164a,b,c,d,e,f) and eight in maize (miR164a,b,c,d,e,f,g,h) [85]. In *Arabidopsis*, the three miR164 family members are functionally redundant but exhibit expression domain differences suggesting some sub-functionalization [86]. miR164b has a role in regulating NAC transcription factor expression during lateral root formation in maize, which indicates a function in patterning lateral outgrowths [87]. The roles of other miRNA164 family members in grasses are yet to be elucidated, especially considering the large size of the NAC transcription family (for example, rice has 149 members) [88]. This expansion of gene families may provide the opportunity for sub-functionalization of key boundary regulatory genes in the grasses.

Forward genetic screens in grasses have identified mutants with tube and fused leaves, which could be indicative of mutations in meristem/organ boundary regulation genes. So far, these mutants with fused leaf phenotypes such as rice *onion-1*, *2*, and *3*, maize *adherant1*, and *fused leaves1* (*fdl1*), have defects in epidermal wax deposition and are not associated with any of the canonical boundary regulatory genes, such as *NAM* or *CUC3* genes [89–94]. The lack of *nam/cuc3* family mutants could suggest functional redundancy in the grass family, or that the mutations are embryo lethal, which implies that the leaf phenotype cannot be observed.

Table 2. Glossary of grass gene names. Where appropriate, the activity relevant to this review is highlighted.

Gene Name	Species	Description
KNOTTED 1 (KN1)	Maize	KNOX Transcription Factor, meristem identity
NO APICAL MERISTEM 1 and 2 (NAM 1,2), CUC3	Maize	NAC domain, transcription factor, expressed in boundary domains
TASSELS REPLACE UPPER EARS 1 (TRU1) and TRU1-like	Maize	Ankyrin repeat domain protein expressed in the sheath and in axillary meristems.
PINFORMED 1a and 1b (PIN1a, PIN1b)	Maize	Auxin transporter
SISTER OF PIN1 (SoPIN1)	Maize	Auxin transporter
RAMOSA 2 (RA2)	Maize	Lateral organ boundary domain transcription factor, involved in axillary meristem development.
SPARSE INFLORESENCE 1 (SPI1)	Maize	YUCCA gene, auxin biosynthesis.
NARROWSHEATH 1 and 2	Maize	WOX genes, involved in leaf development
LIGULELESS1 (LG1)	Maize	Squamosa Binding Protein transcription factor, involved in ligule development.
LIGULELESS2 (LG2)	Maize	BZIP/DOG domain transcription factor, involved in ligule development.
LIGULELESS NARROW (LGN)	Maize	Serine-threonine kinase, involved in ligule development.
LIGULELESS3 (LG3)	Maize	KNOX transcription factor, ectopic expression of LG3 induces ectopic blade/sheath boundaries.
LIGULELESS4 (LG4)	Maize	KNOX transcription factor, ectopic expression of LG4 induces ectopic blade/sheath boundaries.
GNARLEY4 (GN4)	Maize	KNOX transcription factor, ectopic expression of LG4 induces ectopic blade/sheath boundaries
WAVY AURICLES IN BLADE 1 (WAB1)	Maize	TCP transcription factor, ectopic expression of WAB1 induces ectopic blade/sheath boundaries
BEL1-like homeodomain 12 and 14 (BEL12/14)	Maize	BEL1-like homeodomain transcription factors, expressed in the developing ligule
BRASSINOSTEROID INSENSITIVE 1 (BRI1)	Maize	Brassinosteroid receptor, involved in auricle development and leaf angle
BRASSINOSTEROID-DEFICIENT DWARF1 (BRD1)	Maize	Brassinosteroid C6-oxidase, involved in brassino-steroid synthesis, expressed in the base of leaves. Involved in ligule and auricle development.
BETA-D-GULCOSIDASE 1 (GLU1)	Maize	Expressed in developing ligules
UNICULME4 (CUL4)	Barley	Ankyrin repeat domain protein, expressed in the sheath and involved in ligule development
ELIGULUM A (ELIA)	Barley	RNase H domain protein, involved in ligule development

Although no *nam* or *cuc3* mutants have been reported in the grasses, mutants in the orthologues of several boundary genes are known. Two orthologues of *AtBOP1* are found in maize known as *TRU1* and *TRL1*. The maize *tru1* mutant does not have a leaf phenotype, although the protein accumulates in an interesting sheath pattern [84]. *ZmTRL1* has no reported mutant phenotype. The two genes may be partially redundant with respect to vegetative organ boundary specification. In barley, the *AtBOP1* orthologue *HvCUL4*, has a defect in leaf development, with the *cul4* mutant showing a displacement of ligule/ auricle tissue [95]. This may mirror the displacement of distal identities within the proximal tissue observed in *Arabidopsis*.

Arabidopsis *lob* mutants have fusions between cauline leaves and branches but normal vegetative organs. *LOB* is expressed at the base of lateral organs and plays a role in negatively regulating BR signaling in boundaries [33,35]. Double and triple mutant analysis of the homologues of *AtLOB* show no additional phenotypes, but expression analysis highlights distinct expression patterns, suggesting sub-functionalization based on changes in the expression pattern rather than in a coding sequence [96]. In maize, the function of one homolog of *AtLOB* has been examined so far, *RAMOSA2* (*RA2*). RA2 regulates axillary meristem formation during inflorescence development. However, there are no reported organ fusion phenotypes in the *ramosa2* mutant, which contrasts with the AtLOB function [97].

The apparent conservation of boundary gene families suggests a common mechanism for meristem/organ boundary specification in eudicots and grasses, but the exact roles of the genes in grasses are yet to be understood. Some examples studied so far, such as *RA2*, illustrate diversity in gene function.

4.2. Intra-whorl Boundaries (the Boundary Between the Overlapping Margins of the Sheath) in the Grasses

The sheath arises from an overlapping region in the disc of insertion during early P3/ late P4 development, requiring the formation of a new boundary between the two sheath margins (intra-whorl boundary) (Figure 2C, P4 dotted line). The delineation is shown clearly by the expression of adaxial and abaxial markers in the region of the incipient sheath margins [98]. Separation of sheath margins is dependent on auxin since the sheath remains fused and tube-like when plants are cultured in the presence of the auxin inhibitor NPA. In support of this dependency, expression of auxin biosynthesis genes such as *SPARSE INFLORESCENCE 1* (*SPI1*, a *YUCCA* gene) is observed at the incipient sheath boundary. *ZmNAM2* (also called *ZmCUC2*) is also expressed in this region, which suggests the recapitulation of the meristem-organ boundary specification at this location and stage in grass leaf development [98].

What specifies or activates this intra-whorl boundary pathway forming the sheath margins is not clear. The *narrowsheath1/2* double mutant in maize lacks this region, suggesting a role for NS1/2 in patterning or growth of this region [99]. Comparisons of monocots with fused sheaths, such as seen in some members of the sedges, could help elucidate this component in grass sheath development, highlighting factors involved in the evolution of the grass leaf.

4.3. Within-Organ Boundaries (the Blade/Sheath Boundary and the Development of the Ligule and Auricle) in the Grasses

The boundary between the grass leaf sheath and the blade develops characteristic structures; the ligule and the auricle; which directly influence the leaf angle, and can be used to define different species.

The first indication of the ligule during maize leaf development is an apparent increase in cell divisions in both a transverse and longitudinal direction in the adaxial epidermis to form the pre-ligule band [77]. Shortly thereafter, a reoriented accumulation of ZmPIN1a in the epidermis is observed, suggesting that, like organ initiation in the meristem periphery, auxin signaling is important in ligule formation and outgrowth [100]. Laser capture RNAseq of developing ligules found that ligule development involves the recapitulation of the meristem/organ boundary network within the developing leaf [101], highlighting roles for transcription factors such as *ZmNAM2* in addition to auxin, giberellic acid (GA), cytokinin (CK), and brassinosteroid (BR) signaling. This RNAseq dataset suggests that, like eudicot leaf margin modification, the grasses have recruited a common boundary specification network in the development of a novel leaf morphology.

Mirroring the diversification observed in leaf margin development in eudicots, analysis of grass mutants with defects in the ligule/auricle boundary have identified species-specific components. The many blade/sheath boundary mutants in maize, barley, and rice highlight the role of different genes. Some appear to be specific to the outgrowth of the ligule, while others influence the specification of the blade/sheath boundary.

4.3.1. *Liguleless* Mutants and the Patterning of the Ligule

LIGULELESS1 and 2 (LG1 and LG2) are grass-specific transcription factors belonging to the squamosa binding transcription factor and BZIP/DOG domain transcription factor families, respectively. In maize, *lg1* mutants retain a clear blade sheath boundary, but lack the ligule and auricle [102] (Figure 3). In rice and barley, *lg1* mutants are more severe than in maize, completely lacking the ligule region in all leaves [103,104]. The milder phenotype of maize may be explained by duplicates of *ZmLG1*. *LG1* is expressed in the pre-ligular band [101] and acts cell autonomously, which suggests that LG1 functions to specify the ligule [105]. *lg2* mutants, in contrast, have a diffuse blade/sheath boundary and retain reduced auricles at the margins which are displaced vertically relative to each other (Figure 3). *lg2* mutant phenotypes are yet to be described in other grasses. *ZmLG2* has a broad expression pattern but a specific protein localization, and it is able to act non-cell autonomously. The phenotype of *lg2* has led to the hypothesis that LG2 may have a role in defining the blade/sheath boundary itself [105,106]. Double mutant analysis in maize has suggested that both LG1 and LG2 act in the same pathway [105], with *LG2* being expressed earlier than *LG1* [100,106,107].

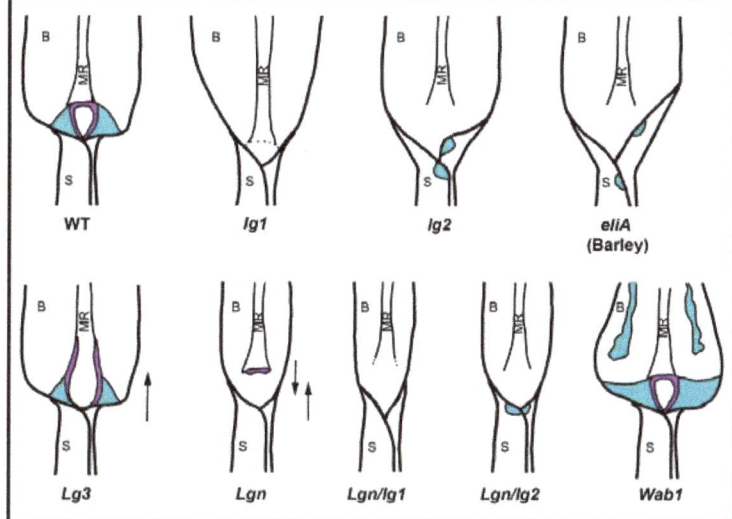

Figure 3. The morphology of the leaf blade/sheath boundary in reported ligule mutants. Cartoons depicting the typical morphology of the blade/sheath boundary in mature leaves of wild-type (WT), *liguleless1* (*lg1*), *liguleless2* (*lg2*), *eligulumA* (*eliA*, a barley mutant), *Liguleless3* (*Lg3*), *Liguleless narrow* (*Lgn*), double *Lgn/lg1*, double *Lgn/lg2*, and *Wavy Auricles in Blade* (*Wab1*) plants. In each cartoon, the blade (B), midrib (MR), and sheath (S) are labelled. The ligule (purple) and the auricles (cyan) are also highlighted. Where the mutant leaf lacks a ligule, but retains a clear boundary between the blade and sheath, the boundary is indicated by a dotted line. The arrows indicate the direction of the displacement of the sheath tissue in the mutant.

RNAseq of *lg1* mutants showed an enrichment of differentially expressed genes involved in auxin signaling, in addition to MYB and SBP transcription factors [101]. The directly bound and modulated targets of LG1 and LG2, however, are yet to be identified. Given data from other species, LG1 and LG2 may form heterodimers with other transcription factors. For example, in *Arabidopsis*, the BZIP DOG domain transcription factor PERIANTHIA (a member of the same clade of BZIP transcription factors as LG2 [108]) is involved in floral development, and interacts with BOP1 and 2 in yeast [49]. The barley orthologue of *AtBOP2*, *UNICULME4* (*HvCUL4*) functions in axillary meristem development and in ligule specification [95]. These observations lead to the hypothesis that LG2 may interact with BOP homologues in the grasses to pattern the blade/sheath boundary.

In addition to homo-dimerization and hetero-dimerization, BZIP transcription factor activity has been shown to be post-translationally regulated via phosphorylation [109]. LIGULELESS NARROW (LGN) is a serine-threonine kinase that is non-functional in the dominant mutant, *Lgn-R*. *Lgn-R* mutants have a pleiotropic phenotype including narrower leaves, the loss of the ligule except at the midrib, and a diffuse blade/sheath boundary (Figure 3). This mutant has led to the hypothesis that a phosphorylation cascade propagates the ligule signal from the midrib to the margins of the leaf. A role for phosphorylation was also highlighted by network analysis where the authors proposed that a membrane associated kinase regulator (MPKR) could act with bHLH transcription factors to influence brassinosteroid (BR) signaling in the ligule [110]. Mutants in rice with reduced BR synthesis such as *dwarf4-1*, *ebisu dwarf* (*d2*), *brassinosteroid-deficient dwarf 1* (*brd1*), or BR signaling, such as *d61*, have more upright leaves [111–114]. Similarly, RNAi knock-down of the BR signaling components, *OsBAK1* in rice and *ZmBRI1* in maize, have reduced BR signaling and more upright leaves with reduced auricles [115,116]. The maize *brd1* mutant has reduced BR synthesis with defects in ligule and auricle development [117]. *ZmBRD1* is expressed in the base of P3 leaves [101], which overlaps with the localization of TRU1 [84]. These results suggest that phosphorylation cascades and BR may be involved in mediolateral patterning of the blade/sheath boundary.

The barley liguleless mutant *eligulumA* has a diffuse blade/sheath boundary (Figure 3) and carries a mutation in a gene that encodes a protein with an RNaseH domain but otherwise of unknown function [118]. In barley, *ELIA* is expressed in an overlapping domain with *LG1*. Although no *eligulum* mutant has yet been reported in maize, gene network analyses highlight a module expressed in the pre-ligule band that includes both maize homologues of ELIA [110]. These results suggest that ELIA may play an, as yet, unknown role in the blade/sheath boundary specification and ligule development across the grasses.

4.3.2. Ectopic Induction of New Blade/Sheath Boundaries

Several dominant maize mutants exhibit ectopic formation of new blade/sheath boundaries, suggesting an additional regulatory network involved in initiating blade/sheath boundary patterning. In support of this, the genes able to trigger ectopic blade/sheath boundaries form a distinct module from the pre-ligule patterning genes (those genes outlined in Section 4.3.1) in gene network analyses [110]. Genes able to ectopically induce new blade/sheath boundaries include the homeobox genes *KNOTTED1* (KN1), *GNARLEY1* (KNOX4), *LIGULELESS 3*, *LIGULELESS 4*, and the TCP transcription factor *WAVY AURICLES IN BLADE1* (*Wab1*) [2,107,119–126]. An additional ectopic blade/sheath boundary mutant, *Hairy sheath frayed* (*Hsf*), has also been identified. *Hsf* develops sheath-like prongs on the blade of the leaf [127,128] and is involved in cytokinin (CK) signaling [Michael Muszynski, Personal Communication]. These mutants suggest that KNOXs, TCPs, and CK signaling could be involved in proximal patterning of the grass leaf before ligule and auricle outgrowth occurs.

In support of the hypothesis that *KNOX* genes are involved in this proximal/distal patterning, KNOX protein accumulates at the base of developing grass leaves, suggesting that KNOXs could provide a "proximal" patterning signal. *KNOX* expression in this boundary may provide competency to respond to the ligule and auricle patterning factors. Interestingly, the KNOX interacting factors, BEL12 and 14, are expressed in the developing ligule [101,129] and are bound and modulated by KN1 [44]. LG3, which is also expressed at the ligule, interacts with both BEL12 and 14 [Aromdee and Hake, unpublished data]. Ectopic expression of *KNOX* genes in other systems also triggers morphological changes and outgrowths. For example, ectopic expression of the *KNOX* gene *BKN3* in the barley lemma/awn boundary triggers the formation of an ectopic floral meristem and triangular marginal outgrowths. This dramatic morphological change correlates with an induced re-orientation of tissue cell polarity (as shown by the localization of SoPIN1) and the ectopic expression of boundary genes such as *NAM* [31], lending further support to the hypothesis that *KNOX* genes are able to pattern new boundary regions and morphological changes.

WAB1 is normally expressed in developing inflorescences and is required for branch initiation in the tassel [121]. In the dominant gain of function mutant, *WAB1* is ectopically expressed in the leaf blade and induces the ectopic expression of *LG1*, which leads to auricle-like outgrowths in the blade (Figure 3). Although WAB1 does not play a role in normal leaf development, it could indicate a possible role for other TCP transcription factors in the regulation of LG1 expression in the leaf.

The recessive mutant *extended auricles 1* (*eta1*) develops ectopic auricle tissue, and has a diffuse blade/sheath boundary. The causal mutation of *eta1* has not been identified, but it has been shown to be involved in the same pathway as LG1 and LG2 [130,131]. ETA1 is proposed to be a possible component of the bridge between the blade/sheath boundary patterning network and the pre-ligule patterning network.

4.3.3. A Proposed Model of Blade/Sheath Boundary Specification

Given that liguleless mutants maintain a blade and a sheath, it is likely that the blade/sheath boundary specification can be separated into two distinct phases.

First, a broad domain boundary between the sheath and blade is specified early in the leaf primordium. Since there are no reported mutants which are only sheath, only blade, or a hybrid of the two identities, it is likely that this stage is genetically redundant. This phase involves factors such as *KNOX* genes, and genes associated with the sheath such as *BOP*, as well as phytohormone gradients such as auxin and cytokinin. Although *KNOX* gene expression is excluded from developing leaf primordia, the accumulation of KNOX protein in the base of the developing leaf could promote the expression of *BOP* genes, specifying the sheath domain. This would predict that the loss of function of multiple *BOP* genes with overlapping functions in the grasses would result in a loss of sheath identity, and that ectopic *KNOX* expression would induce *BOP* expression. Similarly, overexpression of a BOP gene in the developing grass leaf would increase the proportion of sheath to blade. Based on the RNAseq work by Johnston et al. and the mutant phenotypes of *Hsf*, converging gradients of auxin (distal signal) and cytokinin (proximal signal) could contribute to patterning the boundary between the sheath and blade. Early studies that added auxin transport inhibitors to maize seedlings showed a disruption of the blade sheath boundary [132]. It would be of great interest to explore the distribution of auxin and cytokinin in the developing leaf primordium using reporters, as well as to test the effects of differential hormone treatments on the ratio of sheath to blade.

The second phase of boundary development involves the refinement of the blade sheath boundary and the ultimate specification of the pre-ligule band at P6 (Figure 4). This phase likely involves genes expressed at the ligule and those that have liguleless phenotypes. Within this stage, we can predict factors involved in refining the boundary, and those important for ligule specification and outgrowth to function. The *lg2* mutant has a diffuse boundary, which suggests that it is involved in refining the boundary region. *lg1* has a distinct blade/sheath boundary, and is therefore likely specific to the specification and outgrowth of the ligule. The displacement of the ligule and the blade/sheath boundary in the *Lgn* mutant suggests that a phosphorylation cascade and BR signaling may be involved in propagation of the "ligule signal" out from the midrib to the margins of the leaf. It will be of great interest to look at the relative timing of ligule specific gene expression alongside PIN orientations to determine how the ligule region is defined.

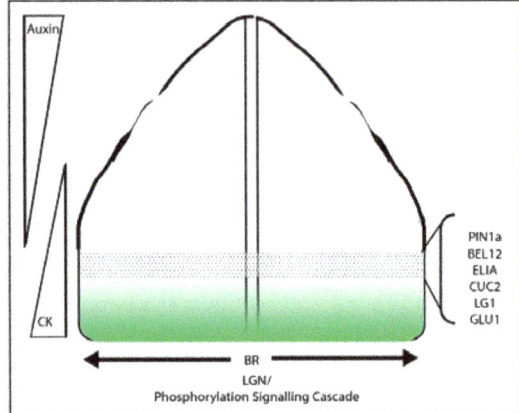

Figure 4. How is a ligule patterned? Summary of the known gene expression patterns in the developing grass leaf at P6 and the hypothetical gradients of phytohormone signaling. Green represents the expression pattern of the *BOP* homologues, which overlaps with *LG3*, *BRD1*, and *BEL14* in the base of the leaf. The dotted region represents the pre-ligule band, where *PIN1a*, *BEL12*, *ELIA*, *CUC2*, *LG1*, and *GLU1* are expressed. Hypothesized gradients are also illustrated for Auxin, CK, and BR.

5. Pleiotropy and Boundaries

Given the profound effects on morphology, manipulation of genes involved in boundary specification could lead to modulation of the leaf phenotype, providing a rich resource for phenotypic plasticity to be tested in different environmental and field conditions. For example, in dense planting fields, more upright leaves, especially in the upper canopy are important, whereas more sparse, inter-cropped fields, may benefit from a wider leaf angle. Many of the existing mutants, however, have pleiotropic effects, which can negatively impact yield. For example, both the maize *lg1* and *lg2* mutants have upright leaves, but also have severely reduced tassel branch numbers [102,133,134]. Similarly, the barley mutant *eliA* is pleiotropic with a shorter stature, ligule defects, and compact inflorescence spikes [118]. This pleiotropy is not unique to the grasses. For example, the *cuc2*, *cuc3*, and *lof1/2* mutants in *Arabidopsis* all have defects in branching [16,135–137]. The combined effects on both leaf architecture, branching and inflorescence architecture of many of these mutants, often leads to a reduced yield.

To explore whether individual phenotypic components could modulate yield, the pleiotropy needs to be broken. Pleiotropic effects could be modulated through: (1) changes in cis-regulatory elements that influence the timing or spatial distribution of expression, (2) altering tissue specific partners, and (3) modulating different tissue-specific downstream elements. For example, DELLA mutants have pleiotropic defects, affecting both stem growth and meristem size. DELLA's effect on stem growth has been linked to direct regulation of the cell cycle inhibitor KRP2 and is independent of meristem size regulation. The genetic uncoupling of stem elongation and meristem size via modulation of KRP2 was effective in both *Arabidopsis* and barley, generating semi-dwarf plants [138].

In model eudicots like *Arabidopsis*, transgenics are used to overcome pleiotropic effects of key regulatory genes. The extensive transgenic toolkits in *Arabidopsis* enable cell-type specific and inducible expression systems [139] to modulate gene expression in a precise manner. For example, conditional dsRNAi silencing of CLV3 allowed identification of the specific function of CLV3 in the meristem, which uncoupled the effects from the severe global changes caused in the full *clv3* mutant [140]. In transformation tractable species, the use of genome editing via CRISPR/cas9, for example, can also be used to alter cis-regulatory elements to uncouple phenotypes. This technique has already been used successfully in tomato to combine alleles that were selected during domestication and more recent breeding for distinct flower morphology and fruit retention traits. Combining the two traits highlighted a negative epistatic relationship, which could be overcome by varying the dosage of the relevant genes using homo/heterozygote mutants, and through CRISPR/cas9 to introduce allelic variation [141]. The use of genome editing to introduce allelic variation in cis-regulatory sequences can also lead to an increase in phenotypic variation [142], which could be used as a resource to break pleiotropy.

Transgenic approaches can be more difficult in grass crops due to the expense and time of transgenics, difficulty in transgenerational maintenance of the transgene, and public opinions regarding genetic modification. Alternatively, rich natural diversity in species such as maize, can be taken advantage of to break links in pleotropic defects.

6. Conclusions

A common underlying mechanism for boundary specification exists between eudicots and grasses, specifying meristem/organ and intra-whorl boundaries during vegetative development. In both eudicots and grass crops, this mechanism has been co-opted to specify within organ boundaries to generate morphological diversity. In both cases, however, there are species and family-specific elements that modulate the core boundary network and highlight the importance of studying boundary specification in both eudicot models and grass crops. The dynamic regulation of these boundary regulatory networks could yield rich phenotypic diversity in agronomically important traits such as leaf angle, making use of targeted natural variation or genome editing in key nodes of the network.

Author Contributions: A.E.R. wrote the manuscript and S.H. helped edit.

Funding: The work was supported by NSF/BIOBBSRC 1547062 and NSF ECA-PGRP 1733606.

Acknowledgments: The work was supported by NSF/BIOBBSRC 1547062 and NSF ECA-PGRP 1733606.

Conflicts of Interest: The authors declare no conflict of interest.

References

1. Jackson, D.; Veit, B.; Hake, S. Expression of maize *KNOTTED1* related homeobox genes in the shoot apical meristem predicts patterns of morphogenesis in the vegetative shoot. *Development* **1994**, *120*, 405–413.
2. Smith, L.G.; Greene, B.; Veit, B.; Hake, S. A dominant mutation in the maize homeobox gene, *Knotted-1*, causes its ectopic expression in leaf cells with altered fates. *Development* **1992**, *116*, 21–30. [PubMed]
3. Long, J.A.; Moan, E.I.; Medford, J.I.; Barton, M.K. A member of the *KNOTTED* class of homeodomain proteins encoded by the *STM* gene of *Arabidopsis*. *Nature* **1996**, *379*, 66–69. [CrossRef] [PubMed]
4. Heisler, M.G.; Hamant, O.; Krupinski, P.; Uyttewaal, M.; Ohno, C.; Jönsson, H.; Traas, J.; Meyerowitz, E.M. Alignment between *PIN1* polarity and microtubule orientation in the shoot apical meristem reveals a tight coupling between morphogenesis and auxin transport. *PLoS Biol.* **2010**, *8*, e1000516. [CrossRef] [PubMed]
5. Abley, K.; De Reuille, P.B.; Strutt, D.; Bangham, A.; Prusinkiewicz, P.; Marée, A.F.M.; Grieneisen, V.A.; Coen, E. An intracellular partitioning-based framework for tissue cell polarity in plants and animals. *Development* **2013**, *140*, 2061–2074. [CrossRef]
6. Bayer, E.M.; Smith, R.S.; Mandel, T.; Nakayama, N.; Sauer, M.; Prusinkiewicz, P.; Kuhlemeier, C. Integration of transport-based models for phyllotaxis and midvein formation. *Genes Dev.* **2009**, *23*, 373–384. [CrossRef]
7. Bhatia, N.; Bozorg, B.; Larsson, A.; Ohno, C.; Jönsson, H.; Heisler, M.G. Auxin acts through *MONOPTEROS* to regulate plant cell polarity and pattern phyllotaxis. *Curr. Biol.* **2016**, *26*, 3202–3208. [CrossRef]
8. Jönsson, H.; Heisler, M.G.; Shapiro, B.E.; Meyerowitz, E.M.; Mjolsness, E. An auxin-driven polarized transport model for phyllotaxis. *Proc. Natl. Acad. Sci. USA* **2006**, *103*, 1633–1638. [CrossRef]
9. Smith, R.S.; Guyomarc'h, S.; Mandel, T.; Reinhardt, D.; Kuhlemeier, C.; Prusinkiewicz, P. A plausible model of phyllotaxis. *Proc. Natl. Acad. Sci. USA* **2006**, *103*, 1301–1306. [CrossRef]
10. Stoma, S.; Lucas, M.; Chopard, J.; Schaedel, M.; Traas, J.; Godin, C. Flux-based transport enhancement as a plausible unifying mechanism for auxin transport in meristem development. *PLoS Comput. Biol.* **2008**, *4*, e1000207. [CrossRef]

11. Benková, E.; Michniewicz, M.; Sauer, M.; Teichmann, T.; Seifertová, D.; Jürgens, G.; Friml, J. Local, efflux-dependent auxin gradients as a common module for plant organ formation. *Cell* **2003**, *115*, 591–602. [CrossRef]
12. Heisler, M.G.; Ohno, C.; Das, P.; Sieber, P.; Reddy, G.V.; Long, J.A.; Meyerowitz, E.M. Patterns of Auxin Transport and Gene Expression during Primordium Development Revealed by Live Imaging of the *Arabidopsis* Inflorescence Meristem. *Curr. Biol.* **2005**, *15*, 1899–1911. [CrossRef] [PubMed]
13. Reinhardt, D.; Mandel, T.; Kuhlemeier, C. Auxin Regulates the Initiation and Radial Position of Plant Lateral Organs. *Plant Cell* **2000**, *12*, 507–518. [CrossRef] [PubMed]
14. Reinhardt, D.; Pesce, E.-R.; Stieger, P.; Mandel, T.; Baltensperger, K.; Bennett, M.; Traas, J.; Friml, J.; Kuhlemeier, C. Regulation of phyllotaxis by polar auxin transport. *Nature* **2003**, *426*, 255. [CrossRef] [PubMed]
15. Okada, K.; Ueda, J.; Komaki, M.K.; Bell, C.J.; Shimura, Y. Requirement of the auxin polar transport system in early stages of *Arabidopsis* floral bud formation. *Plant Cell* **1991**, *3*, 677–684. [CrossRef] [PubMed]
16. Hibara, K.; Karim, M.R.; Takada, S.; Taoka, K.; Furutani, M.; Aida, M.; Tasaka, M. *Arabidopsis CUP-SHAPED COTYLEDON3* Regulates Postembryonic Shoot Meristem and Organ Boundary Formation. *Plant Cell* **2006**, *18*, 2946–2957. [CrossRef] [PubMed]
17. Hamant, O.; Heisler, M.G.; Jönsson, H.; Krupinski, P.; Uyttewaal, M.; Bokov, P.; Corson, F.; Sahlin, P.; Boudaoud, A.; Meyerowitz, E.M. Developmental patterning by mechanical signals in *Arabidopsis*. *Science* **2008**, *322*, 1650–1655. [CrossRef] [PubMed]
18. Abad, U.; Sassi, M.; Traas, J. Flower development: From morphodynamics to morphomechanics. *Philos. Trans. R. Soc. Lond. B. Biol. Sci.* **2017**, *372*. [CrossRef] [PubMed]
19. Green, P.B. Mechanism for Plant Cellular Morphogenesis. *Science* **1962**, *138*, 1404–1405. [CrossRef]
20. Hamant, O.; Traas, J. The mechanics behind plant development. *New Phytol.* **2010**, *185*, 369–385. [CrossRef]
21. Coen, E.; Rebocho, A.B. Resolving conflicts: Modeling genetic control of plant morphogenesis. *Dev. Cell* **2016**, *38*, 579–583. [CrossRef] [PubMed]
22. Rebocho, A.B.; Kennaway, J.R.; Bangham, J.A.; Coen, E. Formation and Shaping of the *Antirrhinum* Flower through Modulation of the *CUP* Boundary Gene. *Curr. Biol.* **2017**, *27*, 2610–2622. [CrossRef] [PubMed]
23. Juarez, M.T.; Twigg, R.W.; Timmermans, M.C.P. Specification of adaxial cell fate during maize leaf development. *Development* **2004**, *131*, 4533–4544. [CrossRef] [PubMed]
24. McConnell, J.R.; Emery, J.; Eshed, Y.; Bao, N.; Bowman, J.; Barton, M.K. Role of *PHABULOSA* and *PHAVOLUTA* in determining radial patterning in shoots. *Nature* **2001**, *411*, 709. [CrossRef] [PubMed]
25. Waites, R.; Hudson, A. Phantastica: A gene required for dorsoventrality of leaves in *Antirrhinum majus*. *Development* **1995**, *121*, 2143–2154.
26. Kumar, S.; Mishra, R.K.; Kumar, A.; Srivastava, S.; Chaudhary, S. Regulation of stipule development by *COCHLEATA* and *STIPULE-REDUCED* genes in pea *Pisum sativum*. *Planta* **2009**, *230*, 449–458. [CrossRef] [PubMed]
27. Harlan, H.V. The origin of hooded barley. *J. Hered.* **1931**, *22*, 265–272. [CrossRef]
28. Müller, K.J.; Romano, N.; Gerstner, O.; Garcia-Marotot, F.; Pozzi, C.; Salamini, F.; Rohde, W. The barley Hooded mutation caused by a duplication in a homeobox gene intron. *Nature* **1995**, *374*, 727. [CrossRef]
29. Stebbins, G.L.; Yagil, E. The morphogenetic effects of the hooded gene in barley I: The course of development in hooded and awned genotypes. *Genetics* **1966**, *54*, 727.
30. Williams-Carrier, R.E.; Lie, Y.S.; Hake, S.; Lemaux, P.G. Ectopic expression of the maize *kn1* gene phenocopies the Hooded mutant of barley. *Development* **1997**, *124*, 3737–3745.
31. Richardson, A.E.; Rebocho, A.B.; Coen, E.S. Ectopic KNOX expression affects plant development by altering tissue cell polarity and identity. *Plant Cell* **2016**. [CrossRef]
32. Chuck, G.; Lincoln, C.; Hake, S. *KNAT1* induces lobed leaves with ectopic meristems when overexpressed in *Arabidopsis*. *Plant Cell* **1996**, *8*, 1277–1289. [CrossRef] [PubMed]
33. Shuai, B.; Reynaga-Peña, C.G.; Springer, P.S. The Lateral Organ Boundaries Gene Defines a Novel, Plant-Specific Gene Family. *Plant Physiol.* **2002**, *129*, 747–761. [CrossRef] [PubMed]
34. Lee, B.H.; Jeon, J.O.; Lee, M.M.; Kim, J.H. Genetic interaction between *GROWTH-REGULATING FACTOR* and *CUP-SHAPED COTYLEDON* in organ separation. *Plant Signal. Behav.* **2015**, *10*, e988071. [CrossRef] [PubMed]

35. Bell, E.M.; Lin, W.; Husbands, A.Y.; Yu, L.; Jaganatha, V.; Jablonska, B.; Mangeon, A.; Neff, M.M.; Girke, T.; Springer, P.S. *Arabidopsis LATERAL ORGAN BOUNDARIES* negatively regulates brassinosteroid accumulation to limit growth in organ boundaries. *Proc. Natl. Acad. Sci. USA* **2012**, *109*, 21146–21151. [CrossRef] [PubMed]
36. Vroemen, C.W.; Mordhorst, A.P.; Albrecht, C.; Kwaaitaal, M.A.C.J.; de Vries, S.C. The *CUP-SHAPED COTYLEDON3* gene is required for boundary and shoot meristem formation in *Arabidopsis*. *Plant Cell* **2003**, *15*, 1563–1577. [CrossRef]
37. Takeda, S.; Hanano, K.; Kariya, A.; Shimizu, S.; Zhao, L.; Matsui, M.; Tasaka, M.; Aida, M. *CUP-SHAPED COTYLEDON1* transcription factor activates the expression of *LSH4* and *LSH3*, two members of the *ALOG* gene family, in shoot organ boundary cells. *Plant J.* **2011**, *66*, 1066–1077. [CrossRef]
38. Tian, C.; Zhang, X.; He, J.; Yu, H.; Wang, Y.; Shi, B.; Han, Y.; Wang, G.; Feng, X.; Zhang, C.; et al. An organ boundary-enriched gene regulatory network uncovers regulatory hierarchies underlying axillary meristem initiation. *Mol. Syst. Biol.* **2014**, *10*, 755. [CrossRef]
39. Borghi, L.; Bureau, M.; Simon, R. *Arabidopsis JAGGED LATERAL ORGANS* is Expressed in Boundaries and Coordinates *KNOX* and *PIN* Activity. *Plant Cell* **2007**, *19*, 1795–1808. [CrossRef]
40. Spinelli, S.V.; Martin, A.P.; Viola, I.L.; Gonzalez, D.H.; Palatnik, J.F. A mechanistic link between STM and CUC1 during Arabidopsis development. *Plant Physiol.* **2011**. [CrossRef]
41. Ha, C.M.; Jun, J.H.; Fletcher, J.C. Control of *Arabidopsis* Leaf Morphogenesis Through Regulation of the *YABBY* and *KNOX* Families of Transcription Factors. *Genetics* **2010**, *186*, 197–206. [CrossRef] [PubMed]
42. Ichihashi, Y.; Aguilar-Martínez, J.A.; Farhi, M.; Chitwood, D.H.; Kumar, R.; Millon, L.V.; Peng, J.; Maloof, J.N.; Sinha, N.R. Evolutionary developmental transcriptomics reveals a gene network module regulating interspecific diversity in plant leaf shape. *Proc. Natl. Acad. Sci. USA* **2014**. [CrossRef]
43. Norberg, M.; Holmlund, M.; Nilsson, O. The *BLADE ON PETIOLE* genes act redundantly to control the growth and development of lateral organs. *Development* **2005**, *132*, 2203–2213. [CrossRef] [PubMed]
44. Bolduc, N.; Yilmaz, A.; Mejia-Guerra, M.K.; Morohashi, K.; O'connor, D.; Grotewold, E.; Hake, S. Unraveling the *KNOTTED1* regulatory network in maize meristems. *Genes Dev.* **2012**, *26*, 1685–1690. [CrossRef] [PubMed]
45. Žádníková, P.; Simon, R. How boundaries control plant development. *Curr. Opin. Plant Biol.* **2014**, *17*, 116–125. [CrossRef] [PubMed]
46. Scofield, S.; Murison, A.; Jones, A.; Fozard, J.; Aida, M.; Band, L.R.; Bennett, M.; Murray, J.A.H. Coordination of meristem and boundary functions by transcription factors in the *SHOOT MERISTEMLESS* regulatory network. *Development* **2018**. [CrossRef] [PubMed]
47. Clouse, S.D.; Langford, M.; McMorris, T.C. A Brassinosteroid-Insensitive Mutant in *Arabidopsis thaliana* Exhibits Multiple Defects in Growth and Development. *Plant Physiol.* **1996**, *111*, 671–678. [CrossRef] [PubMed]
48. Ha, C.M.; Jun, J.H.; Nam, H.G.; Fletcher, J.C. *BLADE-ON-PETIOLE1* and *2* Control *Arabidopsis* Lateral Organ Fate through Regulation of LOB Domain and Adaxial-Abaxial Polarity Genes. *Plant Cell* **2007**, *19*, 1809–1825. [CrossRef]
49. Hepworth, S.R.; Zhang, Y.; McKim, S.; Li, X.; Haughn, G.W. BLADE-ON-PETIOLE–dependent signaling controls leaf and floral patterning in *Arabidopsis*. *Plant Cell* **2005**, *17*, 1434–1448. [CrossRef]
50. Bar, M.; Ori, N. Compound leaf development in model plant species. *Curr. Opin. Plant Biol.* **2015**, *23*, 61–69. [CrossRef]
51. Blein, T.; Pulido, A.; Vialette-Guiraud, A.; Nikovics, K.; Morin, H.; Hay, A.; Johansen, I.E.; Tsiantis, M.; Laufs, P. A Conserved Molecular Framework for Compound Leaf Development. *Science* **2008**, *322*, 1835–1839. [CrossRef] [PubMed]
52. Barkoulas, M.; Hay, A.; Kougioumoutzi, E.; Tsiantis, M. A developmental framework for dissected leaf formation in the *Arabidopsis* relative *Cardamine hirsuta*. *Nat. Genet.* **2008**, *40*, 1136. [CrossRef] [PubMed]
53. Hay, A.; Tsiantis, M. The genetic basis for differences in leaf form between *Arabidopsis thaliana* and its wild relative *Cardamine hirsuta*. *Nat. Genet.* **2006**, *38*, 942. [CrossRef] [PubMed]
54. Bilsborough, G.D.; Runions, A.; Barkoulas, M.; Jenkins, H.W.; Hasson, A.; Galinha, C.; Laufs, P.; Hay, A.; Prusinkiewicz, P.; Tsiantis, M. Model for the regulation of *Arabidopsis thaliana* leaf margin development. *Proc. Natl. Acad. Sci. USA* **2011**, *108*, 3424–3429. [CrossRef] [PubMed]
55. Nikovics, K.; Blein, T.; Peaucelle, A.; Ishida, T.; Morin, H.; Aida, M.; Laufs, P. The balance between the *MIR164A* and *CUC2* genes controls leaf margin serration in *Arabidopsis*. *Plant Cell* **2006**, *18*, 2929–2945. [CrossRef] [PubMed]

56. Hareven, D.; Gutfinger, T.; Parnis, A.; Eshed, Y.; Lifschitz, E. The Making of a Compound Leaf: Genetic Manipulation of Leaf Architecture in Tomato. *Cell* **1996**, *84*, 735–744. [CrossRef]
57. Gourlay, C.W.; Hofer, J.M.I.; Ellis, T.H.N. Pea Compound Leaf Architecture Is Regulated by Interactions among the Genes *UNIFOLIATA, COCHLEATA, AFIL,* and *TENDRIL-LESS*. *Plant Cell* **2000**, *12*, 1279–1294. [CrossRef] [PubMed]
58. Vlad, D.; Kierzkowski, D.; Rast, M.I.; Vuolo, F.; Ioio, R.D.; Galinha, C.; Gan, X.; Hajheidari, M.; Hay, A.; Smith, R.S. Leaf shape evolution through duplication, regulatory diversification, and loss of a homeobox gene. *Science* **2014**, *343*, 780–783. [CrossRef]
59. Chaw, S.-M.; Chang, C.-C.; Chen, H.-L.; Li, W.-H. Dating the monocot-dicot divergence and the origin of core eudicots using whole chloroplast genomes. *J. Mol. Evol.* **2004**, *58*, 424–441.
60. Sharman, B.C. Developmental anatomy of the shoot of *Zea mays* L. *Ann. Bot.* **1942**, *6*, 245–282. [CrossRef]
61. Poethig, R.S.; Szymkowiak, E.J. Clonal analysis of leaf development in maize. Available online: http://agris.fao.org/agris-search/search.do?recordID=IT9561182 (accessed on 24 December 2018).
62. Scanlon, M.J. The Polar Auxin Transport Inhibitor N-1-Naphthylphthalamic Acid Disrupts Leaf Initiation, KNOX Protein Regulation, and Formation of Leaf Margins in Maize. *Plant Physiol.* **2003**, *133*, 597–605. [CrossRef] [PubMed]
63. O'Connor, D.L.; Elton, S.; Ticchiarelli, F.; Hsia, M.M.; Vogel, J.P.; Leyser, O. Cross-species functional diversity within the PIN auxin efflux protein family. *Elife* **2017**, *6*, e31804. [CrossRef] [PubMed]
64. O'Connor, D.L.; Runions, A.; Sluis, A.; Bragg, J.; Vogel, J.P.; Prusinkiewicz, P.; Hake, S. A Division in PIN-Mediated Auxin Patterning during Organ Initiation in Grasses. *PLOS Comput. Biol.* **2014**, *10*, e1003447. [CrossRef] [PubMed]
65. Martinez, C.C.; Koenig, D.; Chitwood, D.H.; Sinha, N.R. A sister of *PIN1* gene in tomato (*Solanum lycopersicum*) defines leaf and flower organ initiation patterns by maintaining epidermal auxin flux. *Dev. Biol.* **2016**, *419*, 85–98. [CrossRef] [PubMed]
66. Zhou, C.; Han, L.; Wang, Z.-Y. Potential but limited redundant roles of *MtPIN4, MtPIN5* and *MtPIN10/SLM1* in the development of Medicago truncatula. *Plant Signal. Behav.* **2011**, *6*, 1834–1836. [CrossRef] [PubMed]
67. Pepper, G.E.; Pearce, R.B.; Mock, J.J. Leaf Orientation and Yield of Maize1. *Crop Sci.* **1977**, *17*, 883–886. [CrossRef]
68. Araus, J.L.; Reynolds, M.P.; Acevedo, E. Leaf Posture, Grain Yield, Growth, Leaf Structure, and Carbon Isotope Discrimination in Wheat. *Crop Sci.* **1993**, *33*, 1273–1279. [CrossRef]
69. Nan, S.S.; Ootsuki, Y.; Adachi, S.; Yamamoto, T.; Ueda, T.; Tanabata, T.; Motobayashi, T.; Ookawa, T.; Hirasawa, T. A near-isogenic rice line carrying a QTL for larger leaf inclination angle yields heavier biomass and grain. *Field Crop. Res.* **2018**, *219*, 131–138. [CrossRef]
70. Pendleton, J.W.; Smith, G.E.; Winter, S.R.; Johnston, T.J. Field Investigations of the Relationships of Leaf (Angle in Corn (*Zea mays* L.) to Grain Yield and Apparent Photosynthesis1. *Agron. J.* **1968**, *60*, 422–424. [CrossRef]
71. Duvick, D.N. The contribution of breeding to yield advances in maize (*Zea mays* L.). *Adv. Agron.* **2005**, *86*, 83–145.
72. Mantilla-Perez, M.B.; Salas Fernandez, M.G. Differential manipulation of leaf angle throughout the canopy: Current status and prospects. *J. Exp. Bot.* **2017**, *68*, 5699–5717. [CrossRef] [PubMed]
73. Lambert, R.J.; Johnson, R.R. Leaf Angle, Tassel Morphology, and the Performance of Maize Hybrids 1. *Crop Sci.* **1978**, *18*, 499–502. [CrossRef]
74. Duncan, W.G. Leaf Angles, Leaf Area, and Canopy Photosynthesis 1. *Crop Sci.* **1971**, *11*, 482–485. [CrossRef]
75. Tian, F.; Bradbury, P.J.; Brown, P.J.; Hung, H.; Sun, Q.; Flint-Garcia, S.; Rocheford, T.R.; McMullen, M.D.; Holland, J.B.; Buckler, E.S. Genome-wide association study of leaf architecture in the maize nested association mapping population. *Nat. Genet.* **2011**, *43*, 159. [CrossRef] [PubMed]
76. Hernandez, M.L.; Passas, H.J.; Smith, L.G. Clonal Analysis of Epidermal Patterning during Maize Leaf Development. *Dev. Biol.* **1999**, *216*, 646–658. [CrossRef] [PubMed]
77. Sylvester, A.W.; Cande, W.Z.; Freeling, M. Division and differentiation during normal and liguleless-1 maize leaf development. *Development* **1990**, *110*, 985–1000. [PubMed]
78. Weir, I.; Lu, J.; Cook, H.; Causier, B.; Schwarz-Sommer, Z.; Davies, B. *CUPULIFORMIS* establishes lateral organ boundaries in *Antirrhinum*. *Development* **2004**, *131*, 915–922. [CrossRef] [PubMed]

79. Souer, E.; van Houwelingen, A.; Kloos, D.; Mol, J.; Koes, R. The no apical meristem gene of *Petunia* is required for pattern formation in embryos and flowers and is expressed at meristem and primordia boundaries. *Cell* **1996**, *85*, 159–170. [CrossRef]
80. Ha, C.M.; Kim, G.-T.; Kim, B.C.; Jun, J.H.; Soh, M.S.; Ueno, Y.; Machida, Y.; Tsukaya, H.; Nam, H.G. The *BLADE-ON-PETIOLE 1* gene controls leaf pattern formation through the modulation of meristematic activity in *Arabidopsis*. *Development* **2003**, *130*, 161–172. [CrossRef]
81. Zimmermann, R.; Werr, W. Pattern Formation in the Monocot Embryo as Revealed by *NAM* and *CUC3* Orthologues from *Zea mays* L. *Plant Mol. Biol.* **2005**, *58*, 669–685. [CrossRef]
82. Adam, H.; Marguerettaz, M.; Qadri, R.; Adroher, B.; Richaud, F.; Collin, M.; Thuillet, A.-C.; Vigouroux, Y.; Laufs, P.; Tregear, J.W.; et al. Divergent Expression Patterns of *miR164* and *CUP-SHAPED COTYLEDON* Genes in Palms and Other Monocots: Implication for the Evolution of Meristem Function in Angiosperms. *Mol. Biol. Evol.* **2011**, *28*, 1439–1454. [CrossRef] [PubMed]
83. Vialette-Guiraud, A.C.M.; Adam, H.; Finet, C.; Jasinski, S.; Jouannic, S.; Scutt, C.P. Insights from ANA-grade angiosperms into the early evolution of *CUP-SHAPED COTYLEDON* genes. *Ann. Bot.* **2011**, *107*, 1511–1519. [CrossRef] [PubMed]
84. Dong, Z.; Li, W.; Unger-Wallace, E.; Yang, J.; Vollbrecht, E.; Chuck, G. Ideal crop plant architecture is mediated by tassels replace upper ears1, a BTB/POZ ankyrin repeat gene directly targeted by TEOSINTE BRANCHED1. *Proc. Natl. Acad. Sci. USA* **2017**, *114*, E8656–E8664. [CrossRef] [PubMed]
85. Sunkar, R.; Zhou, X.; Zheng, Y.; Zhang, W.; Zhu, J.-K. Identification of novel and candidate miRNAs in rice by high throughput sequencing. *BMC Plant Biol.* **2008**, *8*, 25. [CrossRef] [PubMed]
86. Sieber, P.; Wellmer, F.; Gheyselinck, J.; Riechmann, J.L.; Meyerowitz, E.M. Redundancy and specialization among plant microRNAs: Role of the *MIR164* family in developmental robustness. *Development* **2007**, *134*, 1051–1060. [CrossRef]
87. Li, J.; Guo, G.; Guo, W.; Guo, G.; Tong, D.; Ni, Z.; Sun, Q.; Yao, Y. miRNA164-directed cleavage of *ZmNAC1* confers lateral root development in maize (*Zea mays* L.). *BMC Plant Biol.* **2012**, *12*, 220. [CrossRef]
88. Xiong, Y.; Liu, T.; Tian, C.; Sun, S.; Li, J.; Chen, M. Transcription Factors in Rice: A Genome-wide Comparative Analysis between Monocots and Eudicots. *Plant Mol. Biol.* **2005**, *59*, 191–203. [CrossRef]
89. Tsuda, K.; Akiba, T.; Kimura, F.; Ishibashi, M.; Moriya, C.; Nakagawa, K.; Kurata, N.; Ito, Y. ONION2 fatty acid elongase is required for shoot development in rice. *Plant Cell Physiol.* **2013**, *54*, 209–217. [CrossRef]
90. Tsuda, K.; Ito, Y.; Yamaki, S.; Miyao, A.; Hirochika, H.; Kurata, N. Isolation and mapping of three rice mutants that showed ectopic expression of *KNOX* genes in leaves. *Plant Sci.* **2009**, *177*, 131–135. [CrossRef]
91. Takasugi, T.; Ito, Y. Altered expression of auxin-related genes in the fatty acid elongase mutant *oni1* of rice. *Plant Signal Behav.* **2011**, *6*. [CrossRef]
92. Akiba, T.; Hibara, K.I.; Kimura, F.; Tsuda, K.; Shibata, K.; Ishibashi, M.; Moriya, C.; Nakagawa, K.; Kurata, N.; Itoh, J.I.; et al. Organ fusion and defective shoot development in *oni3* mutants of rice. *Plant Cell Physiol.* **2014**, *55*, 42–51. [CrossRef] [PubMed]
93. Sinha, N.; Lynch, M. Fused organs in the *adherent1* mutation in maize show altered epidermal walls with no perturbations in tissue identities. *Planta* **1998**, *206*, 184–195. [CrossRef]
94. La Rocca, N.; Manzotti, P.S.; Cavaiuolo, M.; Barbante, A.; Dalla Vecchia, F.; Gabotti, D.; Gendrot, G.; Horner, D.S.; Krstajic, J.; Persico, M.; et al. The maize fused leaves1 (*fdl1*) gene controls organ separation in the embryo and seedling shoot and promotes coleoptile opening. *J. Exp. Bot.* **2015**, *66*, 5753–5767. [CrossRef] [PubMed]
95. Tavakol, E.; Okagaki, R.; Verderio, G.; Shariati, V.; Hussien, A.; Bilgic, H.; Scanlon, M.J.; Todt, N.R.; Close, T.J.; Druka, A. The barley *Uniculme4* gene encodes a BLADE-ON-PETIOLE-like protein that controls tillering and leaf patterning. *Plant Physiol.* **2015**, *168*, 164–174. [CrossRef] [PubMed]
96. Mangeon, A.; Lin, W.; Springer, P.S. Functional divergence in the Arabidopsis LOB-domain gene family. *Plant Signal. Behav.* **2012**, *7*, 1544–1547. [CrossRef] [PubMed]
97. Bortiri, E.; Chuck, G.; Vollbrecht, E.; Rocheford, T.; Martienssen, R.; Hake, S. *Ramosa2* Encodes a LATERAL ORGAN BOUNDARY Domain Protein That Determines the Fate of Stem Cells in Branch Meristems of Maize. *Plant Cell* **2006**, *18*, 574–585. [CrossRef] [PubMed]
98. Johnston, R.; Leiboff, S.; Scanlon, M.J. Ontogeny of the sheathing leaf base in maize (*Zea mays*). *New Phytol.* **2014**, 306–315. [CrossRef]

99. Scanlon, M.J.; Chen, K.D.; McKnight, C.C. The *narrow sheath* Duplicate Genes: Sectors of Dual Aneuploidy Reveal Ancestrally Conserved Gene Functions During Maize Leaf Development. *Genetics* **2000**, *155*, 1379–1389.
100. Moon, J.; Candela, H.; Hake, S. The Liguleless narrow mutation affects proximal-distal signaling and leaf growth. *Development* **2013**, *140*, 405–412. [CrossRef]
101. Johnston, R.; Wang, M.; Sun, Q.; Sylvester, A.W.; Hake, S.; Scanlon, M.J. Transcriptomic Analyses Indicate That Maize Ligule Development Recapitulates Gene Expression Patterns That Occur during Lateral Organ Initiation. *Plant Cell Online* **2014**. [CrossRef]
102. Moreno, M.A.; Harper, L.C.; Krueger, R.W.; Dellaporta, S.L.; Freeling, M. *liguleless1* encodes a nuclear-localized protein required for induction of ligules and auricles during maize leaf organogenesis. *Genes Dev.* **1997**, *11*, 616–628. [CrossRef] [PubMed]
103. Lee, J.; Park, J.J.; Kim, S.L.; Yim, J.; An, G. Mutations in the rice liguleless gene result in a complete loss of the auricle, ligule, and laminar joint. *Plant Mol. Biol.* **2007**, *65*, 487–499. [CrossRef] [PubMed]
104. Rossini, L.; Vecchietti, A.; Nicoloso, L.; Stein, N.; Franzago, S.; Salamini, F.; Pozzi, C. Candidate genes for barley mutants involved in plant architecture: An *in silico* approach. *Theor. Appl. Genet.* **2006**, *112*, 1073–1085. [CrossRef] [PubMed]
105. Harper, L.; Freeling, M. Interactions of *liguleless1* and *liguleless2* function during ligule induction in maize. *Genetics* **1996**, *144*, 1871–1882. [PubMed]
106. Walsh, J.; Waters, C.A.; Freeling, M. The maize gene *liguleless2* encodes a basic leucine zipper protein involved in the establishment of the leaf blade-sheath boundary. *Genes Dev.* **1998**, *12*, 208–218. [CrossRef] [PubMed]
107. Hay, A.; Hake, S. The dominant mutant Wavy auricle in blade1 disrupts patterning in a lateral domain of the maize leaf. *Plant Physiol.* **2004**, *135*, 300–308. [CrossRef] [PubMed]
108. Jakoby, M.; Weisshaar, B.; Droge-Laser, W.; Vicente-Carbajosa, J.; Tiedemann, J.; Kroj, T.; Parcy, F. bZIP transcription factors in *Arabidopsis*. *Trends Plant Sci.* **2002**, *7*, 106–111. [CrossRef]
109. Schütze, K.; Harter, K.; Chaban, C. Post-translational regulation of plant bZIP factors. *Trends Plant Sci.* **2008**, *13*, 247–255. [CrossRef] [PubMed]
110. Ma, S.; Ding, Z.; Li, P. Maize network analysis revealed gene modules involved in development, nutrients utilization, metabolism, and stress response. *BMC Plant Biol.* **2017**, *17*, 131. [CrossRef] [PubMed]
111. Sakamoto, T.; Morinaka, Y.; Ohnishi, T.; Sunohara, H.; Fujioka, S.; Ueguchi-Tanaka, M.; Mizutani, M.; Sakata, K.; Takatsuto, S.; Yoshida, S. Erect leaves caused by brassinosteroid deficiency increase biomass production and grain yield in rice. *Nat. Biotechnol.* **2006**, *24*, 105. [CrossRef]
112. Hong, Z.; Ueguchi-Tanaka, M.; Umemura, K.; Uozu, S.; Fujioka, S.; Takatsuto, S.; Yoshida, S.; Ashikari, M.; Kitano, H.; Matsuoka, M. A rice brassinosteroid-deficient mutant, *ebisu dwarf* (*d2*), is caused by a loss of function of a new member of cytochrome P450. *Plant Cell* **2003**, *15*, 2900–2910. [CrossRef] [PubMed]
113. Hong, Z.; Ueguchi-Tanaka, M.; Shimizu-Sato, S.; Inukai, Y.; Fujioka, S.; Shimada, Y.; Takatsuto, S.; Agetsuma, M.; Yoshida, S.; Watanabe, Y. Loss-of-function of a rice brassinosteroid biosynthetic enzyme, C-6 oxidase, prevents the organized arrangement and polar elongation of cells in the leaves and stem. *Plant J.* **2002**, *32*, 495–508. [CrossRef] [PubMed]
114. Yamamuro, C.; Ihara, Y.; Wu, X.; Noguchi, T.; Fujioka, S.; Takatsuto, S.; Ashikari, M.; Kitano, H.; Matsuoka, M. Loss of function of a rice brassinosteroid insensitive1 homolog prevents internode elongation and bending of the lamina joint. *Plant Cell* **2000**, *12*, 1591–1605. [CrossRef] [PubMed]
115. Kir, G.; Ye, H.; Nelissen, H.; Neelakandan, A.K.; Kusnandar, A.S.; Luo, A.; Inzé, D.; Sylvester, A.W.; Yin, Y.; Becraft, P.W. RNA interference knockdown of *BRASSINOSTEROID INSENSITIVE1* in maize reveals novel functions for brassinosteroid signaling in controlling plant architecture. *Plant Physiol.* **2015**, *169*, 826–839. [CrossRef] [PubMed]
116. Li, D.; Wang, L.; Wang, M.; Xu, Y.; Luo, W.; Liu, Y.; Xu, Z.; Li, J.; Chong, K. Engineering *OsBAK1* gene as a molecular tool to improve rice architecture for high yield. *Plant Biotechnol. J.* **2009**, *7*, 791–806. [CrossRef]
117. Makarevitch, I.; Thompson, A.; Muehlbauer, G.J.; Springer, N.M. Brd1 Gene in Maize Encodes a Brassinosteroid C-6 Oxidase. *PLoS ONE* **2012**, *7*, e30798. [CrossRef]
118. Okagaki, R.J.; Haaning, A.; Bilgic, H.; Heinen, S.; Druka, A.; Bayer, M.; Waugh, R.; Muehlbauer, G.J. ELIGULUM-A regulates lateral branch and leaf development in barley. *Plant Physiol.* **2018**. [CrossRef]
119. Vollbrecht, E.; Veit, B.; Sinha, N.; Hake, S. The developmental gene *Knotted-1* is a member of a maize homeobox gene family. *Nature* **1991**, *350*, 241. [CrossRef]

120. Ramirez, J.; Bolduc, N.; Lisch, D.; Hake, S. Distal expression of knotted1 in maize leaves leads to reestablishment of proximal/distal patterning and leaf dissection. *Plant Physiol.* **2009**, *151*, 1878–1888. [CrossRef]
121. Lewis, M.W.; Bolduc, N.; Hake, K.; Htike, Y.; Hay, A.; Candela, H.; Hake, S. Gene regulatory interactions at lateral organ boundaries in maize. *Development* **2014**, *141*, 4590–4597. [CrossRef]
122. Foster, T.; Yamaguchi, J.; Wong, B.C.; Veit, B.; Hake, S. Gnarley1 is a dominant mutation in the knox4 homeobox gene affecting cell shape and identity. *Plant Cell* **1999**, *11*, 1239–1252. [CrossRef] [PubMed]
123. Foster, T.; Veit, B.; Hake, S. Mosaic analysis of the dominant mutant, Gnarley1-R, reveals distinct lateral and transverse signaling pathways during maize leaf development. *Development* **1999**, *126*, 305–313. [PubMed]
124. Muehlbauer, G.J.; Fowler, J.E.; Girard, L.; Tyers, R.; Harper, L.; Freeling, M. Ectopic Expression of the Maize Homeobox Gene Liguleless3 Alters Cell Fates in the Leaf 1. *Plant Physiol.* **1999**, *119*, 651–662. [CrossRef] [PubMed]
125. Fowler, J.E.; Muehlbauer, G.J.; Freeling, M. Mosaic analysis of the liguleless3 mutant phenotype in maize by coordinate suppression of mutator-insertion alleles. *Genetics* **1996**, *143*, 489–503. [PubMed]
126. Muehlbauer, G.J.; Fowler, J.E.; Freeling, M. Sectors expressing the homeobox gene *liguleless3* implicate a time-dependent mechanism for cell fate acquisition along the proximal-distal axis of the maize leaf. *Development* **1997**, *124*, 5097–5106. [PubMed]
127. Bertrand-Garcia, R.; Freeling, M. Hairy-Sheath Frayed # 1-0: A Systemic, Heterochronic Mutant of Maize that Specifies Slow Developmental Stage Transitions. *Am. J. Bot.* **1991**, *78*, 747–765. [CrossRef]
128. Saberman, J.; Bertrand-Garcia, R. Hairy-sheath-frayed 1-O is a Non-Cell-Autonomous Mutation That Regulates Developmental Stage Transitions in Maize. *J. Hered.* **1997**, *88*, 549–553. [CrossRef]
129. Tsuda, K.; Abraham-Juarez, M.J.; Maeno, A.; Dong, Z.; Aromdee, D.; Meeley, R.; Shiroishi, T.; Nonomura, K.; Hake, S. KNOTTED1 cofactors, BLH12 and BLH14, regulate internode patterning and vein anastomosis in maize. *Plant Cell* **2017**. [CrossRef]
130. Osmont, K.S.; Sadeghian, N.; Freeling, M. Mosaic analysis of extended *auricle1* (*eta1*) suggests that a two-way signaling pathway is involved in positioning the blade/sheath boundary in *Zea mays*. *Dev. Biol.* **2006**, *295*, 1–12. [CrossRef]
131. Osmont, K.S.; Jesaitis, L.A.; Freeling, M. The *extended auricle1* (*eta1*) Gene Is Essential for the Genetic Network Controlling Postinitiation Maize Leaf Development. *Genetics* **2003**, *165*, 1507–1519.
132. Tsiantis, M.; Brown, M.I.N.; Skibinski, G.; Langdale, J.A. Disruption of Auxin Transport Is Associated with Aberrant Leaf Development in Maize. *Plant Physiol.* **1999**, *121*, 1163–1168. [CrossRef] [PubMed]
133. Lewis, M.W.; Bolduc, N.; Hake, K.; Htike, Y.; Hay, A.; Candela, H. Recruitment of regulatory interactions from the inflorescence to the leaf in the dominant Wavy auricle in blade mutant. *Development* **2014**, *141*, 4590–4597. [CrossRef] [PubMed]
134. Walsh, J.; Freeling, M. The *liguleless2* gene of maize functions during the transition from the vegetative to the reproductive shoot apex. *Plant J.* **1999**, *19*, 489–495. [CrossRef] [PubMed]
135. Busch, B.L.; Schmitz, G.; Rossmann, S.; Piron, F.; Ding, J.; Bendahmane, A.; Theres, K. Shoot branching and leaf dissection in tomato are regulated by homologous gene modules. *Plant Cell* **2011**. [CrossRef] [PubMed]
136. Raman, S.; Greb, T.; Peaucelle, A.; Blein, T.; Laufs, P.; Theres, K. Interplay of *miR164*, *CUP-SHAPED COTYLEDON* genes and *LATERAL SUPPRESSOR* controls axillary meristem formation in *Arabidopsis thaliana*. *Plant J.* **2008**, *55*, 65–76. [CrossRef] [PubMed]
137. Lee, D.-K.; Geisler, M.; Springer, P.S. *LATERAL ORGAN FUSION1* and *LATERAL ORGAN FUSION2* function in lateral organ separation and axillary meristem formation in *Arabidopsis*. *Development* **2009**, *136*, 2423–2432. [CrossRef] [PubMed]
138. Serrano-Mislata, A.; Bencivenga, S.; Bush, M.; Schiessl, K.; Boden, S.; Sablowski, R. *DELLA* genes restrict inflorescence meristem function independently of plant height. *Nat. Plants* **2017**, *3*, 749–754. [CrossRef] [PubMed]
139. Schürholz, A.-K.; López-Salmerón, V.; Li, Z.; Forner, J.; Wenzl, C.; Gaillochet, C.; Augustin, S.; Barro, A.V.; Fuchs, M.; Gebert, M.; et al. A Comprehensive Toolkit for Inducible, Cell Type-Specific Gene Expression in *Arabidopsis*. *Plant Physiol.* **2018**, *178*, 40–53. [CrossRef] [PubMed]
140. Reddy, G.V.; Heisler, M.G.; Ehrhardt, D.W.; Meyerowitz, E.M. Real-time lineage analysis reveals oriented cell divisions associated with morphogenesis at the shoot apex of *Arabidopsis thaliana*. *Development* **2004**, *131*, 4225–4237. [CrossRef] [PubMed]

141. Soyk, S.; Lemmon, Z.H.; Oved, M.; Fisher, J.; Liberatore, K.L.; Park, S.J.; Goren, A.; Jiang, K.; Ramos, A.; van der Knaap, E.; et al. Bypassing Negative Epistasis on Yield in Tomato Imposed by a Domestication Gene. *Cell* **2017**, *169*, 1142–1155. [CrossRef]
142. Rodríguez-Leal, D.; Lemmon, Z.H.; Man, J.; Bartlett, M.E.; Lippman, Z.B. Engineering Quantitative Trait Variation for Crop Improvement by Genome Editing. *Cell* **2017**, *171*, 470–480. [CrossRef] [PubMed]

© 2018 by the authors. Licensee MDPI, Basel, Switzerland. This article is an open access article distributed under the terms and conditions of the Creative Commons Attribution (CC BY) license (http://creativecommons.org/licenses/by/4.0/).

Review

Plant Vascular Tissues—Connecting Tissue Comes in All Shapes

Eva Hellmann [1,†], Donghwi Ko [1,†], Raili Ruonala [1,2,†] and Ykä Helariutta [1,2,*]

[1] The Sainsbury Laboratory, University of Cambridge, Cambridge CB2 1LR, UK; eva.hellmann@slcu.cam.ac.uk (E.H.); donghwi.ko@slcu.cam.ac.uk (D.K.); raili.ruonala@slcu.cam.ac.uk (R.R.);
[2] Institute of Biotechnology, Department of Biological and Environmental Sciences, University of Helsinki, FI-00014 Helsinki, Finland
[*] Correspondence: yrjo.helariutta@slcu.cam.ac.uk (Y.H.)
[†] These authors contributed equally to this work

Received: 2 November 2018; Accepted: 7 December 2018; Published: 13 December 2018

Abstract: For centuries, humans have grown and used structures based on vascular tissues in plants. One could imagine that life would have developed differently without wood as a resource for building material, paper, heating energy, or fuel and without edible tubers as a food source. In this review, we will summarise the status of research on *Arabidopsis thaliana* vascular development and subsequently focus on how this knowledge has been applied and expanded in research on the wood of trees and storage organs of crop plants. We will conclude with an outlook on interesting open questions and exciting new research opportunities in this growing and important field.

Keywords: Vasculature; Organogenesis; Development

1. Vasculature and Its Arrangement

In the 19th century, the variety of vascular arrangements in form of different stele types attracted the interest of researchers. From their analyses, they could conclude that different forms of steles can specialize in supporting different functions and their different shapes are specific for plant groups, enabling them to draw phylogenetic connections between groups [1]. Even within one plant, various stele types occur. The different stele types vary not only with developmental stages, but also within different mature organs such as leaves, stem, hypocotyl, and roots. Although the structures in different species and organs are of diverse build, they share some of the underlying regulatory mechanisms and their main functions for the plant. Generally, they enable plants to transport water, nutrients, assimilates, as well as signalling molecules, and provide stability to the plant body.

In this short review, we will focus on *Arabidopsis thaliana* as an example of an herbaceous species and as a commonly used model plant, in which many of the regulatory pathways for vascular development and arrangement have been elucidated. Furthermore, we will look at angiosperm trees, as they are a model for economically important wood production and tubers, which are essential agronomical food sources all over the world. As many processes underlying wood and tuber formation are shared, research on vascular development in *Arabidopsis* has and will inspire discoveries and development in economically and agronomically important vascular structures. Research on vascular development and expansion involving various species and growth forms is an excellent example of how basic research and applied research can work hand in hand to promote the growth of scientific knowledge and its application.

2. Vascular Development in *Arabidopsis thaliana*

Vascular development in *Arabidopsis thaliana* has been a topic of intensive research for decades. Basic principles of vascular development in roots, hypocotyl, leaves, and stems have been elucidated and gene regulatory networks have been inferred. In the following chapters, we will introduce the primary and secondary development of *Arabidopsis thaliana* root, hypocotyl, and stems, with its main regulators, and subsequently look at wood development and tuber formation.

2.1. Vascular Development in the Root

Arabidopsis root vascular development initiates during embryogenesis. Provascular tissue is specified by a spatially and temporally confined auxin maximum established by the PIN-FORMED (PIN) auxin transport function (Figure 1) [2,3]. *MONOPTEROS* (*MP*) expression, which marks future veins, is induced by auxin [4–7] and provides feedback on the auxin status by promoting *PIN1* expression [5,8–10]. Another component of auxin signalling, BODENLOS (BDL), was found to regulate *TAGRET OF MONOPTEROS* (*TMO*) 3, 5, 6, and 7 upwards [10], which proved to be essential for proper MP function [11]. The MP–TMO5–LONESOME HIGHWAY (LHW) module, regulating cell division in the whole plant, was also found to play a role in the definition of the provasculature [8,11,12]. Among other factors, cytokinin is important for provascular development. The TMO5–LHW module induces cytokinin biosynthesis via activation of *LONELY GUY* (*LOG*) genes [13] and the cytokinin transporter PURINE PERMEASE 14 (PUP14) is required for early vascular development [14].

The postembryonic root vasculature in *Arabidopsis* consists of a xylem strand that is surrounded by procambial cells and two opposing phloem poles. Layers of pericycle and endodermis enclose the vascular cylinder (Figure 2A). As is the case during embryogenesis, auxin and cytokinin play a major role in postembryonic development (Figure 3). Cytokinin reporters are expressed in the procambium, whereas auxin reporters mark the xylem cells [15,16]. The dominant negative cytokinin receptor mutant *wooden leg* (*wol*) shows a reduced number of vascular cell files and all inner cell types differentiate into protoxylem [17,18]. The lack of all three receptor kinases for cytokinin perception leads to a similar phenotype [19] as does the overexpression of a cytokinin degrading enzyme of the CYTOKININE OXIDASE (CKX) family [20,21]. The inhibitor of cytokinin signalling ARABIDOPSIS HISTDINE PHOSPHOTRANSFER PROTEIN 6 (AHP6) plays an important role in protoxylem differentiation [20]. It is upregulated by auxin and is a major component of the mutual inhibitory cytokinin–auxin feedback loop regulating procambium maintenance versus xylem differentiation [22]. Another interconnection between auxin and cytokinin regulation is the TMO5–LHW pathway. In postembryonic development, the TMO5–LHW dimer is, as in provascular development, induced by auxin via MP and activates cytokinin biosynthesis via upregulation of *LOG* genes [13]. Aside from auxin and cytokinin, phytohormone jasmonic acid has also been shown to regulate xylem development. An increase of jasmonic acid levels leads to extra xylem vessels, but this is abolished in jasmonic acid receptor mutants. Jasmonic acid function in vessel development is linked to cytokinin signalling via regulation of *AHP6* by the jasmonic acid regulated transcription factor MYC2 [23]. Further regulators of xylem differentiation include HD-ZIP IIIs that promote metaxylem development [24] and are regulated via the SHORTROOT (SHR)–SCARECROW (SCR) pathway [25,26] via the levels of the inhibitory miRNAs mi165/166 [24,27]. The metaxylem cell fate is also characterised by the expression of the thermospermine biosynthesis gene *ACAULIS 5* (*ACL5*) [28,29]. Thermospermine regulates the translation of the SUPRESSOR OF ACAULIS LIKE (SACL) protein family, which then affects the TMO5–LHW interaction that acts on xylem differentiation and cytokinin biosynthesis [12,13,30–33].

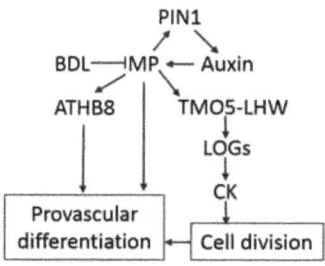

Provascular differentiation

Figure 1. *Arabidopsis* provascular differentiation. MONOPTEROS (MP) is a central regulator in provascular development. It is induced by auxin and promotes auxin flow by induction of *PIN-FORMED (PINs)*. MP function is also modified by BODENLOS (BDL). MP enhances *ATHB8* expression, which contributes to provascular differentiation. It also regulates the TAGRET OF MONOPTEROS 5 (TMO5)–LONESOME HIGHWAY (LHW) dimer, which activates CK (cytokinin) biosynthesis and promotes cell division. LOG—LONELY GUY.

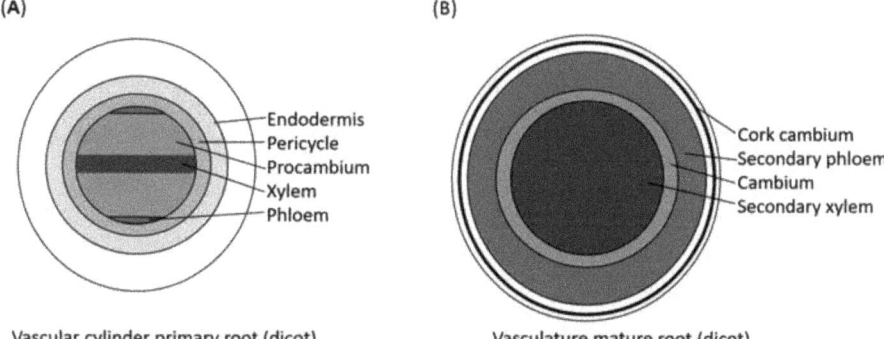

Figure 2. Schematic cross-sections of a primary root (**A**) and a mature root with secondary growth (**B**). In the primary root, two phloem poles are separated by procambium surrounding the central xylem axis. Around this structure, a ring of pericycle cells and endodermis cells can be found (**A**). In roots that have gone through secondary growth, there is a central secondary xylem cylinder surrounded by a continuous cambium and a ring of secondary phloem. Further out, a cork cambium can serve as a lateral meristem giving rise to cork and phelloderm (**B**).

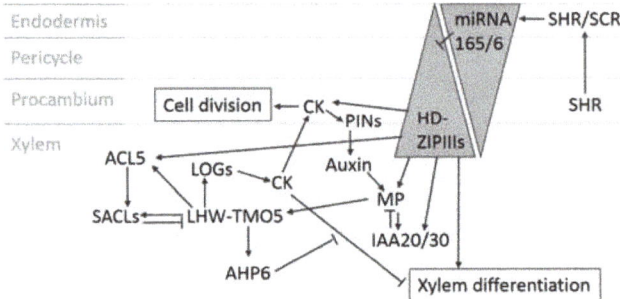

Xylem differentiation

Figure 3. Xylem differentiation in the primary root. HD-ZIP IIIs are important promotors of xylem differentiation. Their level is regulated by a gradient of inhibitory miRNA165/6. miRNA165/6 levels are defined by SHORTROOT (SHR)/SCARECROW (SCR) dimer presence. The gradient is established by SHR diffusion from its production area in the procambium to the endodermis, where it forms the dimer with SCR that promotes miRNA165/6 expression. This results in a miRNA165/6 gradient with highest levels in the endodermis and an inverse gradient for HD-ZIP IIIs that promote xylem differentiation. The HD-ZIP IIIs induce *MP* and *IAA20/30*. They also enhance *ACL5* expression and CK (cytokinin) production. ACL5 induces translation of SUPRESSOR OF ACAULIS LIKE (SACL) genes that inhibit LHW–TMO5 dimerization and thus *LOG* expression, lowering the CK levels. The dimer also induces the CK signalling inhibitor *AHP6*, inhibiting the negative effect of CK on xylem differentiation. In the procambium, the CK inhibitory effects mediated by the HD-ZIP IIIs are not present, which leads to higher CK levels and signalling, resulting in cell division rather than xylem differentiation. CK induces PIN activity, pumping auxin out of the procambium. This causes an auxin maximum in the xylem axis, which subsequently induces *MP* expression.

Protophloem differentiation in *Arabidopsis thaliana* is dependent on the OCTOPUS (OPS)–BRASSINOSTEROID INSENSITIVE 2 (BIN2)–BRASSINOSTEROID INSENSITIVE 1 (BIN1) cascade, on COTYLEDON VASCULAR PATTERN 2 (CVP2), and on the BREVIS RADIX (BRX)–BARELY ANY MERISTEM 3 (BAM3)–CLVATA3/ESR-related (CLE45) module. OPS represses *BIN2* [34,35]. BRX acts in a similar way and restricts *BAM3* expression confining CLE45 perception spatially [36,37]. Recently, other receptors for the CLE peptides have been identified that act independently of CORYNE (CRN)–CLAVATA 2 (CLV2). CLE-RESISTANT RECEPTOR KINASE (CLERK) and its homologues SENESCENCE-ASSOCIATED RECEPTOR-LIKE KINASE (SARK) and NSP-INTERACTING KINASE 1 (NIK1) represent a new module for CLE sensing in protophloem development [38,39]. The CLE45 signal was shown to be enhanced by MEMBRANE-ASSOCIATED KINASE REGULATOR 5 (MAKR5) action [40]. For sieve element differentiation, SUPPRESSOR OF MAX1-LIKE (SMXL) 3, 4, and 5 are required [41]. In contrast to the regulation of procambium proliferation and xylem differentiation, not cytokinin or auxin, but brassinosteroids are the most influential phytohormones for phloem differentiation [40,42,43]. Protophloem sieve element development is modulated by interaction with BRX and PROTEIN KINASE ASSOCIATED WITH BRX (PAX) with PIN1. Whereas BRX inhibits PIN1 mediated auxin efflux, PAX enhances it, leading to a balanced and ordered regulation of auxin distribution that is necessary for protophloem development [44].

ALTERED PHLOEM DEVELOPMENT (APL) regulates phloem differentiation [45,46]. NO APICAL MERISTEM, ATAF, CUP-SHAPED COTYLEDON (NACs), and NAC45/86-DEPENDENT EXONUCLEASE-DOMAIN PROTEINs (NENs) are involved in phloem maturation, which culminates in enucleation and the presence of fully developed sieve pores [47]. Furthermore, NAC20 was found to negatively regulate *APL* in phloem development [48].

Secondary growth in herbaceous dicotyledonous species such as *Arabidopsis* is characterised by the build-up of secondary cell walls in the xylem and lateral growth via a continuous cambium. These events are prominent in the *Arabidopsis* stem and hypocotyl, which are discussed next.

2.2. Vascular Development in Shoot and Hypocotyl

The elongation of the *Arabidopsis* inflorescence stem (bolting) coincides with the transition from the vegetative to the reproductive stage. The primary shoot apical meristem is committed to producing flowers and the rib meristem is activated to push the newly forming flowers upwards from the vegetative rosette. The molecular mechanisms regulating the primary vascular patterning in the extending tip of the young stem are poorly understood [49]. The stem vasculature is organized in separate bundles that eventually become connected by a so-called interfascicular cambium [50]. In the basal part of the stem, in the vicinity of the rosette, the activity of the interfascicular cambium results in complete cylindrical rings of the vascular tissues: phloem, cambium, and xylem, one inside another. Like the primary vascular organization in the *Arabidopsis* root described in the previous section, the *Arabidopsis* hypocotyl (embryonic stem) develops a xylem axis in the centre of the stele and two phloem poles, which are intervened by procambial cells, during the primary growth [51,52]. Common molecular factors modulate the primary vascular development in the root and the hypocotyl. Mutants defected in the primary vascular patterning in the root also exhibit similar flaws in the hypocotyl vasculature. For instance, *MP* and *WOL* are expressed in the root and the hypocotyl vascular tissues during embryogenesis and post-embryonic development, and the mutants are impaired in the vascular patterning of both organs [6,18,53]. In contrast to the *Arabidopsis* root, which has been a representative system to study the primary growth, the *Arabidopsis* hypocotyl and inflorescence stem have been useful model systems to scrutinize the molecular processes underlying secondary growth [52,54,55]. Especially, the hypocotyl undergoes substantial secondary thickening by the activity of vascular cambium and cork cambium, similar to wood formation in trees. The hypocotyl does not grow longitudinally during secondary growth, which makes it easier to observe the progression of radial thickening in a time-dependent manner [52,56,57]. Indeed, multiple molecular components such as phytohormones, transcription factors, peptides, and receptors, orchestrating the secondary growth in the *Arabidopsis* hypocotyls and the inflorescence stem, have been characterised [52,54,55]. In this section, we will mainly introduce the signalling networks underlying the secondary development in the hypocotyl and the inflorescence stem.

The radial secondary growth of the hypocotyl starts after the cambium forms and can be divided into two distinct phases, characterised by the xylem expansion accompanied by a fibre differentiation [56–58]. In phase I, the early phase, xylem vessel elements emerge and the surrounding cells remain as xylem parenchyma cells [56]. Similarly, during the early phase in the phloem, sieve elements, companion cells and parenchyma cells differentiated, but not fibres [52]. The expansion rates of the two conducting tissues are comparable in the early stage; thereby leaving the proportions of xylem and phloem to the total transverse area of the hypocotyls roughly constant [57,58]. In contrast, in phase II, parenchyma cells in the xylem and the phloem differentiate into xylem or phloem fibres with thick secondary cell walls, providing mechanical strength to the plants. The xylem area expands faster than the phloem, which leads to a higher ratio of xylem to phloem, like wood [56–58] (Figure 2B). According to studies done by Ragni and co-workers [59], the transition from phase I to II in hypocotyls concurs with the development of the inflorescence stem (conversion from vegetative to reproductive growth) in various rosette plants including *Arabidopsis thaliana*, *Cardamine hirsute*, *Barberea verna*, and *Taraxacum officinalis* [57,59]. However, this seems to be characteristic to rosette plants as the non-rosette plants (*Arabis alpine*, *Aster alpinus*, *Nicotiana benthamiana*, and *Solanum lycopersicum*) examined undergo the xylem expansion during vegetative growth [59]. Ragni et al. also found that xylem expansion is not regulated by floral specification, bolting, or age of the plants, but by gibberellin (GA), a phytohormone that is produced in the shoot upon flowering induction [59]. The detailed molecular mechanism underlying the GA signalling-mediated fibre differentiation remains to be unveiled, but recently, it was reported that the GA increases the expression of *NAC SECONDARY WALL THICKENING PROMOTING FACTOR 1* (*NST1*) and *NST3*, the master transcription factors implicated in secondary cell wall thickening of xylem fibres [51,60]. They are homologous to the VND6 and VND7 factors, which are sufficient to guide secondary cell wall formation during xylem vessel formation [61]. In addition, it was shown that the leucine-rich receptor-like kinases (LRR-RLKs) ERECTA (ER) and its paralogue ER-LIKE1 (ERL1) prevent the premature GA-induced fibre differentiation in *Arabidopsis* hypocotyls upon the floral transition by suppressing the expression of *NST1* and *NST3* [51]. Not only GA-induced xylem fibre differentiation, but also the suppression of the two *NSTs* by ER and ERL1 are largely dependent on the class I KNOTTED1-like homeobox (KNOX) transcription factor 1 (KNAT1)/BREVIPEDICELLUS (BP), which was previously shown to regulate xylem fibre differentiation in the inflorescence stem [51,62]. Furthermore, KNAT1/BP and another class I KNOX transcription factor, SHOOT MERISTEMLESS (STM), were shown to repress the transcription of *BLADE-ON-PETIOLE 1* (*BOP1*) and *BOP2*. Both encode BTB/POZ domain and ankyrin repeat-containing proteins, which negatively regulate xylem fibre differentiation in the hypocotyl [63]. Recently, Aurora kinases were identified as additional regulators of vascular development. They inhibit xylem and phloem formation via the transcriptional regulation of *ALTERED PHLOEM DEVELOPMENT* (*APL*), *VASCULAR-RELATED NAC-DOMAIN 6* (*VND6*), and *VND7* [64].

In addition to the genetic interactions implicated in fibre differentiation, a few other transcription factors involved in the cambial activity in hypocotyls and stems have been identified (Figure 4). For instance, *WUSHEL-related HOMEOBOX 4* (*WOX4*) and *WOX14* are upregulated by the CLE41/44/TRACHEARY ELEMENT DIFFERENTIATION INHIBITORY FACTOR (TDIF) (peptide ligands)-PHLOEM INTERCALATED WITH XYLEM (PXY)/TDIF RECEPTOR (TDR) (LRR-RLK) module in the cambium and play a part in cambial proliferation [65–70]. In parallel to the CLE41/44/TDIF-PXY/TDR module, the signalling by the phytohormone ethylene facilitates cambial cell division by inducing *ETHYLENE RESPONSE FACTOR*s (*ERF*s), such as *ERF109*, *ERF018*, and *ERF1* [71]. It was suggested that the two signal cascades interact with each other via ethylene, inducing the expression of *PXY/TDR* but WOX4 suppressing ethylene signalling [71]. Two more receptor-like kinases, REDUCED IN LATERAL GROWTH1 (RUL1) and MORE LATERAL GROWTH1 (MOL1), are also involved in regulation of cambial activity [72,73]. There seem to be complex interactions between hormonal pathways, the LRR-RLKs and the transcription factors to fine-tune vascular development. For example, *WOX4* is also shown to be upregulated by auxin and the induction is stabilized in a PXY/TDR-dependent manner [74]. Recently, it was reported that WOX14 is also involved in the xylem differentiation by inducing the expression of GA3-oxidase, which catalyses the production of bioactive GAs in the vascular bundle of the inflorescence stem [75]. Furthermore, in the stem, ER is shown to suppress the expression of *PXY-LIKE 1* (*PXL1*) and *PXL2*, while PXY, PXL1, PXL2, and ER upregulate the expression of *ERL1* and *ERL2* [76]. Interestingly, the interactions in the hypocotyl are distinct from those in the stem. In the hypocotyl, PXY, PXL1, PXL2, and ER repress the expression of *ERL1* and *ERL2* [76].

Furthermore, other phytohormones, such as auxin, cytokinin, strigolactone, and jasmonic acid, positively regulate cambial activity [77–79] and the interactions between key regulators during the secondary growth were recently analysed by network modelling [80]. Recently, the molecular interactions between auxin, cytokinin, and PXY signalling have been elucidated. Han and co-workers demonstrated that the CLE41/44/TDIF-PXY/TDR module regulates cambial proliferation by inhibiting BIN2-LIKE 1 (BIL1). BIL1 phosphorylates MP, which, upon phosphorylation, enhances the expression of *ARABIDOPSIS RESPONSE REGULATOR* (*ARR*) *7* and *15*, resulting in suppression of cambial activity [81]. Moreover, it was reported that auxin signalling in the *Arabidopsis* inflorescence stem not only promotes cambial activity by inducing *AUXIN RESPONSE FACTOR* (*ARF*) *3* and *4* expression outside of the stem cell domain in the cambium, but also facilitates xylem differentiation of cambial cells through MP suppression of *WOX4* activity and direct activation of xylem-related genes [82]. Interestingly, *WOX4* expression is not altered in the *bil1* mutant, suggesting that the suppression of *WOX4* by MP would be independent of the BIL1-mediated phosphorylation [81]. In addition to promoting the cambial proliferation, the CLE41/44/TDIF-PXY/TDR module also represses xylem differentiation of cambial cells by stimulating the activity of BIN2. BIN2 inhibits *BRI1-EMS-SUPPRESSOR (BES1)*, a downstream transcription factor of brassinosteroid signalling [83]. Not much is known about upstream acting factors, but it was shown that *KANADI* genes, GARP family transcription factors, negatively regulate cambial activity by disrupting expression and polar localization of PIN1 [84]. More recently, a novel regulator involved in phloem differentiation has been characterised. The zinc-finger RNA-binding protein JULGI binds to the 5' UTR of SMXL4/5 mRNA, inhibiting their translation and suppressing phloem development [85].

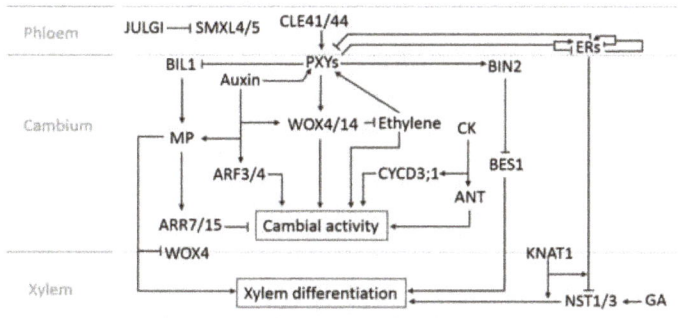

Xylem differentiation in secondary growth

Figure 4. CLAVATA3/ESR-related (CLE)41/44/TRACHEARY ELEMENT DIFFERENTIATION INHIBITORY FACTOR (TDIF) generated in the phloem is perceived by the cambium-localised PHLOEM INTERCALATED WITH XYLEM (PXY)/TDIF RECEPTOR (TDR), which induces expression of *WUSCHEL-related HOMEOBOX (WOX)4/14* and activates BRASSINOSTEROID INSENSITIVE 2 (BIN2). WOX4/14 promotes cambial activity and BIN2 inhibits BRI1-EMS-SUPPRESSOR (BES1), which facilitates xylem differentiation. In addition, the CLE41/44/TDIF-PXY/TDR module enhances cambial activity by suppressing BIN2-LIKE 1 (BIL1)-mediated phosphorylation of MP that induces *ARR7/15* inhibition of cambial activity. The positive role of the auxin on cambium activity involves PXY and WOX4. In addition, auxin signalling upregulates the expression of *ARF3* and *ARF4* outside of the stem cell domain in the cambium, which facilitates the cambial proliferation. MP is induced by auxin and contributes to xylem differentiation via repressing *WOX4* but activating xylem-related genes in the cambium. Ethylene induces the expression of *PXY* and *ERF109*, *ERF018*, and *ERF1*, which enhances the cambial activity. WOX4 suppresses ethylene signalling. Cytokinin upregulates the expression of the D-type cyclin *CYCD3;1* and *AINTEGUMENTA (ANT)* to enhance cambial activity [86]. Gibberellin (GA) signalling facilitates xylem fibre differentiation by elevating the expression of *NST1* and *NST3* in a *KNAT1/BP*-dependent manner. In contrast, ERECTA (ER) and ER-LIKE1 (ERL1) inhibit the expression of *NST1* and *NST3* in a *KNAT1/BP*-dependent manner and suppress xylem differentiation. The two families of leucine-rich receptor-like kinases (LRR-RLKs), PXYs (PXY, PXL1, PXL2) and ERs (ER, ERL1, ERL2), mutually regulate their expressions. In the stem, ER suppresses the expression of *PXL1* and *PXL2*, whereas PXYs and ER upregulate the expression of *ERL1* and *ERL2*. However, in the hypocotyl, PXYs and ER repress the expression of *ERL1* and *ERL2*. Please note that we describe ER in the phloem section of the figure for simplicity, but it was shown that ER is expressed in the epidermis, phloem, and xylem of inflorescence stems [87], and ER and ERL1 are expressed in the stele of hypocotyls [51]. JULGI, which is expressed in the phloem and cambium, inhibits translation of SUPPRESSOR OF MAX1-LIKE (SMXL)4/5 by binding to the 5' UTR of their mRNAs, and thereby suppresses phloem differentiation.

3. Agronomically Important Structures Derived from Plant Vasculatures

As summarised above, a substantial amount of knowledge has been gained by examining vascular development in *Arabidopsis*. In the next two sections, we will focus on how this knowledge can and has been applied to agronomically important plants, especially to wood producing trees, and to species that produce edible tubers as storage organs. On the other hand, research in these fields has provided new insight that is feeding back into research on *Arabidopsis*.

3.1. Wood Development—Secondary Growth of Trees

Spontaneously, one might not consider *Arabidopsis*, a small inconspicuous weed, to be beneficial for studies on secondary growth. However, at a miniature scale, many developmental events found in *Arabidopsis* can mimic the same principal features that are landmarks for trees, even down to a molecular level. One such event characteristic of trees is the extensive formation of woody tissues in the trunk. A multitude of factors, for example, cytokinin, auxin, gibberellin and ethylene, HD-ZIP IIIs, as well as the PXY-CLE41/44 signalling pathway and its target *WOX4*, have been shown to influence secondary growth in trees [88–95] in a manner similar to *Arabidopsis*. These aspects have been extensively reviewed (e.g., [96–98]; also, see above). In this section, we provide an overview of wood (secondary xylem) characteristics in angiosperm trees, and highlight some recent advances in this research field.

Secondary growth relies on closely coordinated cell division in the meristematic zone (the cambium); subsequent expansion; secondary cell wall development; and, in some cases, programmed cell death, all of which finally result in differentiated daughter cells serving their function. In a tree trunk during the active growth season, the cambial zone is composed of several layers of thin-walled cells that appear alike in histological cross-sections (Figure 5A). Recently, Bossinger et al. [99] performed an interesting somatic sector analysis in the *Populus* stem, suggesting the existence of a single cell layer of cambial initials, thought of as stem cells, that can divide in both anticlinal and periclinal orientations, and independently give rise to xylem or phloem. With their system, the authors succeeded in visualizing cell fate during wood development deep inside the trunk over the course of several months, providing insight into the cambial dynamics in a mature tree trunk. Another recently reported toolkit that may be expected to advance our understanding of wood development is the protein–protein and protein–DNA interactome, covering a set of genes expressed in the secondary tissues of *Populus* trunk [100,101]. On top of the high-resolution transcriptomics, hormonal profiling, and proteomics data accumulating from *Populus* ([102–105]), this adds to the growing body of resources available from this prominent tree model species.

Cambium produces secondary xylem, wood, towards the pith of the stem. Wood appears heterogeneous in a sense that it is composed of several cell types with a variable size and function, however, the majority of them are hollow and heavily lignified when mature ([107]). Besides lignin, cellulose and hemicellulose are major components of the secondary cell wall [108,109]. Such solid structures are necessary to support the weight of the plant tissues, including various substances within these tissues, as well as to provide protection against parasites and bacteria. The water-conducting cells are commonly known as tracheary elements (vessels and tracheids). Of these, vessels are the primary conduits for long-distance water transport in the angiosperm wood, while tracheids are predominant in gymnosperms. Typically, vessel elements are decorated by secondary cell wall thickenings and connected at their ends by perforated cell plates to allow a continuum throughout the plant. Vessels are outstanding by terms of a large diameter when compared with any other xylem cell type, which contributes to high efficiency in water transport. On the other hand, the width of the vessels increases the risk of embolism induced by freeze–thaw cycling at temperate regions or during drought (see [110]). Correlations between embolism resistance and lignin contents of wood have been indicated, suggesting that both the herbaceous, including *Arabidopsis*, and tree species with a high lignin content are more resistant to embolism [111,112]. Factors underlying the spatial patterning of vessels, or any other cell type, within the wood are poorly understood, however, a recent report suggests a role for basipetal auxin transport in *Populus* vessel distribution [113].

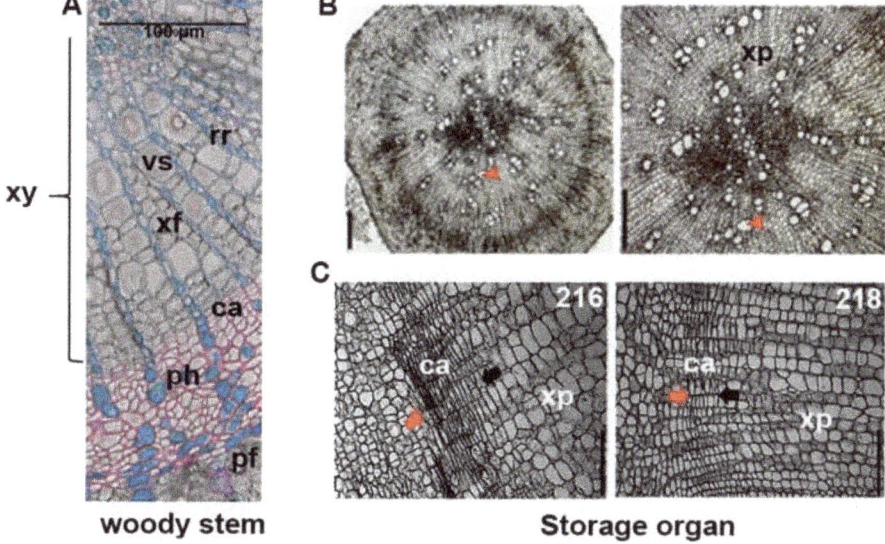

woody stem   **Storage organ**

Figure 5. Cross-sections of birch (*Betula pendula*) stem (**A**) and storage organs of radish (*Raphanus sativus*) (**B**,**C**). Angiosperm wood is composed of xylem fibres and vessels to a large extent (**A**), while there is a substantial number of xylem parenchyma cells in the radish of three-week-old line 216 (**B**). The red arrow head indicates one of the xylem vessel cells. The size bar is 200 µm (**B**). The line 216 (**left**), which generates the larger storage organ, harbours a wider cambial zone when compared with the line 218 (**right**). Cambium zones are marked by red and black arrows with the red arrows on the side of the cortex region. The size bar is 100 µm (**C**). Abbreviations: xy—xylem; vs—vessel; xf—xylem fibre; rr—radial ray; ca—cambium; ph—phloem; pf—phloem fibre; xp—xylem parenchyma. (**A**) By courtesy of Chang Su, University of Helsinki; (**B**) and (**C**) adapted with permission from Jang et al. (2015) [106] and http://www.biologists.com/journal-of-experimental-biology/ doi:10.1093/jxb/erv220.

In addition to the vessels, wood contains two other cell types: fibres and parenchyma. The thick-walled fibres constitute the bulk, up to 80% of the angiosperm wood (commonly named hardwood) volume [110] and provide mechanical support to the plant. In *Arabidopsis*, xylem fibres are found in all organs undergoing secondary growth upon induction of flowering. The hormonal and molecular basis of vessel and fibre differentiation processes has been studied extensively in various plant systems such as *Arabidopsis* and *Populus*, and important discoveries regarding secondary cell wall formation and programmed cell death have also arisen from *Zinnia* and *Arabidopsis* suspension cell culture systems. In previous sections, we discussed some factors involved in fibre and vessel differentiation, however, further perspectives on this topic are provided in a number of recent reviews (e.g., [107,108]). While the vessels and fibres are programmed to die, the xylem parenchyma remain as living cells. In trees, parenchyma cells form rays that facilitate radial transport of water and solutes across the vascular tissues. Furthermore, rays function in carbohydrate storage and protection from embolism within the xylem [114]. As the secondary growth in *Arabidopsis* is limited, rays apparently do not develop spontaneously and seem to represent a rare aspect of wood development that, in addition to seasonality, requires a long-living woody species for functional studies. However, formation of ray-like cells has been reported in *Arabidopsis* stems, where secondary growth was induced under weight stress [115], further highlighting the amenable nature of this little weed for a wide array of manipulations.

It is notable that wood is porous yet stiff, and typically requires drying as well as chemical processing prior to use as a construction material or pulp [116]. Wood processing methods are constantly optimized; for example, Song et al. [117] reported a compression method that, in combination with a carefully designed chemical treatment to partially remove lignin and hemicellulose, increased wood stiffness and strength by an impressive factor of 11. The authors were able to increase the strength of the cellulose component and, in fact, modify the wood structure and composition such that one might draw an imaginary analogue to tension wood (dried and flattened to an extreme). In nature, tension wood develops in the upper side of a tree branch or as a result of bending, to support the weight of the leaning structure. When compared with normal wood, tension wood appears more compact with smaller cell sizes. Furthermore, the cellulose content is higher and the mechanical strength is increased in this special type of angiosperm wood. Various phytohormones, including auxin, gibberellin and ethylene, have been implicated in the formation of tension wood-like features and transcriptomics analyses in *Populus* have shed light on the molecular control of this phenomenon (reviewed by Groover [118]). In the next sections, we move on to different types of special structures, edible storage organs, which, however, are also composed of vascular tissues.

3.2. Tubers—Edible Storage Organs

Various plant species have evolved secondary growth mechanisms specialized to produce storage organs composed of many parenchyma cells that accumulate photosynthates in the form of sucrose or starch. For instance, *Raphanus sativus* (radish), *Brassica rapa* (turnip), *Daucus carota* (carrot), and *Beta vulgaris* (sugar beet) generate storage organs (tubers) from their taproots. *Manihot esculenta* (cassava) and *Ipomoea batatas* (sweet potato) produce them from their fibrous roots and *Solanum tuberosum* (potato) forms tubers from stolons, underground stems [119–123]. The formation of tubers occurs when plants are exposed to the certain conditions, like short days, or when they reach a certain age, and they start to enlarge dramatically upon their initiation [119,123–125]. The initiation of potato tuberisation has been well characterised and has been reviewed [119]. Here, we mainly focus on the bulking stage of storage organs and putative approaches to enhance the secondary growth via modulation of the underlying signalling pathways.

The transverse structure of tubers varies but as a common feature they possess a high number of parenchyma cells for storage [106,126–131] (Figure 5B,C). To generate such structure during organogenesis, high cambial activity is needed to increment the number of cells and inhibit the differentiation of the parenchyma cells to the xylem fibres at the same time. This suggests that engineering tubers to reinforce cambial activity and to sustain the cells as parenchyma cells could increase the capacity and/or size of the storage organs. For this, knowledge about xylem differentiation in *Arabidopsis* secondary growth is of great value. One approach to improve cambial activity in the storage organs could be to engineer cytokinin biogenesis or signalling as cytokinins are crucial for cambial activity. Similarly, the CLE41/44/TDIF-PXY/TDR module and WOX4/14 transcription factors play a crucial part in the cambial activity and could be manipulated to enhance cell proliferation. GA is one of the key factors inducing xylem fibre differentiation. Inhibiting GA signalling or the downstream transcription factors, such as NST1 and NST3, could prevent xylem fibre differentiation of parenchyma cells and contribute to increasing storage capacity. Indeed, there are a few studies showing that storage organs development involves molecular components similar to modulators of *Arabidopsis* secondary development. Jang and co-workers showed that in a radish inbred line development of a larger storage organ correlates with stronger cambial activities and higher cytokinin responses in the cambium [106]. They demonstrated that exogenous cytokinin treatment can result in a substantial increase in cell proliferation in the cambium zone and overall secondary growth in a dose-dependent manner, suggesting that cytokinin signalling and responses are crucial for the secondary thickening of radish [106]. In addition, cytokinin signalling seems to be important for the initiation of tuberisation as the overexpression a cytokinin biosynthesis gene in tomato or exogenous cytokinin application together with sucrose of potato lead to the storage organ formation from their axillary buds [119,132]. Gancheva and co-workers showed that the transcripts of *RsCLE41*, the *AtCLE41* homologue, is primarily expressed in the cambium and the phloem of the radish. Interestingly, its expression in *Raphanus sativus* is much higher than in the presumably ancestral *Raphanus raphanistrum*, that does not produce the enlarged taproot tubers [133]. Moreover, exogenous treatment or overexpression of RsCLE41 increases the number of meristematic foci in the centre of the secondary xylem and facilitates cell division in the regular cambium and the meristematic foci. This suggests that the RsCLE41-mediated signalling is involved in the secondary growth of radish as well as in *Arabidopsis* [133]. GA induces xylem fibre differentiation in *Arabidopsis*, so exogenous GA treatment might reduce the tuber productivity, whereas treatment with the GA biosynthesis inhibitor paclobutrazol (PBZ) might elevate it. Several studies examined the effect of exogenous GA and PBZ treatment on storage organ development and showed that GA-treated carrot and radish are inhibited in storage organ secondary thickening whereas PBZ-treated carrot, radish and potato exhibited enhanced thickening [131,134–136]. It was shown that the exogenous GA facilitates the xylem differentiation and increases the lignin content in the carrot [131]. In radish, it was shown that PBZ treatment increases the number of cells in the xylem area and the size of xylem vessels [135], suggesting that the suppression GA signalling can be used to increase storage organ productivity. In addition to applying knowledge gained from unravelling the networks regulating secondary growth in *Arabidopsis*, there have been approaches that characterize genome-wide transcriptomic changes during the tuberisation or comparisons between tuberous and non-tuberous roots to understand the bulking processes in the radish, cassava, and sweet potato [120,125,128,137]. Altogether, the application of those advances can contribute to progress in engineering or breeding to enhance tuber productivity.

4. Future Perspectives

Our current understanding of secondary growth provides fundamental knowledge to improve wood formation. On the basis of research on *Arabidopsis* secondary growth, engineering of wood formation in tree species has made great progress in the last decade (e.g., [88,93,102]). It is not yet known exactly how storage organs develop a substantial number of xylem parenchyma cells with high sugar or starch content. The *Arabidopsis* hypocotyl and its underlying regulatory network can be very informative for the examination of storage organ regulation and its engineering for higher productivity. A major question is how the switch between differentiation of fibres versus parenchyma cells is regulated in the hypocotyl of *Arabidopsis*, as well as in its Brassiceae relatives with storage root capacity. Insights into secondary growth regulation, storage root development in crops and in potentially new model species, the identification of potential targets for engineering in those, and the development of adjusted methods are particularly relevant, as crop species exhibiting storage roots are currently not compatible with intensive molecular genetics, thus hampering their genetic analysis and efficient bioengineering.

Author Contributions: E.H., D.K. and R.R. contributed equally to the manuscript. Y.H. supervised and edited the work. All authors wrote the manuscript and approved the draft.

Funding: Y.H. was supported by Finnish Centre of Excellence in Molecular Biology of Primary Producers (Academy of Finland CoE program 2014-2019) decision #271832. Y.H.'s laboratory was funded by the Gatsby Foundation [GAT3395/PR3]; the National Science Foundation Biotechnology and Biological Sciences Research Council grant [BB/N013158/1]; University of Helsinki [award 799992091], the European Research Council Advanced Investigator Grant SYMDEV [No. 323052]. D.K. was funded by an EMBO long-term fellowship ALTF 305-2017.

Acknowledgments: We thank Chang Su (University of Helsinki) for kindly providing an unpublished image of a birch stem section. The images of radish root sections were adapted with the permission of Ji-Young Lee (Seoul National University) and the publisher of the original article (Journal of Experimental Botany; [106]), for which we are thankful.

Conflicts of Interest: The authors declare no conflict of interest.

References

1. De Bary, A.; Bower, F.O.; Scott, D.H. *Comparative Anatomy of the Vegetative Organs of the Phanerogams and Ferns*; Clarendon Press: Oxford, UK.
2. Steinmann, T.; Geldner, N.; Grebe, M.; Mangold, S.; Jackson, C.L.; Paris, S.; Gälweiler, L.; Palme, K.; Jürgens, G. Coordinated Polar Localization of Auxin Efflux Carrier PIN1 by GNOM ARF GEF. *Science* **1999**, *286*, 316. [CrossRef] [PubMed]
3. Friml, J.; Vieten, A.; Sauer, M.; Weijers, D.; Schwarz, H.; Hamann, T.; Offringa, R.; Jürgens, G. Efflux-dependent auxin gradients establish the apical–basal axis of Arabidopsis. *Nature* **2003**, *426*, 147–153. [CrossRef]
4. Mayer, U.; Ruiz, R.A.T.; Berleth, T.; Miséra, S.; Jürgens, G. Mutations affecting body organization in the Arabidopsis embryo. *Nature* **1991**, *353*, 402–407. [CrossRef]
5. Wenzel, C.L.; Schuetz, M.; Yu, Q.; Mattsson, J. Dynamics of MONOPTEROS and PIN-FORMED1 expression during leaf vein pattern formation in Arabidopsis thaliana. *Plant J.* **2007**, *49*, 387–398. [CrossRef] [PubMed]
6. Hardtke, C.S.; Berleth, T. The Arabidopsis gene MONOPTEROS encodes a transcription factor mediating embryo axis formation and vascular development. *EMBO J.* **1998**, *17*, 1405–1411. [CrossRef] [PubMed]
7. Berleth, T.; Jurgens, G. The role of the monopteros gene in organising the basal body region of the Arabidopsis embryo. *Development* **1993**, *118*.
8. Ohashi-Ito, K.; Matsukawa, M.; Fukuda, H. An Atypical bHLH Transcription Factor Regulates Early Xylem Development Downstream of Auxin. *Plant Cell Physiol.* **2013**, *54*, 398–405. [CrossRef]
9. Weijers, D.; Schlereth, A.; Ehrismann, J.S.; Schwank, G.; Kientz, M.; Jürgens, G. Auxin Triggers Transient Local Signaling for Cell Specification in Arabidopsis Embryogenesis. *Dev. Cell* **2006**, *10*, 265–270. [CrossRef]
10. Schlereth, A.; Möller, B.; Liu, W.; Kientz, M.; Flipse, J.; Rademacher, E.H.; Schmid, M.; Jürgens, G.; Weijers, D. MONOPTEROS controls embryonic root initiation by regulating a mobile transcription factor. *Nature* **2010**, *464*, 913–916. [CrossRef]
11. De Rybel, B.; Möller, B.; Yoshida, S.; Grabowicz, I.; Barbier de Reuille, P.; Boeren, S.; Smith, R.S.; Borst, J.W.; Weijers, D. A bHLH complex controls embryonic vascular tissue establishment and indeterminate growth in Arabidopsis. *Dev. Cell* **2013**, *24*, 426–437. [CrossRef]
12. Ohashi-Ito, K.; Saegusa, M.; Iwamoto, K.; Oda, Y.; Katayama, H.; Kojima, M.; Sakakibara, H.; Fukuda, H. A bHLH Complex Activates Vascular Cell Division via Cytokinin Action in Root Apical Meristem. *Curr. Biol.* **2014**, *24*, 2053–2058. [CrossRef] [PubMed]
13. De Rybel, B.; Adibi, M.; Breda, A.S.; Wendrich, J.R.; Smit, M.E.; Novák, O.; Yamaguchi, N.; Yoshida, S.; Van Isterdael, G.; Palovaara, J.; et al. Integration of growth and patterning during vascular tissue formation in Arabidopsis. *Science* **2014**, *345*, 1255215. [CrossRef] [PubMed]
14. Zürcher, E.; Liu, J.; di Donato, M.; Geisler, M.; Müller, B. Plant development regulated by cytokinin sinks. *Science* **2016**, *353*, 1027–1030. [CrossRef] [PubMed]
15. Bishopp, A.; Help, H.; El-Showk, S.; Weijers, D.; Scheres, B.; Friml, J.; Benková, E.; Mähönen, A.P.; Helariutta, Y. A Mutually Inhibitory Interaction between Auxin and Cytokinin Specifies Vascular Pattern in Roots. *Curr. Biol.* **2011**, *21*, 917–926. [CrossRef] [PubMed]
16. Müller, C.J.; Valdés, A.E.; Wang, G.; Ramachandran, P.; Beste, L.; Uddenberg, D.; Carlsbecker, A. PHABULOSA Mediates an Auxin Signaling Loop to Regulate Vascular Patterning in Arabidopsis. *Plant Physiol.* **2016**, *170*, 956–970. [CrossRef] [PubMed]
17. Scheres, B.; Di Laurenzio, L.; Willemsen, V.; Hauser, M.T.; Janmaat, K.; Weisbeek, P.; Benfey, P.N. Mutations affecting the radial organisation of the Arabidopsis root display specific defects throughout the embryonic axis. *Development* **1995**, *121*, 53–62.
18. Mähönen, A.P.; Bonke, M.; Kauppinen, L.; Riikonen, M.; Benfey, P.N.; Helariutta, Y. A novel two-component hybrid molecule regulates vascular morphogenesis of the Arabidopsis root. *Genes Dev.* **2000**, *14*, 2938–2943. [CrossRef]
19. Mähönen, A.P.; Higuchi, M.; Törmäkangas, K.; Miyawaki, K.; Pischke, M.S.; Sussman, M.R.; Helariutta, Y.; Kakimoto, T. Cytokinins Regulate a Bidirectional Phosphorelay Network in Arabidopsis. *Curr. Biol.* **2006**, *16*, 1116–1122. [CrossRef]

20. Mähönen, A.P.; Bishopp, A.; Higuchi, M.; Nieminen, K.M.; Kinoshita, K.; Törmäkangas, K.; Ikeda, Y.; Oka, A.; Kakimoto, T.; Helariutta, Y. Cytokinin signaling and its inhibitor AHP6 regulate cell fate during vascular development. *Science* **2006**, *311*, 94–98. [CrossRef]
21. Werner, T.; Motyka, V.; Strnad, M.; Schmülling, T. Regulation of plant growth by cytokinin. *Proc. Natl. Acad. Sci. USA* **2001**, *98*, 10487–10492. [CrossRef]
22. Bishopp, A.; Lehesranta, S.; Vatén, A.; Help, H.; El-Showk, S.; Scheres, B.; Helariutta, K.; Mähönen, A.P.; Sakakibara, H.; Helariutta, Y. Phloem-transported cytokinin regulates polar auxin transport and maintains vascular pattern in the root meristem. *Curr. Biol.* **2011**, *21*, 927–932. [CrossRef] [PubMed]
23. Jang, G.; Chang, S.H.; Um, T.Y.; Lee, S.; Kim, J.-K.; Choi, Y. Do Antagonistic interaction between jasmonic acid and cytokinin in xylem development. *Sci. Rep.* **2017**, *7*, 10212. [CrossRef] [PubMed]
24. Carlsbecker, A.; Lee, J.-Y.; Roberts, C.J.; Dettmer, J.; Lehesranta, S.; Zhou, J.; Lindgren, O.; Moreno-Risueno, M.A.; Vatén, A.; et al. Cell signalling by microRNA165/6 directs gene dose-dependent root cell fate. *Nature* **2010**, *465*, 316–321. [CrossRef] [PubMed]
25. Helariutta, Y.; Fukaki, H.; Wysocka-Diller, J.; Nakajima, K.; Jung, J.; Sena, G.; Hauser, M.-T.; Benfey, P.N. The SHORT-ROOT Gene Controls Radial Patterning of the Arabidopsis Root through Radial Signaling. *Cell* **2000**, *101*, 555–567. [CrossRef]
26. Nakajima, K.; Sena, G.; Nawy, T.; Benfey, P.N. Intercellular movement of the putative transcription factor SHR in root patterning. *Nature* **2001**, *413*, 307–311. [CrossRef] [PubMed]
27. Muraro, D.; Mellor, N.; Pound, M.P.; Help, H.; Lucas, M.; Chopard, J.; Byrne, H.M.; Godin, C.; Hodgman, T.C.; King, J.R.; et al. Integration of hormonal signaling networks and mobile microRNAs is required for vascular patterning in Arabidopsis roots PLANT BIOLOGY. *PNAS* **2014**, *111*, 857–862. [CrossRef] [PubMed]
28. Muñiz, L.; Minguet, E.G.; Singh, S.K.; Pesquet, E.; Vera-Sirera, F.; Moreau-Courtois, C.L.; Carbonell, J.; Blázquez, M.A.; Tuominen, H. ACAULIS5 controls Arabidopsis xylem specification through the prevention of premature cell death. *Development* **2008**, *135*, 2573–2582. [CrossRef]
29. Yoshimoto, K.; Noutoshi, Y.; Hayashi, K.; Shirasu, K.; Takahashi, T.; Motose, H. Thermospermine suppresses auxin-inducible xylem differentiation in Arabidopsis thaliana. *Plant Signal. Behav.* **2012**, *7*, 937–939. [CrossRef]
30. Vera-Sirera, F.; Minguet, E.G.; Singh, S.K.; Ljung, K.; Tuominen, H.; Blázquez, M.A.; Carbonell, J. Role of polyamines in plant vascular development. *Plant Physiol. Biochem.* **2010**, *48*, 534–539. [CrossRef]
31. Katayama, H.; Iwamoto, K.; Kariya, Y.; Asakawa, T.; Kan, T.; Fukuda, H.; Ohashi-Ito, K. A Negative Feedback Loop Controlling bHLH Complexes Is Involved in Vascular Cell Division and Differentiation in the Root Apical Meristem. *Curr. Biol.* **2015**, *25*, 3144–3150. [CrossRef]
32. Yamamoto, M.; Takahashi, T. Thermospermine enhances translation of SAC51 and SACL1 in Arabidopsis. *Plant Signal. Behav.* **2017**, *12*, e1276685. [CrossRef] [PubMed]
33. Cai, Q.; Fukushima, H.; Yamamoto, M.; Ishii, N.; Sakamoto, T.; Kurata, T.; Motose, H.; Takahashi, T. The *SAC51* Family Plays a Central Role in Thermospermine Responses in Arabidopsis. *Plant Cell Physiol.* **2016**, *57*, 1583–1592. [CrossRef] [PubMed]
34. Bauby, H.; Divol, F.; Truernit, E.; Grandjean, O.; Palauqui, J.-C. Protophloem Differentiation in Early Arabidopsis thaliana Development. *Plant Cell Physiol.* **2007**, *48*, 97–109. [CrossRef] [PubMed]
35. Anne, P.; Azzopardi, M.; Gissot, L.; Beaubiat, S.; Hématy, K.; Palauqui, J.-C. OCTOPUS Negatively Regulates BIN2 to Control Phloem Differentiation in Arabidopsis thaliana. *Curr. Biol.* **2015**, *25*, 2584–2590. [CrossRef] [PubMed]
36. Rodriguez-Villalon, A.; Gujas, B.; Kang, Y.H.; Breda, A.S.; Cattaneo, P.; Depuydt, S.; Hardtke, C.S. Molecular genetic framework for protophloem formation. *Proc. Natl. Acad. Sci. USA* **2014**, *111*, 11551–11556. [CrossRef] [PubMed]
37. Depuydt, S.; Rodriguez-Villalon, A.; Santuari, L.; Wyser-Rmili, C.; Ragni, L.; Hardtke, C.S. Suppression of Arabidopsis protophloem differentiation and root meristem growth by CLE45 requires the receptor-like kinase BAM3. *Proc. Natl. Acad. Sci. USA* **2013**, *110*, 7074–7079. [CrossRef] [PubMed]
38. Anne, P.; Amiguet-Vercher, A.; Brandt, B.; Kalmbach, L.; Geldner, N.; Hothorn, M.; Hardtke, C.S. CLERK is a novel receptor kinase required for sensing of root-active CLE peptides in Arabidopsis. *Development* **2018**, *145*. [CrossRef]

39. Hazak, O.; Brandt, B.; Cattaneo, P.; Santiago, J.; Rodriguez-Villalon, A.; Hothorn, M.; Hardtke, C.S. Perception of root-active CLE peptides requires CORYNE function in the phloem vasculature. *EMBO Rep.* **2017**, *18*, 1367–1381. [CrossRef]
40. Kang, Y.H.; Hardtke, C.S. Arabidopsis MAKR5 is a positive effector of BAM3-dependent CLE45 signaling. *EMBO Rep.* **2016**, *17*, 1145–1154. [CrossRef]
41. Wallner, E.-S.; López-Salmerón, V.; Belevich, I.; Poschet, G.; Jung, I.; Grünwald, K.; Sevilem, I.; Jokitalo, E.; Hell, R.; Helariutta, Y.; et al. Strigolactone- and Karrikin-Independent SMXL Proteins Are Central Regulators of Phloem Formation. *Curr. Biol.* **2017**, *27*, 1241–1247. [CrossRef]
42. Salazar-Henao, J.E.; Lehner, R.; Betegón-Putze, I.; Vilarrasa-Blasi, J.; Caño-Delgado, A.I. BES1 regulates the localization of the brassinosteroid receptor BRL3 within the provascular tissue of the Arabidopsis primary root. *J. Exp. Bot.* **2016**, *67*, 4951–4961. [CrossRef] [PubMed]
43. Kang, Y.H.; Breda, A.; Hardtke, C.S. Brassinosteroid signaling directs formative cell divisions and protophloem differentiation in Arabidopsis root meristems. *Development* **2017**, *144*, 272–280. [CrossRef] [PubMed]
44. Marhava, P.; Bassukas, A.E.L.; Zourelidou, M.; Kolb, M.; Moret, B.; Fastner, A.; Schulze, W.X.; Cattaneo, P.; Hammes, U.Z.; Schwechheimer, C.; Hardtke, C.S. A molecular rheostat adjusts auxin flux to promote root protophloem differentiation. *Nature* **2018**, *558*, 297–300. [CrossRef]
45. Bonke, M.; Thitamadee, S.; Mähönen, A.P.; Hauser, M.-T.; Helariutta, Y. APL regulates vascular tissue identity in Arabidopsis. *Nature* **2003**, *426*, 181–186. [CrossRef] [PubMed]
46. Wisman, E.; Cardon, G.H.; Fransz, P.; Saedler, H. The behaviour of the autonomous maize transposable element En/Spm in Arabidopsis thaliana allows efficient mutagenesis. *Plant Mol. Biol.* **1998**, *37*, 989–999. [CrossRef] [PubMed]
47. Furuta, K.M.; Yadav, S.R.; Lehesranta, S.; Belevich, I.; Miyashima, S.; Heo, J.; Vatén, A.; Lindgren, O.; De Rybel, B.; Van Isterdael, G.; Somervuo, P.; Lichtenberger, R.; Rocha, R.; Thitamadee, S.; Tähtiharju, S.; Auvinen, P.; Beeckman, T.; Jokitalo, E.; Helariutta, Y. Plant development. Arabidopsis NAC45/86 direct sieve element morphogenesis culminating in enucleation. *Science* **2014**, *345*, 933–937. [CrossRef] [PubMed]
48. Kondo, Y.; Nurani, A.M.; Saito, C.; Ichihashi, Y.; Saito, M.; Yamazaki, K.; Mitsuda, N.; Ohme-Takagi, M.; Fukuda, H. Vascular Cell Induction Culture System Using Arabidopsis Leaves (VISUAL) Reveals the Sequential Differentiation of Sieve Element-Like Cells. *Plant Cell* **2016**, *28*, 1250–1262. [CrossRef]
49. Serrano-Mislata, A.; Sablowski, R. The pillars of land plants: new insights into stem development. *Curr. Opin. Plant Biol.* **2018**, *45*, 11–17. [CrossRef]
50. Evert, R.F. *Esau's Plant Anatomy*; John Wiley & Sons, Inc.: Hoboken, NJ, USA, 2006; ISBN 9780470047385.
51. Ikematsu, S.; Tasaka, M.; Torii, K.U.; Uchida, N. ERECTA -family receptor kinase genes redundantly prevent premature progression of secondary growth in the *Arabidopsis* hypocotyl. *New Phytol.* **2017**, *213*, 1697–1709. [CrossRef]
52. Lehmann, F.; Hardtke, C.S. Secondary growth of the Arabidopsis hypocotyl—vascular development in dimensions. *Curr. Opin. Plant Biol.* **2016**, *29*, 9–15. [CrossRef]
53. Kuroha, T.; Ueguchi, C.; Sakakibara, H.; Satoh, S. Cytokinin receptors are required for normal development of auxin-transporting vascular tissues in the hypocotyl but not in the adventious roots. *Plant Cell Physiol.* **2006**, *47*, 234–243. [CrossRef] [PubMed]
54. Zhang, J.; Nieminen, K.; Serra, J.A.A.; Helariutta, Y. The formation of wood and its control. *Curr. Opin. Plant Biol.* **2014**, *17*, 56–63. [CrossRef] [PubMed]
55. Ragni, L.; Hardtke, C.S. Small but thick enough—the Arabidopsis hypocotyl as a model to study secondary growth. *Physiol. Plant.* **2014**, *151*, 164–171. [CrossRef] [PubMed]
56. Chaffey, N.; Cholewa, E.; Regan, S.; Sundberg, B. Secondary xylem development in Arabidopsis: a model for wood formation. *Physiol. Plant.* **2002**, *114*, 594–600. [CrossRef] [PubMed]
57. Sibout, R.; Plantegenet, S.; Hardtke, C.S. Flowering as a Condition for Xylem Expansion in Arabidopsis Hypocotyl and Root. *Curr. Biol.* **2008**, *18*, 458–463. [CrossRef]
58. Sankar, M.; Nieminen, K.; Ragni, L.; Xenarios, I.; Hardtke, C.S. Automated quantitative histology reveals vascular morphodynamics during Arabidopsis hypocotyl secondary growth. *eLIFE* **2014**, *3*, 1567. [CrossRef]
59. Ragni, L.; Nieminen, K.; Pacheco-Villalobos, D.; Sibout, R.; Schwechheimer, C.; Hardtke, C.S. Mobile gibberellin directly stimulates Arabidopsis hypocotyl xylem expansion. *Plant Cell* **2011**, *23*, 1322–1336. [CrossRef]

60. Mitsuda, N.; Iwase, A.; Yamamoto, H.; Yoshida, M.; Seki, M.; Shinozaki, K.; Ohme-Takagi, M. NAC Transcription Factors, NST1 and NST3, Are Key Regulators of the Formation of Secondary Walls in Woody Tissues of Arabidopsis. *Plant Cell.* **2007**, *19*, 270–280. [CrossRef]
61. Kubo, M.; Udagawa, M.; Nishikubo, N.; Horiguchi, G.; Yamaguchi, M.; Ito, J.; Mimura, T.; Fukuda, H.; Demura, T. Transcription switches for protoxylem and metaxylem vessel formation. *Genes Dev.* **2005**, *19*, 1855–1860. [CrossRef]
62. Mele, G.; Ori, N.; Sato, Y.; Hake, S. The knotted1-like homeobox gene BREVIPEDICELLUS regulates cell differentiation by modulating metabolic pathways. *Genes Dev.* **2003**, *17*, 2088–2093. [CrossRef]
63. Liebsch, D.; Sunaryo, W.; Holmlund, M.; Norberg, M.; Zhang, J.; Hall, H.C.; Helizon, H.; Jin, X.; Helariutta, Y.; Nilsson, O.; Polle, A.; Fischer, U. Class I KNOX transcription factors promote differentiation of cambial derivatives into xylem fibers in the Arabidopsis hypocotyl. *Development* **2014**, *141*, 4311–4319. [CrossRef]
64. Lee, K.-H.; Avci, U.; Qi, L.; Wang, H. The α Aurora kinases function in vascular development in Arabidopsis. *Plant Cell Physiol.* **2018**. [CrossRef] [PubMed]
65. Ito, Y.; Nakamomyo, I.; Motose, H.; Iwamoto, K.; Sawa, S.; Dohmae, N.; Fukuda, H. Dodeca-CLE peptides as suppressors of plant stem cell differentiation. *Science* **2006**, *313*, 842–845. [CrossRef] [PubMed]
66. Ji, J.; Strable, J.; Shimizu, R.; Koenig, D.; Sinha, N.; Scanlon, M.J. WOX4 Promotes Procambial Development 1. *Plant Physiol.* **2010**, *152*, 1346–1356. [CrossRef]
67. Etchells, J.P.; Provost, C.M.; Mishra, L.; Turner, S.R. WOX4 and WOX14 act downstream of the PXY receptor kinase to regulate plant vascular proliferation independently of any role in vascular organisation. *Development* **2013**, *140*, 2224–2234. [CrossRef]
68. Hirakawa, Y.; Kondo, Y.; Fukuda, H. TDIF peptide signaling regulates vascular stem cell proliferation via the WOX4 homeobox gene in Arabidopsis. *Plant Cell* **2010**, *22*, 2618–2629. [CrossRef]
69. Fisher, K.; Turner, S. PXY, a Receptor-like Kinase Essential for Maintaining Polarity during Plant Vascular-Tissue Development. *Curr. Biol.* **2007**, *17*, 1061–1066. [CrossRef] [PubMed]
70. Hirakawa, Y.; Shinohara, H.; Kondo, Y.; Inoue, A.; Nakamomyo, I.; Ogawa, M.; Sawa, S.; Ohashi-Ito, K.; Matsubayashi, Y.; Fukuda, H. Non-cell-autonomous control of vascular stem cell fate by a CLE peptide/receptor system. *Proc. Natl. Acad. Sci. USA* **2008**, *105*, 15208–15213. [CrossRef] [PubMed]
71. Etchells, J.P.; Provost, C.M.; Turner, S.R. Plant Vascular Cell Division Is Maintained by an Interaction between PXY and Ethylene Signalling. *PLoS Genet.* **2012**, *8*, e1002997. [CrossRef] [PubMed]
72. Agusti, J.; Lichtenberger, R.; Schwarz, M.; Nehlin, L.; Greb, T. Characterization of Transcriptome Remodeling during Cambium Formation Identifies MOL1 and RUL1 As Opposing Regulators of Secondary Growth. *PLoS Genet.* **2011**, *7*, e1001312. [CrossRef] [PubMed]
73. Gursanscky, N.R.; Jouannet, V.; Grünwald, K.; Sanchez, P.; Laaber-Schwarz, M.; Greb, T. MOL1 is required for cambium homeostasis in Arabidopsis. *Plant J.* **2016**, *86*, 210–220. [CrossRef] [PubMed]
74. Suer, S.; Agusti, J.; Sanchez, P.; Schwarz, M.; Greb, T. WOX4 imparts auxin responsiveness to cambium cells in Arabidopsis. *Plant Cell* **2011**, *23*, 3247–3259. [CrossRef] [PubMed]
75. Denis, E.; Kbiri, N.; Mary, V.; Claisse, G.; Conde e Silva, N.; Kreis, M.; Deveaux, Y. *WOX14* promotes bioactive gibberellin synthesis and vascular cell differentiation in Arabidopsis. *Plant J.* **2017**, *90*, 560–572. [CrossRef] [PubMed]
76. Wang, N.; Bagdassarian, K.S.; Doherty, R.; Wang, X.; Kroon, J.; Wang, W.; Jermyn, I.; Turner, S.; Etchells, P. Paralogues of the PXY and ER receptor kinases enforce radial patterning in plant vascular tissue. *bioRxiv* **2018**, 357244. [CrossRef]
77. Agusti, J.; Herold, S.; Schwarz, M.; Sanchez, P.; Ljung, K.; Dun, E.A.; Brewer, P.B.; Beveridge, C.A.; Sieberer, T.; Sehr, E.M.; Greb, T. Strigolactone signaling is required for auxin-dependent stimulation of secondary growth in plants. *Proc. Natl. Acad. Sci. USA* **2011**, *108*, 20242–20247. [CrossRef] [PubMed]
78. Sehr, E.M.; Agusti, J.; Lehner, R.; Farmer, E.E.; Schwarz, M.; Greb, T. Analysis of secondary growth in the Arabidopsis shoot reveals a positive role of jasmonate signalling in cambium formation. *Plant J.* **2010**, *63*, 811–822. [CrossRef] [PubMed]
79. Matsumoto-Kitano, M.; Kusumoto, T.; Tarkowski, P.; Kinoshita-Tsujimura, K.; Václavíková, K.; Miyawaki, K.; Kakimoto, T. Cytokinins are central regulators of cambial activity. *Proc. Natl. Acad. Sci. USA* **2008**, *105*, 20027–20031. [CrossRef]
80. Oles, V.; Panchenko, A.; Smertenko, A. Modeling hormonal control of cambium proliferation. *PLoS One* **2017**, *12*, e0171927. [CrossRef]

81. Han, S.; Cho, H.; Noh, J.; Qi, J.; Jung, H.-J.; Nam, H.; Lee, S.; Hwang, D.; Greb, T.; Hwang, I. BIL1-mediated MP phosphorylation integrates PXY and cytokinin signalling in secondary growth. *Nat. Plants* **2018**, *4*, 605–614. [CrossRef] [PubMed]
82. Brackmann, K.; Qi, J.; Gebert, M.; Jouannet, V.; Schlamp, T.; Grünwald, K.; Wallner, E.-S.; Novikova, D.D.; Levitsky, V.G.; Agustí, J.; Sanchez, P.; Lohmann, J.U.; Greb, T. Spatial specificity of auxin responses coordinates wood formation. *Nat. Commun.* **2018**, *9*, 875. [CrossRef]
83. Kondo, Y.; Ito, T.; Nakagami, H.; Hirakawa, Y.; Saito, M.; Tamaki, T.; Shirasu, K.; Fukuda, H. Plant GSK3 proteins regulate xylem cell differentiation downstream of TDIF–TDR signalling. *Nat. Commun.* **2014**, *5*, 3504. [CrossRef] [PubMed]
84. Ilegems, M.; Douet, V.; Meylan-Bettex, M.; Uyttewaal, M.; Brand, L.; Bowman, J.L.; Stieger, P.A. Interplay of auxin, KANADI and Class III HD-ZIP transcription factors in vascular tissue formation. *Development* **2010**, *137*, 975–984. [CrossRef] [PubMed]
85. Cho, H.; Cho, H.S.; Nam, H.; Jo, H.; Yoon, J.; Park, C.; Dang, T.V.T.; Kim, E.; Jeong, J.; Park, S.; et al. Translational control of phloem development by RNA G-quadruplex–JULGI determines plant sink strength. *Nat. Plants* **2018**, *4*, 376–390. [CrossRef] [PubMed]
86. Randall, R.S.; Miyashima, S.; Blomster, T.; Zhang, J.; Elo, A.; Karlberg, A.; Immanen, J.; Nieminen, K.; Lee, J.-Y.; Kakimoto, T.; et al. AINTEGUMENTA and the D-type cyclin CYCD3;1 regulate root secondary growth and respond to cytokinins. *Biol. Open* **2015**, *4*, 1229–1236. [CrossRef] [PubMed]
87. Uchida, N.; Lee, J.S.; Horst, R.J.; Lai, H.-H.; Katjita, R.; Kakimoto, T.; Tasaka, M.; Torii, K.U. Regulation of inflorescence architecture by intertissue layer ligand–receptor communication between endodermis and ploem. *PNAS* **2012**, *109*, 6337–6342. [CrossRef] [PubMed]
88. Eriksson, M.E.; Israelsson, M.; Olsson, O.; Moritz, T. Increased gibberellin biosynthesis in transgenic trees promotes growth, biomass production and xylem fiber length. *Nat. Biotechnol.* **2000**, *18*, 784–788. [CrossRef] [PubMed]
89. Nieminen, K.; Immanen, J.; Laxell, M.; Kauppinen, L.; Tarkowski, P.; Dolezal, K.; Tahtiharju, S.; Elo, A.; Decourteix, M.; Ljung, K.; et al. Cytokinin signaling regulates cambial development in poplar. *Proc. Natl. Acad. Sci.* **2008**, *105*, 20032–20037. [CrossRef] [PubMed]
90. Nilsson, J.; Karlberg, A.; Antti, H.; Lopez-Vernaza, M.; Mellerowicz, E.; Perrot-Rechenmann, C.; Sandberg, G.; Bhalerao, R.P. Dissecting the Molecular Basis of the Regulation of Wood Formation by Auxin in Hybrid Aspen. *Plant Cell Online* **2008**, *20*, 843–855. [CrossRef] [PubMed]
91. Love, J.; Bjorklund, S.; Vahala, J.; Hertzberg, M.; Kangasjarvi, J.; Sundberg, B. Ethylene is an endogenous stimulator of cell division in the cambial meristem of Populus. *Proc. Natl. Acad. Sci. USA* **2009**, *106*, 5984–5989. [CrossRef]
92. Du, J.; Miura, E.; Robischon, M.; Martinez, C.; Groover, A. The Populus Class III HD ZIP Transcription Factor POPCORONA Affects Cell Differentiation during Secondary Growth of Woody Stems. *PLoS One* **2011**, *6*, e17458. [CrossRef]
93. Etchells, J.P.; Mishra, L.S.; Kumar, M.; Campbell, L.; Turner, S.R. Wood Formation in Trees Is Increased by Manipulating PXY-Regulated Cell Division. *Curr. Biol.* **2015**, *25*, 1050–1055. [CrossRef] [PubMed]
94. Kucukoglu, M.; Nilsson, J.; Zheng, B.; Chaabouni, S.; Nilsson, O. *WUSCHEL-RELATED HOMEOBOX4 (WOX4)* -like genes regulate cambial cell division activity and secondary growth in *Populus* trees. *New Phytol.* **2017**, *215*, 642–657. [CrossRef] [PubMed]
95. Ramachandran, P.; Carlsbecker, A.; Etchells, J.P. Class III HD-ZIPs govern vascular cell fate: an HD view on patterning and differentiation. *J. Exp. Bot.* **2017**, *68*, 55–69. [CrossRef] [PubMed]
96. Nieminen, K.; Blomster, T.; Helariutta, Y.; Mähönen, A.P. Vascular Cambium Development. *Arab. B.* **2015**, *13*, e0177. [CrossRef] [PubMed]
97. Ragni, L.; Greb, T. Secondary growth as a determinant of plant shape and form. *Semin. Cell Dev. Biol.* **2018**, *79*, 58–67. [CrossRef] [PubMed]
98. Bhalerao, R.P.; Fischer, U. Environmental and hormonal control of cambial stem cell dynamics. *J. Exp. Bot.* **2017**, *68*, 79–87. [CrossRef] [PubMed]
99. Bossinger, G.; Spokevicius, A. V Sector analysis reveals patterns of cambium differentiation in poplar stems. *J. Exp. Bot.* **2018**, *69*, 4339–4348. [CrossRef]

100. Petzold, H.E.; Rigoulot, S.B.; Zhao, C.; Chanda, B.; Sheng, X.; Zhao, M.; Jia, X.; Dickerman, A.W.; Beers, E.P.; Brunner, A.M. Identification of new protein–protein and protein–DNA interactions linked with wood formation in Populus trichocarpa. *Tree Physiol.* **2018**, *38*, 362–377. [CrossRef]
101. Petzold, H.E.; Chanda, B.; Zhao, C.; Rigoulot, S.B.; Beers, E.P.; Brunner, A.M. DIVARICATA AND RADIALIS INTERACTING FACTOR (DRIF) also interacts with WOX and KNOX proteins associated with wood formation in *Populus trichocarpa*. *Plant J.* **2018**, *93*, 1076–1087. [CrossRef]
102. Immanen, J.; Nieminen, K.; Smolander, O.-P.; Kojima, M.; Alonso Serra, J.; Koskinen, P.; Zhang, J.; Elo, A.; Mähönen, A.P.; Street, N.; et al. Cytokinin and Auxin Display Distinct but Interconnected Distribution and Signaling Profiles to Stimulate Cambial Activity. *Curr. Biol.* **2016**, *26*, 1990–1997. [CrossRef]
103. Obudulu, O.; Bygdell, J.; Sundberg, B.; Moritz, T.; Hvidsten, T.R.; Trygg, J.; Wingsle, G. Quantitative proteomics reveals protein profiles underlying major transitions in aspen wood development. *BMC Genomics* **2016**, *17*, 119. [CrossRef] [PubMed]
104. Sundell, D.; Street, N.R.; Kumar, M.; Mellerowicz, E.J.; Kucukoglu, M.; Johnsson, C.; Kumar, V.; Mannapperuma, C.; Delhomme, N.; Nilsson, O.; et al. AspWood: High-Spatial-Resolution Transcriptome Profiles Reveal Uncharacterized Modularity of Wood Formation in *Populus tremula*. *Plant Cell.* **2017**, *29*, 1585–1604. [CrossRef]
105. Jin, F.; Li, J.; Ding, Q.; Wang, Q.-S.; He, X.-Q. Proteomic analysis provides insights into changes in the central metabolism of the cambium during dormancy release in poplar. *J. Plant Physiol.* **2017**, *208*, 26–39. [CrossRef] [PubMed]
106. Jang, G.; Lee, J.-H.; Rastogi, K.; Park, S.; Oh, S.-H.; Lee, J.-Y. Cytokinin-dependent secondary growth determines root biomass in radish (*Raphanus sativus* L.). *J. Exp. Bot.* **2015**, *66*, 4607–4619. [CrossRef] [PubMed]
107. Růžička, K.; Ursache, R.; Hejátko, J.; Helariutta, Y. Xylem development - from the cradle to the grave. *New Phytol.* **2015**, *207*, 519–535. [CrossRef] [PubMed]
108. Kumar, M.; Campbell, L.; Turner, S. Secondary cell walls: biosynthesis and manipulation. *J. Exp. Bot.* **2016**, *67*, 515–531. [CrossRef]
109. Meents, M.J.; Watanabe, Y.; Samuels, A.L. The cell biology of secondary cell wall biosynthesis. *Ann. Bot.* **2018**, *121*, 1107–1125. [CrossRef]
110. Spicer, R. Variation in Angiosperm Wood Structure and Its Physiological and Evolutionary Significance. In *Comparative and Evolutionary Genomics of Angiosperm Trees*; Springer: Cham, Switzerland, 2016; pp. 19–60.
111. Lens, F.; Picon-Cochard, C.; Delmas, C.E.L.; Signarbieux, C.; Buttler, A.; Cochard, H.; Jansen, S.; Chauvin, T.; Doria, L.C.; Del Arco, M.; et al. Herbaceous Angiosperms Are Not More Vulnerable to Drought-Induced Embolism Than Angiosperm Trees. *Plant Physiol.* **2016**, *172*, 661–667. [CrossRef]
112. Pereira, L.; Domingues-Junior, A.P.; Jansen, S.; Choat, B.; Mazzafera, P. Is embolism resistance in plant xylem associated with quantity and characteristics of lignin? *Trees* **2018**, *32*, 349–358. [CrossRef]
113. Johnson, D.; Eckart, P.; Alsamadisi, N.; Noble, H.; Martin, C.; Spicer, R. Polar auxin transport is implicated in vessel differentiation and spatial patterning during secondary growth in *Populus*. *Am. J. Bot.* **2018**, *105*, 186–196. [CrossRef]
114. Morris, H.; Plavcov, L.; Cvecko, P.; Fichtler, E.; Gillingham, M.A.F.; Mart Inez-Cabrera, H.I.; Mcglinn, D.J.; Wheeler, E.; Zheng, J.; Ziemi Nska, K.; et al. A global analysis of parenchyma tissue fractions in secondary xylem of seed plants. *New Phytol.* **2016**, *209*, 1553–1565. [CrossRef] [PubMed]
115. Mazur, E.; Kurczynska, E.U. Rays, intrusive growth, and storied cambium in the inflorescence stems of *Arabidopsis thaliana* (L.) Heynh. *Protoplasma* **2012**, *249*, 217–220. [CrossRef] [PubMed]
116. Ramage, M.H.; Burridge, H.; Busse-Wicher, M.; Fereday, G.; Reynolds, T.; Shah, D.U.; Wu, G.; Yu, L.; Fleming, P.; Densley-Tingley, D.; et al. The wood from the trees: The use of timber in construction. *Renew. Sustain. Energy Rev.* **2017**, *68*, 333–359. [CrossRef]
117. Song, J.; Chen, C.; Zhu, S.; Zhu, M.; Dai, J.; Ray, U.; Li, Y.; Kuang, Y.; Li, Y.; Quispe, N.; et al. Processing bulk natural wood into a high-performance structural material. *Nature* **2018**, *554*, 224–228. [CrossRef]
118. Groover, A. Gravitropisms and reaction woods of forest trees - evolution, functions and mechanisms. *New Phytol.* **2016**, *211*, 790–802. [CrossRef]
119. Navarro, C.; Cruz-Oró, E.; Prat, S. Conserved function of FLOWERING LOCUS T (FT) homologues as signals for storage organ differentiation. *Curr. Opin. Plant Biol.* **2015**, *23*, 45–53. [CrossRef] [PubMed]

120. Yu, R.; Wang, J.; Xu, L.; Wang, Y.; Wang, R.; Zhu, X.; Sun, X.; Luo, X.; Xie, Y.; Everlyne, M.; Liu, L. Transcriptome Profiling of Taproot Reveals Complex Regulatory Networks during Taproot Thickening in Radish (*Raphanus sativus* L.). *Front. Plant Sci.* **2016**, *7*, 1210. [CrossRef] [PubMed]
121. Benjamin, L.R.; Wren, M.J. Root Development and Source-Sink Relations in Carrot, *Daucus carota* L. *J. Exp. Bot.* **1978**, *29*, 425–433. [CrossRef]
122. Tanaka, M. Recent Progress in Molecular Studies on Storage Root Formation in Sweetpotato (*Ipomoea batatas*). *Biotechnology* **2016**, *50*, 293–299. [CrossRef]
123. El-Sharkawy, M.A. Cassava biology and physiology. *Plant Mol. Biol.* **2004**, *56*, 481–501. [CrossRef] [PubMed]
124. Wang, G.-L.; Xiong, F.; Que, F.; Xu, Z.-S.; Wang, F.; Xiong, A.-S. Morphological characteristics, anatomical structure and gene expression: novel insights into gibberellin biosynthesis and perception during carrot growth and development. *Hortic. Res.* **2015**, *2*, 15028. [CrossRef]
125. Mitsui, Y.; Shimomura, M.; Komatsu, K.; Namiki, N.; Shibata-Hatta, M.; Imai, M.; Katayose, Y.; Mukai, Y.; Kanamori, H.; Kurita, K.; et al. The radish genome and comprehensive gene expression profile of tuberous root formation and development. *Sci. Rep.* **2015**, *5*, 10835. [CrossRef]
126. Chaweewan, Y.; Taylor, N. Anatomical Assessment of Root Formation and Tuberization in Cassava (*Manihot esculenta* Crantz). *Trop. Plant Biol.* **2015**, *8*, 1–8. [CrossRef]
127. Akoumianakis, K.A.; Alexopoulos, A.A.; Karapanos, I.C.; Kalatzopoulos, K.; Aivalakis, G.; Passam, H.C. Carbohydrate metabolism and tissue differentiation during potato tuber initiation, growth and dormancy induction. *Aust. J. Crop Sci.* **2016**, *10*, 185–192.
128. Firon, N.; LaBonte, D.; Villordon, A.; Kfir, Y.; Solis, J.; Lapis, E.; Perlman, T.S.; Doron-Faigenboim, A.; Hetzroni, A.; Althan, L.; et al. Transcriptional profiling of sweetpotato (*Ipomoea batatas*) roots indicates down-regulation of lignin biosynthesis and up-regulation of starch biosynthesis at an early stage of storage root formation. *BMC Genomics* **2013**, *14*, 460. [CrossRef]
129. Noh, S.A.; Lee, H.-S.; Huh, E.J.; Huh, G.H.; Paek, K.-H.; Shin, J.S.; Bae, J.M. SRD1 is involved in the auxin-mediated initial thickening growth of storage root by enhancing proliferation of metaxylem and cambium cells in sweetpotato (*Ipomoea batatas*). *J. Exp. Bot.* **2010**, *61*, 1337–1349. [CrossRef]
130. Zhang, N.; Zhao, J.; Lens, F.; de Visser, J.; Menamo, T.; Fang, W.; Xiao, D.; Bucher, J.; Basnet, R.K.; Lin, K.; et al. Morphology, carbohydrate composition and vernalization response in a genetically diverse collection of Asian and European turnips (Brassica rapa subsp. rapa). *PLoS One* **2014**, *9*, e114241. [CrossRef]
131. Wang, G.-L.; Que, F.; Xu, Z.-S.; Wang, F.; Xiong, A.-S. Exogenous gibberellin enhances secondary xylem development and lignification in carrot taproot. *Protoplasma* **2017**, *254*, 839–848. [CrossRef]
132. Eviatar-Ribak, T.; Shalit-Kaneh, A.; Chappell-Maor, L.; Amsellem, Z.; Eshed, Y.; Lifschitz, E. Article A Cytokinin-Activating Enzyme Promotes Tuber Formation in Tomato. *Curr. Biol.* **2013**, *23*, 1057–1064. [CrossRef]
133. Gancheva, M.S.; Dodueva, I.E.; Lebedeva, M.A.; Tvorogova, V.E.; Tkachenko, A.A.; Lutova, L.A. Identification, expression, and functional analysis of CLE genes in radish (*Raphanus sativus* L.) storage root. *BMC Plant Biol.* **2016**, *16*, 7. [CrossRef]
134. Wang, G.-L.; Que, F.; Xu, Z.-S.; Wang, F.; Xiong, A.-S. Exogenous gibberellin altered morphology, anatomic and transcriptional regulatory networks of hormones in carrot root and shoot. *BMC Plant Biol.* **2015**, *15*, 290. [CrossRef]
135. Jabir, B.M.; Kinuthia Karanja, B.; Almahadi Faroug, M.; Nureldin Awad, F. Effects of Gibberellin and Gibberellin Biosynthesis Inhibitor (Paclobutrazol) Applications on Radish (*Raphanus sativus* L.) Taproot Expansion and the Presence of Authentic Hormones. *Artic. Int. J. Agric. Biol.* **2017**, *19*, 779–786. [CrossRef]
136. Mabvongwe, O.; Manenji, B.T.; Gwazane, M.; Chandiposha, M. The Effect of Paclobutrazol Application Time and Variety on Growth, Yield, and Quality of Potato (*Solanum tuberosum* L.). *Adv. Agric.* **2016**, *2016*, 1–5. [CrossRef]
137. Yang, J.; An, D.; Zhang, P. Expression Profiling of Cassava Storage Roots Reveals an Active Process of Glycolysis/Gluconeogenesis. *J. Integr. Plant Biol.* **2011**, *53*, 193–211. [CrossRef]

© 2018 by the authors. Licensee MDPI, Basel, Switzerland. This article is an open access article distributed under the terms and conditions of the Creative Commons Attribution (CC BY) license (http://creativecommons.org/licenses/by/4.0/).

Review

Multiple Pathways in the Control of the Shade Avoidance Response

Giovanna Sessa [1], Monica Carabelli [1], Marco Possenti [2], Giorgio Morelli [2] and Ida Ruberti [1,*]

1. Institute of Molecular Biology and Pathology, National Research Council, 00185 Rome, Italy; giovanna.sessa@uniroma1.it (G.S.); monica.carabelli@uniroma1.it (M.C.)
2. Research Centre for Genomics and Bioinformatics, Council for Agricultural Research and Economics (CREA), 00178 Rome, Italy; marco.possenti@crea.gov.it (M.P.); giorgio.morelli.crea@gmail.com (G.M.)
* Correspondence: ida.ruberti@uniroma1.it; Tel.: +39-06-49912211

Received: 19 October 2018; Accepted: 14 November 2018; Published: 17 November 2018

Abstract: To detect the presence of neighboring vegetation, shade-avoiding plants have evolved the ability to perceive and integrate multiple signals. Among them, changes in light quality and quantity are central to elicit and regulate the shade avoidance response. Here, we describe recent progresses in the comprehension of the signaling mechanisms underlying the shade avoidance response, focusing on Arabidopsis, because most of our knowledge derives from studies conducted on this model plant. Shade avoidance is an adaptive response that results in phenotypes with a high relative fitness in individual plants growing within dense vegetation. However, it affects the growth, development, and yield of crops, and the design of new strategies aimed at attenuating shade avoidance at defined developmental stages and/or in specific organs in high-density crop plantings is a major challenge for the future. For this reason, in this review, we also report on recent advances in the molecular description of the shade avoidance response in crops, such as maize and tomato, and discuss their similarities and differences with Arabidopsis.

Keywords: Arabidopsis; auxin; HD-Zip transcription factors; light environment; photoreceptors

1. Introduction

Plants, as sessile organisms, have evolved complex and sophisticated molecular processes to sense and react to the presence of neighboring plants. Plants can be divided into two groups depending on their response to competition for light: shade tolerance and shade avoidance [1–3]. To detect the presence of plants in close proximity, shade-avoiding plants use multiple cues [4]. Among these cues, changes in light intensity and quality play a central role in the regulation of the shade avoidance response. Light reflected or transmitted through photosynthetic plant tissues is depleted in blue (B), red (R), and UV-B wavelengths. Hence, the reflected or transmitted light is enriched in green (G) and far-red (FR) spectral regions, resulting in lowered ratios of R/FR light and B/G light. Plants perceive these differences through multiple photoreceptors, which in turn trigger signaling cascades to regulate plant growth under suboptimal light environments [5–8].

Arabidopsis is very responsive to FR-enriched light. At the early stage of seedling development, the perception of shade results in hypocotyl elongation, a reduction of cotyledon and leaf lamina expansion, and the diminution of root development (Figure 1). Here, we describe the key pathways underlying the shade avoidance response, focusing mainly on Arabidopsis, because most of the molecular processes regulating this response have been characterized in this model plant.

Figure 1. Shade avoidance phenotypes in Arabidopsis seedlings. Seedlings were grown for four days in high red (R)/far-red (FR) $_{High\ PAR}$ and then either maintained in the same light regime or transferred to low R/FR $_{Low\ PAR}$ for six days in a 16-h light/8-h dark photoperiod to simulate, respectively, sunlight and shade. Light outputs were as previously reported [9]. Scale bar, 2 mm.

2. Photoreceptors in the Control of Shade Avoidance

The R/FR ratio is a highly accurate indicator of plant proximity, and probably for this reason, for many years, shade avoidance research has mostly focused on the phytochrome signaling of changes in the R/FR ratio. However, a large number of evidence points to the reduced irradiance and the blue/green ratio as signals that play important roles in activating plant responses to canopy light [5–8].

2.1. Phytochromes

Phytochromes exist in two photo-convertible isoforms: a R light-absorbing form (Pr) and a FR light-absorbing form (Pfr). In the darkness, phytochromes are synthesized in the Pr form, which is inactive. After triggering with R light, the Pr form is converted into the active Pfr form, which, in turn, can absorb FR and switch back to Pr. The active Pfr form is translocated to the nucleus, giving rise to the responses [5,10].

The phytochrome apoproteins are encoded by a small gene family in the majority of plant species. In Arabidopsis, they are encoded by five genes, *PHYA–PHYE*. *PHYE* likely originated from a duplication within the *PHYB* lineage only in dicotyledonous plants. *PHYD*, which is closely related to *PHYB*, presumably emerged from a gene duplication within Brassicaceae [11]. *PHYC* probably arose from a duplication within the *PHYA* lineage [11]. phyA is rapidly degraded in its Pfr form, and signals during the conversion between the Pr and Pfr form mediated by the R/FR ratio light. phyB–E are all relatively stable in the Pfr form [5,10,12].

Among the light-stable phytochromes, phyB has a predominant role in the regulation of the shade avoidance response. However, evidence exists that phyD and phyE function redundantly with phyB in promoting shade-induced elongation [12,13] (Figure 2). By contrast, phyA attenuates the elongation response induced by low R/FR light [9,14–16] (Figure 2).

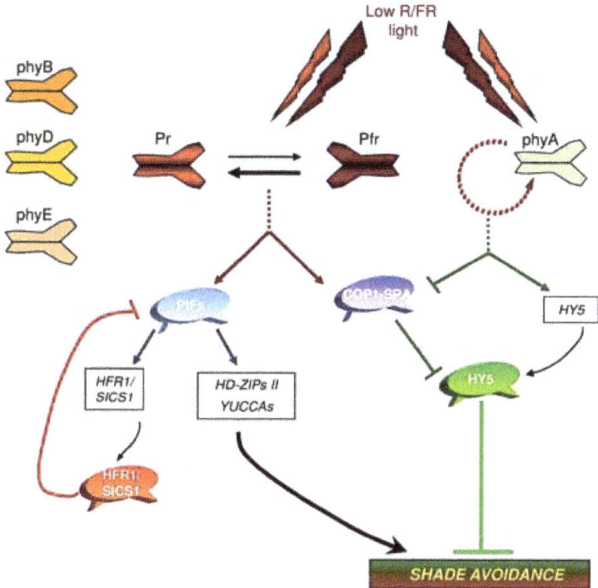

Figure 2. Regulatory routes in the shade avoidance response. Changes in R/FR light causing a shift in the equilibrium between Pr and the FR light-absorbing photo-convertible isoform (Pfr) toward the R light-absorbing photo-convertible isoform (Pr) result in the deactivation of phyB, phyD, and phyE. This, in turn, results in the enhanced stability and/or activity of several phytochrome-interacting transcription factors (PIFs). PIFs, within a few minutes, activate the transcription of *HD-Zips II*, *YUCs*, and *HFR1/SICS1* genes, encoding positive and negative regulators of shade avoidance, respectively. HFR1/SICS1 form non-functional heterodimers with PIF proteins, thereby inhibiting their activity. Shade avoidance is counteracted by the action of phyA, which positively regulates *HY5*, a central regulator of seedling photomorphogenesis. phyA and phyB oppositely affect the activity of COP1/SPA complexes.

In the nucleus, phytochromes directly bind the Phytochrome-Interacting Factors (PIFs), which are a subfamily of basic Helix–Loop–Helix (bHLH) transcription factors involved in the control of plant growth and development [17–19]. The Arabidopsis genome encodes eight PIF/PIF-like proteins—PIF1, PIF3–8, and PIL1/PIF2—all containing a conserved active phytochrome B binding (APB) domain, which is required for the interaction with the Pfr form of phyB. PIF1 and PIF3 also contain an active phytochrome A binding (APA) domain, which is necessary and sufficient for binding the Pfr form of phyA. Most of the PIFs promote growth, whereas PIF6 and PIL1/PIF2 seem to have an opposite function [20]. PIF proteins have both redundant and distinct functions at different stages of plant development, and coherently, only a subset of target genes is regulated by multiple PIFs (PIF1, PIF3–5) [20]. PIFs bind to promoter regions enriched in the cis element G-box and the E-box variant, which is known as the PBE-box (PIF binding E-box) [18]. However, the mechanisms through which different PIF proteins specifically recognize distinct set of target genes are largely unknown. Interestingly, it has been recently shown that the promoters of PIF1 target genes are enriched with G-box coupling elements (GCEs), which bind PIF1-interacting transcription factors (PTFs). These interactions may contribute to the targeting of PIF1 to specific sites in the genome [21].

In most cases, the interaction of PIFs with phyB in the nucleus results in PIF's phosphorylation and ubiquitination, leading to a fast degradation via the 26S proteasome [17]. PIF3, PIF4, and PIF5 protein levels increase rapidly in green seedlings upon inactivation of the phytochromes by simulated shade [22,23]. Instead, PIF7 is not rapidly degraded upon interaction with phyB in high R/FR light, but rather accumulates in a phosphorylated form. Exposure to low R/FR results in a rapid decrease of the amount of phosphorylated PIF7 with a concomitant increase in the level of dephosphorylated PIF7 [24]. PIF1, PIF3, PIF4, PIF5, and PIF7 have all been directly implicated in the shade avoidance response [22–25]. The shade-induced elongation response is indeed reduced in *pif4 pif5*, *pif1*, *pif3*, *pif4*, *pif5*, quadruple (*pifq*), and *pif7* loss-of-function mutants [22–24].

Interestingly, PIF proteins directly control the expression of both positive and negative regulators of the shade avoidance response [5–8,26] (Figure 2).

Among the positive regulators is the *Homeodomain-Leucine Zipper* (*HD-Zip*) *Arabidopsis Thaliana HomeoBox2* (*ATHB2*) transcription factor gene, which is involved in the elongation response induced by light quality changes [27,28]. The *ATHB2* gene is rapidly and reversibly regulated by changes in the R/FR ratio light [29]. phyB, phyD, and phyE have all been implicated in the regulation of *ATHB2* by changes in the ratio of R/FR light [30]. *ATHB2* induction by FR-enriched light does not require de novo protein synthesis [31], and is significantly diminished in loss-of-function *pif* mutants (*pif4 pif5*; *pifq*) [22,32]. Furthermore, there is evidence that *ATHB2* is a direct target of PIF proteins [25]. Relevantly, among the positive regulators are also several auxin biosynthesis *YUCCA* (*YUC*) genes, thus directly linking the perception of shade light to plant growth [24].

Among the negative regulators of shade avoidance controlled by PIF proteins is Long Hypocotyl in Far Red 1/Slender In Canopy Shade 1 (HFR1/SICS1), which is an atypical bHLH protein. *HFR1/SICS1* is rapidly induced by FR-enriched light, and it has been demonstrated that it is recognized in vivo by PIF5 [25,33,34]. Prolonged exposure to Low R/FR leads to the accumulation of HFR1/SICS1 and the formation of non-active heterodimers with PIF4 and PIF5 [33,34]. Consistently, several genes that are rapidly and transiently induced by low R/FR are upregulated in loss-of-function *hfr1/sics1* mutants under persistent shade [33,35]. Moreover, *hfr1/sics1* plants display an exaggerated shade avoidance response, whereas transgenic seedlings overexpressing a stable HFR1/SICS1 protein have suppressed elongation [33,36]. Helix Loop Helix1/Phytochrome Rapidly Regulated1 (HLH1/PAR1) [31,33] is another atypical bHLH protein gene that also acts as a negative regulator of the shade avoidance response. It is rapidly upregulated by low R/FR light, without the requirement of de novo protein synthesis. HLH1/PAR1 has been proposed to act as an antagonist of bHLH transcription factors, including PIF4 [36–39].

The attenuation of shade avoidance responses also involves a low R/FR stimulation of phyA signaling [9,40,41] (Figure 2). The *PHYA* gene is early induced by low R/FR, and phyA is required for the upregulation of the basic leucine zipper (bZIP) transcription factor gene, Elongated Hypocotyl 5 (HY5), which is a central regulator of photomorphogenesis [42]. HY5, on one hand, downregulates genes induced early by low R/FR light, and on the other hand, positively regulates photomorphogenesis-promoting genes under persistent shade [9]. Evidence exists that HY5 binds to PIF proteins [43,44].

phyA in its active Pfr form directly interacts with Suppressor of PhyA-105 (SPA) proteins and inhibits their interaction with Constitutively Photomorphogenic 1 (COP1) [45]. The COP1/SPA complexes are part of the Cullin 4-Damaged DNA Binding 1 ubiquitin E3 ligase complex (CUL4–DDB1$^{COP1/SPA}$), and are required for substrate recognition [46]. Several positive regulators of photomorphogenesis, including HY5 and HFR1/SICS1, are targeted for 26 proteasome-mediated degradation by CUL4–DDB1$^{COP1/SPA}$ [41]. The active form of phyA also interacts with COP1 [45]. Evidence exist that the binding of COP1 and SPA proteins is relevant for the activity of CUL4–DDB1$^{COP1/SPA}$. Therefore, it has been proposed that the direct interaction of phyA and SPA proteins inactivates CUL4–DDB1$^{COP1/SPA}$, which in turn results in the stabilization of positive regulators of photomorphogenesis [41] (Figure 2). phyB in its active form has also been shown to bind to SPAs and inhibit their interaction with COP1 [45] (Figure 2). The analyses of loss-of-function *cop1* and *spa1-4* mutants in low R/FR indicate that the COP1/SPA complex is essential for shade-induced elongation [47,48]. It has been suggested that in low R/FR, reduced levels of the active form of phyB indirectly enhance PIF activity, increasing the COP1/SPA-mediated degradation of negative regulators of the shade avoidance response [48,49]. Together, the data indicate that the phyA and phyB-mediated control of COP1/SPA activity oppositely affect the levels of negative regulators of shade avoidance such as HY5, HFR1/SICS1, HLH1/PAR1, and members of the B-Box (BBX) transcription factor family [50–52].

2.2. Cryptochromes

Cryptochromes are flavoprotein photoreceptors that were originally identified in Arabidopsis, and subsequently found in prokaryotes, archaea, and many eukaryotes [53]. Cryptochromes (CRY) are homologous to photolyases that catalyze light-dependent DNA repair [54]. The Arabidopsis genome encode two cryptochromes, CRY1 and CRY2. They consist of two domains, the PHR (photolyase-homologous region) domain, which is required for photoperception and dimer formation, and the CCE (cryptochrome C-terminal extension) domain, which is involved in signal transduction to downstream factors. It has been proposed that cryptochromes are activated by blue light through conformational changes, mostly in CCE domains [55]. Following blue light activation, CRY2 is rapidly degraded by the 26-proteasome system, whereas CRY1 is stable [54].

Both CRY1 and CRY2 are involved in low blue light (LBL)-induced shade avoidance response [56–58]. Interestingly, it has been recently demonstrated that PIF4 and PIF5 activity is required for LBL-induced hypocotyl growth, and evidence has been provided that these PIFs physically interact with CRY1 and CRY2 [58,59]. Furthermore, chromatin immunoprecipitation sequencing has shown that CRY2 binds to PIF4 and PIF5-regulated gene promoters [58]. Transcriptomic analysis revealed different expression profiles in low R/FR and LBL-treated seedlings. It is relevant that LBL, unlike low R/FR, does not involve changes in auxin levels and sensitivity, further supporting the proposal that phy and CRY photoreceptors control plant responses to shade via largely independent pathways [56–58].

Analogously to the active form of phyB, photoexcited CRY1 has been shown to bind to SPA1, resulting in the suppression of the SPA1–COP1 interaction. This in turn reduces COP1 activity, leading to increased levels of transcription factors such as HY5 [60].

2.3. UVR8

UV-B light is strongly filtered by plant canopies, thus providing further information on plant density [6,61]. In Arabidopsis, the inhibition of hypocotyl elongation by UV-B light depends on the UV-B receptor UVR8 [62,63]. UVR8 in its dimeric form perceives UV-B light; the absorption of UV-B induces the instant monomerization of the photoreceptor followed by interaction with COP1. This, in turn, promotes the accumulation of HY5 and its close relative HY5 Homologue (HYH) [64–66]. UVR8 promotes gibberellic acid (GA) degradation in a HY5/HYH-dependent manner, contributing to the stabilization of DELLA (where D is aspartic acid, E glutamic acid, L leucine, L leucine, A alanine) proteins and the consequent formation of inactive DELLA–PIF complexes [67]. Furthermore, evidence exist that UV-B also enhances the degradation of PIF4 and PIF5 [67]. Together, the data indicate that UV-B light inhibits PIF function, thereby attenuating plant responses to canopy shade [67,68].

3. HD-Zip Transcription Factors in the Control of Shade Avoidance

The HD-Zip class of transcription factors appears to be present exclusively in the plant kingdom [69]. HD-Zip proteins form a dimeric complex that recognize pseudopalindromic DNA elements [70–73], and act as positive or negative regulators of gene expression [74]. The Arabidopsis HD-Zip proteins, on the basis of the sequence homology in the HD-Zip DNA-binding domain, the presence of other conserved motifs, and specific intron and exon positions, have been grouped into four families: HD-Zip I–IV [75–80]. The phylogenetic and bioinformatics analysis of HD-Zip genes using transcriptomic and genomic datasets from a large number of Viridiplantae species indicated that the HD-Zip class of proteins was already present in green algae [81].

All four HD-Zip protein families can be further classified into subfamilies consisting of paralogous genes that have likely originated through genome duplication, considering their association with chromosome-duplicated regions in Arabidopsis and rice [77–80]. Interestingly, members of both the HD-Zip II and HD-Zip III protein families have been implicated in the control of shade avoidance [74,82].

Relevantly, HD-Zip II and HD-Zip III binding sites share the same core sequence [70,76], thereby leading to the hypothesis that members of the two families may control the expression of common target genes [83]. HD-Zip II proteins contain an LxLxL (where L is leucine and x is another amino acid) type of Ethylene-responsive element binding factor-associated amphiphilic repression (EAR) motif [79,84], and there is strong evidence that they function as transcriptional repressors [27,83,85,86]. On the contrary, HD-Zip III transcription factors are considered activators of gene expression [73,83,87–89].

3.1. HD-Zips II

The HD-Zip II protein family includes *ATHB2*, which is the first gene shown to be rapidly and reversibly regulated by light quality changes [29]. phyB, phyD, and phyE are all involved in the regulation of *ATHB2* by low R/FR ratio light [29,30], and it has been shown that *ATHB2* is recognized in vivo by PIF5 [25]. A lack of ATHB2 function results in diminished hypocotyl elongation in low R/FR ratio light, whereas the phenotypes of seedlings with elevated levels of ATHB2 in high R/FR resembles that of wild type in shade [27,28]. The expression of ATHB2, as deduced by the β-glucuronidase (GUS) pattern observed in ATHB2:ATHB2:GUS seedlings, is rapidly and transiently induced by shade in all the cell layers of the hypocotyl [28]. This and other experimental evidence (see below) indicated that ATHB2 acts as a positive regulator of shade avoidance.

The HD-Zip II family consists of 10 genes, five of which [*ATHB2*, *Homeobox Arabidopsis Thaliana (HAT1)*, *HAT2*, *ATHB4* and *HAT3*] are induced by low R/FR ratio light [79]. In the hat3 athb4 double loss-of-function mutant hypocotyl elongation is impaired [90], whereas the overexpression of HAT1, HAT2, HAT3, and ATHB4 causes phenotypes that are analogous to those observed in plants with elevated levels of ATHB2 in high R/FR [26,35,79,86,90], further highlighting the redundancy of these proteins in the regulation of shade avoidance. Relevantly, homologue genes are induced in monocot and dicot plants by low R/FR ratio light, strongly suggesting that the function of HD-Zips II may be conserved through evolution [91–93].

Very recent work has shown that prolonged shade results in an early exit from proliferation in the first pairs of Arabidopsis leaves, and that this process depends on the action of ATHB2 and ATHB4 (Figure 3) [94].

Furthermore, evidence has been provided that ATHB2 and ATHB4 work in concert in the control of leaf development specifically in a low R/FR light environment, likely forming heterodimeric complexes as suggested by yeast two-hybrid assays [94,95]. The data provide novel insights on the molecular mechanisms underlying leaf development in shade. However, further work is needed to uncover the links between the ATHB2 and ATHB4 transcription factors and the known regulatory pathways involved in the control of leaf cell proliferation [96,97].

Links between HD-Zip II proteins and auxin have been established [35,74]. However, how HD-Zips II interact with auxin machineries is still largely unknown.

Interestingly, a growing body of evidence demonstrates that besides their function in plant growth responses to shade, HD-Zips II play a major role in key developmental processes in a sunlight simulated environment, including embryo apical development, shoot apical meristem (SAM) activity, organ polarity, and gynoecium development [74,83,98–101]. These studies suggest that developmental processes and shade avoidance responses, sharing these transcription factors, could be intertwined. Connections between developmental and shade avoidance regulatory networks are further indicated by the recent finding that under shade, PIFs directly suppress multiple *miR156* genes, resulting in the increased expression of the *Squamosa-Promoter Binding Protein-Like* (*SPL*) family of genes [102], which have a role in the regulation of several aspects of plant development [103].

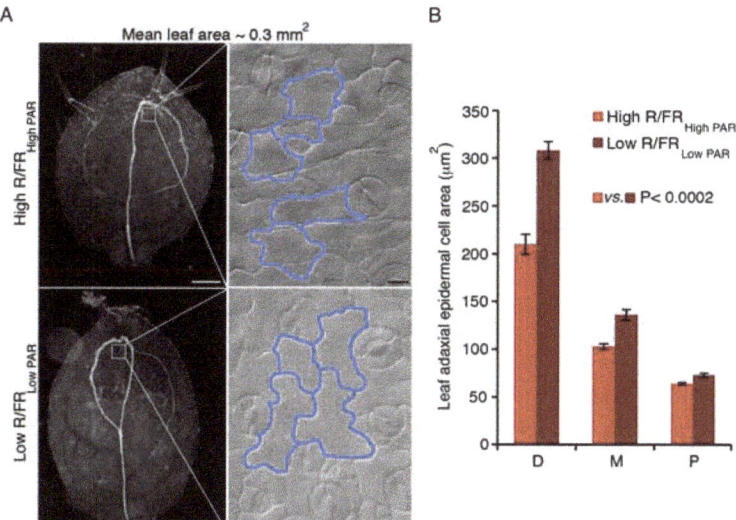

Figure 3. Shade affects adaxial epidermal cell expansion in the Arabidopsis leaf. (**A**) Dark-field images of cleared first/second leaves of wild type grown for eight days in high R/FR $_{High\ PAR}$ (high R/FR $_{High\ PAR}$), or for four days in high R/FR $_{High\ PAR}$ and then for 5.5 days in low R/FR $_{Low\ PAR}$ (low R/FR $_{Low\ PAR}$), respectively. The insets show a paradermal view of leaf adaxial epidermis; the borders of a few cells have been highlighted manually with a blue line. Light outputs were as previously reported [9]. Scale bars: (**A**), 100 μm; insets, 10 μm. (**B**) The graph shows the mean epidermal cell area at three positions along the proximo-distal leaf axis, distal (D), median (M) and proximal (P) in the two light conditions. At least 100 adaxial epidermal cells in 10 leaves were analyzed for each condition. Statistical analysis was performed as described [94].

3.2. HD-Zips III

The HD-Zip III protein family consists of five members: ATHB8, Corona (CNA), Phabulosa (PHB), Phavoluta (PHV), and Revoluta (REV). Several evidence have indicated that HD-Zip III proteins act as master regulators of embryonic apical fate [104], and are required to maintain SAM activity and establish lateral organ polarity [105,106]. The pattern of HD-Zips III expression largely overlaps with that of auxin distribution [89,107–112]. Furthermore, *HD-Zip III* genes are regulated at the post-transcriptional level by the microRNAs miR165/166, which negatively affect their expression through mRNA cleavage [105,113].

Interestingly, there is evidence that REV directly positively regulates *Tryptophan Aminotransferase of Arabidopsis 1* (*TAA1*) and *YUC5*, indicating that at least part of its role in plant development implies the regulation of auxin biosynthesis [73,114]. Relevantly, *TAA1* and *YUC5* are directly negatively regulated by KANADI1 (KAN1), which is a key determinant of abaxial cell fate in the leaf [57,115–117]. Furthermore, it has been recently demonstrated that genes implicated in auxin transport, including the influx carriers *LIKE Auxin Resistant 2* (*LAX2*) and *LAX3*, and response are also direct targets of REV [89,112,114,115].

Among the genes directly regulated by REV are also *HAT3*, *ATHB4*, *ATHB2*, and *HAT2*, and there is evidence that PHB and PHV are involved in the regulation of *HAT3* [73,83]. Coherently, the HAT3 and ATHB4 expression pattern in simulated sunlight essentially coincides with that of PHB, PHV, and REV. *ATHB2* expression is instead restricted to procambial cells early during embryo and leaf development; however, *ATHB2* is expressed in the *HAT3* and *ATHB4* domains in the *hat3 athb4* mutant, compensating in part for the lack of HAT3 and ATHB4 [83].

The direct regulation of *HD-Zip II* genes by HD-Zip III transcription factors and the finding that the phenotypes of *hat3 athb4 athb2* loss-of-function *HD-Zip II* mutants in sunlight resemble those of *rev phb phv* indicate that HD-Zip II and HD-Zip III proteins function in the same pathways under a sun-simulated environment [74,82]. Considering that HD-Zip II proteins work as negative regulators of gene expression [27,83,85], it was proposed that they may restrict HD-Zip III expression [74]. Interestingly, it was recently shown that REV, which is expressed exclusively in the adaxial side of the leaf because of the activity of microRNA (miR) 165/166 in the abaxial leaf domain, physically interacts with HAT3 and ATHB4 to directly repress the expression of *MIR165/166* genes in the adaxial side [118].

The analysis of *HD-Zip III* loss-of-function and gain-of function mutants has uncovered the involvement of REV in shade-induced elongation growth. *rev* loss-of-function mutants as well as plants ectopically expressing *MIR165a* display reduced elongation growth under simulated shade, whereas REV gain-of-function mutants (*rev10D*) show slightly long hypocotyl phenotypes under simulated sunlight [73,82]. It will be of interest in the future to investigate whether HD-Zip II and HD-Zip III proteins act together in the regulation of gene expression under a simulated shade environment.

4. Auxin as a Driver of the Shade Avoidance Response

There is a large body of evidence showing that plant responses to shade involve changes in hormonal pathways. Here, we focus on auxin, whereas for other hormones involved in the shade avoidance response, we recommend recent reviews [119,120]. Auxin has a central role in many responses induced by neighbor detection and canopy shade, such as the increased elongation of hypocotyl and petioles, and reduced leaf and root growth. Auxin homeostasis, transport, and signaling are all regulated in response to shade [35,121]. Interestingly, it has been shown that whereas the increase in auxin synthesis is a major event at the early stages of shade avoidance, the persistence of shade mainly results in the modulation of auxin sensitivity [25,122–124].

4.1. Auxin Homeostasis

Exposure to shade results in a rapid increase in the levels of auxin [24,25,125]. New auxin is synthesized in cotyledons from tryptophan (Trp) through TAA1, which is an enzyme encoded by the *Shade Avoidance3* (*SAV3*) gene [125,126]. Trp is converted to indole-3-pyruvic acid (IPA), and IPA in turn is modified to indole-3-acetic acid (IAA) by the action of the YUC family of flavin monooxygenases [127–130]. *YUC2, YUC5, YUC8,* and *YUC9* are rapidly regulated by low R/FR ratio light through PIF transcription factors [24,125]. Furthermore, the *sav3* mutant and the quadruple *yuc2 yuc3 yuc8 yuc 9* mutant are impaired in low R/FR-induced responses [125,131,132].

Low R/FR ratio light also controls auxin homeostasis by modulating its inactivation. Indeed, a number of auxin-inducible genes of the Gretchen Hagen 3 (GH3) family are quickly upregulated by low R/FR [14,133]. GH3 proteins promote the reduction of the free IAA pool by the conjugation of IAA to different amino acids [134], and it has been reported that GH3 mutants show defects in the elongation responses of the hypocotyl to light [135,136]. Furthermore, it has been recently shown that the loss-of-function of *VAS2* [*IAA-amido synthetase* (*GH3.17*)] results in an increase in free IAA at the expense of IAA-glutamate in the hypocotyl epidermis. Interestingly, the *vas2* mutants display longer hypocotyls in response to low R/FR light largely independently of the novo IAA biosynthesis in cotyledons, demonstrating the relevance of local auxin metabolism to modulate IAA homeostasis in an organ-specific manner in response to shade [137].

The relevance of local responses is also demonstrated by the recent finding that the alteration of the R/FR ratio at the leaf tip induces an upwards leaf movement that is confined to the leaf perceiving the light signal. Evidence have been provided that this hyponastic response depends on the synthesis of auxin in the leaf and its transport to the petiole [138,139].

4.2. Auxin Transport

It has been proposed that auxin that is synthesized in the cotyledons through the TAA1/YUC pathway upon low R/FR exposure is transported to hypocotyls, where it stimulates cell elongation [125]. Consistent with this proposal, auxin transport inhibitors abolish low R/FR-induced elongation, highlighting the relevance of auxin distribution for shade avoidance [27,125].

A large body of evidence indicates that the active transport of auxin is strictly controlled during neighbor detection and canopy shade. A low R/FR light ratio regulates the expression of the polar-auxin-transport efflux carriers PIN-Formed (PIN) 1, PIN3, PIN4, and PIN7 [14,25,133,140,141]. Moreover, the triple loss-of-function *pin3 pin4 pin7* mutant does not elongate under simulated shade [131]. The regulation of ATP-binding cassette B (ABCB) auxin transporters is also important for proper auxin distribution in the hypocotyl in simulated shade [142].

In the hypocotyls, low R/FR ratio light also controls the localization of PIN3 [140], which plays a key role in tropic responses [143,144]. Analogous to tropic responses, it was hypothesized almost 20 years ago that shade-induced elongation could be produced by a laterally symmetric redistribution of auxin [27,145,146]. In accordance, it has been subsequently demonstrated that a low R/FR ratio light leads to PIN3 lateral localization in the hypocotyl endodermal cells toward the cortical and epidermal cells [140].

Interestingly, it has been recently demonstrated that the control of auxin fluxes is essential to coordinate shoot and root growth in response to light cues [141,147]. *PIN1* is expressed at low levels in the hypocotyls of Arabidopsis etiolated seedlings, and it is significantly upregulated upon light exposure, thus suggesting that light may control shoot-to-root polar auxin transport mainly through the regulation of *PIN1* expression in the hypocotyl. Accordingly, it has been shown that *pin1* displays a reduced root length and alterations in the root apical meristem (RAM) that were highly similar to those of plants treated with polar auxin transport inhibitors. Remarkably, the expression of *PIN1* in the hypocotyl is regulated by COP1. Therefore, COP1, whose activity is determined by light, affects shoot-derived auxin levels in the root. This affects root elongation and adapts auxin transport and cell proliferation in the RAM, modulating the intracellular distribution of PIN1 and PIN2 in the root in a COP1-dependent manner [147]. Under simulated shade, a significant downregulation of *PIN1* in the hypocotyl, together with a concomitant reduction in auxin levels in the RAM, has also been observed, indicating that it is likely that a low R/FR light light may activate a PIN1-dependent mechanism, similar to that described in etiolated seedlings [141,147]. Interestingly, it appears that COP1 plays a dual role in the regulation of root growth according to the light present in the environment. Indeed, COP1, on one hand, controls the long-distance transport of auxin, and, on the other hand, regulates local fluxes of auxin in the RAM through different mechanisms [147]. As for the first mechanism, it has been suggested that HY5, which is one of the best characterized targets of COP1, might directly regulate *PIN1* transcription in the hypocotyl [147]. Notably, recent work has shown that HY5 is a shoot-to-root mobile signal involved in the promotion of root growth by light [148,149]. The perception of low R/FR in the shoot also results in a decrease in lateral root (LR) emergence, and it has been proposed that HY5 regulates this process by inhibiting the auxin efflux carrier PIN3 and the influx carrier LIKE-AUX1 3 (LAX3) auxin transporters, which act in concert in the process of LR emergence [149,150].

4.3. Auxin Signaling

The Transport Inhibitor Response 1/Auxin Signaling F-Box (TIR1/AFBs) proteins are auxin receptors and are components of the SKP1 CULLIN–FBOX (SCF)-type E3 ligase complex, SCF$^{TIR1\text{-}AFBs}$. Auxin binding to SCF^{TIR1AFBs} determines the ubiquitination and degradation of the Auxin/Indole-3-Acetic Acid (Aux/IAA) proteins. Aux/IAAs function as repressors by forming dimers with Auxin Response Factors (ARFs), and their degradation releases the inhibition of ARF transcription factors [151,152].

Relevantly, it has been shown that a low R/FR light ratio rapidly and transiently diminishes the frequency of cell division in Arabidopsis leaf primordia through a mechanism that requires

TIR1. Consistent with the role of HFR1/SICS1 in the shade avoidance response, the leaf primordium phenotype is enhanced in *hfr1/sics1* mutant seedlings in a low R/FR light ratio (Figure 4).

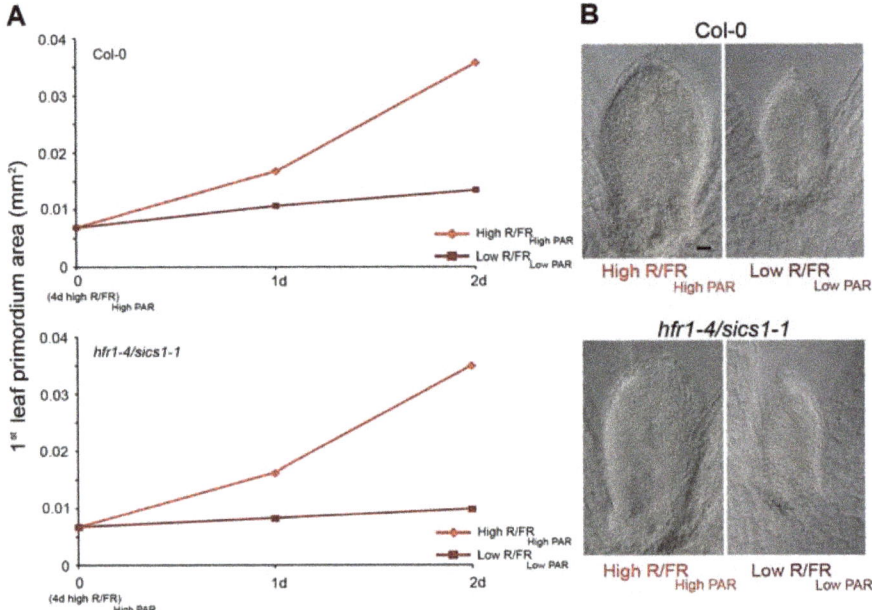

Figure 4. *hfr1/sics1* **mutation causes an exaggerated leaf primordium phenotype in shade.** (**A**) *hfr1/sics1* and control (Col-0) seedlings were grown for four days in high R/FR $_{High\ PAR}$, and then either maintained in the same light regime (red lines) or transferred to low R/FR $_{Low\ PAR}$ for different times (garnet red lines). The mean area of the first/second leaf primordium was calculated by analyzing 50 samples in each condition. (**B**) Leaf primordia, observed under Differential Interference Contrast (DIC) optics, of *hfr1/sics1* and Col-0 grown for four days in high R/FR $_{High\ PAR}$, and then either maintained in the same light regime or transferred for two days to low R/FR $_{Low\ PAR}$. Light outputs were as previously reported [9]. Scale bar, 10 µm.

The auxin increase perceived through TIR1 results in the upregulation of *Cytokinin Oxidase/Dehydrogenase 6* (*CKX6*), which is a gene encoding an enzyme that catalyzes the irreversible degradation of cytokinin [153,154]. This, in turn, lowers local cytokinin levels, and reduces cell proliferation in developing leaf primordia [133,155]. Further studies are needed to identify the specific ARF(s) that are involved in the induction of *CKX6* by a low R/FR light ratio.

A number of studies have identified auxin-related genes as overrepresented among the genes induced by shade in young seedlings [9,14,23,24,33,49,131,156]. Interestingly, a large fraction of these genes are upregulated in both cotyledons and hypocotyl, thus indicating that shade-induced elongation depends not only on the cotyledon-derived auxin, but also on local hypocotyl signals [131]. Among the auxin-related genes rapidly induced by low R/FR are several early auxin response genes, particularly members of the *Aux/IAA* and the *Small Auxin Up RNA* (*SAUR*) gene families, thus indicating that a number of ARF proteins contribute to the shade avoidance response. Recent work indeed provided evidence that three ARF proteins, ARF6, NPH4/ARF7, and ARF8, together play a key role in the regulation of hypocotyl elongation in a low R/FR environment, as well as in response to other signals, including high temperature [157].

5. From Arabidopsis to Crops

The yield of a crop depends to a large degree on its radiation use efficiency and capacity of light interception. At a high planting density, the light interception depends on plant architecture, the degree of mutual shading among plants, and the genetically defined ability of the plant to react to shading, i.e., producing new leaves or reorienting the leaves toward open light [5]. Indeed, several of the effects of the perception of low R/FR signals appear to be negative for yield. Interestingly, despite breeding programs resulting in new cultivars with increased performance under high planting density, many crops still retain the ability to sense and react to canopy shade. For instance, the sensing and reactions to low R/FR, including elongation responses, are present in modern commercial hybrids of maize [158–160]. Similarly, the analysis of 10 modern Argentinian wheat cultivars revealed that the selection for yield did not reduced the ability to respond to a low R/FR ratio and diminish the impact of the negative control of productivity [161]. The reduction of these responses may allow increasing plant productivity at a higher density or may provide higher yield at current densities. This could be realized through the selection of natural variants or mutants, as well as by the generation of mutations in critical factor genes by New Breeding Techniques (NBT) or the production of transgenic plants (a.k.a. Genetically Modified Organism, GMO) expressing specific regulators. The latter two approaches require the identification of key regulatory factors. Arabidopsis is an excellent model system to uncover and dissect mechanisms regulating the shade avoidance response, some of which are likely to be conserved during evolution. However, some important differences are emerging from the analysis of other plant species, which have been recently described in several excellent reviews [162–164]. It is clear that we have to expand our knowledge of other plant species, especially those representing crop model plants, both for food and energy production. Effective approaches for studying the dynamics of shade avoidance and the identification of critical regulators include genome-wide transcriptional analyses, also taking advantage of the genetic diversity of wild and cultivated species and introgression line (IL) populations produced by their crossing. Here, we briefly review the main results obtained in maize and tomato, which are two economically important mono and dicotyledonous crops, respectively.

5.1. Maize

The genome of maize encodes three types of phytochromes (PHYA, PHYB, and PHYC) [165]. PHYB is encoded by two genes (*PHYB1* and *PHYB2*) derived from an ancient tetraploidization event, and both phytochromes contribute differently to distinct physiological aspects of the shade avoidance response [166]. The *phyB1 phyB2* double mutant phenocopies wild-type plants grown in shade, including increased plant height and internode length, reduced tillering, and early flowering [166]. Studies in hybrid maize and teosinte using end-of-day far-red (EOD-FR) light treatments suggested that mesocotyl elongation responses were of the same magnitude [160]. However, a comparison between a modern and an old variety suggested that hybrids that are more productive under high-density plantings may have a reduced auxin response to changes in light quality [159]. The recent data of a genome-wide expression analysis using the maize B73 elite inbred line support this hypothesis [93]. Interestingly, light conditions mimicking canopy shade identical to those utilized by Ruberti et al. to study the process in Arabidopsis [9] were used for the analysis of the shade avoidance response in maize [93]. Consistently, under this light condition, maize seedlings showed an elongated phenotype that was typical of the shade avoidance response. Thereby, the authors were able to compare the dynamics of the transcriptional reprogramming in the two plant species. Two major important differences, among several others, came out from this analysis. First of all, the *YUC* genes, which were strongly induced by low R/FR light in Arabidopsis, were not found regulated in maize. Conversely, *TAA1* was slightly upregulated in maize seedlings, whereas it is downregulated to some extent in Arabidopsis. Coherently, the Gene Ontology (GO) analysis revealed the lack of an enrichment in auxin response genes among those induced by low R/FR light. Furthermore, a genome-wide expression analysis in rice also revealed the lack of induction of auxin response genes in the coleoptile when the seedlings were exposed to low R/FR light [167]. Therefore, it seems possible that the auxin response may have a less important role in monocots, or be a peculiarity of the shade avoidance response in dicotyledonous plants, as confirmed by the large amount of data collected [121,126,168]. A confirmation of such a hypothesis will require a more systematic analysis of monocotyledonous plant species and their undomesticated ancestors, including teosinte. In addition, the comparison of maize and Arabidopsis transcriptional responses also revealed very little overlap between the early response genes, even though hundreds of genes are regulated by low R/FR [93]. In particular, only 20 upregulated and 11 downregulated maize genes have orthologous genes similarly regulated by shade in Arabidopsis. In addition, 19 orthologous gene pairs displayed opposite regulation in response to low R/FR light. Among the upregulated orthologous pairs, there are *ATHB2* and *Gigantea* (*GI*). GI has been implicated in the induction of shade-mediated rapid flowering in low R/FR [169]. The role of ATHB2 in the shade avoidance response has been discussed earlier in this review, and, it is of interest that it is induced by low R/FR light in other plant species [92,167,170,171]. The Arabidopsis *ATHB2* gene is a direct target of the PIF proteins [25,172], and the maize genome encodes for homologs of the Arabidopsis PIF proteins. The constitutive expression of either ZmPIF4 or ZmPIF5 partially rescues the reduced hypocotyl phenotype of the quadruple *pif1 pif3 pif4 pif5* (*pifq*) Arabidopsis mutant, and the overexpression of *ZmPIF5* in Arabidopsis exhibited a constitutive shade avoidance phenotype [173]. Further studies should clarify if the ZmPIFs have any role in the shade avoidance response, including the upregulation of *ATHB2*-like maize genes.

5.2. Tomato

Physiological and molecular studies have begun to dissect the effects of neighbor detection and shade avoidance in tomato [92,171,174–176]. As other plant species, tomato plants exposed to low R/FR elongate both internodes and petioles more. Unlike other species, tomato plants increase the size of the SAM and incipient leaf primordia, and of the leaf blade when exposed to shade. The alteration of leaf morphology has been observed both in cultivated [129] and wild species [177]. Molecular studies have begun to highlight specific patterns of gene expression in the leaf and stem. Particularly significant is the differential regulation of genes involved in photosynthesis in the leaf and stem, being upregulated and downregulated, respectively [170]. As in the case of maize, the domestication of tomato results in plants that exhibit a reduced shade avoidance response compared to wild tomato species. By means of the introgression analysis of a population arising from a cross between the cultivated tomato M82 and the wild relative *Solanum pennellii*, several loci have been found to affect the strength of shade avoidance, either positively or negatively. The expression analysis of the introgressed lines (ILs) confirmed and extended the molecular data obtained by Casal et al. [170]. In particular, this analysis identified a group of auxin-related genes whose expression correlates with the strength of the shade avoidance response, being upregulated in strong responding and downregulated in tolerant lines, respectively [174]. However, prolonged exposure to shade, while still producing shade avoidance responses, results in normal levels of auxin both in the leaf and stem, although auxin-responsive genes are found upregulated [168]. Similar results are also found in Arabidopsis and soybean [124,178,179], indicating that part of the responses to prolonged exposure to shade is produced by an increased sensitivity to auxin [179]. The analysis of ILs also revealed a very limited number of transcription factor genes regulated by shade; among these genes, only three homologs of *ATHB2* and the homolog of *Ethylene and Salt Inducible 3* (*ESE3*) [174] are induced by shade in Arabidopsis, whereas ESE3 is not regulated in maize [93]. Expression profiling studies in the first emerging leaf primordium exposed to shade light for 28 h also revealed a significant upregulation in the expression of the tomato ortholog of *Shootmeristemless* and other *KNOX*-related genes that are known to promote indeterminacy, and the downregulation of genes involved in leaf differentiation [92].

6. Conclusions

Dose-dependent responses to transient and/or persistent stimuli are very common in nature. Generally, a transient behavior with very steep initial upregulation and a subsequent decay region is observed. The overall shape of the response depends on the magnitude of the stimulus received, i.e., it shows a dose-dependent behavior, likely as the product of negative feedback(s). The persistence or the extinction of the response depends on the permanence of the stimulus.

Recent data in Arabidopsis and tomato strongly suggest that the strength of the shade avoidance response depends on auxin. Studies at the molecular level that were conducted mainly in Arabidopsis have highlighted two distinct molecular programs operating in the shade avoidance response. The first one, which is defined as neighbor detection, is characterized by a strong induction of auxin biosynthesis, its accumulation and transport, and transduction of the auxin signal, together with the upregulation of several transcription factor genes and the expression of multiple hormone pathways with distinct and/or overlapping programs taking place in different organs [131]. This molecular response is rapid and transient; it is a "warning signal" that is comparable to a defense response, with the auxin biosynthesis quickly turned off by the intensity of the light reaching the plant, which affects the stability of the negative regulator HRF1/SICS1 [123]. The second program (canopy shade) takes place later on, in part overlaps with the first one, and persists even when the plant is unable to escape shade by the need of the plant to acclimate to the new environmental conditions characterized by a less efficient photosynthetic light. It has been proposed that auxin signaling is also involved in the regulation of this program, likely by a change in the sensitivity to auxin rather than an increase in the concentration of this hormone [25,122–124,178,179]. However, intriguingly, the data accumulating in monocotyledonous plant species seem to indicate a reduced or even the lack of an auxin response(s), in spite of the presence of a characteristic shade avoidance response [93,159,167].

It is worth reminding that neighbor detection and canopy shade are both under the strict control of the phytochrome systems through the PIF proteins, and that the whole processes are rapidly reversed by high R/FR light, eventually just by increased irradiance and/or the altered spectral composition of sunflecks perceived through the canopy [156]. Consistently, *ATHB2*, being a direct target of PIF proteins, is rapidly and reversibly regulated by changes in the R/FR light ratio [29], and it is fully induced even by local irradiation [180]. Evidence is accumulating that ATHB2 and its homologs are key regulators of the shade avoidance response, at least in Arabidopsis. Indeed, the overexpression of different members of the HD-Zip II family phenocopies the effect of shade light on distinct organs and flowering, even when the plants are grown in high R/FR [26,27,35,79,86,90]. On the contrary, single and double loss-of-function *HD-ZIP II* mutants display altered growth responses to shade both in the hypocotyl and in the leaf [28,90,94]. In agreement, the expression of a dominant-negative *athb2* mutation in transgenic Arabidopsis and tomato plants results in phenotypic alterations that are suggestive of an overall attenuation of the shade avoidance response [181]. Unfortunately, multiple loss-of-function *HD-Zip II* mutants are difficult to test in shade, since they are strongly altered in embryo, SAM activity, leaf polarity, and gynoecium and fruit development under simulated sunlight conditions [83,98,100], implying a fundamental role of these proteins in the regulation of plant growth and development. Indeed, there are evidences that the alteration of selected HD-Zip II proteins affects at least a regulatory circuit between HD-Zip II and HD-Zip III transcription factors [73,79,83,98,118] and hormones' signal transduction pathways [101,182]. In addition, evidence exists that a PIF/HD-Zip II genetic module was recruited to carpel development in Arabidopsis [99].

In evolutionary terms, the shade avoidance response appears to be a relatively recent invention that is predominantly found in angiosperms, and it has been considered one of the factors that has contributed to their success [13].

Although the transcriptional program(s) that regulate the developmental responses to shade may be different in distant evolutionary species, it is relevant to emphasize that *ATHB2* and its homologs are the only transcription factor genes regulated by low R/FR light in all of the species that have been analyzed up to today, including poplar [183].

Further work is needed to establish whether ATHB2 and ATHB2-like proteins, together with the PIF proteins, may be considered as the "core regulatory module" recruited to escape and/or adapt to canopy shade.

Author Contributions: G.S., M.C., and M.P. performed the review of the literature; G.M., and I.R. wrote the manuscript. All authors approved the final draft.

Funding: This research was funded by the Italian Ministry of Education, University and Research, PRIN Program (https://www.researchitaly.it/), grant number 2010HEBBB8_004.

Acknowledgments: We thank all our collaborators who made the work on the shade avoidance response an exciting and gratifying experience. Our apologies to the many researchers whose work or original publications has not been cited here because of space limitations.

Conflicts of Interest: The authors declare no conflict of interest.

Abbreviation

BBX	B-BOX
DELLA proteins	Named after their conserved N-terminal
DIC	Differential Interference Contrast
GMO	Genetically Modified Organisms
GUS	β-glucuronidase

References

1. Gommers, C.M.M.; Visser, E.J.W.; St Onge, K.R.; Voesenek, L.A.C.J.; Pierik, R. Shade tolerance: When growing tall is not an option. *Trends Plant Sci.* **2013**, *18*, 65–71. [CrossRef] [PubMed]
2. Jacobs, M.; Lopez-Garcia, M.; Phrathep, O.P.; Lawson, T.; Oulton, R.; Whitney, H.M. Photonic multilayer structure of Begonia chloroplasts enhances photosynthetic efficiency. *Nat. Plants* **2016**, *2*, 16162. [CrossRef] [PubMed]
3. Gommers, C.M.; Keuskamp, D.H.; Buti, S.; van Veen, H.; Koevoets, I.T.; Reinen, E.; Voesenek, L.A.; Pierik, R. Molecular profiles of contrasting shade response strategies in wild plants: Differential control of immunity and shoot elongation. *Plant Cell* **2017**, *29*, 331–344. [CrossRef] [PubMed]
4. Pierik, R.; De Wit, M. Shade avoidance: Phytochrome signalling and other aboveground neighbour detection cues. *J. Exp. Bot.* **2014**, *65*, 2815–2824. [CrossRef] [PubMed]
5. Casal, J.J. Photoreceptor signaling networks in plant responses to shade. *Annu. Rev. Plant Biol.* **2013**, *64*, 403–427. [CrossRef] [PubMed]
6. Fraser, D.P.; Hayes, S.; Franklin, K.A. Photoreceptor crosstalk in shade avoidance. *Curr. Opin. Plant Biol.* **2016**, *33*, 1–7. [CrossRef] [PubMed]
7. Ballaré, C.L.; Pierik, R. The shade-avoidance syndrome: Multiple signals and ecological consequences. *Plant Cell Environ.* **2017**, *40*, 2530–2543. [CrossRef] [PubMed]
8. Fiorucci, A.S.; Fankhauser, C. Plant Strategies for Enhancing Access to Sunlight. *Curr. Biol.* **2017**, *27*, R931–R940. [CrossRef] [PubMed]
9. Ciolfi, A.; Sessa, G.; Sassi, M.; Possenti, M.; Salvucci, S.; Carabelli, M.; Morelli, G.; Ruberti, I. Dynamics of the shade-avoidance response in Arabidopsis. *Plant Physiol.* **2013**, *163*, 331–353. [CrossRef] [PubMed]
10. Bae, G.; Choi, G. Decoding of light signals by plant phytochromes and their interacting proteins. *Annu. Rev. Plant Biol.* **2008**, *59*, 281–311. [CrossRef] [PubMed]
11. Mathews, S.; Sharrock, R. Phytochrome gene diversity. *Plant Cell Environ.* **1997**, *20*, 666–671. [CrossRef]
12. Franklin, K.A.; Quail, P.H. Phytochrome functions in Arabidopsis development. *J. Exp. Bot.* **2010**, *61*, 11–24. [CrossRef] [PubMed]
13. Smith, H.; Whitelam, G.C. The shade avoidance syndrome: Multiple responses mediated by multiple phytochromes. *Plant Cell Environ.* **1997**, *20*, 840–844. [CrossRef]
14. Devlin, P.F.; Yanovsky, M.J.; Kay, S.A. A genomic analysis of the shade avoidance response in Arabidopsis. *Plant Physiol.* **2003**, *133*, 1617–1629. [CrossRef] [PubMed]

15. Johnson, E.; Bradley, J.M.; Harberd, N.P.; Whitelam, G.C. Photoresponses of light-grown phyA mutants of Arabidopsis: Phytochrome A is required for the perception of daylength extensions. *Plant Physiol.* **1994**, *105*, 141–149. [CrossRef] [PubMed]
16. Wang, X.; Roig-Villanova, I.; Khan, S.; Shanahan, H.; Quail, P.H.; Martinez-Garcia, J.F.; Devlin, P.F. A novel high-throughput in vivo molecular screen for shade avoidance mutants identifies a novel phyA mutation. *J. Exp. Bot.* **2011**, *62*, 2973–2987. [CrossRef] [PubMed]
17. Leivar, P.; Quail, P.H. PIFs: Pivotal components in a cellular signaling hub. *Trends Plant Sci.* **2011**, *16*, 19–28. [CrossRef] [PubMed]
18. Leivar, P.; Monte, E. PIFs: Systems integrators in plant development. *Plant Cell* **2014**, *26*, 56–78. [CrossRef] [PubMed]
19. de Lucas, M.; Prat, S. PIFs get BRright: PHYTOCHROME INTERACTING FACTORs as integrators of light and hormonal signals. *New Phytol.* **2014**, *202*, 1126–1141. [CrossRef] [PubMed]
20. Lee, N.; Choi, G. Phytochrome-interacting factor from Arabidopsis to liverwort. *Curr. Opin. Plant Biol.* **2017**, *35*, 54–60. [CrossRef] [PubMed]
21. Kim, J.; Kang, H.; Park, J.; Kim, W.; Yoo, J.; Lee, N.; Yoon, T.Y.; Choi, G. PIF1-interacting transcription factors and their binding sequence elements determine the in vivo targeting sites of PIF1. *Plant Cell* **2016**, *28*, 1388–1405. [CrossRef] [PubMed]
22. Lorrain, S.; Allen, T.; Duek, P.D.; Whitelam, G.C.; Fankhauser, C. Phytochrome-mediated inhibition of shade avoidance involves degradation of growth-promoting bHLH transcription factors. *Plant J.* **2008**, *53*, 312–323. [CrossRef] [PubMed]
23. Leivar, P.; Monte, E.; Cohn, M.M.; Quail, P.H. Phytochrome signaling in green Arabidopsis seedlings: Impact assessment of a mutually negative phyB–PIF feedback loop. *Mol. Plant* **2012**, *5*, 734–749. [CrossRef] [PubMed]
24. Li, L.; Ljung, K.; Breton, G.; Pruneda-Paz, J.; Cowing-Zitron, C.; Cole, B.J.; Ivans, L.J.; Pedmale, U.V.; Jung, H.S.; Ecker, J.R.; et al. Linking photoreceptor excitation to changes in plant architecture. *Genes Dev.* **2012**, *26*, 785–790. [CrossRef] [PubMed]
25. Hornitschek, P.; Kohnen, M.V.; Lorrain, S.; Rougemont, J.; Ljung, K.; López-Vidriero, I.; Franco-Zorrilla, J.M.; Solano, R.; Trevisan, M.; Pradervand, S.; et al. Phytochrome interacting factors 4 and 5 control seedling growth in changing light conditions by directly controlling auxin signaling. *Plant J.* **2012**, *71*, 699–711. [CrossRef] [PubMed]
26. Ruberti, I.; Sessa, G.; Ciolfi, A.; Possenti, M.; Carabelli, M.; Morelli, G. Plant adaptation to dynamically changing environment: The shade avoidance response. *Biotechnol. Adv.* **2012**, *30*, 1047–1058. [CrossRef] [PubMed]
27. Steindler, C.; Matteucci, A.; Sessa, G.; Weimar, T.; Ohgishi, M.; Aoyama, T.; Morelli, G.; Ruberti, I. Shade avoidance responses are mediated by the ATHB-2 HD-Zip protein, a negative regulator of gene expression. *Development* **1999**, *125*, 4235–4245.
28. Carabelli, M.; Turchi, L.; Ruzza, V.; Morelli, G.; Ruberti, I. Homeodomain-Leucine Zipper II family of transcription factors to the limelight: Central regulators of plant development. *Plant Signal. Behav.* **2013**, *8*, e25447. [CrossRef] [PubMed]
29. Carabelli, M.; Morelli, G.; Whitelam, G.; Ruberti, I. Twilight-zone and canopy shade induction of the *ATHB-2* homeobox gene in green plants. *Proc. Natl. Acad. Sci. USA* **1996**, *93*, 3530–3535. [CrossRef] [PubMed]
30. Franklin, K.A.; Praekelt, U.; Stoddart, W.M.; Billingham, O.E.; Halliday, K.J.; Whitelam, G.C. Phytochromes B, D, and E act redundantly to control multiple physiological responses in Arabidopsis. *Plant Physiol.* **2003**, *131*, 1340–1346. [CrossRef] [PubMed]
31. Roig-Villanova, I.; Bou, J.; Sorin, C.; Devlin, P.F.; Martínez-García, J.F. Identification of primary target genes of phytochrome signaling. Early transcriptional control during shade avoidance responses in Arabidopsis. *Plant Physiol.* **2006**, *141*, 85–96. [CrossRef] [PubMed]
32. Leivar, P.; Tepperman, J.M.; Cohn, M.M.; Monte, E.; Al-Sady, B.; Erickson, E.; Quail, P.H. Dynamic antagonism between phytochromes and PIF family basic helix-loop-helix factors induces selective reciprocal responses to light and shade in a rapidly responsive transcriptional network in Arabidopsis. *Plant Cell* **2012**, *24*, 1398–1419. [CrossRef] [PubMed]

33. Sessa, G.; Carabelli, M.; Sassi, M.; Ciolfi, A.; Possenti, M.; Mittempergher, F.; Becker, J.; Morelli, G.; Ruberti, I. A dynamic balance between gene activation and repression regulates the shade avoidance response in Arabidopsis. *Genes Dev.* **2005**, *19*, 2811–2815. [CrossRef] [PubMed]
34. Hornitschek, P.; Lorrain, S.; Zoete, V.; Michielin, O.; Fankhauser, C. Inhibition of the shade avoidance response by formation of non-DNA binding bHLH heterodimers. *EMBO J.* **2009**, *28*, 3893–3902. [CrossRef] [PubMed]
35. Ruzza, V.; Sessa, G.; Sassi, M.; Morelli, G.; Ruberti, I. Auxin coordinates shoot and root development during shade avoidance response. In *Auxin and Its Role in Plant Development*; Zažímalová, E., Petrasek, J., Benková, E., Eds.; Springer-Verlag: Wien, Austria, 2014; pp. 349–412.
36. Galstyan, A.; Cifuentes-Esquivel, N.; Bou-Torrent, J.; Martinez-Garcia, J.F. The shade avoidance syndrome in Arabidopsis: A fundamental role for atypical basic helix-loop-helix proteins as transcriptional cofactors. *Plant J.* **2011**, *66*, 258–267. [CrossRef] [PubMed]
37. Roig-Villanova, I.; Bou-Torrent, J.; Galstyan, A.; Carretero-Paulet, L.; Portolés, S.; Rodríguez-Concepción, M.; Martínez-García, J.F. Interaction of shade avoidance and auxin responses: A role for two novel atypical bHLH proteins. *EMBO J.* **2007**, *26*, 4756–4767. [CrossRef] [PubMed]
38. Hao, Y.; Oh, E.; Choi, G.; Liang, Z.; Wang, Z.Y. Interactions between HLH and bHLH factors modulate light-regulated plant development. *Mol. Plant* **2012**, *5*, 688–697. [CrossRef] [PubMed]
39. Cifuentes-Esquivel, N.; Bou-Torrent, J.; Galstyan, A.; Gallemí, M.; Sessa, G.; Salla Martret, M.; Roig-Villanova, I.; Ruberti, I.; Martínez-García, J.F. The bHLH proteins BEE and BIM positively modulate the shade avoidance syndrome in Arabidopsis seedlings. *Plant J.* **2013**, *75*, 989–1002. [CrossRef] [PubMed]
40. Martínez-García, J.F.; Gallemí, M.; Molina-Contreras, M.J.; Llorente, B.; Bevilaqua, M.R.R.; Quail, P.H. The shade avoidance syndrome in Arabidopsis: The antagonistic role of phytochrome A and B differentiates vegetation proximity and canopy shade. *PLoS ONE* **2014**, *9*, e109275. [CrossRef] [PubMed]
41. Sheerin, D.J.; Hiltbrunner, A. Molecular mechanisms and ecological function of far-red light signalling. *Plant Cell Environ.* **2017**, *40*, 2509–2529. [CrossRef] [PubMed]
42. Lau, O.S.; Deng, X.W. Plant hormone signaling lightens up: Integrators of light and hormones. *Curr. Opin. Plant Biol.* **2010**, *13*, 571–577. [CrossRef] [PubMed]
43. Chen, D.; Xu, G.; Tang, W.; Jing, Y.; Ji, Q.; Fei, Z.; Lin, R. Antagonistic basic helix-loop-helix/bZIP transcription factors form transcriptional modules that integrate light and reactive oxygen species signaling in Arabidopsis. *Plant Cell* **2013**, *25*, 1657–1673. [CrossRef] [PubMed]
44. Toledo-Ortiz, G.; Johansson, H.; Lee, K.P.; Bou-Torrent, J.; Stewart, K.; Steel, G.; Rodríguez-Concepción, M.; Halliday, K.J. The HY5-PIF regulatory module coordinates light and temperature control of photosynthetic gene transcription. *PLoS Genet.* **2014**, *10*, e1004416. [CrossRef] [PubMed]
45. Sheerin, D.J.; Menon, C.; zur Oven-Krockhaus, S.; Enderle, B.; Zhu, L.; Johnen, P.; Schleifenbaum, F.; Stierhof, Y.D.; Huq, E.; Hiltbrunner, A. Light-activated Phytochrome A and B interact with members of the SPA family to promote photomorphogenesis in Arabidopsis by disrupting the COP1-SPA complex. *Plant Cell* **2015**, *27*, 189–201. [CrossRef] [PubMed]
46. Lau, O.S.; Deng, X.W. The photomorphogenic repressors COP1 and DET1: 20 years later. *Trends Plant Sci.* **2012**, *17*, 584–593. [CrossRef] [PubMed]
47. McNellis, T.W.; von Arnim, A.G.; Araki, T.; Komeda, Y.; Miséra, S.; Deng, X.-W. Genetic and molecular analysis of an allelic series of cop1 mutants suggests functional roles for the multiple protein domains. *Plant Cell* **1994**, *6*, 487–500. [CrossRef] [PubMed]
48. Rolauffs, S.; Fackendahl, P.; Sahm, J.; Fiene, G.; Hoecker, U. Arabidopsis COP1 and SPA genes are essential for plant elongation but not for acceleration of flowering time in response to a low red light to far-red light ratio. *Plant Phys.* **2012**, *160*, 2015–2027. [CrossRef] [PubMed]
49. Pacín, M.; Semmoloni, M.; Legris, M.; Finlayson, S.A.; Casal, J.J. Convergence Of Constitutive Photomorphogenesis 1 and Phytochrome Interacting Factor Signalling during Shade Avoidance. *New Phytol.* **2016**, *211*, 967–979. [CrossRef] [PubMed]
50. Crocco, C.D.; Holm, M.; Yanovsky, M.J.; Botto, J.F. AtBBX21 and COP1 genetically interact in the regulation of shade avoidance. *Plant J.* **2010**, *64*, 551–562. [CrossRef] [PubMed]
51. Chang, C.S.J.; Maloof, J.N.; Wu, S.H. COP1 mediated degradation of BBX22/LZF1 optimizes seedling development in Arabidopsis. *Plant Physiol.* **2011**, *156*, 228–239. [CrossRef] [PubMed]

52. Xu, D.; Jiang, Y.; Li, J.; Lin, F.; Holm, M.; Deng, X.W. BBX21, an Arabidopsis B-box protein, directly activates HY5 and is targeted by COP1 for 26S proteasome-mediated degradation. *Proc. Natl. Acad. Sci. USA* **2016**, *113*, 7655–7660. [CrossRef] [PubMed]
53. Mei, Q.; Dvornyk, V. Evolutionary history of the photolyase/cryptochrome superfamily in eukaryotes. *PLoS ONE* **2015**, *10*, e0135940. [CrossRef] [PubMed]
54. Chaves, I.; Pokorny, R.; Byrdin, M.; Hoang, N.; Ritz, T.; Brettel, K.; Essen, L.O.; van der Horst, G.T.; Batschauer, A.; Ahmad, M. The cryptochromes: Blue light photoreceptors in plants and animals. *Annu. Rev. Plant Biol.* **2011**, *62*, 335–364. [CrossRef] [PubMed]
55. Yang, Z.; Liu, B.; Su, J.; Liao, J.; Lin, C.; Oka, Y. Cryptochromes Orchestrate Transcription Regulation of Diverse Blue Light Responses in Plants. *Photochem. Photobiol.* **2017**, *93*, 112–127. [CrossRef] [PubMed]
56. Keller, M.M.; Jaillais, Y.; Pedmale, U.V.; Moreno, J.E.; Chory, J.; Ballaré, C.L. Cryptochrome 1 and phytochrome B control shade-avoidance responses in Arabidopsis via partially independent hormonal cascades. *Plant J.* **2011**, *67*, 195–207. [CrossRef] [PubMed]
57. Keuskamp, D.H.; Sasidharan, R.; Vos, I.; Peeters, A.J.; Voesenek, L.A.C.J.; Pierik, R. Blue-light-mediated shade avoidance requires combined auxin and brassinosteroid action in Arabidopsis seedlings. *Plant J.* **2011**, *67*, 208–217. [CrossRef] [PubMed]
58. Pedmale, U.V.; Huang, S.S.; Zander, M.; Cole, B.J.; Hetzel, J.; Ljung, K.; Reis, P.-A.; Sridevi, P.; Nito, K.; Nery, J.R.; et al. Cryptochromes interact directly with PIFs to control plant growth in limiting blue light. *Cell* **2016**, *164*, 233–245. [CrossRef] [PubMed]
59. Ma, D.; Li, X.; Guo, Y.; Chu, J.; Fang, S.; Yan, C.; Noel, J.P.; Liu, H. Cryptochrome 1 interacts with PIF4 to regulate high temperature-mediated hypocotyl elongation in response to blue light. *Proc. Natl. Acad. Sci. USA* **2016**, *113*, 224–229. [CrossRef] [PubMed]
60. Liu, B.; Zuo, Z.; Liu, H.; Liu, X.; Lin, C. Arabidopsis cryptochrome 1 interacts with SPA1 to suppress COP1 activity in response to blue light. *Genes Dev.* **2011**, *25*, 1029–1034. [CrossRef] [PubMed]
61. Casal, J.J. Shade Avoidance. *Arabidopsis Book* **2012**, *10*, e0157. [CrossRef] [PubMed]
62. Rizzini, L.; Favory, J.-J.; Cloix, C.; Faggionato, D.; O'Hara, A.; Kaiserli, E.; Baumeister, R.; Schäfer, E.; Nagy, F.; Jenkins, G.I.; et al. Perception of UV-B by the Arabidopsis UVR8 protein. *Science* **2011**, *332*, 103–106. [CrossRef] [PubMed]
63. Ulm, R.; Jenkins, G. Q&A: How do plants sense and respond to UV-B radiation? *BMC Biol.* **2015**, *13*, 45.
64. Oravecz, A.; Baumann, A.; Mate, Z.; Brzezinska, A.; Molinier, J.; Oakeley, E.J.; Adam, E.; Schäfer, E.; Nagy, F.; Ulm, R. Constitutively photomorphogenic1 is required for the UV-B response in Arabidopsis. *Plant Cell* **2006**, *18*, 1975–1990. [CrossRef] [PubMed]
65. Brown, B.A.; Jenkins, G.I. UV-B signaling pathways with different fluence-rate response profiles are distinguished in mature Arabidopsis leaf tissue by requirement for UVR8, HY5, and HYH. *Plant Physiol.* **2008**, *146*, 576–588. [CrossRef] [PubMed]
66. Favory, J.J.; Stec, A.; Gruber, H.; Rizzini, L.; Oravecz, A.; Funk, M.; Albert, A.; Cloix, C.; Jenkins, G.I.; Oakeley, E.J.; et al. Interaction of COP1 and UVR8 regulates UV-B-induced photomorphogenesis and stress acclimation in Arabidopsis. *EMBO J.* **2009**, *28*, 591–601. [CrossRef] [PubMed]
67. Hayes, S.; Velanis, C.N.; Jenkins, G.I.; Franklin, K.A. UV-B detected by the UVR8 photoreceptor antagonizes auxin signaling and plant shade avoidance. *Proc. Natl. Acad. Sci. USA* **2014**, *111*, 11894–11899. [CrossRef] [PubMed]
68. Mazza, C.A.; Ballaré, C.L. Photoreceptors UVR8 and phytochrome B cooperate to optimize plant growth and defense in patchy canopies. *New Phytol.* **2015**, *207*, 4–9. [CrossRef] [PubMed]
69. Ruberti, I.; Sessa, G.; Lucchetti, S.; Morelli, G. A novel class of plant proteins containing a homeodomain with a closely linked leucine zipper motif. *EMBO J.* **1991**, *10*, 1787–1791. [CrossRef] [PubMed]
70. Sessa, G.; Morelli, G.; Ruberti, I. The Athb-1 and -2 HD-Zip domains homodimerize forming complexes of different DNA binding specificities. *EMBO J.* **1993**, *12*, 3507–3517. [CrossRef] [PubMed]
71. Ariel, F.D.; Manavella, P.A.; Dezar, C.A.; Chan, R.L. The true story of the HD-Zip family. *Trends Plant Sci.* **2007**, *12*, 419–426. [CrossRef] [PubMed]
72. Tron, A.E.; Bertoncini, C.W.; Palena, C.M.; Chan, R.L.; Gonzales, D.H. Combinatorial interactions of two amino acids with a single base pair define target site specificity in plant dimeric homeodomain proteins. *Nucl. Acids Res.* **2001**, *29*, 4866–4872. [CrossRef] [PubMed]

73. Brandt, R.; Salla-Martret, M.; Bou-Torrent, J.; Musielak, T.; Stahl, M.; Lanz, C.; Ott, F.; Schmid, M.; Greb, T.; Schwarz, M.; et al. Genome-wide binding-site analysis of REVOLUTA reveals a link between leaf patterning and light-mediated growth responses. *Plant J.* **2012**, *72*, 31–42. [CrossRef] [PubMed]
74. Turchi, L.; Baima, S.; Morelli, G.; Ruberti, I. Interplay of HD-Zip II and III transcription factors in auxin-regulated plant development. *J. Exp. Bot.* **2015**, *66*, 5043–5053. [CrossRef] [PubMed]
75. Sessa, G.; Carabelli, M.; Ruberti, I.; Baima, S.; Lucchetti, S.; Morelli, G. Identification of distinct families of HD-Zip proteins in Arabidopsis thaliana. In *Molecular-Genetic Analysis of Plant Development and Metabolism*; Puigdomenech, P., Coruzzi, G., Eds.; NATO-ASI Series; Springer-Verlag: Berlin/Heidelberg, Germany, 1994; Volume 81, pp. 411–426.
76. Sessa, G.; Steindler, C.; Morelli, G.; Ruberti, I. The Arabidopsis Athb-8, -9 and -14 genes are members of a small gene family coding for highly related HD-Zip proteins. *Plant Mol. Biol.* **1998**, *38*, 609–622. [CrossRef] [PubMed]
77. Henriksson, E.; Olsson, A.S.; Johannesson, H.; Johansson, H.; Hanson, J.; Engstrom, P.; Soderman, E. Homeodomain leucine zipper class I genes in Arabidopsis. Expression patterns and phylogenetic relationships. *Plant Physiol.* **2005**, *139*, 509–518. [CrossRef] [PubMed]
78. Agalou, A.; Purwantomo, S.; Overnaes, E.; Johannesson, H.; Zhu, X.; Estiati, A.; de Kam, R.J.; Engström, P.; Slamet-Loedin, I.H.; Zhu, Z.; et al. A genome-wide survey of HD-Zip genes in rice and analysis of drought responsive family members. *Plant Mol. Biol.* **2008**, *66*, 87–103. [CrossRef] [PubMed]
79. Ciarbelli, A.R.; Ciolfi, A.; Salvucci, S.; Ruzza, V.; Possenti, M.; Carabelli, M.; Fruscalzo, A.; Sessa, G.; Morelli, G.; Ruberti, I. The Arabidopsis homeodomain-leucine zipper II gene family: Diversity and redundancy. *Plant Mol. Biol.* **2008**, *68*, 465–478. [CrossRef] [PubMed]
80. Harris, J.C.; Hrmova, M.; Lopato, S.; Langridge, P. Modulation of plant growth by HD-Zip class I and II transcription factors in response to environmental stimuli. *New Phytol.* **2011**, *190*, 823–837. [CrossRef] [PubMed]
81. Romani, F.; Reinheimer, R.; Florent, S.N.; Bowman, J.L.; Moreno, J.E. Evolutionary history of HOMEODOMAIN LEUCINE ZIPPER transcription factors during plant transition to land. *New Phytol.* **2018**, *219*, 408–421. [CrossRef] [PubMed]
82. Merelo, P.; Paredes, E.B.; Heisler, M.G.; Wenkel, S. The shady side of leaf development: The role of the REVOLUTA/KANADI1 module in leaf patterning and auxin-mediated growth promotion. *Curr. Opin. Plant Biol.* **2017**, *35*, 111–116. [CrossRef] [PubMed]
83. Turchi, L.; Carabelli, M.; Ruzza, V.; Possenti, M.; Sassi, M.; Peñalosa, A.; Sessa, G.; Salvi, S.; Forte, V.; Morelli, G.; Ruberti, I. Arabidopsis HD-Zip II transcription factors control embryo development and meristem function. *Development* **2013**, *140*, 2118–2129. [CrossRef] [PubMed]
84. Kagale, S.; Links, M.G.; Rozwadowski, K. Genome-wide analysis of ethylene-responsive element binding factor-associated amphiphilic repression motif-containing transcriptional regulators in Arabidopsis. *Plant Physiol.* **2010**, *152*, 1109–1134. [CrossRef] [PubMed]
85. Ohgishi, M.; Oka, A.; Morelli, G.; Ruberti, I.; Aoyama, T. Negative autoregulation of the Arabidopsis homeobox gene ATHB-2. *Plant J.* **2001**, *25*, 389–398. [CrossRef] [PubMed]
86. Sawa, S.; Ohgishi, M.; Goda, H.; Higuchi, K.; Shimada, Y.; Yoshida, S.; Koshiba, T. The HAT2 gene, a member of the HD-Zip gene family, isolated as an auxin inducible gene by DNA microarray screening, affects auxin response in Arabidopsis. *Plant J.* **2002**, *32*, 1011–1022. [CrossRef] [PubMed]
87. Wenkel, S.; Emercy, J.; Hou, B.; Evans, M.M.S.; Barton, M.K. A feedback regulatory module formed by little zipper and HD-ZIPIII genes. *Plant Cell* **2007**, *19*, 3379–3390. [CrossRef] [PubMed]
88. Kim, Y.-S.; Kim, S.-G.; Lee, M.; Lee, I.; Park, H.Y.; Seo, P.J.; Jung, J.H.; Kwon, E.J.; Suh, S.W.; Paek, K.H.; Park, C.M. HD-ZIP III activity is modulated by competitive inhibitors via a feedback loop in Arabidopsis shoot apical meristem development. *Plant Cell* **2008**, *20*, 920–933. [CrossRef] [PubMed]
89. Baima, S.; Forte, V.; Possenti, M.; Peñalosa, A.; Leoni, G.; Salvi, S.; Felici, B.; Ruberti, I.; Morelli, G. Negative feedback regulation of auxin signaling by ATHB8/ACL5-BUD2 transcription module. *Mol. Plant* **2014**, *7*, 1006–1025. [CrossRef] [PubMed]
90. Sorin, C.; Salla-Martret, M.; Bou-Torrent, J.; Roig-Villanova, I.; Martínez-García, J.F. ATHB4, a regulator of shade avoidance, modulates hormone response in Arabidopsis seedlings. *Plant J.* **2009**, *59*, 266–277. [CrossRef] [PubMed]

91. Ueoka-Nakanishi, H.; Hori, N.; Ishida, K.; Ono, N.; Yamashino, T.; Nakamichi, N.; Mizuno, T. Characterization of shade avoidance responses in Lotus japonicus. *Biosci. Biotechnol. Biochem.* **2011**, *75*, 2148–2154. [CrossRef] [PubMed]
92. Chitwood, D.H.; Kumar, R.; Ranjan, A.; Pelletier, J.M.; Townsley, B.T.; Ichihashi, Y.; Martinez, C.C.; Zumstein, K.; Harada, J.J.; Maloof, J.N.; et al. Light-induced indeterminacy alters shade-avoiding tomato leaf morphology. *Plant Physiol.* **2015**, *169*, 2030–2047. [CrossRef] [PubMed]
93. Wang, H.; Wu, G.; Zhao, B.; Wang, B.; Lang, Z.; Zhang, C.; Wang, H. Regulatory modules controlling early shade avoidance response in maize seedlings. *BMC Genom.* **2016**, *17*, 269. [CrossRef] [PubMed]
94. Carabelli, M.; Possenti, M.; Sessa, G.; Ruzza, V.; Morelli, G.; Ruberti, I. Arabidopsis HD-Zip II proteins regulate the exit from proliferation during leaf development in canopy shade. *J. Exp. Bot.* **2018**. [CrossRef] [PubMed]
95. Trigg, S.A.; Garza, R.M.; MacWilliams, A.; Nery, J.R.; Bartlett, A.; Castanon, R.; Goubil, A.; Feeney, J.; O'Malley, R.; Huang, S.C.; et al. CrY2H-seq: A massively multiplexed assay for deep-coverage interactome mapping. *Nat. Methods* **2017**, *14*, 819–825. [CrossRef] [PubMed]
96. Gonzales, N.; Vanharen, H.; Inzé, D. Leaf size control: Complex coordination of cell division and expansion. *Trends Plant Sci.* **2012**, *17*, 332–340. [CrossRef] [PubMed]
97. Bar, M.; Ori, N. Leaf development and morphogenesis. *Development* **2014**, *141*, 4219–4230. [CrossRef] [PubMed]
98. Bou-Torrent, J.; Salla-Martret, M.; Brandt, R.; Musielak, T.; Palauquim, J.C.; Martinez-Garcia, J.F.; Wenkel, S. ATHB4 and HAT3, two class II HD-ZIP transcription factors, control leaf development in Arabidopsis. *Plant Signal. Behav.* **2012**, *7*, 1382–1387. [CrossRef] [PubMed]
99. Reymond, M.C.; Brunoud, G.; Chauvet, A.; Martínez-Garcia, J.F.; Martin-Magniette, M.L.; Monéger, F.; Scutt, C.P. A light-regulated genetic module was recruited to carpel development in Arabidopsis following a structural change to SPATULA. *Plant Cell* **2012**, *24*, 2812–2825. [CrossRef] [PubMed]
100. Zúñiga-Mayo, V.M.; Marsch-Martínez, N.; de Folter, S. JAIBA, a class-II HD-ZIP transcription factor involved in the regulation of meristematic activity, and important for correct gynoecium and fruit development in Arabidopsis. *Plant J.* **2012**, *71*, 314–326. [CrossRef] [PubMed]
101. Zhang, D.; Ye, H.; Guo, H.; Johnson, A.; Zhang, M.; Lin, H.; Yin, Y. Transcription factor HAT1 is phosphorylated by BIN2 kinase and mediates brassinosteroid repressed gene expression in Arabidopsis. *Plant J.* **2014**, *77*, 59–70. [CrossRef] [PubMed]
102. Xie, Y.; Liu, Y.; Wang, H.; Ma, X.; Wang, B.; Wu, G.; Wang, H. Phytochrome-interacting factors directly suppress *MIR156* expression to enhance shade-avoidance syndrome in *Arabidopsis*. *Nat. Commun.* **2017**, *8*, 348–358. [CrossRef] [PubMed]
103. Xu, M.; Hu, T.; Zhao, J.; Park, M.-Y.; Earley, K.W.; Wu, G.; Yang, L.; Poethig, R.S. Developmental functions of miR156-regulated squamosa promoter binding protein-like (SPL) genes in *Arabidopsis thaliana*. *PLoS Genet.* **2016**, *12*, e1006263. [CrossRef] [PubMed]
104. Smith, Z.R.; Long, J.A. Control of Arabidopsis apical-basal embryo polarity by antagonistic transcription factors. *Nature* **2010**, *464*, 423–426. [CrossRef] [PubMed]
105. Emery, J.F.; Floyd, S.K.; Alvarez, J.; Eshed, Y.; Hawker, N.P.; Izhaki, A.; Baum, S.F.; Bowman, J.L. Radial patterning of Arabidopsis shoots by class III HD-ZIP and KANADI genes. *Curr. Biol.* **2003**, *13*, 1768–1774. [CrossRef] [PubMed]
106. Prigge, M.J.; Otsuga, D.; Alonso, J.M.; Ecker, J.R.; Drews, G.N.; Clark, S.E. Class III homeodomain-leucine zipper gene family members have overlapping, antagonistic, and distinct roles in Arabidopsis development. *Plant Cell* **2005**, *17*, 61–76. [CrossRef] [PubMed]
107. Baima, S.; Nobili, F.; Sessa, G.; Lucchetti, S.; Ruberti, I.; Morelli, G. The expression of the Athb-8 homeobox gene is restricted to provascular cells in Arabidopsis thaliana. *Development* **1995**, *121*, 4171–4182. [PubMed]
108. Heisler, M.G.; Ohno, C.; Das, P.; Sieber, P.; Reddy, G.V.; Long, J.A.; Meyerowitz, E.M. Patterns of auxin transport and gene expression during primordium development revealed by live imaging of the Arabidopsis inflorescence meristem. *Curr. Biol.* **2005**, *15*, 1899–1911. [CrossRef] [PubMed]
109. Ohashi-Ito, K.; Kubo, M.; Demura, T.; Fukuda, H. Class III homeodomain leucine-zipper proteins regulate xylem cell differentiation. *Plant Cell Physiol.* **2005**, *46*, 1646–1656. [CrossRef] [PubMed]
110. Floyd, S.K.; Zalewski, C.S.; Bowman, J.L. Evolution of class III homeodomain-leucine zipper genes in streptophytes. *Genetics* **2006**, *173*, 373–388. [CrossRef] [PubMed]

111. Floyd, S.K.; Bowman, J.L. Distinct developmental mechanisms reflect the independent origins of leaves in vascular plants. *Curr. Biol.* **2006**, *16*, 1911–1917. [CrossRef] [PubMed]
112. Donner, T.J.; Sherr, I.; Scarpella, E. Regulation of preprocambial cell state acquisition by auxin signaling in Arabidopsis leaves. *Development* **2009**, *136*, 3235–3246. [CrossRef] [PubMed]
113. Tang, G.; Reinhart, B.J.; Bartel, D.P.; Zamore, P.D. A biochemical framework for RNA silencing in plants. *Genes Dev.* **2003**, *17*, 49–63. [CrossRef] [PubMed]
114. Huang, T.; Harrar, Y.; Lin, C.; Reinhart, B.; Newell, N.R.; Talavera-Rauh, F.; Hokin, S.A.; Barton, M.K.; Kerstetter, R.A. Arabidopsis KANADI1 acts as a transcriptional repressor by interacting with a specific cis-element and regulates auxin biosynthesis, transport, and signaling in opposition to HD-ZIPIII factors. *Plant Cell* **2014**, *26*, 246–262. [CrossRef] [PubMed]
115. Reinhart, B.J.; Liu, T.; Newell, N.R.; Magnani, E.; Huang, T.; Kerstetter, R.; Michaels, S.; Barton, M.K. Establishing a framework for the Ad/abaxial regulatory network of Arabidopsis: Ascertaining targets of class III homeodomain leucine zipper and KANADI regulation. *Plant Cell* **2013**, *25*, 3228–3249. [CrossRef] [PubMed]
116. Eshed, Y.; Baum, S.F.; Perea, J.V.; Bowman, J.L. Establishment of polarity in lateral organs of plants. *Curr. Biol.* **2001**, *11*, 1251–1260. [CrossRef]
117. Xie, Y.; Straub, D.; Eguen, T.; Brandt, R.; Stahl, M.; Martínez-García, J.F.; Wenkel, S. Meta-Analysis of Arabidopsis KANADI1 Direct Target Genes Identifies a Basic Growth-Promoting Module Acting Upstream of Hormonal Signaling Pathways. *Plant Phys.* **2015**, *169*, 1240–1253. [CrossRef] [PubMed]
118. Merelo, P.; Ram, H.; Pia Caggiano, M.; Ohno, C.; Ott, F.; Straub, D.; Graeff, M.; Cho, S.K.; Yang, S.W.; Wenkel, S.; et al. Regulation of MIR165/166 by class II and class III homeodomain leucine zipper proteins establishes leaf polarity. *Proc. Natl. Acad. Sci. USA* **2016**, *113*, 11973–11978. [CrossRef] [PubMed]
119. de Wit, M.; Galvão, V.C.; Fankhauser, C. Light-Mediated Hormonal Regulation of Plant Growth and Development. *Annu. Rev. Plant Biol.* **2016**, *67*, 513–537. [CrossRef] [PubMed]
120. Yang, C.; Li, L. Hormonal Regulation in Shade Avoidance. *Front. Plant Sci.* **2017**, *8*, 1527. [CrossRef] [PubMed]
121. Iglesias, M.J.; Sellaro, R.; Zurbriggen, M.D.; Casal, J.J. Multiple links between shade avoidance and auxin networks. *J. Exp. Bot.* **2018**, *69*, 213–228. [CrossRef]
122. Nozue, K.; Harmer, S.L.; Maloof, J.N. Genomic analysis of circadian clock-, light-, and growth-correlated genes reveals PIF5 as a modulator of auxin signaling in Arabidopsis. *Plant Physiol.* **2011**, *448*, 358–361.
123. Hersch, M.; Lorrain, S.; De Wit, M.; Trevisan, M.; Ljung, K.; Bergmann, S.; Fankhauser, C. Light intensity modulates the regulatory network of the shade avoidance response in Arabidopsis. *Proc. Natl. Acad. Sci. USA* **2014**, *111*, 6515–6520. [CrossRef] [PubMed]
124. Pucciariello, O.; Legris, M.; Costigliolo Rojas, C.; Iglesias, M.J.; Hernando, C.E.; Dezar, C.; Vazquez, M.; Yanovsky, M.J.; Finlayson, S.A.; Prat, S.; et al. Rewiring of auxin signaling under persistent shade. *Proc. Natl. Acad. Sci. USA.* **2018**, *115*, 5612–5617. [CrossRef] [PubMed]
125. Tao, Y.; Ferrer, J.L.; Ljung, K.; Pojer, F.; Hong, F.; Long, J.A.; Li, L.; Moreno, J.E.; Bowman, M.E.; Ivans, L.J.; et al. Rapid synthesis of auxin via a new tryptophan-dependent pathway is required for shade avoidance in plants. *Cell* **2008**, *133*, 164–176. [CrossRef] [PubMed]
126. Procko, C.; Crenshaw, C.M.; Ljung, K.; Noel, J.P.; Chory, J. Cotyledon-generated auxin is required for shade-induced hypocotyl growth in Brassica rapa. *Plant Physiol.* **2014**, *165*, 1285–1301. [CrossRef] [PubMed]
127. Zhao, Y.; Christensen, S.K.; Fankhauser, C.; Cashman, J.R.; Cohen, J.D.; Weigel, D.; Chory, J. A role for flavin monooxygenase-like enzymes in auxin biosynthesis. *Science* **2001**, *291*, 306–309. [CrossRef] [PubMed]
128. Mashiguchi, K.; Tanaka, K.; Sakai, T.; Sugawara, S.; Kawaide, H.; Natsume, M.; Hanada, A.; Yaeno, T.; Shirasu, K.; Yao, H.; et al. The main auxin biosynthesis pathway in Arabidopsis. *Proc. Natl. Acad. Sci. USA* **2011**, *108*, 18512–18517. [CrossRef] [PubMed]
129. Stepanova, A.N.; Yun, J.; Robles, L.M.; Novak, O.; He, W.; Guo, H.; Ljung, K.; Alonso, J.M. The Arabidopsis YUCCA1 flavin monooxygenase functions in the indole-3-pyruvic acid branch of auxin biosynthesis. *Plant Cell* **2011**, *23*, 3961–3973. [CrossRef] [PubMed]
130. Won, C.; Shen, X.; Mashiguchi, K.; Zheng, Z.; Dai, X.; Cheng, Y.; Kasahara, H.; Kamiya, Y.; Chory, J.; Zhao, Y. Conversion of tryptophan to indole-3-acetic acid by TRYPTOPHAN AMINOTRANSFERASES OF ARABIDOPSIS and YUCCAs in Arabidopsis. *Proc. Natl. Acad. Sci. USA* **2011**, *108*, 18518–18523. [CrossRef] [PubMed]

131. Kohnen, M.V.; Schmid-Siegert, E.; Trevisan, M.; Petrolati, L.A.; Senechal, F.; Muller-Moule, P.; Maloof, J.; Xenarios, I.; Fankhauser, C. Neighbor detection induces organ-specific transcriptomes, revealing patterns underlying hypocotyl-specific growth. *Plant Cell* **2016**, *28*, 2889–2904. [CrossRef] [PubMed]
132. Muller-Moule, P.; Nozue, K.; Pytlak, M.L.; Palmer, C.M.; Covington, M.F.; Wallace, A.D.; Harmer, S.L.; Maloof, J.N. YUCCA auxin biosynthetic genes are required for Arabidopsis shade avoidance. *Peer J.* **2016**, *4*, e2574. [CrossRef] [PubMed]
133. Carabelli, M.; Possenti, M.; Sessa, G.; Ciolfi, A.; Sassi, M.; Morelli, G.; Ruberti, I. Canopy shade causes a rapid and transient arrest in leaf development through auxin-induced cytokinin oxidase activity. *Genes Dev.* **2007**, *21*, 1863–1868. [CrossRef] [PubMed]
134. Staswick, P.E.; Serban, B.; Rowe, M.; Tiryaki, I.; Maldonado, M.T.; Maldonado, M.C.; Suza, W. Characterization of an Arabidopsis enzyme family that conjugates amino acids to indole-3-acetic acid. *Plant Cell* **2005**, *17*, 616–627. [CrossRef] [PubMed]
135. Nakazawa, M.; Yabe, N.; Ichikawa, T.; Yamamoto, Y.Y.; Yoshizumi, T.; Hasunuma, K.; Matsui, M. DFL1, an auxin-responsive GH3 gene homologue, negatively regulates shoot cell elongation and lateral root formation, and positively regulates the light response of hypocotyl length. *Plant J.* **2001**, *25*, 213–221. [CrossRef] [PubMed]
136. Takase, T.; Nakazawa, M.; Ishikawa, A.; Kawashima, M.; Ichikawa, T.; Takahashi, N.; Shimada, H.; Manabe, K.; Matsui, M. ydk1-D, an auxin-responsive GH3 mutant that is involved in hypocotyl and root elongation. *Plant J.* **2004**, *37*, 471–483. [CrossRef] [PubMed]
137. Zheng, Z.; Guo, Y.; Novak, O.; Chen, W.; Ljung, K.; Noel, J.P.; Chory, J. Local auxin metabolism regulates environment-induced hypocotyl elongation. *Nat. Plants* **2016**, *2*, 16025. [CrossRef] [PubMed]
138. Michaud, O.; Fiorucci, A.S.; Xenarios, I.; Fankhauser, C. Local auxin production underlies a spatially restricted neighbor-detection response in *Arabidopsis*. *Proc. Natl. Acad. Sci. USA* **2017**, *114*, 7444–7449. [CrossRef] [PubMed]
139. Pantazopoulou, C.K.; Bongers, F.J.; Küpers, J.J.; Reinen, E.; Das, D.; Evers, J.B.; Anten, N.P.R.; Pierik, R. Neighbor detection at the leaf tip adaptively regulates upward leaf movement through spatial auxin dynamics. *Proc. Natl. Acad. Sci. USA* **2017**, *114*, 7450–7455. [CrossRef] [PubMed]
140. Keuskamp, D.H.; Pollmann, S.; Voesenek, L.A.; Peeters, A.J.; Pierik, R. Auxin transport through PIN-FORMED 3 (PIN3) controls shade avoidance and fitness during competition. *Proc. Natl. Acad. Sci. USA* **2010**, *107*, 22740–22744. [CrossRef] [PubMed]
141. Sassi, M.; Wang, J.; Ruberti, I.; Vernoux, T.; Xu, J. Shedding light on auxin movement: Light-regulation of polar auxin transport in the photocontrol of plant development. *Plant Signal. Behav.* **2013**, *8*, e23355. [CrossRef] [PubMed]
142. Ge, Y.; Yan, F.; Zourelidou, M.; Wang, M.; Ljung, K.; Fastner, A.; Hammes, U.Z.; Di Donato, M.; Geisler, M.; Schwechheimer, C.; et al. SHADE AVOIDANCE 4 is required for proper auxin distribution in the hypocotyl. *Plant Physiol.* **2017**, *173*, 788–800. [CrossRef] [PubMed]
143. Friml, J.; Wiśniewska, J.; Benková, E.; Mendgen, K.; Palme, K. Lateral relocation of auxin efflux regulator PIN3 mediates tropism in Arabidopsis. *Nature* **2002**, *415*, 806–809. [CrossRef] [PubMed]
144. Fankhauser, C.; Christie, J.M. Plant phototropic growth. *Curr. Biol.* **2015**, *25*, R384–R389. [CrossRef] [PubMed]
145. Morelli, G.; Ruberti, I. Shade avoidance responses. Driving auxin along lateral routes. *Plant Physiol.* **2000**, *122*, 621–626. [CrossRef] [PubMed]
146. Morelli, G.; Ruberti, I. Light and shade in the photocontrol of Arabidopsis growth. *Trends Plant Sci.* **2002**, *7*, 399–404. [CrossRef]
147. Sassi, M.; Lu, Y.; Zhang, Y.; Wang, J.; Dhonukshe, P.; Blilou, I.; Dai, M.; Li, J.; Gong, X.; Jaillais, Y.; et al. COP1 mediates the coordination of root and shoot growth by light through modulation of PIN1- and PIN2-dependent auxin transport in Arabidopsis. *Development* **2012**, *139*, 3402–3412. [CrossRef] [PubMed]
148. Chen, X.; Yao, Q.; Gao, X.; Jiang, C.; Harberd, N.P.; Fu, X. Shoot-to-Root Mobile Transcription Factor HY5 Coordinates Plant Carbon and Nitrogen Acquisition. *Curr. Biol.* **2016**, *26*, 640–646. [CrossRef] [PubMed]
149. van Gelderen, K.; Kang, C.; Paalman, R.; Keuskamp, D.; Hayes, S.; Pierik, R. Far-Red Light Detection in the Shoot Regulates Lateral Root Development through the HY5 Transcription Factor. *Plant Cell* **2018**, *30*, 101–116. [CrossRef] [PubMed]

150. Péret, B.; Middleton, A.M.; French, A.P.; Larrieu, A.; Bishopp, A.; Njo, M.; Wells, D.M.; Porco, S.; Mellor, N.; Band, L.R.; et al. Sequential induction of auxin efflux and influx carriers regulates lateral root emergence. *Mol. Syst. Biol.* **2013**, *9*, 699. [CrossRef] [PubMed]
151. Lavy, M.; Estelle, M. Mechanisms of auxin signaling. *Development* **2016**, *143*, 3226–3229. [CrossRef] [PubMed]
152. Leyser, O. Auxin signaling. *Plant Physiol.* **2018**, *176*, 465–479. [CrossRef] [PubMed]
153. Redman, J.C.; Haas, B.J.; Tanimoto, G.; Town, C.D. Development and evaluation of an Arabidopsis whole genome Affymetrix probe array. *Plant J.* **2004**, *38*, 545–561. [CrossRef] [PubMed]
154. Werner, T.; Motyka, V.; Laucou, V.; Smets, R.; Van Onckelen, H.; Schmülling, T. Cytokinin-deficient transgenic Arabidopsis plants show multiple developmental alterations indicating opposite functions of cytokinins in the regulation of shoot and root meristem activity. *Plant Cell* **2003**, *11*, 2532–2550. [CrossRef] [PubMed]
155. Carabelli, M.; Possenti, M.; Sessa, G.; Ciolfi, A.; Sassi, M.; Morelli, G.; Ruberti, I. A novel regulatory circuit underlying plant response to canopy shade. *Plant Signal. Behav.* **2008**, *3*, 137–139. [CrossRef] [PubMed]
156. Sellaro, R.; Yanovsky, M.J.; Casal, J.J. Repression of shade avoidance reactions by sunfleck induction of HY5 expression in Arabidopsis. *Plant J.* **2011**, *68*, 919–928. [CrossRef] [PubMed]
157. Reed, J.W.; Wu, M.F.; Reeves, P.H.; Hodgens, C.; Yadav, V.; Hayes, S.; Pierik, R. Three Auxin Response Factors Promote Hypocotyl Elongation. *Plant Physiol.* **2018**. [CrossRef] [PubMed]
158. Maddonni, G.A.; Otegui, M.E.; Andrieu, B.; Chelle, M.; Casal, J.J. Maize leaves turn away from neighbors. *Plant Physiol.* **2002**, *130*, 1181–1189. [CrossRef] [PubMed]
159. Fellner, M.; Horton, L.A.; Cocke, A.E.; Stephens, N.R.; Ford, E.D.; Van Volkenburgh, E. Light interacts with auxin during leaf elongation and leaf angle development in young corn seedlings. *Planta* **2003**, *216*, 366–376. [PubMed]
160. Dubois, P.G.; Olsefski, G.T.; Flint-Garcia, S.; Setter, T.L.; Hoekenga, O.A.; Brutnell, T.P. Physiological and genetic characterization of end-of-day far-red light response in maize seedlings. *Plant Physiol.* **2010**, *154*, 173–186. [CrossRef] [PubMed]
161. Ugarte, C.C.; Trupkin, S.A.; Ghiglione, H.; Slafer, G.; Casal, J.J. Low red/far-red ratios delay spike and stem growth in wheat. *J. Exp. Bot.* **2010**, *61*, 3151–3162. [CrossRef] [PubMed]
162. Carriedo, L.G.; Maloof, J.N.; Brady, S.M. Molecular control of crop shade avoidance. *Curr. Opin. Plant Biol.* **2016**, *30*, 151–158. [CrossRef] [PubMed]
163. Casal, J.J. Canopy light signals and crop yield in sickness and in health. *ISRN Agron.* **2013**, *2013*, 650439. [CrossRef]
164. Küpers, J.J.; van Gelderen, K.; Pierik, R. Location Matters: Canopy Light Responses over Spatial Scales. *Trends Plant Sci.* **2018**, *23*, 865–873. [CrossRef] [PubMed]
165. Sheehan, M.J.; Farmer, P.R.; Brutnell, T.P. Structure and expression of maize phytochrome family homeologs. *Genetics* **2004**, *167*, 1395–1405. [CrossRef] [PubMed]
166. Sheehan, M.J.; Kennedy, L.M.; Costich, D.E.; Brutnell, T.P. Subfunctionalization of PhyB1 and PhyB2 in the control of seedling and mature plant traits in maize. *Plant J.* **2007**, *49*, 338–353. [CrossRef] [PubMed]
167. Liu, H.; Yang, C.; Li, L. Shade-induced stem elongation in rice seedlings: Implication of tissue-specific phytohormone regulation. *J. Integr. Plant Biol.* **2016**, *58*, 614–617. [CrossRef] [PubMed]
168. Kurepin, L.V.; Emery, R.J.N.; Pharis, R.P.; Reid, D.M. Uncoupling light quality from light irradiance effects in Helianthus annuus shoots: Putative roles for plant hormones in leaf and internode growth. *J. Exp. Bot.* **2007**, *58*, 2145–2157. [CrossRef] [PubMed]
169. Wollenberg, A.C.; Strasser, B.; Cerdán, P.D.; Amasino, R.M. Acceleration of flowering during shade avoidance in Arabidopsis alters the balance between FLOWERING LOCUS C-mediated repression and photoperiodic induction of flowering. *Plant Physiol.* **2008**, *148*, 1681–1694. [CrossRef] [PubMed]
170. Cagnola, J.I.; Ploschuk, E.; Benech-Arnold, T.; Finlayson, S.A.; Casal, J.J. Stem transcriptome reveals mechanisms to reduce the energetic cost of shade-avoidance responses in tomato. *Plant Physiol.* **2012**, *160*, 1110–1119. [CrossRef] [PubMed]
171. Ding, Z.; Zhang, Y.; Xiao, Y.; Liu, F.; Wang, M.; Zhu, X.; Liu, P.; Sun, Q.; Wang, W.; Peng, M.; et al. Transcriptome response of cassava leaves under natural shade. *Sci. Rep.* **2016**, *6*, 31673. [CrossRef] [PubMed]
172. Zhang, Y.; Mayba, O.; Pfeiffer, A.; Shi, H.; Tepperman, J.M.; Speed, T.P.; Quail, P.H. A quartet of PIF bHLH factors provides a transcriptionally centered signaling hub that regulates seedling morphogenesis through differential expression-patterning of shared target genes in Arabidopsis. *PLoS Genet.* **2013**, *9*, e1003244. [CrossRef] [PubMed]

173. Shi, Q.; Zhang, H.; Song, X.; Jiang, Y.; Liang, R.; Li, G. Functional Characterization of the Maize Phytochrome-Interacting Factors PIF4 and PIF5. *Front. Plant Sci.* **2017**, *8*, 2273. [CrossRef] [PubMed]
174. Bush, S.M.; Carriedo, L.; Fulop; Ichihashi, Y.; Covington, M.F.; Kumar, R.; Ranjan, A.; Chitwood, D.H.; Headland, L.; Filiault, D.L.; et al. Auxin signaling is a common factor underlying natural variation in tomato shade avoidance. *bioRxiv* **2015**, 031088. [CrossRef]
175. Chitwood, D.H.; Headland, L.R.; Kumar, R.; Peng, J.; Maloof, J.N.; Sinha, N.R. The developmental trajectory of leaflet morphology in wild tomato species. *Plant Physiol.* **2012**, *158*, 1230–1240. [CrossRef] [PubMed]
176. Chitwood, D.H.; Ranjan, A.; Kumar, R.; Ichihashi, Y.; Zumstein, K.; Headland, L.R.; Ostria-Gallardo, E.; Aguilar-Martínez, J.A.; Bush, S.; Carriedo, L.; et al. Resolving distinct genetic regulators of tomato leaf shape within a heteroblastic and ontogenetic context. *Plant Cell* **2014**, *26*, 3616–3629. [CrossRef] [PubMed]
177. Chitwood, D.H.; Headland, L.R.; Filiault, D.L.; Kumar, R.; Jiménez-Gómez, J.M.; Schrager, A.V.; Park, D.S.; Peng, J.; Sinha, N.R.; Maloof, J.N. Native environment modulates leaf size and response to simulated foliar shade across wild tomato species. *PLoS ONE* **2012**, *7*, e29570. [CrossRef] [PubMed]
178. Bou-Torrent, J.; Galstyan, A.; Gallemí, M.; Cifuentes-Esquivel, N.; Molina-Contreras, M.J.; Salla-Martret, M.; Jikumaru, Y.; Yamaguchi, S.; Kamiya, Y.; Martínez-García, J.F. Plant proximity perception dynamically modulates hormone levels and sensitivity in Arabidopsis. *J. Exp. Bot.* **2014**, *65*, 2937–2947. [CrossRef] [PubMed]
179. de Wit, M.; Ljung, K.; Fankhauser, C. Contrasting growth responses in lamina and petiole during neighbor detection depend on differential auxin responsiveness rather than different auxin levels. *New Phytol.* **2015**, *208*, 198–209. [CrossRef] [PubMed]
180. Kim, S.; Mochizuki, N.; Deguchi, A.; Nagano, A.J.; Suzuki, T.; Nagatani, A. Auxin Contributes to the Intraorgan Regulation of Gene Expression in Response to Shade. *Plant Physiol.* **2018**, *177*, 847–862. [CrossRef] [PubMed]
181. Iannacone, R.; Mittempergher, F.; Morelli, G.; Panio, G.; Perito, A.; Ruberti, I.; Sessa, G.; Cellini, F. Influence of an Arabidopsis dominant negative athb2 mutant on tomato plant development. *Acta Hortic.* **2008**, *789*, 263–276. [CrossRef]
182. Tan, W.; Zhang, D.; Zhou, H.; Zheng, T.; Yin, Y.; Lin, H. Transcription factor HAT1 is a substrate of SnRK2.3 kinase and negatively regulates ABA synthesis and signaling in Arabidopsis responding to drought. *PLoS Genet.* **2018**, *14*, e1007336. [CrossRef] [PubMed]
183. Karve, A.A.; Jawdy, S.S.; Gunter, L.E.; Allen, S.M.; Yang, X.; Tuskan, G.A.; Wullschleger, S.D.; Weston, D.J. *New Phytol.* **2012**, *196*, 726–737. [CrossRef] [PubMed]

 © 2018 by the authors. Licensee MDPI, Basel, Switzerland. This article is an open access article distributed under the terms and conditions of the Creative Commons Attribution (CC BY) license (http://creativecommons.org/licenses/by/4.0/).

Review

Translating Flowering Time from *Arabidopsis thaliana* to Brassicaceae and Asteraceae Crop Species

Willeke Leijten [1], Ronald Koes [2], Ilja Roobeek [1,*] and Giovanna Frugis [3,*]

1. ENZA Zaden Research & Development B.V., Haling 1E, 1602 DB Enkhuizen, The Netherlands; W.Leijten@enzazaden.nl
2. Swammerdam Institute for Life Sciences (SILS), University of Amsterdam, Science Park 904, 1098 XH Amsterdam, The Netherlands; r.e.koes@uva.nl
3. Istituto di Biologia e Biotecnologia Agraria (IBBA), Operative Unit of Rome, Consiglio Nazionale delle Ricerche (CNR), via Salaria Km. 29,300, 00015 Monterotondo Scalo, Roma, Italy
* Correspondence: I.Roobeek@enzazaden.nl (I.R.); giovanna.frugis@cnr.it (G.F.); Tel.: +31-(0)228-350100 (I.R.); +39-06-9067-2889 (G.F.)

Received: 1 November 2018; Accepted: 13 December 2018; Published: 16 December 2018

Abstract: Flowering and seed set are essential for plant species to survive, hence plants need to adapt to highly variable environments to flower in the most favorable conditions. Endogenous cues such as plant age and hormones coordinate with the environmental cues like temperature and day length to determine optimal time for the transition from vegetative to reproductive growth. In a breeding context, controlling flowering time would help to speed up the production of new hybrids and produce high yield throughout the year. The flowering time genetic network is extensively studied in the plant model species *Arabidopsis thaliana*, however this knowledge is still limited in most crops. This article reviews evidence of conservation and divergence of flowering time regulation in *A. thaliana* with its related crop species in the Brassicaceae and with more distant vegetable crops within the Asteraceae family. Despite the overall conservation of most flowering time pathways in these families, many genes controlling this trait remain elusive, and the function of most Arabidopsis homologs in these crops are yet to be determined. However, the knowledge gathered so far in both model and crop species can be already exploited in vegetable crop breeding for flowering time control.

Keywords: Brassicaceae; Asteraceae; flowering time; photoperiod; vernalization; ambient temperature; gibberellins; age; plant breeding

---

1. Introduction

The switch from vegetative stage to flowering is essential for plant reproduction, and flowering time diversity has adaptive value in natural populations [1]. The time at which flowering occurs plays a major role in agricultural production as it affects the quality and quantity of leaf, flower, seed and fruit products, ease of harvest and marketing. Shifting the seasonal timing of reproduction is a major goal of plant breeders to develop novel varieties that are better adapted to local environments and changing climatic conditions [2]. Over the last few years, climate underwent significant changes, such as relatively mild winters, dry and warm summers, and more heavy rain fall in spring and autumn. All of those changes affect plant growth and flowering time. Besides natural occurring climate change, adapting varieties to new environments makes crop production more flexible [2]. To produce varieties that are more robust and predictable in flowering time is also a desirable trait for reliable production. Obtaining varieties with increased yield is also a major breeding goal, and will enhance food production within the same amount of land in a world where the population is growing, and demanding more food production. However, yield is influenced by several factors, including premature bolting (see Glossary in Table 1) during crop production, and therefore, a prolonged vegetative phase will increase yield for leafy crops that are harvested before the transition to the reproductive phase.

Table 1. Glossary of main terms as used in the review.

Term	Definition
flowering time	the switch from plant vegetative growth to reproductive development
bolting	rapid elongation of the inflorescence/flowering stem
annuals	plants that complete their entire life cycle from seed to flower within one year and are characterized by short vegetative phase
biennials	plants which require two years to complete their life cycle,
perennials	plants that survive for several years and restrict the duration of reproduction by cycling between vegetative growth and flowering; perennials are characterized by prolonged vegetative phase that can last from a few weeks to several years
shoot apical meristem (SAM)	population of cells located at the tip of the shoot axis that produce lateral organs, stem tissue and regenerates itself
inflorescence meristem (IM)	a meristem that underwent transition from vegetative to reproductive fate and can produce floral meristems
floral meristem (FM)	group of cells responsible for the formation of floral organs
facultative photoperiod	plants that flower faster under a particular photoperiod but will eventually flower under all photoperiods (also called "quantitative")
obligate photoperiod	plants that flower only under a particular photoperiod (also called "qualitative")
long days	day length more than about 12 h, usually 16 h light and 8 h dark periods
short days	day length less than about 12 h, usually 8 h light and 16 h dark periods
Double Haploid (DH)	chromosome doubling of haploid cells to produce genetically homozygous plants
genome-wide association study (GWAS)	observational study of a genome-wide set of genetic variants in different individuals that occur more frequently in correlation with a specific trait, identifying inherited genetic variants associated with a trait
homolog	a gene related to a second gene by descent from a common ancestral DNA sequence
ortholog	genes in different species that evolved from a common ancestral gene by speciation; normally, orthologs retain the same function in the course of evolution
paralog	genes related by duplication within a genome that may evolve new functions
maturity	the state of being fully developed or full grown
uniformity	a state or condition of the plant in which everything is regular, homogeneous, or unvarying
predictable	always behaving or occurring in the way expected
robust	is a characteristic of being strong that, when transposed into a system, it refers to the ability of tolerating perturbations and remain effective
QTL	(or Quantitative Trait Locus), is a locus (section of DNA) which correlates with variation of a quantitative trait in the phenotype of a population of organisms
vernalization	cold treatment needed to get many perennials to flower; usually the minimum period is six to twelve weeks at 4 °C
spring types	plants which flower early without vernalization
winter types	plants which have an obligate requirement for prolonged periods of cold temperatures
semi-winter types	plants which require mild vernalization and lack frost hardiness

Controlling flowering time would therefore help grow crops in all seasons to speed up the production of new hybrids and produce high yield throughout the year. Early bolting potentially limits vegetative growth and can severely decrease yield, while non-flowering inhibits seed production. The timing of bolting and flowering are especially important for vegetable crops. For cauliflower and broccoli, synchronization of flowering is essential as the plants are harvested in the inflorescence meristems phase (curds). For lettuce, plants flower early when grown at high temperature. Early stages of bolting are not visible, but the flavor changes more towards bitter. Therefore, late bolting is preferred to enhance yield without the bitterness.

In the past, selection for flowering time was based on plant phenotyping in the greenhouse or in the field. The increasing availability of crop genetic and genomic resources, and the current knowledge on both gene function and natural genetic variation, are of great value and can be used in breeding. The development of trait specific markers, e.g., based on QTL (see Glossary in Table 1) analysis, are useful to select for favorable genotypes in a breeding program [3–5]. On the other hand, reverse screening for genetic variation in specific flowering time related genes in wild accessions or mutant populations could be of benefit for the trait. The latter approach is still underexploited as knowledge about the flowering pathways, the molecular mechanisms, and the genes involved is still limited in most crops [6,7]. However, the flowering time genetic network is extensively studied in the model species *Arabidopsis thaliana* [6,8–11] (Figure 1a), which is an annual facultative long-day (LD) plant belonging to the Brassicaceae family. Hence, if the function of the *A. thaliana* flowering time genes would be conserved in the crops of interest, this would provide targets for genetic selection and improvement to speed up breeding and agricultural biotechnology.

Figure 1. Plant species considered in this review. (**a**) *Arabidopsis thaliana*, (**b**) *Brassica napus* (rapeseed), (**c**) *Brassica rapa* (turnip), (**d**) *Raphanus sativus* (radish), (**e**) *Diplotaxis tenuifolia* (wild rocket), (**f**) *Cichorium intybus* (chicory), (**g**) *Brassica oleracea* (cauliflower) in winter, (**h**) *Brassica oleracea* (cauliflower) in summer and (**i**) *Lactuca sativa* (lettuce).

Here, we review current knowledge regarding the conservation and divergence of the mechanisms that regulate flowering time in *A. thaliana* related crop species from the Brassicaceae family and more distantly related leafy crops within the Asteraceae family, that are of great interest for food market and vegetable breeding. We will focus on *Brassica napus* (Figure 1b), *Brassica rapa* (Figure 1c), *Brassica oleracea* (Figure 1g-h), *Raphanus sativus* (Figure 1d and *Diplotaxis tenuifolia* (Figure 1e) for the Brassicaceae and on *Lactuca sativa* (Figure 1i) and *Cichorium intybus* (Figure 1f) as key leafy crops within the Asteraceae family. The possible exploitation of this knowledge in vegetable crop breeding, and the potential of translational biology and genomics to crops, will be discussed. The different flowering time pathways and genes explored in these crops will be discussed, and gene function will be compared to the knowledge acquired in *A. thaliana*.

2. Flowering Requirements of Brassicaceae and Asteraceae Species

Optimal conditions for flowering vary between and within species, as plants respond and adapt to specific combination of light (quality and day length) and temperature (cold, warm, hot) to undergo floral transition [11]. Plants can be either long-day or short-day if they flower when exposed to light periods longer (as in summer) or shorter (as in winter) of a certain critical length. Many plant species require prolonged exposure to low temperatures (vernalization) to flower, while others flower independently of cold conditions [8]. Plants have annual, biennial or perennial life cycles depending on the number of growing seasons required to complete their life cycle [11]. Further classification of plant types can be made based on geographical origin and growing season.

The plant model species *Arabidopsis thaliana* (Figure 1a, Table 2) can be annual or biennial. Annual plants flowers earlier in response to long days (facultative long-day), and natural accessions are classified into summer annuals and winter annuals [11]. Summer annuals flower rapidly when grown under long days, whereas in winter annuals flowering is not induced until they are exposed to low temperature (4 °C) for several weeks (vernalization), followed by warmer temperatures in spring [11].

Table 2. Brief description of the plant species.

Species	Chr.	Life Span	Vernalization	Types	Breeding Goal	Day Length [1]	Ref.
A. thaliana	2n = 10	Annual/biennial	Yes/no	Spring/semi-winter/winter	none	facultative LD	[12]
B. napus	2n = 38 (AACC)	Annual/biennial	Yes/no	Spring/semi-winter/winter	flowering time adaptation	LD	[13,14]
B. rapa	2n = 20 (AA)	Annual/biennial	Yes/no	Spring/semi-winter/winter	Late bolting	LD	[15]
B. oleracea	2n = 18 (CC)	Annual/biennial	Yes/no	Spring/semi-winter/winter	Predictable harvest time	LD	[16]
R. sativus	2n = 18	Annual	No		Late bolting	facultative LD	[17,18]
D. tenuifolia	2n = 22	Annual	No		Late bolting	LD	[19]
L. sativa	2n = 18	Annual	No		Heat resistance	facultative LD	[20]
C. intybus	2n = 18	Biennial/perennial	Yes		Resistance to bolting	LD	[21,22]

[1] All species flower early under long-day (LD, 16 h/8 h of light/dark) conditions.

B. napus, *B. rapa* and *B. oleracea*, closely related to *A. thaliana*, share similar life cycles (annuals and biennials) and have spring, semi-winter and winter types: Spring types flower early without vernalization and are grown in geographical regions with strong winters or in subtropical climates; winter types have an obligate requirement for prolonged periods of cold temperatures and are grown in moderate temperate climates; semi-winter types, which are sown before winter, flower after winter and are grown in geographical regions with moderate winter temperatures (>0 °C) (Figure 2) [14]. *B. napus* is a domesticated allotetraploid species with two genomes, AA and CC, derived from *B. rapa* and *B. oleracea*, respectively. Two other Brassicaceae species, *R. sativus* and *D. tenuifolia*, and the Asteraceae species *Lactuca sativa*, are annuals that do not require vernalization to flower. In contrast, the Asteraceae species *Cichorium intybus* is biennial or perennial and does require a cold period for flowering induction. All species considered in this review flower faster in response to longer photoperiods and warmer temperatures (more details about species characteristics are in Table 2). Despite similarities amongst these species, breeding goals with regard to plant growth and flowering time differ for each crop.

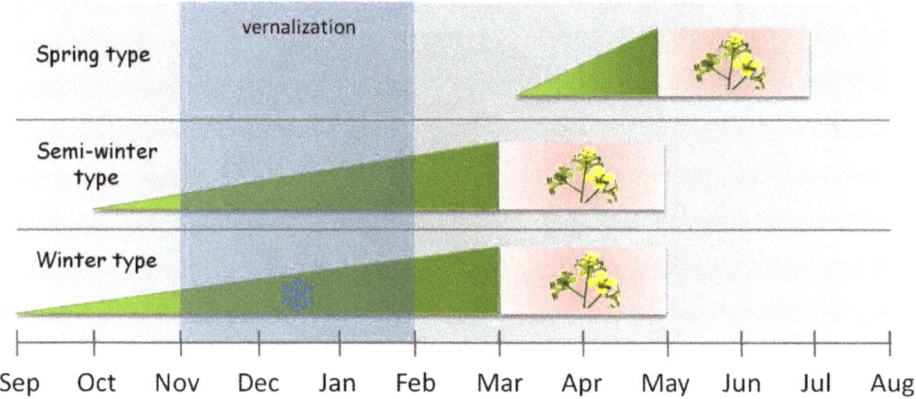

Figure 2. Schematic representation of the life cycles of annual Brassica species. Green triangles represent vegetative growth, pink boxes plant flowering. Periods of cold required for vernalization are indicated by a blue box. Frost symbols indicate frost hardiness in winter types that does not occur in semi-winter plants.

3. Breeding Goals

Breeders aim to improve varieties by adapting them to climate changes, new environments or increasing yield in general, and flowering time affects all these traits [2].

B. napus L. (Figure 1b) is one of the most important oilseed crops worldwide and includes oilseed rape and rapeseed (Figure 3). The yield potential of rapeseed largely depends on flowering time, thus creating lines with optimal flowering time is a major breeding goal [14].

B. rapa (Figure 1c) is cultivated worldwide, particularly in Asia, and includes the vegetable crops Chinese cabbage, pak choi, turnip and cime di rapa (Figure 3). Premature bolting is a severe problem as it reduces yield of the harvested crops, e.g., for the spring cultivation of Chinese cabbage. Extremely late bolting is a major breeding goal in this crop as unexpectedly low temperatures can induce flowering and so yield loss [23].

B. oleracea (Figure 1g–h) encompasses multiple cultivar groups (Figure 3) that are classified based on the morphology of their edible structures: Kohlrabi, kales and cabbages are harvested at vegetative stage; broccoli and cauliflower are cultivated for their curd (the edible flower head of the plant) that is harvested at the transition to reproductive phase. Cultivars and wild species accumulate anti-carcinogenic and antioxidants, which are beneficial for human health [16]. Breeding strategies for broccoli and cauliflower include uniformity in time to curd production for easy crop handling during production. In cauliflower, slight deviations from optimal growth temperature, either lower or higher, lead to uneven curd formation and therefore less predictable harvest times. On one hand, vernalization is required to produce a harvestable curd, while on the other hand, high temperatures in spring result in prolonged vegetative growth before the curd is produced. Adjusting the vernalization and temperature sensitivity of plants will help to create cauliflower varieties with a predictable curd formation, for example by exploring genetic variation in temperature-dependent flowering time genes. Early prediction of the thermal time to curd induction in untested genotypes and environments can be achieved by using the genome-based model proposed by Rosen et al. [24], making this a good tool for early selection of the desired genotypes to be incorporated into breeding material.

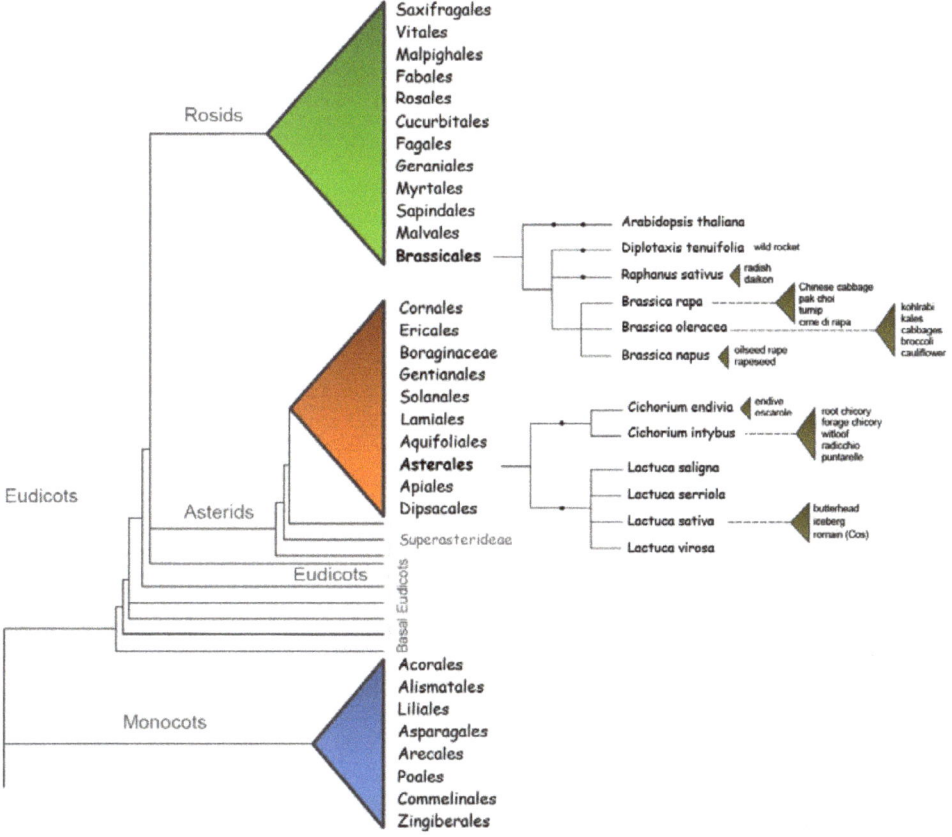

Figure 3. Schematic phylogenetic tree of eudicot and monocot species. Plant families are indicated for the main phylogenetic groups (http://science.kennesaw.edu/jmcneal7/plantsys/index.html). Phylogenetic relationships within Brassicaceae and Asteraceae species of interest were obtained using phyloT, a phylogenetic tree generator based on NCBI taxonomy (https://phylot.biobyte.de/). Cultivated crops for the different plant species are shown.

R. sativus (Figure 1d), including radish and daikon (Figure 3), are important vegetable root crops with large variation in root size and shapes [18,25]. Late flowering is a relevant breeding goal as some varieties are sensitive to premature bolting. In radish, when plants are grown under LD conditions, premature bolting reduces yield and quality of the harvested product. Enhancing the quality of radish can be achieved by late flowering, combined with increased post-harvest shelf life through delayed leaf senescence, as the whole plant (bulb and leaves) is harvested and the leaves are used as an indication for the post-harvest quality of the plant.

Wild rocket (*D. tenuifolia*) (Figures 1e and 3) is a popular salad leaf that has a similar shape and taste as rucola (*Eruca sativa*), but with a stronger peppery flavor. The crop can be harvested multiple times. For cultivation, *D. tenuifolia* is selected against pre-harvest flowering as it leads to unsaleable crops.

L. sativa (Figure 1i) encompasses multiple lettuce cultivars that are classified based on the morphology of the leafy or head type: Iceberg types form a close head resembling that of a cabbage; butterhead types form a head with large ruffled outer leafs; and romaine (Cos) types do not form a head but have long, broad and upright leaves. Wild relatives used in breeding include *L. virosa*, *L. serriola* and *L. saligna* (Figure 3). While cultivated lettuce is annual, the wild species *L. virosa* is biennial and does require vernalization for flower initiation. High temperature during the cultivation of lettuce results in heat-induced early bolting. Heat resistance is therefore a major breeding goal to produce better tasting lettuce when grown at high temperature. Although early bolting is beneficial for fast seed production in the creation of varieties, it does reduce the quality of harvestable crops. The early stages of bolting are not visible when the crop is harvested, but the flavor changes towards an undesirable bitter taste [26]. Exploring the genetic determinants of this response will help understand the mechanism of heat-induced early flowering, and enable breeders to produce better tasting lettuce when grown at high temperature.

C. intybus (Figure 1f) includes multiple cultivar groups that are classified based on purpose and use of the harvested product: Root and forage chicory is used for inulin extraction and grown for live stock, respectively; witloof and radicchio are leafy vegetables that can be cooked or eaten fresh [27]. Among the leaf chicory groups, several "Catalogna" landraces are cultivated in Italy for both leaves and stems/buds, the latter appreciated for the bitter and crispy taste (puntarelle) (Figure 3) [28]. If sown too early in spring, the plants could be vernalized and flower during the first growing period [22]. Breeding goals include uniformity in crop yield and maturity, and resistance to bolting [22,29]. For the production of root chicory, a cold season during growth induces early bolting and therefore decreases yield. Investigating the cold-response of root chicory in more detail is needed to delay bolting under these conditions.

4. Conserved and Divergent Flowering Time Genes in Brassicaceae and Asteraceae

The switch from plant vegetative growth to reproductive development (transition to flowering), is a critical stage in the life cycle of a plant. Plants need to coordinate their developmental programs precisely in response to seasonal changes and in an ecological context in order to ensure their reproductive success. As such, flowering is tightly controlled by diverse developmental, hormonal and environmental cues, day length and temperature being the most important of these environmental signals. [10]. Six major genetic pathways, converging to a small number of floral integrator genes, have been described in *Arabidopsis thaliana* (Figure 4): The vernalization and photoperiod pathways, which control flowering in response to seasonal changes; the ambient temperature pathway, which regulates flowering time in response to changing ambient temperature; the age, autonomous, and gibberellin pathways, acting more independently of environmental stimuli [9,10]. When the switch towards flowering is made, the shoot apical meristem (SAM) transforms into an inflorescence meristem (IM) as an intermediate step. From the IM, floral meristems (FM) are initiated that can produce floral primordia [30]. The transition to reproduction is accompanied by shoot stem elongation (bolting).

4.1. Floral Integrator Genes: An Overview

In plants, the signaling pathways that are activated by various endogenous and environmental cues ultimately converge to a few floral integrator genes to control flowering time, leading to the activation of floral meristem identity genes, the first step in the formation of a flower [10]. In Arabidopsis, two floral integrators play a major role in the transition to flowering, FLOWERING

LOCUS T (FT), which belongs to the PEBP (phosphatidylethanolamine-binding protein) family, and SUPPRESSOR OF OVEREXPRESSION OF CONSTANS 1 (SOC1/AGL20), a MADS-box transcription factor [9] (Figure 4). Two homologs of FT, TWIN SISTER OF FT (TSF) and TERMINAL FLOWER 1 (TFL1), act redundantly or antagonistically to *FT*, respectively [31,32]. If floral integrator genes were conserved between Arabidopsis and crops, mutations reducing *FT*, *TSF* and *SOC1* orthologs activity would result in late flowering plants, whereas increased expression of the corresponding genes should induce early flowering. The opposite would occur for mutations or increased expression of *TFL1*.

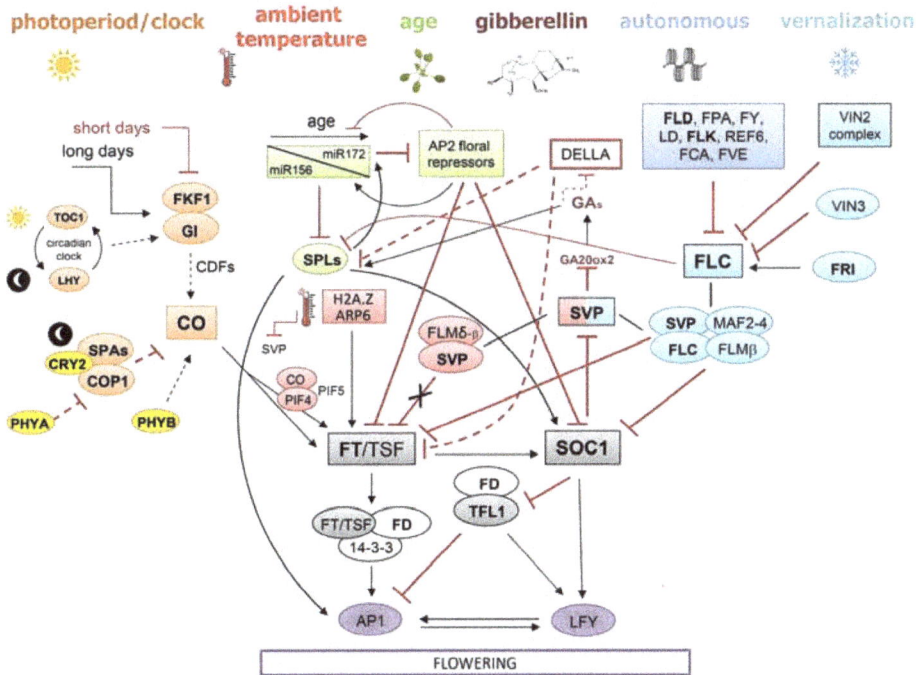

Figure 4. Main flowering time pathways acting in *Arabidopsis thaliana*: Photoperiod (orange and yellow), ambient temperature (red), age (green), gibberellins (brown), autonomous (sky blue), vernalization (light blue). Grey boxes represent the main floral integrators FT/TSF and SOC1. The two main genes conferring inflorescence meristem identity, *AP1* and *LFY*, are indicated in purple. Squared boxes indicate genes having a pivotal role in the specific pathway. Boxes with rounded corners represent several genes or complexes. Solid and dotted lines indicate either direct or indirect regulation, black arrows and red T-ends indicate positive or negative regulation, respectively. The cartoon represents only the main regulatory genes in the different pathways, whereas the complete flowering time network, involving more than 300 genes, is available at the WikiPathways Web Site [33].

The *FT/TFL* family and *SOC1* have a major role in flowering time response that seems to be conserved across different species [34]. However, due to Brassica genus evolutionary history, with genome triplication within diploid species of Brassica and polyploidism, *B. napus*, *B. oleracea* and *B. napus* contain several copies of floral integrator homologs. Among these, only some conserve a key role in flowering time whereas others may have been either inactivated or undergone a process of neofunctionalization [35]. Lettuce *SOC1* shows a unique role in heat-promoted bolting [36], whereas its role downstream of *FT* to induce flowering transition in natural conditions is yet to be determined.

4.1.1. Floral Integrator Genes in Brassicaceae

In Arabidopsis, FT plays a key role in the floral transition process, since it is the mobile signal moving from the leaves, through the phloem, to promote flowering at the SAM [37,38]. In the leaves, the circadian clock-associated gene *CONSTANS* (*CO*) enhances *FT* expression under long-day (LD) conditions, while temperature-dependent genes such as *SHORT VEGETATIVE PHASE* (*SVP*) and *FLOWERING LOCUS C* (*FLC*) repress *FT* expression independently of day length [39–41]. FT protein is produced in the leaves and transported through the phloem to the SAM, with FT-INTERACTING PROTEIN 1 (FTIP1) assisting with FT protein transport [37]. At the SAM, FT interacts via a 14-3-3 protein with FLOWERING LOCUS D (FD) [42] to activate the floral promoter *SOC1* and the downstream floral meristem identity genes *APETALA 1* (*AP1*) and the AP1 paralog *CAULIFLOWER* (*CAL*) [38]. SOC1 also activates floral meristem identity through *LEAFY* (*LFY*) and *AGAMOUS-LIKE 24* (*AGL24*). LFY, AP1 and CAL direct certain groups of cells in the flanks of the SAM to differentiate into floral meristems, leading to the transition from vegetative to reproductive phase [43]. A close homolog of *FT*, *TWIN SISTER OF FT* (*TSF*) with 82% amino acid identity, seems to act redundantly to *FT*. Overexpression of either *FT* or *TSF* results in early flowering, a mutation in *FT* results in late flowering under LD conditions and a mutation in *TSF* shows a greater effect under short-day (SD) conditions [31]. A more distantly related homolog of *FT*, *TERMINAL FLOWER 1* (*TFL1*) with 59% amino acid identity, acts antagonistically to *FT*. Plants overexpressing *TFL1* are late flowering with an extended first inflorescence phase, during which they form cauline leaves and branches [44]. TFL1 is a mobile signal like FT [45], but acts as a transcription repressor rather than a transcriptional activator as FT [46]. The antagonistic activity of TFL1 and FT originates from an external loop in the protein [32] and interchanging one specific residue in the loop (Y85 in FT and H88 in TFL1) is sufficient to convert TFL1 into FT function and vice versa [44].

B. napus contains six paralogs of *FT*, four of *TFL1* [47,48] and four paralogs of *SOC1* (Table 3) [35]. *FT* paralogs map to six distinct regions of conserved blocks of the A and C genomes homologous to a common ancestral block of *Arabidopsis* chromosome 1. *BnFT* gene coding sequences show 92% to 99% identities to each other and 85% to 87% identity with those of Arabidopsis. *Bna.FT.C02* and the corresponding ortholog in *B. oleracea* are not expressed, possibly due to the presence of a transposable element (TE) causing high cytosine methylation at the promoter [47]. Differently, three *Bna.FT* paralogs, *Bna.FT.A02*, *BnaFT.C06a* and *Bna.FT.C06b*, are expressed and were associated with two major QTL clusters for flowering time, one of which encompasses two *Bna.FT* paralogs *Bna.FT-C06a* and *Bna.FT.C06b*. Their function in flowering time variation was confirmed by association analysis in vernalization-free conditions in both spring and winter type cultivars of rapeseed. *Bna.FT.A02* is expressed in leaves of both winter and spring type plants, with and without vernalization [47,49], and was found to associate with flowering time in a panel of 188 *Brassica* spp. accessions collected from different geographic locations worldwide [50]. *Bna.FT.C06* and *Bna.FT.A07* are expressed in winter type plants after vernalization and spring type plants, but not in winter types without vernalization [47,49]. EMS (ethyl methanesulfonate) lines harboring different mutant alleles of *Bna.FT.C06b* were late flowering and displayed reduced fertility [51]. Plants harboring different mutant alleles of *Bna.TFL1* paralogs were not affected in flowering time [51], whereas amongst the four paralogs of *SOC1*, only *Bna.SOC1.A03* was associated with flowering time and seed yield-related QTLs on chromosome A03, and its expression was induced by vernalization [52]. These data point to a function of *Bna.FT.A02* and *Bna.FT.C06* in controlling flowering time, the latter also in response to vernalization similarly to their paralog in Arabidopsis, whereas *B. napus* paralogs of *TFL1* seem to affect seed yield but not flowering time [51]. *Bna.SOC1.A03* might play a role in flowering time control, but this is yet to be explored in *B. napus* species. These data are consistent with the association of *Bna.FT-A02*, but not *B. napus* TLF1 and *SOC1* paralogs, with a spring-environment specific flowering QTL in double haploid populations grown in different environmental conditions [53].

Table 3. Flowering time genes in Brassica species.

Pathway	Gene	Arabidopsis	B. napus gene ID	B. napus Chr position	B. rapa gene ID	B. rapa Chr position	B. oleracea gene ID	B. oleracea Chr position
Floral integrators	FT	AT1G65480	GSBRNA2T00090951001 Bna.FT.A02	A02:6375936.6379058	Bra022475 BrFT1	A02:8551268.8553758		
			GSBRNA2T00030311001 Bna.FT.C02	C02:996695.998788			Bol045330	Scaffold000001_P2: 1990327.1992083
	TSF	AT4G20370	GSBRNA2T00124448001 Bna.FT.A07	A07:18855196.18857952	Bra004117 BrFT2	A07:20213069.20215397		
			GSBRNA2T00146560001	A07:22787807.22790354	Bra015710	A07:24515213.24516895	Bol039209	C02:19458855.19452577
			GSBRNA2T00077948001	C02:20907503.20909228			Bol017639	C04:17148775.17151658
			GSBRNA2T00011334200001	C04:12435074.12437644			Bol012573	C07:9349005.9351279
			GSBRNA2T00067517001 Bna.FT.C06	C06:28352966.28355216			Bol027595	C07:1423408.1425133
			GSBRNA2T00050890001	Cnn:48285424.48286397				
	TFL1	AT5G03840	GSBRNA2T00136426001	A10:16767409.16768474	Bra009508	A10:15774055.15775120	Bol005471	C02:1447642.1448756
			GSBRNA2T00119620001	Ann:609805.611005	Bra028815	A02:545667.546787	Bol015337	C03:438359.439413
			GSBRNA2T00078527001	C02:1320757.1321835			Bol010027	C09:39511589.39512660
			GSBRNA2T00134290001	C03:673349.674628				
			GSBRNA2T00073025001	Cnn:9572005.9573076				
	SOC1	AT2G45660	GSBRNA2T00011646001	A03:901877.905188 BnSOC1-A3	Bra005783	A03:603455.604516		
			GSBRNA2T00006326001	A04:18732428.18735897	Bra000393	A03:109118286.10920672		
			GSBRNA2T00116723001	A05:2627051.2630394	Bra039324	A04:18723546.18725960		
			GSBRNA2T00037309001	C04:48074887.48078345	Bra004928	A05:2530305.2532747	Bol021742	C04:40413670.40414880
			GSBRNA2T00008301001	C14:867297.870707			Bol030200	C04:2998426.2999594
			GSBRNA2T00029970001	Cnn:35198162.35204681			Bol029556	C10:34211271.34232327
Vernalization	FLC	AT5G10140	GSBRNA2T00143535001 Bna.FLC.A02	A02:134962.138212	Bra028599 BrFLC2	A02:1524995.1528254		
			GSBRNA2T00129741001	A03:1360971.1364359	Bra006051 BrFLC3	A03:1764912.1767856		
			GSBRNA2T00142187001	A03:6240056.6245305	Bra022771 BrFLC5	A03:6971946.6976797		
			GSBRNA2T00135921001	A10:14998617.15003197	Bra009055 BrFLC1	A10:13856133.13860473		
			GSBRNA2T00068991001 Bna.FLC.C02	C02:208562.212139			Bol024642	C02:2720826.2721596
							BoFLC4	C02:2722189.2724345
							BoFLC2	
			GSBRNA2T00134620001	C03:20011058.2004465			Bol008758	C03:1890867.1890743
							BoFLC3	
			GSBRNA2T00024568001	C03:84033312.84110062			BoFLC5	C03:49708405.49709316
			GSBRNA2T00016124001	C09:46345350.46350092			Bol043693	C09:37175182.37179020
			GSBRNA2T00016119001	C09:46366645.46371180			BoFLC1	

Table 3. *Cont.*

Pathway	Gene	Arabidopsis	*B. napus* gene ID	*B. napus* Chr position	*B. rapa* gene ID	*B. rapa* Chr position	*B. oleracea* gene ID	*B. oleracea* Chr position
Ambient temperature	FRI	AT4G00650	GSBRNA2T00006686001 Bna.FRl.Xa	A03:6053113..6055294	Bra029192 BrFRla	A03:6784863..6787013		
			GSBRNA2T00120967001 Bna.FRl.Xb	A10:4019556..4021675	Bra035723 BrFRlb	A10:4133444..4134764		
			GSBRNA2T00052682001 Bna.FRl.Xd	C03:8149599..8151810			Bol028107 BoFRla	C03:7962008..7964180
			GSBRNA2T00015226001 Bna.FRl.Xc	C09:29041826..29043953			Bol004294 BoFRlb	Scaffold000327:204688..206816
	SVP	AT2G22540	GSBRNA2T00032884001	A04:10961147..10963402	Bra030228	A04:10192172..10194736		
			GSBRNA2T00078179001	A09:29590705..29594744	Bra038511	A09:33434743..33437921		
			GSBRNA2T00114975201	C04:36478652..36481951			Bol031759	Scaffold000153:1406474..1408404
			GSBRNA2T00127429001	C08:32995398..32998881	Bol044741	C08:35213085..35214818		
Photoperiod	CO	AT5G15840	GSBRNA2T00135488001	A10:13358777..13360064	Bra008669	A10:12117648..12118929	Bol030488	C09:33143053..33144339
	GI	AT1G22770	GSBRNA2T00035272001	C09:43745679..43747139				
			GSBRNA2T00015763001	A09:22588149..22593013	Bra024536	A09:25756404..25760934		
			GSBRNA2T00119480001	C05:11778931..11784461			Bol023541	Scaffold000099_P1:794479..799157
Age	SPL3	AT2G33810	GSBRNA2T00064576001	A04:15462653..15463366	Bra021880	A04:15123762..15124274		
			GSBRNA2T00095270001	A05:5425249..5426076	Bra005470	A05:5668800..5669314		
			GSBRNA2T00132295001	C03:9629272..9630113			Bol036997	C06:40526300..40526809
			GSBRNA2T00020688001	C04:44354526..44355241			Bol037895	C04:35340992..35341501
			GSBRNA2T00038835001	Cnn:4854484..4855112			Bol027299	C04:20510435..20510961
	SPL9	AT2G42200	GSBRNA2T00123166001	A04:17845227..17847617	Bra016891	A04:17839490..17841541		
			GSBRNA2T00132740001	A05:1443071..1445187	Bra004674	A05:1325605..1327387		
			GSBRNA2T00108440001	C04:1886612..1888780			Bol004847	C04:9669229..968875
			GSBRNA2T00084688001	C04:46904649..46905101				
			GSBRNA2T00084692001	C04:46915351..46917939			Bol012678	Scaffold000379:152205..154297
	SPL15	AT3G57920	GSBRNA2T00034335001	A04:1548881..156614	Bra014599	A04:1684031..1685273		
			GSBRNA2T00089000001	A07:14658857..14660105	Bra003305	A07:15783674..15784920	Bol011022	C04:9176952..9178238
			GSBRNA2T00087887001	C04:25001142..25003655			Bol007052	C07:28887260..28888520
			GSBRNA2T00105779001	C06:19172101..19173360				
GA20OX1		AT4G25420			Bra013890	A01:8279446..8280885	Bol039527	C01:11622628..11624071
					Bra019165	A03:25974634..25976038	Bol042237	C06:43862307..43863764
							Bol041615	Scaffold000009_P1:317473..317667
GA20OX2		AT5G51810	GSBRNA2T00036929001	A02:5851980..5853392	Bra022565	A02:7878457..7879856		
			GSBRNA2T00110217001	A10:6243369..6244766	Bra028277	A10:4556457..4557854		
			GSBRNA2T00153037001	C02:11109632..11111035			Bol045266	Scaffold000001_P2:739593..740993
			GSBRNA2T00108658001	C09:30358841..30359437				
			GSBRNA2T00108686001	C09:30368578..30369977			Bol029404	C09:18392061..18393124
			GSBRNA2T00025818001	Cnn:77727013..77728046				

Table 3. Cont.

Pathway	Gene	Arabidopsis	B. napus gene ID	B. napus Chr position	B. rapa gene ID	B. rapa Chr position	B. oleracea gene ID	B. oleracea Chr position
Gibberellin	GA20OX3	AT5G07200			Bra028706 Bra005927 Bra010064 Bra009285	A02:1065763..1067251 A03:1251910..1253082 A06:14066314..14068214 A10:14501966..14503398	Bol024532 Bol008872 Bol043862 Bol041616 Bol024814	C02:2131323..2133012 C03:1262233..1263937 C09:38109369..38110750 Scaffold000009_P1:331428..332030 Scaffold000091:1192479..1195919
	GA20OX4	AT1G60980	CSBRNA2T00070537001 CSBRNA2T00043758001 CSBRNA2T00043759001 CSBRNA2T00080857001 CSBRNA2T00028142001	A09:7950969..7952898 Ann:21491303..21492435 Ann:21493033..21494525 Cnn:11274341..11276251 Cnn:34189212..34190700	Bra027106 Bra039251 Bra031467	A09:8952409..8954196 Scaffold000162:173188..176593 A01:17039613..17041240	Bol014320 Bol044153	C03:52855996..52857792 Scaffold000003_P1:2530105..2531509
	GA2OX5	AT1G44090	CSBRNA2T00097054001 CSBRNA2T00060328001 CSBRNA2T00013490001	A08:969550..970698 C08:6199021..6199389 Cnn:72340763..72341689	Bra014019	A08:4525615..4527790	Bol007374 Bol021441	Scaffold000262:280424..282063 C07:30202821..30203904

Note: Gene symbols, gene names and position in the chromosome of the paralogs for *B. napus*, *B. rapa* and *B. oleracea*, and corresponding *A. thaliana* genes, are shown for each pathway. Black text means the same chromosome in all Brassica species, grey text means best *B. napus* hit from *B. rapa* or *B. oleracea* protein sequence; brown text means no direct homolog available in the list with *A. thaliana* syntheny (http://brassicadb.org/brad/searchAll.php).

Three paralogs of *SOC1* (*Br004928*, *Br000393* and *Br009324*) and two paralogs of *FT* (*BrFT*) are found and expressed in *B. rapa* (Table 3) [47,54]. *BrFT1* and *BrFT2* show a similar expression as their corresponding *B. napus* orthologs *Bna.FT.A02* and *Bna.FT.A07*, respectively. *BrFT1* is expressed in all plant types and diurnally regulated [47,54]. *BrFT2* is only expressed in winter type plants after vernalization and spring-type plants [47]. A TE in exon 2 of *BrFT2* causes plants to flower 4.9 days later in spring and 14.7 days later in autumn. Due to the bigger effect under SD conditions, it was suggested that *BrFT2* might be an ortholog of *AtTSF* [49], but this still needs to be confirmed. Overexpression of a *B. rapa SOC1* ortholog (*BrAGL20*) in *B. napus* causes early flowering [55], suggesting that the function of this gene may be conserved in Brassicaceae. Moreover, association between flowering time and expression of the two *SOC1* paralogs *Br004928* and *Br000393* was found in a natural population of *B. rapa* [56].

Four paralogs of *FT* (*BoFT*) (two copies on C02 and one on C04 and C06), and three homologs of *SOC1* (*BoSOC1*) (C03 and two copies on C04) have been identified in the genomes of *B. oleracea* (Table 3) [35], but no functional studies are available so far.

One *FT*, *TFL1*, *TSF* and two *SOC1* genes, sharing 82.58%, 89.47%, 83.3%, 85.49% and 88.82% of nucleotide homology with their Arabidopsis homologs, can be found in the de novo assembled transcriptome of *D. tenuifolia* [57] that was obtained from leaves of stressed young plants. However, no characterization of floral integrator genes is available for this species.

4.1.2. Floral Integrator Genes in Asteraceae

In *L. sativa*, an *FT* homolog (*LsFT*) was characterized [58] and shown to express in the largest lettuce leaves, stems and flower bud in controlled high temperature (35/25 °C) conditions which induce lettuce flowering [58]. *LsFT* overexpression could induce early flowering in transgenic *A. thaliana*, although the phenotype was less strong compared to *AtFT* overexpressing plants. However, other studies showed that expression of *LsFT* under the viral 35S constitutive promoter control could fully complement *Arabidopsis ft* null mutant [59]. Correlation between *LsFT* expression and lettuce bolting (measured as the days to the first visible elongated stem) was further analyzed in nine lettuce varieties, which were selected amongst 705 lettuce accessions, with either late, middle and early bolting times [59]. Heat treatment (35 °C day/ 25 °C night) for 48 h also promoted expression of *LsFT* in all lettuce varieties. RNAi-mediated knockdown of *LsFT* in *L. sativa* results in a late bolting phenotype, lack of response to heat treatment and reduced levels of *LsLFY* and *LsAP1* [59], which expression is most abundant at the onset of bolting [58,60]. Induction of high *LsFT* expression during the transition to reproductive growth and activation of *LsLFY* and *LsAP1* was also observed in three heading and non-heading lettuce varieties grown in the field in natural conditions [60].

Transcriptomic data from lines that are either bolting resistant or sensitive to high temperature, identified floral integrator genes like *LsSOC1*, *LsFT* and *LsAP1* as upregulated in the bolting sensitive line. [61]. Gene expression analysis of shoot apical meristem cells undergoing flowering transition in response to high temperature on the bolting-sensitive lettuce line S39, and further gene function studies, confirmed a role of *LsSOC1* in heat-promoted bolting in lettuce [36]. When expressed from the 35S promoter, *LsSOC1* acts as an activator of flowering in *A. thaliana* and can fully complement the Arabidopsis *soc1* null mutant [36]. RNAi-mediated knockdown of *LsSOC1* in *L. sativa* results in late flowering plants with reduced *LsLFY* expression [36]. The important function of *LsSOC1* in heat induced bolting in lettuce was further supported by the identification of two heat shock transcription factors that bind to the promoter of *LsSOC1* [36].

Overall, *LsFT* and *LsSOC1*, and their putative floral meristem identity targets *LsAP1* and *LsLFY* seem to play a key role in flowering transition in *L. sativa*, similar to other plant species. However, the key role of *LsSOC1* in promoting heat-induced flowering was not observed in other species so far, and may constitute a unique feature whose conservation amongst other Asteraceae species should be investigated.

4.2. Overview of the Vernalization and Autonomous Pathways

Vernalization refers to a process by which prolonged period of cold (winter) renders plants competent to flower, often many weeks later when other conditions, like day length or ambient temperature, are favorable [62]. Duration of cold exposure and the optimal temperature for vernalization vary among species, and among ecotypes of a given species, as plants adapt to periods of cold that are typical of a winter season in their natural habitat [8]. Plants can be either annual, biennial or perennial depending on the time required to complete their life cycles, from germination to seed setting, and the length of vegetative phase. Perennial plants can reproduce several times with recurrent vegetative to flowering cycles, and often do not respond to vernalization in the first year(s) of life. In annuals and biennials, vegetative to reproductive transition occurs once and flowering is associated with senescence and death of the whole plant [63].

In Arabidopsis, two genes are responsible for much of the variation in flowering time among natural population, *FLOWERING LOCUS C* (*FLC*), which acts as a repressor of flowering, and *FRIGIDA* (*FRI*), which promotes expression of *FLC*. In response to prolonged exposure to low temperatures, *FLC* is progressively repressed through epigenetic and silencing mechanisms, leading to flowering response. *VERNALIZATION INSENSITIVE3* (*VIN3*), a factor needed for epigenetic silencing of *FLC*, was recently found to have a key and complex role in vernalization and response to different temperatures [64]. These studies indicate that the absence of warmth rather than the presence of cold might be necessary for vernalization. Pivotal roles of *FLC* and *VIN3* in flowering time adaptation to natural environments were also confirmed by genome-wide association studies with nearly complete genotype information from 1135 Arabidopsis accessions [65].

The vernalization response is largely conserved within the Brassicaceae species due to conserved function of the main regulators FLC and FRI. However, the complex rearrangements occurred in the Brassica genomes [66–68] likely led to neofunctionalization processes of some *FLC* and *FRI* paralogs, which have lost their role in flowering control in response to vernalization.

In perennial Brassicaceae (e.g., *Arabis alpine*), orthologs of *FLC* are repressed by winter cold and reactivated in spring conferring seasonal flowering patterns, differently from annuals where they are stably repressed by cold as in Arabidopsis. Sequence comparisons of *FLC* orthologs from annuals and perennials identified two regulatory regions in the first intron whose sequence variation correlates with divergence of the annual and perennial expression patterns [69]. Unstable repression of a *C. intybus FLC* homolog during the cold season was also confirmed in root chicory that is perennial [22]. This points to key role of *FLC* regulation in evolutionary transitions between perenniality and annuality that seems to have occurred often among higher plants.

Questions regarding flowering response to vernalization in *Lactuca* species remain open as cultivated plants seem to have lost the need for the vernalization that is present in wild relatives. More generally, several species in the Asteraceae family require vernalization to flower, however molecular mechanisms underlying this trait have been poorly investigated.

Overall, null mutations or decreased expression of either *FLC* or *FRI*, as well increased expression of *FLC* negative regulators, would result in early and vernalization-independent flowering induction.

4.2.1. Vernalization and Autonomous Pathway in Brassicaceae

In *A. thaliana*, winter annuals contain active alleles at two loci, *FLOWERING LOCUS C* (*FLC*) and *FRIGIDA* (*FRI*), whereas summer annuals harbor inactivating mutations in one or both of these genes [70–73]. FLC is a MADS box transcription factor that acts as a repressor of flowering by directly binding to the floral promoting genes *FT*, *SOC1* and *SQUAMOSA PROMOTER-BINDNG PROTEIN-LIKE* 15 (*SPL15*) to block their transcription [40,70] (Figure 4). FRIGIDA encodes a coiled-coil protein that promotes *FLC* transcription, probably by affecting its chromatin structure [72]. During cold treatment, *FLC* is repressed through chromatin remodeling [74], and epigenetic mechanisms maintain the repressed state of *FLC* upon return to higher temperatures. [75]. During vernalization, transcription of several long noncoding RNAs (lncRNAs) starts from sites within the intron

(COLDAIR) and promoter of FLC (COLDWRAP) and a set of antisense transcripts of *FLC*, collectively named COOLAIR, are induced and physically associate with the *FLC* locus. This accelerates the transcriptional shutdown of *FLC* by recruitment of chromatin remodelers and switching of chromatin states [76–79]. Histone modifications mediated by genes like *VERNALIZATION 1* (*VRN1*), *VRN2*, *VERNALIZATION INSENSITIVE 3* (*VIN3*), cooperate to repress *FLC* at chromatin level [80–84]. FLOWERING LOCUS CA (FCA), FLOWERING LOCUS D (FLD), FLOWERING LOCUS KH DOMAIN (FLK), FLOWERING LOCUS PA (FPA), FLOWERING LOCUS VE (FVE), FLOWERING LOCUS Y (FY), and LUMINIDEPENDENS (LD) also repress *FLC* to accelerate flowering independently of vernalization. The corresponding genes are part of the so-called autonomous flowering pathway and act through repressive chromatin remodeling complexes and small RNAs to negatively regulate *FLC* [8]. FLC-like proteins form a specific phylogenetic clade, some members of which (MADS AFFECTING FLOWERING, MAF) can form protein complexes with FLC and redundantly affect flowering in response to vernalization [85].

Many Brassica species are biennial and require vernalization at seedling or mature plant stage. The temperature and duration of vernalization varies between spring, semi-winter and winter type plants: Flowering occurs without vernalization in spring types, with low vernalization (exposure to cold for shorter periods) in semi-winter types and with longer exposure to cold temperature in winter types. Rapid cycling populations, with extremely short reproductive cycles and which flower early independent of vernalization, have been developed in different Brassica species [86]. Comparative phylogenetic analysis of *B. napus*, *B. rapa* and *B. oleracea* identified three FLC clades which reflects the whole-genome triplication events that occurred during the evolution of the Brassica genome [66,87]. Four *FLC* paralogs in *B. rapa* (*BrFLC*) [67], five in *B. olearacea* (*BoFLC*) [68] and nine in *B. napus* (*BnaFLC*) [66] were identified (Table 3). *FLC* homologs in the chromosome A10 and C02 of *B. napus*, and an additional one in A03 (*Bna.FLC.A03b*), were initially associated with flowering time in *B. napus*. However, genome-wide association studies of flowering time and vernalization response in 188 different accessions demonstrated that *Bna.FLC.A02* and *Bna.FLC.C02* account for a significant proportion (22%) of natural variation in diverse accessions [50]. Expression of eight out of nine *BnaFLC* genes were downregulated during vernalization. This suggests that vernalization modulates *FLC* expression levels in a similar manner as in Arabidopsis. A cold-responsive *FLC-FRI-CBF1* cluster including *Bna.FLC.A03b* and *Bna.FRI.A3/Bna.FRI.Xa* was identified. It has been shown in other species that gene clusters with functionally related genes might be maintained by selection pressure to enable adaptation to extremely diverse environments in a similar manner as the cold-responsive cluster *FLC-FRI-CBF1* [88,89]. *Bna.FLC.A03b* shows enhanced expression levels in winter compared to semi-winter type plants [66]. Four *FRI* possible orthologs were identified in *B. napus*. [14]. Association analysis in a double-haploid population revealed that six SNPs (Single Nucleotide Polymorphism) in *Bna.FRI.A03* are associated with flowering time variation in 248 accessions, and that specific haplotypes are over-represented in semi-winter or winter types, while spring type plants did not show this correlation [3,14,90]. These data suggest that *Bna.FLC.A03b* and *Bna.FRI.A03* are functionally related, similar to *FLC* and *FRI* in *A. thaliana*, and have a key role in *B. napus* flowering response to vernalization.

In *B. rapa*, both *Bra.FLC.A10* (*BrFLC1*) and *Bra.FLC.A02* (*BrFLC2*), were found to underlie QTLs for flowering time in different studies [49,91–94], possibly due to alternative splicing and a 57 InDel (INsertion/DELetion) leading to a non-functional allele [91], respectively. However, the most similar *B. rapa* homolog of *Bna.FLC.A03b*, *BrFLC5*, is truncated and is expected to be not functional [95]. A recent study showed that the reference genome sequence indeed contains a truncated *BrFLC5* sequence, while other accessions contain functional genes with different splicing patterns resulting from a single nucleotide mutation. Genetic variation within the *BrFLC5* locus indicates that *BrFLC5* is not a major regulator of flowering time [96]. *BrFLC2* acts as a repressor of flowering when overexpressed in *A. thaliana* and shows early flowering when silenced in *B. rapa ssp. chinensis* (Pak-choi) [95,97]. *BrFLC2* seems to negatively regulate flowering by enhancing *MADS AFFECTING FLOWERING2* (*BrMAF2*) expression, while inhibiting expression of *BrSOC1* and *BrSPL15* [97]. In *B. rapa* seedlings, *BrFLC2* expression levels decrease upon vernalization treatment and remained low after return to higher temperatures. Contrarily to *BrFLC2*, expression of *BrVIN3*, a negative regulator of *FLC*, is very low in 14-day-old seedlings without vernalization, activates after four-week vernalization treatment on seeds and decreases again after transfer to higher temperature [95].

At chromatin level, *BrFLC* genes contained active chromatin marks H3K4me3 and H3K37me3 under normal growth conditions. During vernalization, alternative splicing of five *BrCOOLAIR* transcripts (*BrFLC2as406*, *-477*, *-599*, *-755* and *-816*) reduced H3K37me3 levels of *BrFLC1*, *BrFLC2* and *BrFLC3*. Differently from the Arabidopsis *COOLAIR*, *BrCOOLAIR* is located further downstream of *BrFLC2* and, during vernalization, class II transcripts, which are polyadenylated in the region complementary to the *BrFLC* promoter, are more abundant than class I, which are polyadenylated in the region complementary to the last intron of *BrFLC* [98]. Together with reduced H3K37me3 levels, an increase of H3K27me3 was detected in *BrFLC1*, *BrFLC2* and *BrFLC3* upon vernalization, which was maintained when plants were transferred to higher temperatures [95]. Besides affecting *FLC*, vernalization also resulted in enhanced H3K27 methylation in *BrMAF1* and DNA demethylation of two subunits of *casein kinase II* (CK2), *BrCKA2* and *BrCKB4*, altering daily expression period of clock-related gene *CIRCADIAN CLOCK-ASSOCIATED1* (*BrCCA1*) [95,99]. These findings indicate that the mechanisms underlying vernalization in *B. rapa* are very similar to those of Arabidopsis and involve chromatin modifications and COOLAIR antisense transcription.

In *B. oleracea*, expression levels of *BoFLC2* and *BoVRN* are enhanced in early compared to late flowering *B. oleracea* genotypes when grown at ambient temperature (22.5 °C and 12/12 h light/dark period) [100]. Two alleles for *BoFLC4* are described, which both confer a requirement for vernalization but respond with different kinetics to temperature shifts. Plants containing allele E9 require longer cold periods and flower late compared to those harboring allele E5. Introduction of genomic fragments containing the *BoFLC4^{E5}* or *BoFLC^{E9}* allele complemented an Arabidopsis *flc* null mutant, with a stronger effect for *BoFLC^{E9}* [101]. The closest *B. oleracea* ortholog of *BnaXFRId*, *BoFRIa*, also acts as a repressor of flowering when transformed into an Arabidopsis *fri* null mutant [90,101,102]. This indicates that FLC and FRI function in vernalization is also conserved in *B. olearacea*.

R. sativus is not a vernalization-requiring plant, but cold treatment does accelerate flowering. Radish transcriptome analysis during vernalization resulted in the identification of several vernalization-related differentially expressed genes [18]. Three copies of *RsFLC* were detected and all three act as flowering repressors when overexpressed in *A. thaliana* [103]. *RsFLC* expression before vernalization was enhanced in a late- compared to an early-bolting *R. sativus* inbred line, and reduced during vernalization or after GA treatment [18,104,105]. Overall, negative regulators of the vernalization pathway, such as *RsFLC*, *RsMAF2*, *RsSPA1*, and *RsAGL18*, were highly expressed in the late-bolting line, whereas positive regulators of vernalization, such as *RsVRN1*, *RsVIN3*, and *RsAGL19* were relatively highly expressed in the early-bolting line [104]. These results suggest that the vernalization pathway is conserved between radish and Arabidopsis.

D. tenuifolia is not a vernalization-requiring plant, and cold treatment of either seeds or plantlets does not accelerate flowering. Hence, even though *DtFLC* acts as repressor of flowering when overexpressed in *A. thaliana* and can complement the Arabidopsis *flc* null mutant, its role as a regulator of flowering time in wild rocket has to be further investigated [19,106].

4.2.2. Vernalization and Autonomous Pathway in Asteraceae

Wild lettuce-related species like *L. virosa* require vernalization to induce flowering. The cultivated *L. sativa* does not require a cold treatment for flowering, but a few days of cold does result in a better germination. Expression of the lettuce homolog of *FVE*, *FLD* and *LD* of the autonomous pathways were found to correlate with *LsFT* expression and flowering induction in two early or late *L. sativa* varieties grown in the field [60]. This finding suggests the existence and function of the autonomous pathway in lettuce flowering induction. However, the expression of lettuce *FLC* homologous genes was not analyzed either in this or in other studies, which impedes any further consideration about a possible role of *FLC*-like genes in cultivated lettuce.

A *FLC*-like gene, *CiFL1*, was identified and studied in *C. intybus*, which is biennial and requires vernalization at seedling or mature plant stage. Overexpression of *CiFL1* in Arabidopsis causes late flowering and prevents upregulation of the *AtFLC* target *FLOWERING LOCUS T* by photoperiod, suggesting functional conservation between root chicory and Arabidopsis [107]. *CiFL1* was repressed during vernalization of seeds or plantlets of chicory, like *AtFLC* in Arabidopsis. However, *CiFL1* repression was not maintained when plants were returned to warmer temperatures. This may be linked to the perenniality of root chicory compared with the annual life cycle of Arabidopsis. [22]. Indeed, recent studies on the divergence of seasonal flowering behavior among annual and perennial species in Brassicaceae showed that in perennial Brassicaceae orthologs of *FLC* are repressed by winter cold and reactivated in spring conferring seasonal flowering patterns, whereas in annuals, they are stably repressed by cold [69].

4.3. Overview of the Ambient Temperature Pathways

Responsiveness to ambient temperature is an adaptive trait and varies widely between and within species and accessions [108]. Besides extreme changes in temperature (e.g., vernalization), small changes in ambient temperature can also have an effect on flowering time. In *A. thaliana* plants grown under controlled laboratory conditions, a shift to lower (23 °C to 16 °C) and higher (23 °C to 27 °C) temperature delays and enhances flowering time, respectively [109]. The MADS box transcription factor *SHORT VEGETATIVE PHASE* (*SVP*) and most genes from the *FLC* clade, such as *FLOWERING LOCUS M* (*FLM/MAF1*) and *MADS AFFECTING FLOWERING-2-4* (*MAF2–MAF4*), have been implicated in the thermosensory pathway [85,107,110], with *SVP* and *FLM* having key roles in this process in Arabidopsis (Figure 4). SVP represses *FT* transcription at lower temperatures, but the levels of *FT* mRNA increase at higher temperatures. The control of floral transition in response to ambient temperature seems to differ among plant species, and many important questions concerning the regulation of flowering time by ambient temperature in Arabidopsis remain unsolved. However, *FT*-like genes seem to integrate the response to changes in ambient temperature in many species [111].

Although flowering induction in response to temperature changes may greatly affect yield and product quality in both Brassicaceae and Asteraceae crop species, insufficient work has been done to identify the genes responsible for this trait, especially in Brassica species. The floral integrator SOC1 was suggested to mediate heat-promoted bolting in lettuce, but further studies are needed to establish the exact mechanisms of flowering induction under these conditions, and whether this role is conserved in other Asteraceae [36].

Mutations that increase or decrease the expression of *MAFs* and *SVP* genes, known to be negative regulators of flowering time in Arabidopsis, may delay or speed up flowering time, respectively, if molecular mechanisms were conserved in crop species. On the other hand, reduction of *SOC1* in Asteraceae would potentially result in delayed timing of bolting and insensitivity to high temperature.

4.3.1. Ambient Temperature Pathways in Brassicaceae

Ambient temperature affects the deposition of the histone variant H2A.Z by the chromatin remodeling factor ACTIN RELATED PROTEIN 6 (ARP6). H2A.Z has been proposed to compact DNA in a temperature-dependent manner, thereby functioning as a temperature sensor in *A. thaliana* [112]. Accordingly, *arp6* mutants display a constitutive warm temperature response, but are still temperature responsive, indicating that H2A.Z is not the only thermosensor that mediates flowering. Recently, the basic helix-loop-helix (bHLH) transcription factor PHYTOCHROME INTERACTING PROTEIN 4 (PIF4) was shown to mediate flowering in response to temperature downstream of H2A.Z [113]. Mutations in *PIF4* suppress the induction of flowering by high ambient temperature only in SD, whereas the *pif4* mutant flowers normally in inductive LD [114]. The response to 27 °C-SD in the leaves was found to depend on the coordinate functions of CO, PIF4 and PIF5, as well as SVP, providing a genetic and molecular framework for the interaction between the photoperiod and thermosensory pathways [115].

SVP is directly activated by the chromatin remodeler BRAHMA (BRM) during the vegetative phase, whereas *FLM* is also regulated by the vernalization and photoperiodic pathways (reviewed in [116]). SVP can interact with FLC or FLM to form a repressor complex to prevent the expression of *FT* and *SOC1* [117,118]. Loss-of-function of *SVP* or *FLM* results in early and temperature-insensitive flowering, although *flm* loss-of-function plants retain some temperature sensitivity below 10 °C [118]. *FLM* is subject to temperature-dependent alternative splicing [110]. Two most abundant splice forms of *FLM*, *FLMβ* and *FLMδ*, which differ in the incorporation of either the second or third cassette exon, are both translated into proteins and their splicing pattern changes in response to changes in ambient temperature [110,118–121]. Different studies have shown that the abundance of *FLM-β* and *FLM-δ* splicing variants is regulated by temperature in an opposite fashion, with *FLM-β* enhanced at low temperature (16 °C) and *FLM-δ* increased at high temperature (27 °C) [118,120]. Overexpression of either *FLM-β* or *FLM-δ* results in opposite phenotypes, with *FLM-β* overexpression delaying flowering, as expected for a floral repressor, and overexpression of *FLM-δ* accelerating the transition to flowering [116,120]. A model was proposed in which only the incorporation of the FLM-β protein in the SVP–FLM complex would result in active repression of flowering targets, whereas incorporation of FLM-δ would form an inactive complex, indirectly promoting the transition to flowering [120]. More recent studies have shown that splice variant *FLM-β* has a stronger effect on flowering time compared to *FLM-δ* and therefore the function of *FLM-δ* under natural conditions is a matter of debate [108,122]. SVP and FLM contribute to the variation of flowering time among natural accessions of *A. thaliana* [73,123]. Alternative splicing is an important mechanism in sensing and adapting to changes in ambient temperature, and several genes in the thermosensory pathway undergo alternative splicing in response to temperature changes [121]. *MAF2*, *MAF3*, and circadian clock associated genes *PRR7* and *CCA1*, showed alternative splicing variants after a temperature shift.

Genes homologous to *SVP* and *FLM/MAF1* have been identified in *B. napus*, *B. rapa*, *B. oleracea* and *R. sativus*. In *B. rapa*, BrSVP and BcMAF1, a MAF-related Pak-choi (*B. rapa ssp chinensis*) gene, cause late flowering when transformed individually into *A. thaliana* [3,107]. Silencing of BcMAF1 in Pak-choi resulted in enhanced expression of BcFT1, BcFT2 and BcSOC1, reduced expression of BcMAF2 and early flowering compared to control plants [3]. These findings point to a function of SVP and FLM/MAF1 in the regulation of flowering time, but their role in ambient temperature response was not explored.

R. sativus plants flower early in spring, with LD conditions and higher temperature, compared to autumn. Vernalization and LD conditions reduces *RsSVP* expression, while expression is enhanced in SD conditions [105], indicating that RsSVP may act as a repressors of flowering in radish, as in Arabidopsis.

In *B. oleracea*, shifting plants to higher (23 °C to 27 °C) temperature results in differential splicing of about 156 genes. However, only 1% to 2.2% of those overlap with transcripts that are differentially expressed in the two investigated *A. thaliana* accessions (Gy-0 and Col-0). In contrast to *A. thaliana*, no differential splicing in flowering time genes was described in *B. oleracea* in response to high temperature [121], indicating that alternative splicing may not be a general regulatory mechanism by which ambient temperature regulates flowering response in Brassica species other than Arabidopsis.

4.3.2. Ambient Temperature Pathways in Asteraceae

L. sativa plants grown at high temperatures (35/25 °C) flower early compared to plants grown at lower temperatures (25/15 °C) [58]. RNA-seq analysis revealed 1443 and 1216 genes that were upregulated respectively in leaves and stems of plants that had been shifted to 37 °C for one week compared to control plants that were maintained at 25 °C [124]. Among these genes were homologs of *AP2*, *AP2-like*, *SOC1* and *FLM* in the leaves and homologs of *AP2-like*, *FLC* and *FLM* in the stem. The shift to 37 °C resulted in the downregulation of 1038 genes in leaves and of 933 genes in stems, as compared to the controls at 25 °C. These included photoperiod-related genes in both leaf and stem, and two *LsFLC*-like homologs in leaf. Unexpectedly, *SVP*-like genes were not present in the sets of differentially regulated transcripts [124].

In *C. intybus*, treatment of non-vernalized plants with elevated temperatures (increase of 6 °C) in the field resulted in a variety of phenotypic differences like more leaves, reduced mean leaf area, decreased root weight and early flowering [125]. The severity of these heat stress-induced phenotypic changes was cultivar dependent. Early flowering in response to elevated temperature seems to be conserved in *L. sativa* and *C. intybus*. However, no genetic or molecular data are available in *Cichorium* spp. for heat-induced bolting response.

4.4. Overview of the Photoperiodic Pathway

Day length is an important factor for a plant to track seasonal changes, where short days (SD, 8/16 h light and dark) indicate winter and long days (LD, 16/8 h light and dark) indicate spring or summer. Plants can be divided into three major groups on the basis of their responses to photoperiod: Long-day plants flower when the day exceeds a critical length, short-day plants flower when the day is shorter than a critical length and day-neutral plants flower independently of day length [126]. As plants aim to flower in the optimal season, most plants show a delay in bolting when grown under SD conditions and early bolting under LD conditions. The mechanism behind light perception and integration has been intensively studied in *A. thaliana* over the past 15 years (reviewed in [11] and [127]). The circadian clock and photoreceptors influences transcription and protein stability of the transcriptional activator *CONSTANS* (*CO*) which, in a signaling cascade involving GIGANTEA (GI), in turn activates the floral integrator *FT* in a long-day afternoon [39,128].

Photoperiod and circadian rhythm are involved in many processes of adaptive response to environmental conditions, including flowering time. Their molecular mechanisms are widely conserved amongst plant species to such an extent that mechanisms of photoperiod measurement are more diverse between long-day and short-day plants than between eudicots and monocots [129]. Based on gene expression, it is suggested that the photoperiod pathway is conserved between the Brassicaceae and Asteraceae family, which include mainly plants requiring long days to flower. Despite our knowledge on the genetic control of flowering time in response to different light conditions is quite limited in the species we are reviewing, preliminary studies suggest a key role of CO, GI and photoreceptors in adaptation to different environments [54,106,130].

4.4.1. The Photoperiodic Pathway in Brassicaceae

CONSTANS promotes flowering by initiating transcription of the *FT* and *TSF* genes (Figure 4). The blue light receptor FLAVIN-BINDING KELCH REPEAT F-BOX 1 (FKF1) and the clock-associated protein GI form a complex to degrade transcriptional repressors of *CO*, *CYCLING DOF FACTORs* (*CDFs*), and to stabilize the CO protein [131–135]. Post-translational regulation of CO is essential for a flowering response to long days. The CO protein is ubiquitylated by a ubiquitin ligase complex that includes CONSTITUTIVE PHOTOMORPHOGENIC 1 (COP1) and SUPPRESSOR OF PHYTOCHROME A (SPA1), facilitating CO degradation by the 26S proteasome [136–138]. Activity of this complex is repressed by light so that it mainly promotes the degradation of CO protein in the dark. Thus, only the peak of *CO* mRNA that occurs in the light at the end of a long day after degradation of the CDFs by GI–FKF1 leads to CO protein accumulation (Figure 4).

The circadian clock is a time-keeping mechanism with a periodicity of 24 h. In Arabidopsis, the circadian clock confers diurnal patterns of gene expression on roughly one-third of the genes, and comprises interlocked feed-back loops [139,140]. Core clock components include the morning phased genes *CCA1*, *LATE ELONGATED HYPOCOTYL* (*LHY*), *REVEILLE8* (*RVE8*) and *PSEUDO-RESPONSE REGULATOR 9* (*PRR9*) [141–147]; the afternoon phased genes *PRR5*, *PRR7*, *GI* [145,147]; and the evening phased genes *EARLY FLOWERING 3* (*ELF3*), *ELF4*, *LUX ARRHYTHMO* (*LUX*) and *TIMING OF CAB EXPRESSION1* (*TOC1*) [142,147–150].

Homologs of all genes involved in photoperiodic response were identified in *B. napus* and shown to be highly variable in studies of targeted deep-sequencing of essential flowering time regulators [35] in a panel of 280 inbred lines. Four *CO* and four *CO*-like genes are present in the genome of *B. napus*, including those initially characterized by Robert et al. [151], one of which shown to complement *co* mutants in *A. thaliana* [151]. One *BnPHYA* gene has undergone two coupled duplication-deletion events (HNRTs), where one region of the genome replaces a respective homeologous genome region. It was suggested that such rearrangements may represent a necessary co-adaptation of the photoperiodic pathway to the strong vernalization requirement in winter inbred lines [4].

Compared to *A. thaliana*, several duplicated or triplicated photoperiod genes, such as *BrCO*, *BrFKF1*, *BrCDF1*, *BrLHY* and *BrTOC1*, were detected in *B. rapa* [35,54]. Expression of these genes throughout the day differed when plants were grown under LD or SD conditions, only *BrCDF1* showed a similar trend under both growth conditions [54]. So far, no complementation or other functional studies for the core clock components are available in Brassica species. *BrGI* was identified as an important component for circadian rhythm and multiple abiotic stress responses and acts as an activator of flowering when transformed into an Arabidopsis *gi* null mutant [152]. Two putative null alleles of *BrGI* resulted in late flowering when homozygous in *B. rapa* [152,153]. Furthermore, BrGI protein physically interacts with GI-interacting partners, like BrFKF1, suggesting a conserved function with Arabidopsis [152].

In both *B. oleracea* and *R. sativus*, silencing of *GI* resulted in delayed bolting and flowering, with a correlation between *GI* expression levels and days to flowering [154,155].

D. tenuifolia plants flower later under SD compared to LD conditions, with 50 and 20 days to flowering, respectively [106]. *DtCO* and *DtGI* are both diurnally regulated. Under LD conditions, *DtCO* acts as activator of flowering when transformed into *A. thaliana* and could complement the *co* null mutant [106].

4.4.2. The Photoperiodic Pathway in Asteraceae

Lou et al. [156] hypothesized that *CCA1*, *RVE2*, *RVE4* and *RVE5* function might be restricted to the Brassicaceae family. However, Higashi et al. [130] later described 215 common oscillating transcripts in *L. sativa*, including *LsCCA1*, *LsGI*, *LsLHY*, *LsFKF1*, *LsTOC1*, *LsPRR7* and *LsCO*-like. The expression pattern of these genes show a large degree of overlap with those of *A. thaliana* [130], indicating a possible functional conservation in Asteraceae. Despite the great importance of photoperiodic control of flowering time for vegetable crop production and adaptation to different cultivation environments, no further molecular and genetic data are available for either lettuce or chicory species.

4.5. Overview of the Age Pathway

Plants go through developmental phases such as juvenile-to-adult transition and floral induction during their life cycle. As the plant ages, concentrations of the SQUAMOSA PROMOTER BINDING LIKE (SPL) transcription factors (also known as SQUAMOSA promoter binding protein, box family, SBP) increase. SPLs promote flowering by initiating the expression of several other transcription factors, such as LEAFY (LFY), FRUITFULL (FUL), and SOC1 [157,158]. SPL proteins are negatively regulated by the microRNAs [158]. MicroRNAs (miRNA) are key regulators of the age pathway, preventing precocious flowering when the plant is too young. Two major miRNAs, *miR156* and *miR172*, have an antagonistic effect on flowering time by downregulating their own set of target genes. *miR156* expression is high in young plant stage, decreases over time and is low at the onset of flowering [159,160].

The involvement of miRNAs in flowering time and the important role of *miR156* and *miR172* and their corresponding targets, is widely conserved across plant species [161]. As expected, both *miR156* and *miR172* seem to be conserved between the Brassicaceae and Asteraceae families, although very few reports are available in Asteraceae. The miR156/SPL module plays a central function in age-dependent competence to flowering, but seems to be even more fundamental in perennial Brassica species that undergo reiterative flowering induction cycles and do not respond to vernalization in the first year of life. Therefore, miR156/SPL may play a key role in flowering control in biennial crops [162]. Other miRNA like *miR824* and *miR5227*, the latter only detected in *R. sativus* [163], are less conserved and seem to be newly evolved Brassica-specific miRNAs as they were not found in families other than Brassicaceae so far. In *L. sativa*, a homolog of the Arabidopsis DELAY OF GERMINATION1 (DOG1) seems to have acquired a novel function in the miRNA-mediated response to flowering time, but further studies are needed to investigate DOG1 role in other Asteraceae and in other plant families [164].

4.5.1. Age Pathway in Brassicaceae

A. thaliana contains eight *miR156* members (*miR156a* to *miR156h*) which target different *SPL* genes (Figure 4) [159,165]. Besides enhancing expression of floral meristem identify genes, *SPL* genes also promote *miR172* expression [166]. *miR172* shows an inverse expression pattern with increasing expression over time [159,160]. *A. thaliana* contains five *miR172* members (*miR172a* to *miR172e*) which target *AP2* and the *AP2*-like genes *TARGET OF EAT1* (*TOE1*), *TOE2*, *TOE3*, *SCHLAFMÜTZE* (*SMZ*) and *SCHNARCHZAPFEN* (*SNZ*) [167–169]. *AP2* and *AP2*-like genes inhibit the onset of flowering by repressing expression of *SOC1*, *FUL* and *AGAMOUS* (*AG*) (Figure 4, [170]). Another miRNA, *miR824*, targets *AGL16*, which encodes a MADS-box repressor of flowering time that interacts with SVP and FLC to regulate *FT* expression levels [171]. SPL15 cooperates with SOC1 to coordinate the basal floral promotion pathways required for flowering in non-inductive environments by directly activating transcription of *FUL* and *miR172* in the SAM [162]. The capacity of SPL15 to promote flowering is regulated by age through *miR156* that targets *SPL15* mRNA. Strong evidence is emerging that miR156/SPL control competence to flower as well as vegetative phase change [162]. Several studies point to a major role of SPL9 and SPL15, with SPL15 playing the larger role in floral induction, particularly under noninductive short days, and SPL9 acting in floral primordia after the floral induction. The miR156/SPL module is of special interest for the acquisition of competence to flowering in biennial and perennial Brassicaceae relatives of Arabidopsis, where *miR156* levels act as the timer in controlling competence to flower, and often make plants insensitive to vernalization when too young (Figure 4, [172,173]). It was suggested that the miR156/SPL module, which is evolutionarily conserved in all flowering plants, might have acquired increased dependency for flowering in perennials, whereas annuals would have evolved genetic mechanisms to bypass this module by alternative inductive pathways such as light/photoperiod [162].

B. napus contains 36 copies of *miR156*, of which 17 located on the A genome and 19 on the C genome, and 14 copies of *miR172*, with eight located on the A genome and six on the C genome [174]. A total of 58 genes encoding putative SPL/SBP proteins are present in the *B. napus* genome, 44 of which harboring *miR156* binding sites [175]. This suggests that relationship between *miR156* and *SBP* genes is conserved across species, although distinct regulation pattern of the homologous genes exist between *B. napus* and Arabidopsis that may reveal some divergence of the SBP-box genes in oilseed rape.

B. rapa contains 17 copies of *miR156* and 11 of *miR172* [174]. BrmiR156 is highly expressed in early plant stages and expression decreases during plant development. BrSPL9-2 and SPL15-1 show an opposite expression pattern compared to BrmiR156, with increasing expression over time. Cabbage plants expressing a mutated BrSPL9 (mBrSPL9) allele, resistant to BrmiR156, showed enhanced BrSPL9 and BrmiR172 expression. In the field, mBrSPL9 plants had dark green leaves with enhanced chlorophyll content and a prolonged heading stage with delayed flowering, but no significant change in head weight, size or shape. Overexpression of BrmiR156 in cabbage resulted in decreased BrSPL9-2 transcript levels and a prolonged seedling and rosette stage [176], pointing to conservation of the miR156/SPL module in *B. rapa*.

B. oleracea contains 15 copies of *miR156*, where BomiR156c is known to target BoSPL9 while BomiR156g targets BoSPL3. The *miR172* family contains nine copies and targets BoAP2 and BoTOE2 [174,177,178]. A newly evolved *miR824*, which seems specific for Brassicaceae, was also identified and targets *BoAGL16* [178]. This function is conserved with Arabidopsis where the miR824/AGL16 quantitatively modulate the extent of flowering time repression in a long-day photoperiod through *FT* [171].

In *R. sativus*, 11 members of *miR156/miR157*, five members of *miR172*, two members of *miR824* and one member of *miR5227* are detected. Different *RsmiR156* copies target *RsSPLs* and *RsmiR156a* also *RsTOC1*, *RsmiR172a* targets *RsAP2*, *RsmiR824* targets *RsAGL16* and *RsmiR5227* targets *RsVRN1*. Expression of *RsmiR156a*, *RsmiR824* and *RsmiR5227* decreased when plants shifted from vegetative to reproductive phase [163], strongly indicating that these miRNAs and their corresponding target genes might play important roles during bolting and flowering processes of radish.

4.5.2. Age Pathway in Asteraceae

LsmiR156 and *LsmiR172* act as repressor and activator of flowering, respectively, when expressed in *A. thaliana* [164], and targets *LsSPLs* and *LsAP2* in *L. sativa* [179]. In Arabidopsis, expression of the *DELAY OF GERMINATION1* (*DOG1*) gene responds to seed maturation temperature and determines the depth of seed dormancy [180]. Huo et al. showed that DOG1 could regulate seed dormancy and flowering times in lettuce through the modulation of *miR156* and *miR172* levels [164]. *LsDOG1* silencing lines flowered early compared to control *L. sativa* plants, with an enhanced effect in autumn, and showed reduced expression of *LsmiR156*, enhanced expression of *LsmiR172*, *LsFT*, *LsSPL3* and *LsSPL4* and no difference in transcript levels of *LsSPL9*. This would suggest that *LsDOG1* has an additive role in *LsmiR156*- and *LsmiR172*-mediated flowering time, besides the thermo-inhibition of seed germination described in *A. thaliana* [164].

Srivastava et al. [181] has predicted two copies for *miR156* and one copy for *miR157* in *C. intybus*. For the miRNA targets, only *CiSPL3* and *CiSPL12* were detected and confirmed as targets of *CimiR156* [181].

4.6. Overview of the Hormonal Pathway

Gibberellins (GAs) are growth regulators involved in plant developmental processes that promote transition to flowering in several plant species [61,124,182–185]. In Arabidopsis, GA contributes to flowering under inductive long days (LDs) through the activation of *SOC1* and *LFY* in the inflorescence and floral meristems, and of *FT* in leaves. Under non-inductive short days (SDs) conditions, the GA pathway assumes a major role as under SDs flowering is delayed and correlates with a gradual increase in bioactive GA at the shoot apex [186]. Mutations that impair GA biosynthesis prevent flowering under SDs [183].

Besides GA, it has been suggested that cytokinins, major growth regulators in plants, also participate in the regulation of flowering time (reviewed in [187]). For a long time, it has been known that exogenously applied cytokinin can promote flowering in Arabidopsis [188–190]. However, it is unclear whether endogenous cytokinins can also have the same inductive activity.

Regulation of GA and its involvement in the switch to flowering seems conserved between the Brassicaceae and Asteraceae families. In grasses and cereals, GAs are similarly regulated and also involved in flowering time [191], suggesting that the GA role in promoting flowering is widely conserved in plants. Genes involved in GA metabolism or sensitivity may constitute good targets to modulate flowering time in crops, as enhanced GA content or signaling can induce early flowering whereas low GA amount or signal can delay bolting and flowering.

4.6.1. Hormonal Pathway in Brassicaceae

The GA pathway is well described in *A. thaliana* (reviewed in [192]). In brief, the last steps of the GA pathway involves the conversion of GA_{12} into GA_9 and GA_{53} into GA_{20}, by GA 20-oxidases (GA20ox1-5), the conversion of GA_9 and GA_{20} into bioactive GA_4 and GA_1, respectively, by GA 3-oxidases (GA3ox1-3) and the deactivation of GA_4 and GA_1 by GA 2-oxidases (GA2ox1-5) [193–198]. Bioactive GAs binds to *GIBBERELLIN INSENSITIVE DWARF1* (*GID1a*, *-b* and *-c*) to promote degradation of DELLA proteins [199–201], negative regulators of gibberellin signaling that act immediately downstream of the GA receptor. DELLA proteins repress transcription of many genes, including *FT*, *TSF*, and some *SPL* genes [202]. Low levels of bioactive GA result in the accumulation of DELLA proteins, which delay flowering independent of photoperiod [202,203]. The MADS box transcription factor SVP, besides repressing floral integrator gene expression, regulates bioactive GAs at the shoot apex by repressing the *GA20ox2* gene [204]. In response to inductive photoperiods, repression of SVP contributes to the increase of GAs at the shoot apex, promoting rapid induction of flowering. The ambient temperature and GA pathways are tightly linked (Figure 4, [205]).

Cytokinins (CK) were also proposed to affect flowering time as exogenous application of CKs can promote flowering in Arabidopsis. It has been shown that exogenous cytokinins promote flowering independently of *FT*, but through the transcriptional activation of its paralog *TSF* [189]. Cytokinins are perceived by membrane-located receptors called *A. THALIANA HISTIDINE KINASE2* (*AHK2*), *AHK3* and *AHK4* and are involved in many plant processes during plant development. Gain-of-function variants of *AHK2*, with enhanced cytokinin signaling, showed either early or late flowering [206]. Furthermore, it has been suggested that there is a cross-talk between cytokinins and GA, mediated by *SPINDLY* (*SPY*) [207].

In *B. napus*, genes encoding DELLA proteins and genes of the GA metabolism have been identified [208], however their role in flowering was not explored. During *B. napus* vernalization, the content of cytokinins increases significantly and reaches a maximum during reproductive transitions. Cis-Zeatin riboside accounted for ca. 87% to 89% of the total isoprenoid cytokinin content in control and vernalized plants, whilst isopentenyladenosine and cis-zeatin were the next most abundant cytokinins. In the post-vernalization period, endogenous cytokinin levels decreased, but remained significantly higher in the reproductive plants than in the vegetative controls. Changes in cytokinin accumulation during vernalization-induced reproductive development may suggest a possible role of CK in this process. [209].

In *B. rapa*, low-temperature treatment increases the GA content, and enhanced GA accumulation initiates floral bud differentiation [210]. Expression patterns of most genes involved in GA metabolism, particularly those of four genes including one *GA20ox* were consistent with observed GA levels [210].

In *B. oleracea*, treatment with bioactive GA_3 and GA_{4+7} result in early curd formation in cauliflower and broccoli plants [184]. GA treatment induces bracting and stem elongation, but not flower initiation, when cauliflower and broccoli are at the IM or floral bud stage, respectively. As confirmation, treatment with GA does not show differences in the expression of *BoAP1-a*, *BoAP1-c*, *BoLFY* and *BoSOC1* in cauliflower plants at the IM stage. These results suggest that GA has an effect on vegetative-to-reproductive transition and another pathway is responsible for the IM-to-FM transition [184].

In *R. sativus*, two homologs for *GID1a*, one for *GID1b*, one for *GID1c* and three for *GA2ox* have been described. Before vernalization, expression level of one *RsGA2ox* homolog was upregulated in a late compared to early bolting line. Expression level of one homolog of *RsGID1a* was induced by vernalization treatment [104].

4.6.2. Hormonal Pathway in Asteraceae

In *L. sativa*, plants treated with exogenous GA have enhanced levels of GA_3 and GA_4 in the leaves and flower early, with an enhanced effect in early flowering varieties [61,124]. Early flowering plants treated with CCC (a GA inhibitor) have reduced GA_3, GA_4 and IAA levels in the leaves and stem, are compact and do not bolt. Transferring plants from ambient (25/15°C) to higher temperature (35/25 °C) results in enhanced expression level of *LsGA2ox1*, *LsGA3ox1* with corresponding enhanced endogenous levels GA_8 and GA_1, respectively. Expression level of *LsGA20ox1* and corresponding endogenous level of GA_{20} was unaffected by the transfer to higher temperature. Therefore, it is suggested that *LsGA3ox1* might be responsible for enhanced bioactive GA_1 levels in plants grown at higher temperatures [211]. With transcriptome analysis of a bolting resistant and sensitive line, Han et al. [61] have shown that *LsGA3ox1*, *LsGA20ox1*, *LsGA20ox2* and 28 out of 41 auxin-related genes were upregulated in leaves of a bolting sensitive line. Liu et al. [124] showed that heat treatment of bolting sensitive plants results in early bolting, enhanced GA_3 and GA_4 levels in the leaves, reduced IAA levels in the leaves and enhanced IAA levels in the stem. Transcriptomic analysis of a bolting sensitive line has shown that, out of 1443 and 1038 differentially up and down regulated genes, *LsGA20ox* was upregulated in leaves and a gibberellin-regulated family protein upregulated in the stem tip after heat treatment [124]. *L. sativa* plants overexpressing Arabidopsis *KNAT1*, a KNOTTED1-like homeobox (KNOX) transcription factor, show altered plant architecture and early flowering compared to control plants. Their striking leaf morphology phenotype was associated to a consistent increase in cytokinin content. Based on these results, correlation between temperature, GA levels and flowering time is suggested, together with a role of KNAT1 in flowering time, directly or indirectly, through cytokinins [212]. It has been proposed that the KNOX transcription factor *KNAT1* could regulate flowering by increasing cytokinin biosynthesis [212], and *ISOPENTENYL TRANSFERASE (IPT)* biosynthetic genes were shown to be downstream targets of KNOX transcription factors [213,214]. However, there is no direct evidence that the early flowering phenotype observed in *KNAT1* overexpressing lettuce plants depends on CK increase as KNOXs also control other hormonal and metabolic pathways, including GA biosynthesis (reviewed in [215]). Hence, the observed early flowering phenotype may depend on mis-regulation of as of yet unknown targets in the flowering time genetic network.

5. Quantitative Trait Loci (QTL)

The identification and functional characterization of genes controlling different pathways of flowering time has increased the knowledge about this complex trait. In parallel, the genetic basis of natural variation in flowering time has been investigated by quantitative trait loci (QTL) analysis. Salomé et al. [216] and Brachi et al. [217] have described the QTL mapping of 17 F2 populations and 13 RIL (recombinant inbred line) families in *A. thaliana*, which has led to the identification of many QTLs. Most of the QTLs are located in five genomic regions (region At1–5) and contain flowering time genes previously described in this review (Figure 4). All the five QTL regions described contain large-effect alleles [216,217]. Within the detected QTL regions, epistatic interaction between *FLC* and *FRI* alleles is highly associated with flowering time and could explain up to 70% of the variation [218,219]. Recently, a genome wide association map of flowering time, with nearly complete genotype information, was obtained taking advantage of the genomic sequencing and phenotype information from different environments (10 °C and 16 °C) of 1135 natural inbred lines of *Arabidopsis thaliana* [65]. The identified peaks from the genome wide association study (GWAS) contained *VIN3*, *FT*, *SVP*, *FLC* and *DOG1*, all previously linked to flowering time [81,123,164,220,221].

In *B. napus*, a mapping population made from a cross between Tapidor (winter type) and Ningyou7 (semi-winter type) is the most used for the identification of QTLs affecting flowering time [5,66,222–225]. Other analyses include different mapping populations or a broad set of accessions and inbred lines. Overall, phenotyping was performed in field trials in different locations and over multiple years, and flowering time was scored when 25% or 50% of the plants within a plot had an open flower. Many QTLs have been discovered in the different populations, with 23 genomic regions (Bn1-23) overlapping between at least two QTL analyses (Table 4). Of the flowering related genes within these genomic regions, *Bna.FRI.Xa* (region Bn5) is shown to have specific haplotypes overrepresented in either semi-winter or winter type plants [3,14]. Long et al. [222] have shown that genomic region Bn13 explains 50% of the variation in flowering time, is specific for spring environments and suggested that *Bna.FLC.A10* might control flowering time in non-vernalization environments. Later, Hou et al. [224] observed that one of the polymorphic sites upstream of *Bna.FLC.A10* is strongly associated with vernalization requirement of rapeseed. For *Bna.FT.A07b* (region Bn11), differential expression between types or treatments has been described [47], but no haplotype information is available so far.

In *B. rapa*, QTL analyses were performed on mapping populations mainly involving Yellow Sarson or a rapid cycling line 09A001. Phenotyping was scored based on flowering time (days to first open flower) or bolting time (days to first internode elongation). Of the detected QTLs in different populations, six genomic regions (Br1–6) were overlapping in at least two QTL analyses (Table 4). Of the flowering related genes within these genomic regions, *BrFLC2* (region Br2) is a major factor in determining flowering time. A 57 bp deletion on the exon4/intron4 border of *BrFLC2*, resulting in alternative splicing, is significantly associated with flowering time [91]. Zhang et al. [49] showed that a transposon insertion in exon 2 of *BrFT2* (region Br5) results in late flowering, and that there is a correlation between flowering time and different *BrFLC2* and *BrFT2* alleles. Plants with functional or non-functional alleles for both genes result in similar flowering time. However, a non-functional allele of either *BrFLC2* or *BrFT2* results in early or late flowering, respectively [49]. Besides the QTLs detected in multiple analysis, Xie et al. [153] has described one QTL (ChrA09:25634145.25774304), containing *BrGI* as a candidate gene responsible for circadian period determination. Two detected *BrGI* alleles ($BrGI^{imb211}$ and $BrGI^{500}$) could complement the late flowering phenotype of the Arabidopsis *GI* null mutant, but plants with allele $BrGI^{500}$ showed a shorter circadian period and could not (fully) complement the response to red and blue light [153].

In *B. oleracea*, different mapping populations and commercial parents have been used for QTL analysis. Phenotyping was performed in the greenhouse and was scored as days to flowering or days to curd initiation (curd larger than 1 cm). Of the detected QTLs in different populations, six genomic regions (Bo1–6) were overlapping in at least two QTL analyses (Table 4). One of the candidate genes in region Bo1 is *BoFLC4* (Table 4). The two main alleles $BoFLC4^{E5}$ and $BoFLC4^{E9}$ both confer a requirement for vernalization, but differ with regard to their transcription regulation in response to temperature shifts, due to cis-regulatory differences [101]. One of the candidate genes in region Bo1 is *BoFRIa* (Table 4). Sequencing of *BoFRIa* from 55 accessions detected six different alleles with numerous substitutions and InDels. Expression of the two most abundant alleles from the *AtFRI* promoter prolonged the time to flower equally when overexpressed in *A. thaliana*, suggesting that the potential effect of these alleles on flowering time in *B. oleracea* may result from differences in their expression [90].

For *L. sativa*, a RIL population of cultivar *L. sativa* cv. Salinas (Crisphead) and Californian *L. serriola* unveiled two QTLs (Ls1 and Ls3) for days to flowering [226,227]. Furthermore, backcrossed lines selfed for one generation (BC1S1) from a cross between cultivar *L. sativa* cv. Dynamite (Butterhead) and a *L. serriola* uncovered four additional QTLs (Ls2, Ls4–6) [228]. A few flowering time related genes are located within QTL regions Ls1, Ls5 and Ls6. However, it remains an open question whether polymorphisms in these candidate flowering time related genes underlie the detected QTLs. Recently the *L. sativa* genome sequence [20] and RNA-seq data from 240 wild and cultivated lettuce accessions were realized, which will provide valuable tools to explore genetic variations contributing to flowering time and other traits in *L. sativa* [229].

In general, *FLC* and *FRI* seem to be overlapping in QTL analyses between different Brassica species. This provides more evidence that indeed these are key regulators of flowering time in many Brassicaceae. However, the specific genes and alleles responsible for the other QTLs remain unexplored. Identification of the causal genes and genetic variation for all QTLs would help to further understand the regulation of flowering time in the different crops. More QTL analyses have been performed for some species other than those discussed here. Even though these data are of great value, it is difficult to determine if the QTLs overlap with the reported QTLs, as reported positions cannot be related to the physical map (e.g., QTL analysis from [230,231]). For *L. sativa*, only two populations have been used for QTL analysis, both involving wild source *L. serriola*. It might be worthwhile exploring other wild sources such as *L. virosa* or *L. saligna* to expand the number of currently known QTLs. The availability of new genetic and genomics resources will consistently speed up genetic studies to unravel the key regulatory nodes of flowering time pathways in Asteraceae leafy crops.

Table 4. Flowering-time related QTL regions for *A. thaliana*, *B. napus*, *B. rapa*, *B. oleracea* and *L. sativa* with candidate flowering-time genes within these QTL regions.

QTL Region	Species	Region [1]	Candidate Genes	References
At1	*A. thaliana*	Chr1:24500000-29000000	FT, FKF1, AP1, FLM	[216,217]
At2	*A. thaliana*	Chr4:300000-1900000	FRI	[216,217]
At3	*A. thaliana*	Chr4:8000000-12000000	VRN2, TSF, GA2ox	[216,217]
At4	*A. thaliana*	Chr5:2700000-8100000	FLC, CO, TFL2	[216,217]
At5	*A. thaliana*	Chr5:21500000-26000000	VIN3, PRR3, TOE2, LFY, CDF1, MAF2-5	[216,217]
Bn1	*B. napus*	chrA02:114931.1575498	Bna.FLC.A2, CO-like, RVE1	[232,233]
Bn2	*B. napus*	chrA02:1575449.4330821	AP2-like, TOE2, PRR3	[225,232,233]
Bn3	*B. napus*	chrA02:5233136.8233310	GA20ox, Bna.FT.A02	[223,225,232-235]
Bn4	*B. napus*	chrA02:8776742.9248051		[5,222,234,236]
Bn5	*B. napus*	chrA03:5046910.6515058	**Bna.FRI.Xa**, SPL13, CBF1, Bna.FLC.A03b	[66,222,234,236]
Bn6	*B. napus*	chrA03:18872718.20131639	AP2-like, FUL, TOC1	[223,232,233,236]
Bn7	*B. napus*	chrA04:257040.4734286	AP2-like	[233,234,236]
Bn8	*B. napus*	chrA04:7743947.10942653		[233,234]
Bn9	*B. napus*	chrA04:11898475.13460703	CO-like, ELF3	[234,236]
Bn10	*B. napus*	chrA06:23330530.23617143		[232,236]
Bn11	*B. napus*	chrA07:14463578.18916565	SPL15, AP2-like, GID1, AP1, **Bna.FT.A07b**	[232-234,236]
Bn12	*B. napus*	chrA10:9835903.10695100	PRR3, TOE2, AP2-like	[222,234]
Bn13	*B. napus*	chrA10:13375104.15191366	**Bna.FLC.A10**	[66,222,224,233,234]
Bn14	*B. napus*	chrC01:27417076.34893173	FRI-like, VRN1	[232,233]
Bn15	*B. napus*	chrC02:6956919.13653054	GA20ox, SPL	[222,232,234]
Bn16	*B. napus*	chrC02:22287455.22560553		[222,234]
Bn17	*B. napus*	chrC02:44366336.45788246	FUL, MAF2, MAF3	[225,232]
Bn18	*B. napus*	chrC03:58161161.58296560		[233,234]
Bn19	*B. napus*	chrC04:40003810.41181656		[222,234]
Bn20	*B. napus*	chrC06:21784608.29654361	ELF4, AP1	[225,232,233,236]
Bn21	*B. napus*	chrC07:26989258.31787256	SEP4	[225,232,234]
Bn22	*B. napus*	chrC09:39312343.43429210	SPL7, AP2-like, TFL2, RVE	[234,236]
Bn23	*B. napus*	chrC09:45206288.47504024	Bna.FLC.C09b, Ga20ox	[225,232]
Br1	*B. rapa*	A01:81263.3282650	AP2-like	[237,238]
Br2	*B. rapa*	A02:1244721.4284193	**BrFLC2**, AP2-like, CO-like, SPL7	[49,92,237-239]
Br3	*B. rapa*	A03:14357780.27239372	CO-like, AP2-like, GA2ox, AGL24	[237-239]
Br4	*B. rapa*	A06:13769411.18840509	LFY, GA20ox, CDF1, FLM, MAF4, VIN3-like, CO-like,	[238,240]
Br5	*B. rapa*	A07:12545242.20240840	AP2-like, SPL15, ELF4-like, AP1, BrFT2	[49,238,240]
Br6	*B. rapa*	A10:12936259.13856133	BrFLC1	[237,238,241]
Bo1	*B. oleracea*	C02:900000.2900000	GRF6, **BoFLC4**	[101,242]
Bo2	*B. oleracea*	C03:1800000.20000000	BoFLC3, SOC1, **BoFRIa**, ELF4, GA20ox	[100,242,243]
Bo3	*B. oleracea*	C04:10726862.16070000	TOE2	[100,242]
Bo4	*B. oleracea*	C04:32446947.35540000		[100,244]
Bo5	*B. oleracea*	C06:2396965.6360269	TOE1, VIN3	[242,243]
Bo6	*B. oleracea*	C06:22550000.32446947		[100,243]
Ls1	*L. sativa*	LG2:163353056.165477161	CDF1, CO, FLC, PRR5, VRN1	[226,227]
Ls2	*L. sativa*	LG6:140450832.140481276		[228]
Ls3	*L. sativa*	LG7:158780460.159063877		[226,227]
Ls4	*L. sativa*	LG7:172306237.193636147		[228]
Ls5	*L. sativa*	LG8:25874939.47456612	PRR3, PRR5, PRR7, PRR9	[228]
Ls6	*L. sativa*	LG8:63537238.76202393	FKF1	[228]

[1] Regions on genomes of *A. thaliana* (Tair10), *B. napus* (Brassica_napus_v4.1.chromosomes), *B. rapa* (Brapa_genome_sequence_v1.5), *B. oleracea* (B. oleracea var. capitate V1.0) and *L. sativa* (lettuce genome V8.1). [2] For Brassica species, only QTLs detected in more than one study, encompassing different mapping populations and/or varieties, are shown. For lettuce, only two populations have been used for QTL mapping. Flowering time genes with described allelic variation are highlighted in bold.

6. Perspectives for Breeding Strategies

Knowledge about conservation and divergence of *A. thaliana* flowering time with its related crop species, and with more distant leafy crops within the Asteraceae family, is of great value to select candidate genes for the improvement of flowering time in commercial varieties. Introducing genetic variation in those candidate genes can be achieved by identifying novel alleles from wild relatives, the production of mutant populations or, when allowed, via a transgenic or genome editing approach.

6.1. Environmental Changes

Breeders aim to produce commercial varieties that are more robust and predictable in flowering time to adapt to climate change and new environments.

In cauliflower, exploring genetic variation in temperature-dependent flowering time genes such as *SVP*, *FLM* and *FLC* [3,101,107] would help in adjusting the vernalization and temperature sensitivity of plants for a predictable curd formation.

In lettuce, exploring genetic variation in the floral integrator genes *FT* and *SOC1* will help to understand the mechanism of heat-induced early flowering and can therefore be used to produce better tasting lettuce when grown at high temperatures. Different studies have described that silencing of either *LsFT* or *LsSOC1* results in late flowering and heat insensitive lettuce plants [36,59]. *LsSOC1* expression was enhanced in both heat-treated wild type and *LsFT* silenced lines, indicating that *LsSOC1* can induce bolting independent of *LsFT* upon heat treatment [36]. Heat shock elements (HSE1 and HSE2) are detected in the promoter of *LsSOC1* and two heat shock proteins (LsHsfA1e and LsHsfA4c) bind to these elements to induce flowering [36]. Genetic variation at the heat-responsive promoter elements of *SOC1* might selectively affect heat sensitivity rather than flowering time in general.

6.2. Yield Increase

Prolonged vegetative phase can increase yield in leafy crops that are harvested before the transition to the reproductive phase.

In radish, premature bolting under LD conditions reduces yield and quality of the harvested product. Delayed bolting is described for *RsGI* loss-of-function mutants in *R. sativus* [154], while silencing of *BoGI* in *B. oleracea* also resulted in delayed post-harvest leaf senescence. Based on the phenotype of the *B. oleracea* silencing line, it is worthwhile to test the effect of genetic variation in *RsGI* as added value of delayed leaf senescence together with delayed bolting.

Cold season during growth induces early bolting and decreases yield in root chicory. A *C. intybus* homolog of the Arabidopsis *FLC*, *CiFL1*, was characterized and seems conserved in the vernalization response [22]. However, it remains to be demonstrated that the high expression level of *CiFL1* in non-vernalized chicory plants is the cause of the absolute vernalization requirement for flowering. This indicates that more research about the vernalization response of chicory is required to achieve late bolting plants when grown at low temperature.

6.3. Genetic Resources

In the past centuries, domestication has led to the creating of edible vegetables from their wild relatives. During this domestication process, plants are selected for specific desirable traits, thereby losing some of the genetic variation in the current germplasm. As a result, some variation in flowering time genes, producing crops that are adapting to specific environments, are not present in our current breeding material. Exploring phenotypic and genotypic differences in closely related (wild) species, and introducing desired traits back into breeding material, will help create new varieties that are adapted to climate change and produce higher yield.

The Brassicaceae family contains both annual, biennial and perennial species, and spring, semi-winter and winter type plants within a species, indicating that this family varies greatly in flowering time response [3,86]. Within the family, different family members are closely related and can be crossed through interspecific crosses, making it easier to introduce new genetics. Hybridization between *R. sativus* and Brassica species *B. napus*, *B. rapa* and *B. oleracea*, and between *D. tenuifolia* and *B. rapa* has been proven to be successful even though the number of successful hybridizations might be rather low [244]. Schiessl et al. [35] have described the amount of copies of 35 flowering time regulatory genes and their genetic variation between *B. rapa* and *B. oleracea*. This genetic information could be used as a basis to look for candidate genes to follow in an interspecific cross. As an example, Shea et al. [245] have developed late flowering *B. rapa* plants by replacing the *BrFLC2* genomic region with a 6.5 Mb region containing *BoFLC2* from *B. oleracea*. As many of the flowering pathways are conserved within the Brassicaceae family, it is worthwhile to explore introgression of flowering time genes from Brassica species into *R. sativus* or *D. tenuifolia* to alter flowering time.

Introducing genetic variation in flowering time genes from wild material into cultivated lettuce and chicory is possible [246,247], however, the flowering pathway is largely undiscovered in these species. Recently, high quality transcriptomes of both *C. intybus* and *C. endivia* were obtained by de novo assembly using RNA of several organs and Illumina HiSeq2000 technology [248,249], paving the way to the identification of flowering time transcripts in *Cichorium* spp. More research is required before specific candidate genes can be selected to introduce from wild material into breeding lines.

6.4. Speeding up Breeding

From a breeding perspective, introducing genetic variation from wild relatives or mutant populations into a new variety will take up to years. Speeding up this breeding process, using early flowering plants to grow more generations in one year would be of added value for the breeding companies. Similar to adapting plants to climate change or increasing yield, generating early flowering plants is possible by the use of genetic variation in flowering time genes. Besides exploring the genetics of wild material, it has also been shown for lettuce that screening mutant populations are a great source to discover plants with an altered flowering time. In Brassica spp., rapid-cycling lines and RIL populations have been obtained [86], which can be used to speed up breeding and for rapid analysis of QTL.

In Arabidopsis, winter type plants that require vernalization contain functional alleles for both *FLC* and *FRI*, while summer type plants lack a functional allele for either *FLC* or *FRI* [70–73]. With this system, early bolting parental lines can be created, while the F1 hybrids are late flowering. As an example, by producing a female line containing an *FLC* knock-out and a male line with FRI knock-out. The parents do not require vernalization to initiate flowering, as both parents lack a functional allele for either *FLC* or *FRI*. In the F1 hybrid, both genes are heterozygous, resulting in winter type plants that do require vernalization.

Different articles have shown that treatment with bioactive GAs can induce early flowering in both bolting sensitive and bolting resistance lettuce lines [61,124]. The benefit of GA application is that it will speed up the breeding process, when this is desired, but will not have a negative influence on flowering time during crop production.

7. Conclusions

Overall, most flowering time pathways seem to be genetically conserved between Brassicaceae and Asteraceae families, paving the way for exploitation of the fundamental knowledge acquired in the Brassica model species Arabidopsis to closely or more distantly related vegetable crops. This is highlighted in Figure 5, which represents a simplified model of the main regulatory genes shown to have a function in the various species within the Brassicaceae or Asteraceae family. However, a comprehensive comparison of the different flowering time pathways between Brassicaceae and Asteraceae is impaired by the poor knowledge available about molecular biology and gene function in *D. tenuifolia*, *L. sativa* and *C. intybus*. Fundamental biology studies in crop species to identify casual genes of advantageous traits is advisable to apply candidate gene approaches for successful breeding strategies. An increasing number of tools for molecular marker assisted breeding is expected to come in the near future from genomic and transcriptomic studies. With the rapid development of sequencing technology, whole genome sequences assembly and resequencing from crop plants is becoming routine, enabling genome-wide investigations into fundamental genetic pathways that underlie important agricultural traits. In addition, generating a pan-genome, capturing the genomic diversity of ecotypes, geographical isolates, and domesticated crop varieties, will make comparative approaches and association studies possible to identify the genetic components of adaptive and domestication traits. Increasing "omics" information (e.g., genomics, transcriptomics, metabolomics,

SNP-omics) will enable systems biology approaches to understand complex traits, such as flowering time, and identify hub/master gene regulators for the so-called "smart" or "precision breeding," which aims to develop new varieties more precisely and rapidly.

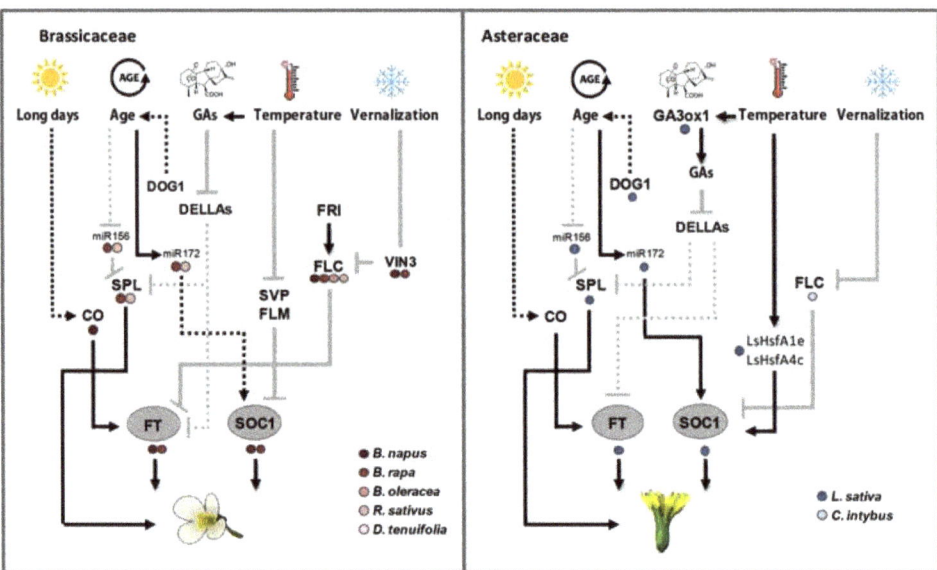

Figure 5. Simplified model of the main regulatory genes and flowering pathways acting in various crops within the Brassicaceae and Asteraceae family. Grey and black lines represent repression and induction, respectively, dotted lines indicate indirect regulation.

Author Contributions: Conceptualization, R.K., I.R. and G.F.; Investigation, W.L.; Writing-Original Draft Preparation, W.L.; Writing-Review & Editing, W.L., R.K., I.R. and G.F.; Visualization, W.L., I.R. and G.F.; Supervision, I.R. and G.F.

Funding: This research received no external funding.

Acknowledgments: The authors thank the lettuce and brassica breeding team of Enza Zaden Research & Development B.V. for providing information about crops; the manager of the Biotechnology Research Department, G.J. de Boer for his input in the breeding strategy session; S.A. Raccuia for providing the picture of the *B. napus* field in Figure 1; Rita Romani for providing the pictures of her own cultivated field of *B. rapa*, and delicious "broccoletti" product, in Figure 1; the Communication Department of Enza Zaden Marketing and Sales B.V. for providing the remaining pictures in Figure 1; Giulio Testone for helping to prepare Figure 3.

Conflicts of Interest: The authors declare no conflict of interest.

References

1. Blackman, B.K. Changing Responses to Changing Seasons: Natural Variation in the Plasticity of Flowering Time. *Plant Physiol.* **2017**, *173*, 16–26. [CrossRef] [PubMed]
2. Jung, C.; Muller, A.E. Flowering time control and applications in plant breeding. *Trends Plant Sci.* **2009**, *14*, 563–573. [CrossRef] [PubMed]
3. Huang, F.; Liu, T.; Hou, X. Isolation and Functional Characterization of a Floral Repressor, BcMAF1, from Pak-choi (Brassica rapa ssp. Chinensis). *Front. Plant Sci.* **2018**, *9*, 290. [CrossRef] [PubMed]
4. Schiessl, S.; Huettel, B.; Kuehn, D.; Reinhardt, R.; Snowdon, R.J. Targeted deep sequencing of flowering regulators in Brassica napus reveals extensive copy number variation. *Sci. Data* **2017**, *4*, 170013. [CrossRef] [PubMed]

5. Shi, J.; L, R.; Qiu, D.; Jiang, C.; Long, Y.; Morgan, C.; Bancroft, I.; Zhao, J.; Meng, J. Unraveling the complex trait of crop yield with quantitative trait loci mapping in Brassica napus. *Genetics* **2009**, *182*, 851–861. [CrossRef] [PubMed]
6. Bluemel, M.; Dally, N.; Jung, C. Flowering time regulation in crops—What did we learn from Arabidopsis? *Curr. Opin. Biotechnol.* **2015**, *32*, 121–129. [CrossRef] [PubMed]
7. Higuchi, Y. Florigen and anti-florigen: Flowering regulation in horticultural crops. *Breed. Sci.* **2018**, *68*, 109–118. [CrossRef] [PubMed]
8. Amasino, R. Seasonal and developmental timing of flowering. *Plant J.* **2010**, *61*, 1001–1013. [CrossRef] [PubMed]
9. Fornara, F.; de Montaigu, A.; Coupland, G. SnapShot: Control of flowering in Arabidopsis. *Cell* **2010**, *141*, 550-550.e2. [CrossRef] [PubMed]
10. Srikanth, A.; Schmid, M. Regulation of flowering time: All roads lead to Rome. *Cell. Mol. Life Sci.* **2011**, *68*, 2013–2037. [CrossRef] [PubMed]
11. Andrés, F.; Coupland, G. The genetic basis of flowering responses to seasonal cues. *Nat. Rev. Genet.* **2012**, *13*, 627–639. [CrossRef] [PubMed]
12. Koornneef, M.; Alonso-Blanco, C.; Vreugdenhil, D. Naturally occurring genetic variation in Arabidopsis thaliana. *Annu. Rev. Plant. Biol.* **2004**, *55*, 141–172. [CrossRef] [PubMed]
13. Gulden, R.H.; Warwick, S.I.; Thomas, A.G. The biology of Canadian weeds. 137. Brassica napus L. and B. rapa L. *Can. J. Plant Sci.* **2008**, *88*, 951–996. [CrossRef]
14. Wang, N.; Qian, W.; Suppanz, I.; Wei, L.; Mao, B.; Long, Y.; Meng, J.; Müller, A.E.; Jung, C. Flowering time variation in oilseed rape (Brassica napus L.) is associated with allelic variation in the FRIGIDA homologue BnaA. FRI. a. *J. Exp. Bot.* **2011**, *62*, 5641–5658. [CrossRef] [PubMed]
15. Wang, X.; Wang, H.; Wang, J.; Sun, R.; Wu, J.; Liu, S.; Bai, Y.; Mun, J.-H.; Bancroft, I.; Cheng, F. The genome of the mesopolyploid crop species Brassica rapa. *Nat. Genet.* **2011**, *43*, 1035–1039. [CrossRef] [PubMed]
16. Parkin, I.A.; Koh, C.; Tang, H.; Robinson, S.J.; Kagale, S.; Clarke, W.E.; Town, C.D.; Nixon, J.; Krishnakumar, V.; Bidwell, S.L. Transcriptome and methylome profiling reveals relics of genome dominance in the mesopolyploid Brassica oleracea. *Genome Boil.* **2014**, *15*, R77. [CrossRef] [PubMed]
17. Kitashiba, H.; Li, F.; Hirakawa, H.; Kawanabe, T.; Zou, Z.; Hasegawa, Y.; Tonosaki, K.; Shirasawa, S.; Fukushima, A.; Yokoi, S. Draft sequences of the radish (Raphanus sativus L.) genome. *DNA Res.* **2014**, *21*, 481–490. [CrossRef] [PubMed]
18. Liu, C.; Wang, S.; Xu, W.; Liu, X. Genome-wide transcriptome profiling of radish (Raphanus sativus L.) in response to vernalization. *PLoS ONE* **2017**, *12*, e0177594. [CrossRef] [PubMed]
19. Taylor, J.L.; Massiah, A.; Kennedy, S.; Hong, Y.; Jackson, S.D. FLC expression is down-regulated by cold treatment in Diplotaxis tenuifolia (wild rocket), but flowering time is unaffected. *J. Plant Physiol.* **2017**, *214*, 7–15. [CrossRef] [PubMed]
20. Reyes-Chin-Wo, S.; Wang, Z.; Yang, X.; Kozik, A.; Arikit, S.; Song, C.; Xia, L.; Froenicke, L.; Lavelle, D.O.; Truco, M.J.; et al. Genome assembly with in vitro proximity ligation data and whole-genome triplication in lettuce. *Nat. Commun.* **2017**, *8*, 14953. [CrossRef] [PubMed]
21. Gonthier, L.; Bellec, A.; Blassiau, C.; Prat, E.; Helmstetter, N.; Rambaud, C.; Huss, B.; Hendriks, T.; Bergès, H.; Quillet, M.-C. Construction and characterization of two BAC libraries representing a deep-coverage of the genome of chicory (Cichorium intybus L., Asteraceae). *BMC Res. Notes* **2010**, *3*, 225. [CrossRef] [PubMed]
22. Périlleux, C.; Pieltain, A.; Jacquemin, G.; Bouché, F.; Detry, N.; D'aloia, M.; Thiry, L.; Aljochim, P.; Delansnay, M.; Mathieu, A.S. A root chicory MADS box sequence and the Arabidopsis flowering repressor FLC share common features that suggest conserved function in vernalization and de-vernalization responses. *Plant J.* **2013**, *75*, 390–402. [CrossRef] [PubMed]
23. Kitamoto, N.; Nishikawa, K.; Tanimura, Y.; Urushibara, S.; Matsuura, T.; Yokoi, S.; Takahata, Y.; Yui, S. Development of late-bolting F1 hybrids of Chinese cabbage (Brassica rapa L.) allowing early spring cultivation without heating. *Euphytica* **2017**, *213*, 292. [CrossRef]
24. Rosen, A.; Hasan, Y.; Briggs, W.; Uptmoor, R. Genome-based prediction of time to curd induction in cauliflower. *Front. Plant Sci.* **2018**, *9*, 78. [CrossRef] [PubMed]
25. Kitamoto, N.; Yui, S.; Nishikawa, K.; Takahata, Y.; Yokoi, S. A naturally occurring long insertion in the first intron in the Brassica rapaFLC2 gene causes delayed bolting. *Euphytica* **2014**, *196*, 213–223. [CrossRef]

26. Simonne, A.; Simonne, E.; Eitenmiller, R.; Coker, C.H. Bitterness and composition of lettuce varieties grown in the southeastern United States. *HortTechnology* **2002**, *12*, 721–726. [CrossRef]
27. Street, R.A.; Sidana, J.; Prinsloo, G. Cichorium intybus: Traditional uses, phytochemistry, pharmacology, and toxicology. *Evid.-Based Complement. Altern. Med.* **2013**, *2013*, 579319. [CrossRef] [PubMed]
28. Renna, M.; Gonnella, M.; Giannino, D.; Santamaria, P. Quality evaluation of cook-chilled chicory stems (Cichorium intybus L., Catalogna group) by conventional and sous vide cooking methods. *J. Sci. Food Agric.* **2014**, *94*, 656–665. [CrossRef] [PubMed]
29. Barcaccia, G.; Ghedina, A.; Lucchin, M. Current advances in genomics and breeding of leaf Chicory (Cichorium intybus L.). *Agriculture* **2016**, *6*, 50. [CrossRef]
30. Kwiatkowska, D. Flowering and apical meristem growth dynamics. *J. Exp. Bot.* **2008**, *59*, 187–201. [CrossRef] [PubMed]
31. Yamaguchi, A.; Kobayashi, Y.; Goto, K.; Abe, M.; Araki, T. TWIN SISTER OF FT (TSF) acts as a floral pathway integrator redundantly with FT. *Plant Cell Physiol.* **2005**, *46*, 1175–1189. [CrossRef] [PubMed]
32. Ahn, J.H.; Miller, D.; Winter, V.J.; Banfield, M.J.; Lee, J.H.; Yoo, S.Y.; Henz, S.R.; Brady, R.L.; Weigel, D. A divergent external loop confers antagonistic activity on floral regulators FT and TFL1. *EMBO J.* **2006**, *25*, 605–614. [CrossRef] [PubMed]
33. Flowering Time Pathway (Arabidopsis thaliana). Available online: https://www.wikipathways.org/index.php/Pathway:WP2312 (accessed on 19 September 2018).
34. Wickland, D.P.; Hanzawa, Y. The FLOWERING LOCUS T/TERMINAL FLOWER 1 gene family: Functional evolution and molecular mechanisms. *Mol. Plant* **2015**, *8*, 983–997. [CrossRef] [PubMed]
35. Schiessl, S.V.; Huettel, B.; Kuehn, D.; Reinhardt, R.; Snowdon, R.J. Flowering time gene variation in Brassica species shows evolutionary principles. *Front. Plant Sci.* **2017**, *8*, 1742. [CrossRef] [PubMed]
36. Chen, Z.; Zhao, W.; Ge, D.; Han, Y.; Ning, K.; Luo, C.; Wang, S.; Liu, R.; Zhang, X.; Wang, Q. LCM-seq reveals the crucial role of Ls SOC 1 in heat-promoted bolting of lettuce (Lactuca sativa L.). *Plant J.* **2018**. [CrossRef] [PubMed]
37. Liu, L.; Liu, C.; Hou, X.; Xi, W.; Shen, L.; Tao, Z.; Wang, Y.; Yu, H. FTIP1 is an essential regulator required for florigen transport. *PLoS Boil.* **2012**, *10*, e1001313. [CrossRef] [PubMed]
38. Abe, M.; Kobayashi, Y.; Yamamoto, S.; Daimon, Y.; Yamaguchi, A.; Ikeda, Y.; Ichinoki, H.; Notaguchi, M.; Goto, K.; Araki, T. FD, a bZIP protein mediating signals from the floral pathway integrator FT at the shoot apex. *Science* **2005**, *309*, 1052–1056. [CrossRef] [PubMed]
39. Suárez-López, P.; Wheatley, K.; Robson, F.; Onouchi, H.; Valverde, F.; Coupland, G. CONSTANS mediates between the circadian clock and the control of flowering in Arabidopsis. *Nature* **2001**, *410*, 1116–1120. [CrossRef] [PubMed]
40. Deng, W.; Ying, H.; Helliwell, C.A.; Taylor, J.M.; Peacock, W.J.; Dennis, E.S. FLOWERING LOCUS C (FLC) regulates development pathways throughout the life cycle of Arabidopsis. *Proc. Natl. Acad. Sci. USA* **2011**, *108*, 6680–6685. [CrossRef] [PubMed]
41. Mateos, J.L.; Madrigal, P.; Tsuda, K.; Rawat, V.; Richter, R.; Romera-Branchat, M.; Fornara, F.; Schneeberger, K.; Krajewski, P.; Coupland, G. Combinatorial activities of SHORT VEGETATIVE PHASE and FLOWERING LOCUS C define distinct modes of flowering regulation in Arabidopsis. *Genome Boil.* **2015**, *16*, 31. [CrossRef] [PubMed]
42. Taoka, K.-i.; Ohki, I.; Tsuji, H.; Furuita, K.; Hayashi, K.; Yanase, T.; Yamaguchi, M.; Nakashima, C.; Purwestri, Y.A.; Tamaki, S. 14-3-3 proteins act as intracellular receptors for rice Hd3a florigen. *Nature* **2011**, *476*, 332–335. [CrossRef] [PubMed]
43. Liu, C.; Chen, H.; Er, H.L.; Soo, H.M.; Kumar, P.P.; Han, J.-H.; Liou, Y.C.; Yu, H. Direct interaction of AGL24 and SOC1 integrates flowering signals in Arabidopsis. *Development* **2008**, *135*, 1481–1491. [CrossRef] [PubMed]
44. Hanzawa, Y.; Money, T.; Bradley, D. A single amino acid converts a repressor to an activator of flowering. *Proc. Natl. Acad. Sci. USA* **2005**, *102*, 7748–7753. [CrossRef] [PubMed]
45. Conti, L.; Bradley, D. TERMINAL FLOWER1 is a mobile signal controlling Arabidopsis architecture. *Plant Cell* **2007**, *19*, 767–778. [CrossRef] [PubMed]
46. Hanano, S.; Goto, K. Arabidopsis TERMINAL FLOWER1 is involved in the regulation of flowering time and inflorescence development through transcriptional repression. *Plant Cell* **2011**, *23*, 3172–3184. [CrossRef] [PubMed]

47. Wang, J.; Hopkins, C.J.; Hou, J.; Zou, X.; Wang, C.; Long, Y.; Kurup, S.; King, G.J.; Meng, J. Promoter variation and transcript divergence in Brassicaceae lineages of FLOWERING LOCUS T. *PLoS ONE* **2012**, *7*, e47127. [CrossRef] [PubMed]
48. Wang, J.; Long, Y.; Wu, B.; Liu, J.; Jiang, C.; Shi, L.; Zhao, J.; King, G.J.; Meng, J. The evolution of Brassica napus FLOWERING LOCUST paralogues in the context of inverted chromosomal duplication blocks. *BMC Evol. Boil.* **2009**, *9*, 271. [CrossRef] [PubMed]
49. Zhang, X.; Meng, L.; Liu, B.; Hu, Y.; Cheng, F.; Liang, J.; Aarts, M.G.; Wang, X.; Wu, J. A transposon insertion in FLOWERING LOCUS T is associated with delayed flowering in Brassica rapa. *Plant Sci.* **2015**, *241*, 211–220. [CrossRef] [PubMed]
50. Raman, H.; Raman, R.; Coombes, N.; Song, J.; Prangnell, R.; Bandaranayake, C.; Tahira, R.; Sundaramoorthi, V.; Killian, A.; Meng, J. Genome-wide association analyses reveal complex genetic architecture underlying natural variation for flowering time in canola. *Plant Cell Environ.* **2016**, *39*, 1228–1239. [CrossRef] [PubMed]
51. Guo, Y.; Hans, H.; Christian, J.; Molina, C. Mutations in single FT-and TFL1-paralogs of rapeseed (Brassica napus L.) and their impact on flowering time and yield components. *Front. Plant Sci.* **2014**, *5*, 282. [CrossRef] [PubMed]
52. Shah, S.; Weinholdt, C.; Jedrusik, N.; Molina, C.; Zou, J.; Große, I.; Schiessl, S.; Jung, C.; Emrani, N. Whole transcriptome analysis reveals genetic factors underlying flowering time regulation in rapeseed (Brassica napus L.). *Plant Cell Environ.* **2018**. [CrossRef] [PubMed]
53. Li, B.; Zhao, W.; Li, D.; Chao, H.; Zhao, X.; Ta, N.; Li, Y.; Guan, Z.; Guo, L.; Zhang, L. Genetic dissection of the mechanism of flowering time based on an environmentally stable and specific QTL in Brassica napus. *Plant Sci.* **2018**, *277*, 296–310. [CrossRef] [PubMed]
54. Song, X.; Duan, W.; Huang, Z.; Liu, G.; Wu, P.; Liu, T.; Li, Y.; Hou, X. Comprehensive analysis of the flowering genes in Chinese cabbage and examination of evolutionary pattern of CO-like genes in plant kingdom. *Sci. Rep.* **2015**, *5*, 14631. [CrossRef] [PubMed]
55. Hong, J.K.; Kim, S.-Y.; Kim, K.-S.; Kwon, S.-J.; Kim, J.S.; Kim, J.A.; Lee, S.I.; Lee, Y.-H. Overexpression of a Brassica rapa MADS-box gene, BrAGL20, induces early flowering time phenotypes in Brassica napus. *Plant. Biotechnol. Rep.* **2013**, *7*, 231–237. [CrossRef]
56. Franks, S.J.; Perez-Sweeney, B.; Strahl, M.; Nowogrodzki, A.; Weber, J.J.; Lalchan, R.; Jordan, K.P.; Litt, A. Variation in the flowering time orthologs BrFLC and BrSOC1 in a natural population of Brassica rapa. *PeerJ* **2015**, *3*, e1339. [CrossRef] [PubMed]
57. Cavaiuolo, M.; Cocetta, G.; Spadafora, N.D.; Müller, C.T.; Rogers, H.J.; Ferrante, A. Gene expression analysis of rocket salad under pre-harvest and postharvest stresses: A transcriptomic resource for Diplotaxis tenuifolia. *PLoS ONE* **2017**, *12*, e0178119. [CrossRef] [PubMed]
58. Fukuda, M.; Matsuo, S.; Kikuchi, K.; Kawazu, Y.; Fujiyama, R.; Honda, I. Isolation and functional characterization of the FLOWERING LOCUS T homolog, the LsFT gene, in lettuce. *J. Plant Physiol.* **2011**, *168*, 1602–1607. [CrossRef] [PubMed]
59. Chen, Z.; Han, Y.; Ning, K.; Ding, Y.; Zhao, W.; Yan, S.; Luo, C.; Jiang, X.; Wang, Q.; Zhang, X. Inflorescence development and the role of LsFT in regulating bolting in lettuce (Lactuca sativa L.). *Front. Plant Sci.* **2017**, *8*, 2248. [CrossRef] [PubMed]
60. Fukuda, M.; Yanai, Y.; Nakano, Y.; Sasaki, H.; Uragami, A.; Okada, K. Isolation and Gene Expression Analysis of Flowering-related Genes in Lettuce (Lactuca sativa L.). *Hortic. J.* **2017**, *86*, 340–348. [CrossRef]
61. Han, Y.; Chen, Z.; Lv, S.; Ning, K.; Ji, X.; Liu, X.; Wang, Q.; Liu, R.; Fan, S.; Zhang, X. MADS-box genes and gibberellins regulate bolting in Lettuce (Lactuca sativa L.). *Front. Plant Sci.* **2016**, *7*, 1889. [CrossRef] [PubMed]
62. Chouard, P. Vernalization and its relations to dormancy. *Annu. Rev. Plant Physiol.* **1960**, *11*, 191–238. [CrossRef]
63. Albani, M.C.; Coupland, G. Comparative analysis of flowering in annual and perennial plants. In *Current Topics in Developmental Biology*; Elsevier: Amsterdam, The Netherlands, 2010; Volume 91, pp. 323–348.
64. Hepworth, J.; Antoniou-Kourounioti, R.L.; Bloomer, R.H.; Selga, C.; Berggren, K.; Cox, D.; Collier Harris, B.R.; Irwin, J.A.; Holm, S.; Sall, T.; et al. Absence of warmth permits epigenetic memory of winter in Arabidopsis. *Nat. Commun* **2018**, *9*, 639. [CrossRef] [PubMed]

65. Alonso-Blanco, C.; Andrade, J.; Becker, C.; Bemm, F.; Bergelson, J.; Borgwardt, K.M.; Cao, J.; Chae, E.; Dezwaan, T.M.; Ding, W. 1,135 genomes reveal the global pattern of polymorphism in Arabidopsis thaliana. *Cell* **2016**, *166*, 481–491. [CrossRef] [PubMed]
66. Zou, X.; Suppanz, I.; Raman, H.; Hou, J.; Wang, J.; Long, Y.; Jung, C.; Meng, J. Comparative analysis of FLC homologues in Brassicaceae provides insight into their role in the evolution of oilseed rape. *PLoS ONE* **2012**, *7*, e45751. [CrossRef] [PubMed]
67. Schranz, M.E.; Quijada, P.; Sung, S.-B.; Lukens, L.; Amasino, R.; Osborn, T.C. Characterization and effects of the replicated flowering time gene FLC in Brassica rapa. *Genetics* **2002**, *162*, 1457–1468. [PubMed]
68. Razi, H.; Howell, E.C.; Newbury, H.J.; Kearsey, M.J. Does sequence polymorphism of FLC paralogues underlie flowering time QTL in Brassica oleracea? *Theor. Appl. Genet.* **2008**, *116*, 179–192. [CrossRef] [PubMed]
69. Kiefer, C.; Severing, E.; Karl, R.; Bergonzi, S.; Koch, M.; Tresch, A.; Coupland, G. Divergence of annual and perennial species in the Brassicaceae and the contribution of cis-acting variation at FLC orthologues. *Mol. Ecol.* **2017**, *26*, 3437–3457. [CrossRef] [PubMed]
70. Michaels, S.D.; Amasino, R.M. FLOWERING LOCUS C encodes a novel MADS domain protein that acts as a repressor of flowering. *Plant Cell* **1999**, *11*, 949–956. [CrossRef] [PubMed]
71. Sheldon, C.C.; Burn, J.E.; Perez, P.P.; Metzger, J.; Edwards, J.A.; Peacock, W.J.; Dennis, E.S. The FLF MADS box gene: A repressor of flowering in Arabidopsis regulated by vernalization and methylation. *Plant Cell* **1999**, *11*, 445–458. [CrossRef] [PubMed]
72. Johanson, U.; West, J.; Lister, C.; Michaels, S.; Amasino, R.; Dean, C. Molecular analysis of FRIGIDA, a major determinant of natural variation in Arabidopsis flowering time. *Science* **2000**, *290*, 344–347. [CrossRef] [PubMed]
73. Werner, J.D.; Borevitz, J.O.; Uhlenhaut, N.H.; Ecker, J.R.; Chory, J.; Weigel, D. FRIGIDA-independent variation in flowering time of natural Arabidopsis thaliana accessions. *Genetics* **2005**, *170*, 1197–1207. [CrossRef] [PubMed]
74. Liu, F.; Marquardt, S.; Lister, C.; Swiezewski, S.; Dean, C. Targeted 3′ processing of antisense transcripts triggers Arabidopsis FLC chromatin silencing. *Science* **2010**, *327*, 94–97. [CrossRef] [PubMed]
75. Yang, H.; Howard, M.; Dean, C. Antagonistic roles for H3K36me3 and H3K27me3 in the cold-induced epigenetic switch at Arabidopsis FLC. *Curr. Boil.* **2014**, *24*, 1793–1797. [CrossRef] [PubMed]
76. Csorba, T.; Questa, J.I.; Sun, Q.; Dean, C. Antisense COOLAIR mediates the coordinated switching of chromatin states at FLC during vernalization. *Proc. Natl. Acad. Sci. USA* **2014**, *111*, 16160–16165. [CrossRef] [PubMed]
77. Heo, J.B.; Sung, S. Vernalization-mediated epigenetic silencing by a long intronic noncoding RNA. *Science* **2011**, *331*, 76–79. [CrossRef] [PubMed]
78. Kim, D.-H.; Sung, S. Vernalization-triggered intragenic chromatin loop formation by long noncoding RNAs. *Dev. Cell* **2017**, *40*, 302.e4–312.e4. [CrossRef] [PubMed]
79. Kim, D.-H.; Xi, Y.; Sung, S. Modular function of long noncoding RNA, COLDAIR, in the vernalization response. *PLoS Genet.* **2017**, *13*, e1006939. [CrossRef] [PubMed]
80. Bastow, R.; Mylne, J.S.; Lister, C.; Lippman, Z.; Martienssen, R.A.; Dean, C. Vernalization requires epigenetic silencing of FLC by histone methylation. *Nature* **2004**, *427*, 164–167. [CrossRef] [PubMed]
81. Sung, S.; Amasino, R.M. Vernalization in Arabidopsis thaliana is mediated by the PHD finger protein VIN3. *Nature* **2004**, *427*, 159. [CrossRef] [PubMed]
82. Ausín, I.; Alonso-Blanco, C.; Jarillo, J.A.; Ruiz-García, L.; Martínez-Zapater, J.M. Regulation of flowering time by FVE, a retinoblastoma-associated protein. *Nat. Genet.* **2004**, *36*, 162–166. [CrossRef] [PubMed]
83. Gendall, A.R.; Levy, Y.Y.; Wilson, A.; Dean, C. The VERNALIZATION 2 gene mediates the epigenetic regulation of vernalization in Arabidopsis. *Cell* **2001**, *107*, 525–535. [CrossRef]
84. Levy, Y.Y.; Mesnage, S.; Mylne, J.S.; Gendall, A.R.; Dean, C. Multiple roles of Arabidopsis VRN1 in vernalization and flowering time control. *Science* **2002**, *297*, 243–246. [CrossRef] [PubMed]
85. Gu, X.; Le, C.; Wang, Y.; Li, Z.; Jiang, D.; Wang, Y.; He, Y. Arabidopsis FLC clade members form flowering-repressor complexes coordinating responses to endogenous and environmental cues. *Nat. Commun.* **2013**, *4*, 1947. [CrossRef] [PubMed]
86. Williams, P.H.; Hill, C.B. Rapid-cycling populations of Brassica. *Science* **1986**, *232*, 1385–1389. [CrossRef] [PubMed]

87. Okazaki, K.; Sakamoto, K.; Kikuchi, R.; Saito, A.; Togashi, E.; Kuginuki, Y.; Matsumoto, S.; Hirai, M. Mapping and characterization of FLC homologs and QTL analysis of flowering time in Brassica oleracea. *Theor. Appl. Genet.* **2007**, *114*, 595–608. [CrossRef] [PubMed]
88. Al-Shahrour, F.; Minguez, P.; Marqués-Bonet, T.; Gazave, E.; Navarro, A.; Dopazo, J. Selection upon genome architecture: Conservation of functional neighborhoods with changing genes. *PLoS Comput. Boil.* **2010**, *6*, e1000953. [CrossRef] [PubMed]
89. Lee, J.M.; Sonnhammer, E.L. Genomic gene clustering analysis of pathways in eukaryotes. *Genome Res.* **2003**, *13*, 875–882. [CrossRef] [PubMed]
90. Irwin, J.A.; Lister, C.; Soumpourou, E.; Zhang, Y.; Howell, E.C.; Teakle, G.; Dean, C. Functional alleles of the flowering time regulator FRIGIDA in the Brassica oleracea genome. *BMC Plant Boil.* **2012**, *12*, 21. [CrossRef] [PubMed]
91. Wu, J.; Wei, K.; Cheng, F.; Li, S.; Wang, Q.; Zhao, J.; Bonnema, G.; Wang, X. A naturally occurring InDel variation in BraA. FLC. b (BrFLC2) associated with flowering time variation in Brassica rapa. *BMC Plant Boil.* **2012**, *12*, 151.
92. Xiao, D.; Zhao, J.J.; Hou, X.L.; Basnet, R.K.; Carpio, D.P.; Zhang, N.W.; Bucher, J.; Lin, K.; Cheng, F.; Wang, X.W. The Brassica rapa FLC homologue FLC2 is a key regulator of flowering time, identified through transcriptional co-expression networks. *J. Exp. Bot.* **2013**, *64*, 4503–4516. [CrossRef] [PubMed]
93. Yuan, Y.-X.; Wu, J.; Sun, R.-F.; Zhang, X.-W.; Xu, D.-H.; Bonnema, G.; Wang, X.-W. A naturally occurring splicing site mutation in the Brassica rapa FLC1 gene is associated with variation in flowering time. *J. Exp. Bot.* **2009**, *60*, 1299–1308. [CrossRef] [PubMed]
94. Zhao, J.; Kulkarni, V.; Liu, N.; Pino Del Carpio, D.; Bucher, J.; Bonnema, G. BrFLC2 (FLOWERING LOCUS C) as a candidate gene for a vernalization response QTL in Brassica rapa. *J. Exp. Bot.* **2010**, *61*, 1817–1825. [CrossRef] [PubMed]
95. Kawanabe, T.; Osabe, K.; Itabashi, E.; Okazaki, K.; Dennis, E.S.; Fujimoto, R. Development of primer sets that can verify the enrichment of histone modifications, and their application to examining vernalization-mediated chromatin changes in Brassica rapa L. *Genes Genet. Syst.* **2016**, *91*, 1–10. [CrossRef] [PubMed]
96. Xi, X.; Wei, K.; Gao, B.; Liu, J.; Liang, J.; Cheng, F.; Wang, X.; Wu, J. BrFLC5: A weak regulator of flowering time in Brassica rapa. *Theor. Appl. Genet.* **2018**, 1–10. [CrossRef] [PubMed]
97. Huang, F.; Liu, T.; Wang, J.; Hou, X. Isolation and functional characterization of a floral repressor, BcFLC2, from Pak-choi (Brassica rapa ssp. chinensis). *Planta* **2018**, *248*, 423–435. [CrossRef] [PubMed]
98. Li, X.; Zhang, S.; Bai, J.; He, Y. Tuning growth cycles of Brassica crops via natural antisense transcripts of Br FLC. *Plant. Biotechnol. J.* **2016**, *14*, 905–914. [CrossRef] [PubMed]
99. Duan, W.; Zhang, H.; Zhang, B.; Wu, X.; Shao, S.; Li, Y.; Hou, X.; Liu, T. Role of vernalization-mediated demethylation in the floral transition of Brassica rapa. *Planta* **2017**, *245*, 227–233. [CrossRef] [PubMed]
100. Matschegewski, C.; Zetzsche, H.; Hasan, Y.; Leibeguth, L.; Briggs, W.; Ordon, F.; Uptmoor, R. Genetic variation of temperature-regulated curd induction in cauliflower: Elucidation of floral transition by genome-wide association mapping and gene expression analysis. *Front. Plant Sci.* **2015**, *6*, 720. [CrossRef] [PubMed]
101. Irwin, J.A.; Soumpourou, E.; Lister, C.; Ligthart, J.D.; Kennedy, S.; Dean, C. Nucleotide polymorphism affecting FLC expression underpins heading date variation in horticultural brassicas. *Plant J.* **2016**, *87*, 597–605. [CrossRef] [PubMed]
102. Fadina, O.; Pankin, A.; Khavkin, E. Molecular characterization of the flowering time gene FRIGIDA in Brassica genomes A and C. *Russ. J. Plant Physiol.* **2013**, *60*, 279–289. [CrossRef]
103. Yi, G.; Park, H.; Kim, J.-S.; Chae, W.B.; Park, S.; Huh, J.H. Identification of three FLOWERING LOCUS C genes responsible for vernalization response in radish (Raphanus sativus L.). *Hortic. Environ. Biotechnol.* **2014**, *55*, 548–556. [CrossRef]
104. Jung, W.Y.; Park, H.J.; Lee, A.; Lee, S.S.; Kim, Y.-S.; Cho, H.S. Identification of flowering-related genes responsible for differences in bolting time between two radish inbred lines. *Front. Plant Sci.* **2016**, *7*, 1844. [CrossRef] [PubMed]
105. Li, C.; Wang, Y.; Xu, L.; Nie, S.; Chen, Y.; Liang, D.; Sun, X.; Karanja, B.K.; Luo, X.; Liu, L. Genome-wide characterization of the MADS-Box gene family in Radish (Raphanus sativus L.) and assessment of its roles in flowering and floral organogenesis. *Front. Plant Sci.* **2016**, *7*, 1390. [CrossRef] [PubMed]

106. Taylor, J.L. Delayed Bolting in Rocket for Improved Quality and GREATER Sustainability. Ph.D. Thesis, University of Warwick, Coventry, UK, 2015.
107. Lee, J.H.; Park, S.H.; Lee, J.S.; Ahn, J.H. A conserved role of SHORT VEGETATIVE PHASE (SVP) in controlling flowering time of Brassica plants. *Biochim. Biophys. Acta (BBA)-Gene Struct. Expr.* **2007**, *1769*, 455–461. [CrossRef] [PubMed]
108. Lutz, U.; Nussbaumer, T.; Spannagl, M.; Diener, J.; Mayer, K.F.; Schwechheimer, C. Natural haplotypes of FLM non-coding sequences fine-tune flowering time in ambient spring temperatures in Arabidopsis. *eLife* **2017**, *6*, e22114. [CrossRef] [PubMed]
109. Balasubramanian, S.; Sureshkumar, S.; Lempe, J.; Weigel, D. Potent induction of Arabidopsis thaliana flowering by elevated growth temperature. *PLoS Genet.* **2006**, *2*, e106. [CrossRef] [PubMed]
110. Balasubramanian, S.; Weigel, D. Temperature induced flowering in Arabidopsis thaliana. *Plant Signal. Behav.* **2006**, *1*, 227–228. [CrossRef] [PubMed]
111. Capovilla, G.; Schmid, M.; Pose, D. Control of flowering by ambient temperature. *J. Exp. Bot* **2015**, *66*, 59–69. [CrossRef] [PubMed]
112. Kumar, S.V.; Wigge, P.A. H2A. Z-containing nucleosomes mediate the thermosensory response in Arabidopsis. *Cell* **2010**, *140*, 136–147. [CrossRef] [PubMed]
113. Kumar, S.V.; Lucyshyn, D.; Jaeger, K.E.; Alós, E.; Alvey, E.; Harberd, N.P.; Wigge, P.A. Transcription factor PIF4 controls the thermosensory activation of flowering. *Nature* **2012**, *484*, 242–245. [CrossRef] [PubMed]
114. Thines, B.C.; Youn, Y.; Duarte, M.I.; Harmon, F.G. The time of day effects of warm temperature on flowering time involve PIF4 and PIF5. *J. Exp. Bot.* **2014**, *65*, 1141–1151. [CrossRef] [PubMed]
115. Fernández, V.; Takahashi, Y.; Le Gourrierec, J.; Coupland, G. Photoperiodic and thermosensory pathways interact through CONSTANS to promote flowering at high temperature under short days. *Plant J.* **2016**, *86*, 426–440. [CrossRef] [PubMed]
116. Capovilla, G.; Pajoro, A.; Immink, R.G.; Schmid, M. Role of alternative pre-mRNA splicing in temperature signaling. *Curr. Opin. Plant Boil.* **2015**, *27*, 97–103. [CrossRef] [PubMed]
117. Li, D.; Liu, C.; Shen, L.; Wu, Y.; Chen, H.; Robertson, M.; Helliwell, C.A.; Ito, T.; Meyerowitz, E.; Yu, H. A repressor complex governs the integration of flowering signals in Arabidopsis. *Dev. Cell* **2008**, *15*, 110–120. [CrossRef] [PubMed]
118. Lee, J.H.; Ryu, H.-S.; Chung, K.S.; Posé, D.; Kim, S.; Schmid, M.; Ahn, J.H. Regulation of temperature-responsive flowering by MADS-box transcription factor repressors. *Science* **2013**, *342*, 628–632. [CrossRef] [PubMed]
119. Jiao, Y.; Meyerowitz, E.M. Cell-type specific analysis of translating RNAs in developing flowers reveals new levels of control. *Mol. Syst. Boil.* **2010**, *6*, 419. [CrossRef] [PubMed]
120. Posé, D.; Verhage, L.; Ott, F.; Yant, L.; Mathieu, J.; Angenent, G.C.; Immink, R.G.; Schmid, M. Temperature-dependent regulation of flowering by antagonistic FLM variants. *Nature* **2013**, *503*, 414–417. [CrossRef] [PubMed]
121. Verhage, L.; Severing, E.I.; Bucher, J.; Lammers, M.; Busscher-Lange, J.; Bonnema, G.; Rodenburg, N.; Proveniers, M.C.; Angenent, G.C.; Immink, R.G. Splicing-related genes are alternatively spliced upon changes in ambient temperatures in plants. *PLoS ONE* **2017**, *12*, e0172950. [CrossRef] [PubMed]
122. Capovilla, G.; Symeonidi, E.; Wu, R.; Schmid, M. Contribution of major FLM isoforms to temperature-dependent flowering in Arabidopsis thaliana. *J. Exp. Bot.* **2017**, *68*, 5117–5127. [CrossRef] [PubMed]
123. Méndez-Vigo, B.; Martínez-Zapater, J.M.; Alonso-Blanco, C. The flowering repressor SVP underlies a novel Arabidopsis thaliana QTL interacting with the genetic background. *PLoS Genet.* **2013**, *9*, e1003289. [CrossRef] [PubMed]
124. Yi, L.; Chen, C.; Yin, S.; Li, H.; Li, Z.; Wang, B.; King, G.J.; Wang, J.; Liu, K. Sequence variation and functional analysis of a FRIGIDA orthologue (BnaA3. FRI) in Brassica napus. *BMC Plant Boil.* **2018**, *18*, 32. [CrossRef] [PubMed]
125. Mathieu, A.-S.; Lutts, S.; Vandoorne, B.; Descamps, C.; Périlleux, C.; Dielen, V.; Van Herck, J.-C.; Quinet, M. High temperatures limit plant growth but hasten flowering in root chicory (Cichorium intybus) independently of vernalisation. *J. Plant Physiol.* **2014**, *171*, 109–118. [CrossRef] [PubMed]
126. Garner, W.W.; Allard, H.A. Effect of the relative length of day and night and other factors of the environment on growth and reproduction in plants. *Mon. Weather. Rev.* **1920**, *48*, 415. [CrossRef]

127. Hernando, C.E.; Romanowski, A.; Yanovsky, M.J. Transcriptional and post-transcriptional control of the plant circadian gene regulatory network. *Biochim. Biophys. Acta (BBA)-Gene Regul. Mech.* **2017**, *1860*, 84–94. [CrossRef] [PubMed]
128. Valverde, F.; Mouradov, A.; Soppe, W.; Ravenscroft, D.; Samach, A.; Coupland, G. Photoreceptor regulation of CONSTANS protein in photoperiodic flowering. *Science* **2004**, *303*, 1003–1006. [CrossRef] [PubMed]
129. Song, Y.H.; Ito, S.; Imaizumi, T. Similarities in the circadian clock and photoperiodism in plants. *Curr. Opin. Plant Biol.* **2010**, *13*, 594–603. [CrossRef] [PubMed]
130. Higashi, T.; Aoki, K.; Nagano, A.J.; Honjo, M.N.; Fukuda, H. Circadian Oscillation of the Lettuce Transcriptome under Constant Light and Light–Dark Conditions. *Front. Plant Sci.* **2016**, *7*, 1114. [CrossRef] [PubMed]
131. Fornara, F.; Panigrahi, K.C.; Gissot, L.; Sauerbrunn, N.; Rühl, M.; Jarillo, J.A.; Coupland, G. Arabidopsis DOF transcription factors act redundantly to reduce CONSTANS expression and are essential for a photoperiodic flowering response. *Dev. Cell* **2009**, *17*, 75–86. [CrossRef] [PubMed]
132. Imaizumi, T.; Schultz, T.F.; Harmon, F.G.; Ho, L.A.; Kay, S.A. FKF1 F-box protein mediates cyclic degradation of a repressor of CONSTANS in Arabidopsis. *Science* **2005**, *309*, 293–297. [CrossRef] [PubMed]
133. Song, Y.H.; Smith, R.W.; To, B.J.; Millar, A.J.; Imaizumi, T. FKF1 conveys timing information for CONSTANS stabilization in photoperiodic flowering. *Science* **2012**, *336*, 1045–1049. [CrossRef] [PubMed]
134. Sawa, M.; Nusinow, D.A.; Kay, S.A.; Imaizumi, T. FKF1 and GIGANTEA complex formation is required for day-length measurement in Arabidopsis. *Science* **2007**, *318*, 261–265. [CrossRef] [PubMed]
135. Song, Y.H.; Estrada, D.A.; Johnson, R.S.; Kim, S.K.; Lee, S.Y.; MacCoss, M.J.; Imaizumi, T. Distinct roles of FKF1, GIGANTEA, and ZEITLUPE proteins in the regulation of CONSTANS stability in Arabidopsis photoperiodic flowering. *Proc. Natl. Acad. Sci. USA* **2014**, *111*, 17672–17677. [CrossRef] [PubMed]
136. Jang, S.; Marchal, V.; Panigrahi, K.C.; Wenkel, S.; Soppe, W.; Deng, X.W.; Valverde, F.; Coupland, G. Arabidopsis COP1 shapes the temporal pattern of CO accumulation conferring a photoperiodic flowering response. *EMBO J.* **2008**, *27*, 1277–1288. [CrossRef] [PubMed]
137. Laubinger, S.; Marchal, V.; Gentilhomme, J.; Wenkel, S.; Adrian, J.; Jang, S.; Kulajta, C.; Braun, H.; Coupland, G.; Hoecker, U. Arabidopsis SPA proteins regulate photoperiodic flowering and interact with the floral inducer CONSTANS to regulate its stability. *Development* **2006**, *133*, 3213–3222. [CrossRef] [PubMed]
138. Liu, L.-J.; Zhang, Y.-C.; Li, Q.-H.; Sang, Y.; Mao, J.; Lian, H.-L.; Wang, L.; Yang, H.-Q. COP1-mediated ubiquitination of CONSTANS is implicated in cryptochrome regulation of flowering in Arabidopsis. *Plant Cell* **2008**, *20*, 292–306. [CrossRef] [PubMed]
139. Covington, M.F.; Maloof, J.N.; Straume, M.; Kay, S.A.; Harmer, S.L. Global transcriptome analysis reveals circadian regulation of key pathways in plant growth and development. *Genome Boil.* **2008**, *9*, R130. [CrossRef] [PubMed]
140. Nohales, M.A.; Kay, S.A. Molecular mechanisms at the core of the plant circadian oscillator. *Nat. Struct. Mol. Boil.* **2016**, *23*, 1061–1069. [CrossRef] [PubMed]
141. Lu, S.X.; Knowles, S.M.; Andronis, C.; Ong, M.S.; Tobin, E.M. CIRCADIAN CLOCK ASSOCIATED1 and LATE ELONGATED HYPOCOTYL function synergistically in the circadian clock of Arabidopsis. *Plant Physiol.* **2009**, *150*, 834–843. [CrossRef] [PubMed]
142. Huang, W.; Pérez-García, P.; Pokhilko, A.; Millar, A.; Antoshechkin, I.; Riechmann, J.L.; Mas, P. Mapping the core of the Arabidopsis circadian clock defines the network structure of the oscillator. *Science* **2012**, *336*, 75–79. [CrossRef] [PubMed]
143. Kamioka, M.; Takao, S.; Suzuki, T.; Taki, K.; Higashiyama, T.; Kinoshita, T.; Nakamichi, N. Direct repression of evening genes by CIRCADIAN CLOCK-ASSOCIATED 1 in Arabidopsis circadian clock. *Plant Cell* **2016**, *28*, 696–711. [CrossRef] [PubMed]
144. Lu, S.X.; Webb, C.J.; Knowles, S.M.; Kim, S.H.; Wang, Z.-Y.; Tobin, E.M. CCA1 and ELF3 Interact in the control of hypocotyl length and flowering time in Arabidopsis. *Plant Physiol.* **2012**, *158*, 1079–1088. [CrossRef] [PubMed]
145. Farré, E.M.; Harmer, S.L.; Harmon, F.G.; Yanovsky, M.J.; Kay, S.A. Overlapping and distinct roles of PRR7 and PRR9 in the Arabidopsis circadian clock. *Curr. Boil.* **2005**, *15*, 47–54. [CrossRef] [PubMed]
146. Hsu, P.Y.; Devisetty, U.K.; Harmer, S.L. Accurate timekeeping is controlled by a cycling activator in Arabidopsis. *eLife* **2013**, *2*, e00473. [CrossRef] [PubMed]

147. Nakamichi, N.; Kiba, T.; Henriques, R.; Mizuno, T.; Chua, N.-H.; Sakakibara, H. PSEUDO-RESPONSE REGULATORS 9, 7, and 5 are transcriptional repressors in the Arabidopsis circadian clock. *Plant Cell* **2010**, *22*, 594–605. [CrossRef] [PubMed]
148. Herrero, E.; Kolmos, E.; Bujdoso, N.; Yuan, Y.; Wang, M.; Berns, M.C.; Uhlworm, H.; Coupland, G.; Saini, R.; Jaskolski, M. EARLY FLOWERING4 recruitment of EARLY FLOWERING3 in the nucleus sustains the Arabidopsis circadian clock. *Plant Cell* **2012**, *24*, 428–443. [CrossRef] [PubMed]
149. Chow, B.Y.; Helfer, A.; Nusinow, D.A.; Kay, S.A. ELF3 recruitment to the PRR9 promoter requires other Evening Complex members in the Arabidopsis circadian clock. *Plant Signal. Behav.* **2012**, *7*, 170–173. [CrossRef] [PubMed]
150. Gendron, J.M.; Pruneda-Paz, J.L.; Doherty, C.J.; Gross, A.M.; Kang, S.E.; Kay, S.A. Arabidopsis circadian clock protein, TOC1, is a DNA-binding transcription factor. *Proc. Natl. Acad. Sci. USA* **2012**, *109*, 3167–3172. [CrossRef] [PubMed]
151. Robert, L.S.; Robson, F.; Sharpe, A.; Lydiate, D.; Coupland, G. Conserved structure and function of the Arabidopsis flowering time gene CONSTANS in Brassica napus. *Plant Mol. Boil.* **1998**, *37*, 763–772. [CrossRef]
152. Kim, J.A.; Jung, H.-E.; Hong, J.K.; Hermand, V.; McClung, C.R.; Lee, Y.-H.; Kim, J.Y.; Lee, S.I.; Jeong, M.-J.; Kim, J. Reduction of GIGANTEA expression in transgenic Brassica rapa enhances salt tolerance. *Plant Cell Rep.* **2016**, *35*, 1943–1954. [CrossRef] [PubMed]
153. Xie, Q.; Lou, P.; Hermand, V.; Aman, R.; Park, H.J.; Yun, D.-J.; Kim, W.Y.; Salmela, M.J.; Ewers, B.E.; Weinig, C. Allelic polymorphism of GIGANTEA is responsible for naturally occurring variation in circadian period in Brassica rapa. *Proc. Natl. Acad. Sci. USA* **2015**, *112*, 3829–3834. [PubMed]
154. Curtis, I.S.; Nam, H.G.; Yun, J.Y.; Seo, K.-H. Expression of an antisense GIGANTEA (GI) gene fragment in transgenic radish causes delayed bolting and flowering. *Transgenic Res.* **2002**, *11*, 249–256. [CrossRef] [PubMed]
155. Thiruvengadam, M.; Shih, C.-F.; Yang, C.-H. Expression of an antisense Brassica oleracea GIGANTEA (BoGI) gene in transgenic broccoli causes delayed flowering, leaf senescence, and post-harvest yellowing retardation. *Plant Mol. Boil. Rep.* **2015**, *33*, 1499–1509. [CrossRef]
156. Lou, P.; Wu, J.; Cheng, F.; Cressman, L.G.; Wang, X.; McClung, C.R. Preferential retention of circadian clock genes during diploidization following whole genome triplication in Brassica rapa. *Plant Cell* **2012**, *24*, 2415–2426. [CrossRef] [PubMed]
157. Yamaguchi, A.; Wu, M.-F.; Yang, L.; Wu, G.; Poethig, R.S.; Wagner, D. The microRNA-regulated SBP-Box transcription factor SPL3 is a direct upstream activator of LEAFY, FRUITFULL, and APETALA1. *Dev. Cell* **2009**, *17*, 268–278. [CrossRef] [PubMed]
158. Wang, J.-W.; Czech, B.; Weigel, D. miR156-regulated SPL transcription factors define an endogenous flowering pathway in Arabidopsis thaliana. *Cell* **2009**, *138*, 738–749. [CrossRef] [PubMed]
159. Wu, G.; Poethig, R.S. Temporal regulation of shoot development in Arabidopsis thaliana by miR156 and its target SPL3. *Development* **2006**, *133*, 3539–3547. [CrossRef] [PubMed]
160. Wu, G.; Park, M.Y.; Conway, S.R.; Wang, J.-W.; Weigel, D.; Poethig, R.S. The sequential action of miR156 and miR172 regulates developmental timing in Arabidopsis. *Cell* **2009**, *138*, 750–759. [CrossRef] [PubMed]
161. Teotia, S.; Tang, G. To bloom or not to bloom: Role of microRNAs in plant flowering. *Mol. Plant* **2015**, *8*, 359–377. [CrossRef] [PubMed]
162. Hyun, Y.; Richter, R.; Coupland, G. Competence to flower: Age-controlled sensitivity to environmental cues. *Plant Physiol.* **2017**, *173*, 36–46. [CrossRef] [PubMed]
163. Nie, S.; Xu, L.; Wang, Y.; Huang, D.; Muleke, E.M.; Sun, X.; Wang, R.; Xie, Y.; Gong, Y.; Liu, L. Identification of bolting-related microRNAs and their targets reveals complex miRNA-mediated flowering-time regulatory networks in radish (Raphanus sativus L.). *Sci. Rep.* **2015**, *5*, 14034. [CrossRef] [PubMed]
164. Huo, H.; Wei, S.; Bradford, K.J. DELAY OF GERMINATION1 (DOG1) regulates both seed dormancy and flowering time through microRNA pathways. *Proc. Natl. Acad. Sci. USA* **2016**, *113*, E2199–E2206. [CrossRef] [PubMed]
165. Schwarz, S.; Grande, A.V.; Bujdoso, N.; Saedler, H.; Huijser, P. The microRNA regulated SBP-box genes SPL9 and SPL15 control shoot maturation in Arabidopsis. *Plant Mol. Boil.* **2008**, *67*, 183–195. [CrossRef] [PubMed]
166. Xu, M.; Hu, T.; Zhao, J.; Park, M.-Y.; Earley, K.W.; Wu, G.; Yang, L.; Poethig, R.S. Developmental functions of miR156-regulated SQUAMOSA PROMOTER BINDING PROTEIN-LIKE (SPL) genes in Arabidopsis thaliana. *PLoS Genet.* **2016**, *12*, e1006263. [CrossRef] [PubMed]

167. Aukerman, M.J.; Sakai, H. Regulation of flowering time and floral organ identity by a microRNA and its APETALA2-like target genes. *Plant Cell* **2003**, *15*, 2730–2741. [CrossRef] [PubMed]
168. Jung, J.-H.; Seo, Y.-H.; Seo, P.J.; Reyes, J.L.; Yun, J.; Chua, N.-H.; Park, C.-M. The GIGANTEA-regulated microRNA172 mediates photoperiodic flowering independent of CONSTANS in Arabidopsis. *Plant Cell* **2007**, *19*, 2736–2748. [CrossRef] [PubMed]
169. Mathieu, J.; Yant, L.J.; Mürdter, F.; Küttner, F.; Schmid, M. Repression of flowering by the miR172 target SMZ. *PLoS Boil.* **2009**, *7*, e1000148. [CrossRef] [PubMed]
170. Yant, L.; Mathieu, J.; Dinh, T.T.; Ott, F.; Lanz, C.; Wollmann, H.; Chen, X.; Schmid, M. Orchestration of the floral transition and floral development in Arabidopsis by the bifunctional transcription factor APETALA2. *Plant Cell* **2010**, *22*, 2156–2170. [CrossRef] [PubMed]
171. Hu, J.-Y.; Zhou, Y.; He, F.; Dong, X.; Liu, L.-Y.; Coupland, G.; Turck, F.; de Meaux, J. miR824-regulated AGAMOUS-LIKE16 contributes to flowering time repression in Arabidopsis. *Plant Cell* **2014**, *26*, 2024–2037. [CrossRef] [PubMed]
172. Bergonzi, S.; Albani, M.C.; van Themaat, E.V.L.; Nordström, K.J.V.; Wang, R.; Schneeberger, K.; Moerland, P.D.; Coupland, G. Mechanisms of age-dependent response to winter temperature in perennial flowering of Arabis alpina. *Science* **2013**, *340*, 1094–1097. [CrossRef] [PubMed]
173. Zhou, M.; Luo, H. MicroRNA-mediated gene regulation: Potential applications for plant genetic engineering. *Plant Mol. Boil.* **2013**, *83*, 59–75. [CrossRef] [PubMed]
174. Shen, E.; Zou, J.; Hubertus Behrens, F.; Chen, L.; Ye, C.; Dai, S.; Li, R.; Ni, M.; Jiang, X.; Qiu, J. Identification, evolution, and expression partitioning of miRNAs in allopolyploid Brassica napus. *J. Exp. Bot.* **2015**, *66*, 7241–7253. [CrossRef] [PubMed]
175. Cheng, H.; Hao, M.; Wang, W.; Mei, D.; Tong, C.; Wang, H.; Liu, J.; Fu, L.; Hu, Q. Genomic identification, characterization and differential expression analysis of SBP-box gene family in Brassica napus. *BMC Plant Boil.* **2016**, *16*, 196. [CrossRef] [PubMed]
176. Wang, J.; Hou, X.; Yang, X. Identification of conserved microRNAs and their targets in Chinese cabbage (Brassica rapa subsp. pekinensis). *Genome* **2011**, *54*, 1029–1040. [CrossRef] [PubMed]
177. Wang, J.; Yang, X.; Xu, H.; Chi, X.; Zhang, M.; Hou, X. Identification and characterization of microRNAs and their target genes in Brassica oleracea. *Gene* **2012**, *505*, 300–308. [CrossRef] [PubMed]
178. Geng, M.; Li, H.; Jin, C.; Liu, Q.; Chen, C.; Song, W.; Wang, C. Genome-wide identification and characterization of miRNAs in the hypocotyl and cotyledon of cauliflower (Brassica oleracea L. var. botrytis) seedlings. *Planta* **2014**, *239*, 341–356. [CrossRef] [PubMed]
179. Han, Y.; Zhu, B.; Luan, F.; Zhu, H.; Shao, Y.; Chen, A.; Lu, C.; Luo, Y. Conserved miRNAs and their targets identified in lettuce (Lactuca) by EST analysis. *Gene* **2010**, *463*, 1–7. [CrossRef] [PubMed]
180. Footitt, S.; Huang, Z.; Clay, H.A.; Mead, A.; Finch-Savage, W.E. Temperature, light and nitrate sensing coordinate Arabidopsis seed dormancy cycling, resulting in winter and summer annual phenotypes. *Plant J.* **2013**, *74*, 1003–1015. [CrossRef] [PubMed]
181. Srivastava, S.; Singh, N.; Srivastava, G.; Sharma, A. MiRNA mediated gene regulatory network analysis of Cichorium intybus (chicory). *Agric. Gene* **2017**, *3*, 37–45. [CrossRef]
182. Bernier, G.; Havelange, A.E.; Houssa, C.; Petitjean, A.; Lejeune, P. Physiological signals that induce flowering. *Plant Cell* **1993**, *5*, 1147–1155. [CrossRef] [PubMed]
183. Wilson, R.N.; Heckman, J.W.; Somerville, C.R. Gibberellin is required for flowering in Arabidopsis thaliana under short days. *Plant Physiol.* **1992**, *100*, 403–408. [CrossRef] [PubMed]
184. Duclos, D.V.; Björkman, T. Gibberellin Control of Reproductive Transitions in Brassica oleracea Curd Development. *J. Am. Soc. Hortic. Sci.* **2015**, *140*, 57–67.
185. Bernier, G. The control of floral evocation and morphogenesis. *Annu. Rev. Plant Physiol. Plant Mol. Boil.* **1988**, *39*, 175–219. [CrossRef]
186. Eriksson, S.; Böhlenius, H.; Moritz, T.; Nilsson, O. GA4 is the active gibberellin in the regulation of LEAFY transcription and Arabidopsis floral initiation. *Plant Cell* **2006**, *18*, 2172–2181. [CrossRef] [PubMed]
187. Bernier, G. My favourite flowering image: The role of cytokinin as a flowering signal. *J. Exp. Bot.* **2011**, *64*, 5795–5799. [CrossRef] [PubMed]
188. Besnard-Wibaut, C. Effectiveness of gibberellins and 6-benzyladenine on flowering of Arabidopsis thaliana. *Physiol. Plant.* **1981**, *53*, 205–212. [CrossRef]

189. D'aloia, M.; Bonhomme, D.; Bouché, F.; Tamseddak, K.; Ormenese, S.; Torti, S.; Coupland, G.; Périlleux, C. Cytokinin promotes flowering of Arabidopsis via transcriptional activation of the FT paralogue TSF. *Plant J.* **2011**, *65*, 972–979. [CrossRef] [PubMed]
190. Michniewicz, M.; Kamieńska, A. Studies on the role of kinetin and vitamin E in the flowering of the cold requiring plant Cichorium intybus) and the long-day plant (Arabidopsis thaliana) grown in non-inductive. *Acta Soc. Bot. Pol.* **1967**, *36*, 67–72. [CrossRef]
191. King, R.W.; Evans, L.T. Gibberellins and flowering of grasses and cereals: Prizing open the lid of the "florigen" black box. *Annu. Rev. Plant. Boil.* **2003**, *54*, 307–328. [CrossRef] [PubMed]
192. Hedden, P.; Phillips, A.L. Gibberellin metabolism: New insights revealed by the genes. *Trends Plant Sci.* **2000**, *5*, 523–530. [CrossRef]
193. Rieu, I.; Ruiz-Rivero, O.; Fernandez-Garcia, N.; Griffiths, J.; Powers, S.J.; Gong, F.; Linhartova, T.; Eriksson, S.; Nilsson, O.; Thomas, S.G. The gibberellin biosynthetic genes AtGA20ox1 and AtGA20ox2 act, partially redundantly, to promote growth and development throughout the Arabidopsis life cycle. *Plant J.* **2008**, *53*, 488–504. [CrossRef] [PubMed]
194. Plackett, A.R.; Powers, S.J.; Fernandez-Garcia, N.; Urbanova, T.; Takebayashi, Y.; Seo, M.; Jikumaru, Y.; Benlloch, R.; Nilsson, O.; Ruiz-Rivero, O. Analysis of the developmental roles of the Arabidopsis gibberellin 20-oxidases demonstrates that GA20ox1,-2, and-3 are the dominant paralogs. *Plant Cell* **2012**, *24*, 941–960. [CrossRef] [PubMed]
195. Mitchum, M.G.; Yamaguchi, S.; Hanada, A.; Kuwahara, A.; Yoshioka, Y.; Kato, T.; Tabata, S.; Kamiya, Y.; Sun, T.p. Distinct and overlapping roles of two gibberellin 3-oxidases in Arabidopsis development. *Plant J.* **2006**, *45*, 804–818. [CrossRef] [PubMed]
196. Chen, M.-L.; Su, X.; Xiong, W.; Liu, J.-F.; Wu, Y.; Feng, Y.-Q.; Yuan, B.-F. Assessing gibberellins oxidase activity by anion exchange/hydrophobic polymer monolithic capillary liquid chromatography-mass spectrometry. *PLoS ONE* **2013**, *8*, e69629. [CrossRef] [PubMed]
197. Porri, A.; Torti, S.; Romera-Branchat, M.; Coupland, G. Spatially distinct regulatory roles for gibberellins in the promotion of flowering of Arabidopsis under long photoperiods. *Development* **2012**, *139*, 2198–2209. [CrossRef] [PubMed]
198. Rieu, I.; Eriksson, S.; Powers, S.J.; Gong, F.; Griffiths, J.; Woolley, L.; Benlloch, R.; Nilsson, O.; Thomas, S.G.; Hedden, P. Genetic analysis reveals that C19-GA 2-oxidation is a major gibberellin inactivation pathway in Arabidopsis. *Plant Cell* **2008**, *20*, 2420–2436. [CrossRef] [PubMed]
199. Ariizumi, T.; Murase, K.; Sun, T.-p.; Steber, C.M. Proteolysis-independent downregulation of DELLA repression in Arabidopsis by the gibberellin receptor GIBBERELLIN INSENSITIVE DWARF1. *Plant Cell* **2008**, *20*, 2447–2459. [CrossRef] [PubMed]
200. Willige, B.C.; Ghosh, S.; Nill, C.; Zourelidou, M.; Dohmann, E.M.; Maier, A.; Schwechheimer, C. The DELLA domain of GA INSENSITIVE mediates the interaction with the GA INSENSITIVE DWARF1A gibberellin receptor of Arabidopsis. *Plant Cell* **2007**, *19*, 1209–1220. [CrossRef] [PubMed]
201. Nakajima, M.; Shimada, A.; Takashi, Y.; Kim, Y.C.; Park, S.H.; Ueguchi-Tanaka, M.; Suzuki, H.; Katoh, E.; Iuchi, S.; Kobayashi, M. Identification and characterization of Arabidopsis gibberellin receptors. *Plant J.* **2006**, *46*, 880–889. [CrossRef] [PubMed]
202. Galvão, V.C.; Horrer, D.; Küttner, F.; Schmid, M. Spatial control of flowering by DELLA proteins in Arabidopsis thaliana. *Development* **2012**, *139*, 4072–4082. [CrossRef] [PubMed]
203. Yu, S.; Galvão, V.C.; Zhang, Y.-C.; Horrer, D.; Zhang, T.-Q.; Hao, Y.-H.; Feng, Y.-Q.; Wang, S.; Markus, S.; Wang, J.-W. Gibberellin regulates the Arabidopsis floral transition through miR156-targeted SQUAMOSA PROMOTER BINDING–LIKE transcription factors. *Plant Cell* **2012**, *24*, 3320–3332. [CrossRef] [PubMed]
204. Andrés, F.; Porri, A.; Torti, S.; Mateos, J.; Romera-Branchat, M.; García-Martínez, J.L.; Fornara, F.; Gregis, V.; Kater, M.M.; Coupland, G. SHORT VEGETATIVE PHASE reduces gibberellin biosynthesis at the Arabidopsis shoot apex to regulate the floral transition. *Proc. Natl. Acad. Sci. USA* **2014**, 201409567.
205. Galvão, V.C.; Collani, S.; Horrer, D.; Schmid, M. Gibberellic acid signaling is required for ambient temperature-mediated induction of flowering in Arabidopsis thaliana. *Plant J.* **2015**, *84*, 949–962. [CrossRef] [PubMed]
206. Bartrina, I.; Jensen, H.; Novak, O.; Strnad, M.; Werner, T.; Schmülling, T. Gain-of-function mutants of the cytokinin receptors AHK2 and AHK3 regulate plant organ size, flowering time and plant longevity. *Plant Physiol.* **2017**. [CrossRef] [PubMed]

207. Greenboim-Wainberg, Y.; Maymon, I.; Borochov, R.; Alvarez, J.; Olszewski, N.; Ori, N.; Eshed, Y.; Weiss, D. Cross talk between gibberellin and cytokinin: The Arabidopsis GA response inhibitor SPINDLY plays a positive role in cytokinin signaling. *Plant Cell* **2005**, *17*, 92–102. [CrossRef] [PubMed]
208. Zhao, B.; Li, H.; Li, J.; Wang, B.; Dai, C.; Wang, J.; Liu, K. Brassica napus DS-3, encoding a DELLA protein, negatively regulates stem elongation through gibberellin signaling pathway. *Theor. Appl. Genet.* **2017**, *130*, 727–741. [CrossRef] [PubMed]
209. Tarkowská, D.; Filek, M.; Biesaga-Kościelniak, J.; Marcińska, I.; Macháčková, I.; Krekule, J.; Strnad, M. Cytokinins in shoot apices of Brassica napus plants during vernalization. *Plant Sci.* **2012**, *187*, 105–112. [CrossRef] [PubMed]
210. Shang, M.; Wang, X.; Zhang, J.; Qi, X.; Ping, A.; Hou, L.; Xing, G.; Li, G.; Li, M. Genetic Regulation of GA Metabolism during Vernalization, Floral Bud Initiation and Development in Pak Choi (Brassica rapa ssp. chinensis Makino). *Front. Plant Sci.* **2017**, *8*, 1533. [CrossRef] [PubMed]
211. Fukuda, M.; Matsuo, S.; Kikuchi, K.; Mitsuhashi, W.; Toyomasu, T.; Honda, I. The endogenous level of GA1 is upregulated by high temperature during stem elongation in lettuce through LsGA3ox1 expression. *J. Plant Physiol.* **2009**, *166*, 2077–2084. [CrossRef] [PubMed]
212. Frugis, G.; Giannino, D.; Mele, G.; Nicolodi, C.; Chiappetta, A.; Bitonti, M.B.; Innocenti, A.M.; Dewitte, W.; Van Onckelen, H.; Mariotti, D. Overexpression of KNAT1 in lettuce shifts leaf determinate growth to a shoot-like indeterminate growth associated with an accumulation of isopentenyl-type cytokinins. *Plant Physiol.* **2001**, *126*, 1370–1380. [CrossRef] [PubMed]
213. Jasinski, S.; Piazza, P.; Craft, J.; Hay, A.; Woolley, L.; Rieu, I.; Phillips, A.; Hedden, P.; Tsiantis, M. KNOX action in Arabidopsis is mediated by coordinate regulation of cytokinin and gibberellin activities. *Curr. Boil.* **2005**, *15*, 1560–1565. [CrossRef] [PubMed]
214. Yanai, O.; Shani, E.; Dolezal, K.; Tarkowski, P.; Sablowski, R.; Sandberg, G.; Samach, A.; Ori, N. Arabidopsis KNOXI proteins activate cytokinin biosynthesis. *Curr. Boil.* **2005**, *15*, 1566–1571. [CrossRef] [PubMed]
215. Di Giacomo, E.; Iannelli, M.A.; Frugis, G. TALE and shape: How to make a leaf different. *Plants* **2013**, *2*, 317–342. [CrossRef] [PubMed]
216. Salomé, P.A.; Bomblies, K.; Laitinen, R.A.; Yant, L.; Mott, R.; Weigel, D. Genetic architecture of flowering time variation in Arabidopsis thaliana. *Genetics* **2011**, *188*, 421–433. [CrossRef] [PubMed]
217. Brachi, B.; Faure, N.; Horton, M.; Flahauw, E.; Vazquez, A.; Nordborg, M.; Bergelson, J.; Cuguen, J.; Roux, F. Linkage and association mapping of Arabidopsis thaliana flowering time in nature. *PLoS Genet.* **2010**, *6*, e1000940. [CrossRef] [PubMed]
218. Shindo, C.; Aranzana, M.J.; Lister, C.; Baxter, C.; Nicholls, C.; Nordborg, M.; Dean, C. Role of FRIGIDA and FLOWERING LOCUS C in determining variation in flowering time of Arabidopsis. *Plant Physiol.* **2005**, *138*, 1163–1173. [CrossRef] [PubMed]
219. Caicedo, A.L.; Stinchcombe, J.R.; Olsen, K.M.; Schmitt, J.; Purugganan, M.D. Epistatic interaction between Arabidopsis FRI and FLC flowering time genes generates a latitudinal cline in a life history trait. *Proc. Natl. Acad. Sci. USA* **2004**, *101*, 15670–15675. [CrossRef] [PubMed]
220. Li, P.; Filiault, D.; Box, M.S.; Kerdaffrec, E.; van Oosterhout, C.; Wilczek, A.M.; Schmitt, J.; McMullan, M.; Bergelson, J.; Nordborg, M. Multiple FLC haplotypes defined by independent cis-regulatory variation underpin life history diversity in Arabidopsis thaliana. *Genes Dev.* **2014**, *28*, 1635–1640. [CrossRef] [PubMed]
221. Schwartz, C.; Balasubramanian, S.; Warthmann, N.; Michael, T.P.; Lempe, J.; Sureshkumar, S.; Kobayashi, Y.; Maloof, J.N.; Borevitz, J.O.; Chory, J. Cis-regulatory changes at FLOWERING LOCUS T mediate natural variation in flowering responses of Arabidopsis thaliana. *Genetics* **2009**, *183*, 723–732. [CrossRef] [PubMed]
222. Long, Y.; Shi, J.; Qiu, D.; Li, R.; Zhang, C.; Wang, J.; Hou, J.; Zhao, J.; Shi, L.; Park, B.-S. Flowering time quantitative trait loci analysis of oilseed Brassica in multiple environments and genomewide alignment with Arabidopsis. *Genetics* **2007**, *177*, 2433–2444. [PubMed]
223. Li, L.; Long, Y.; Zhang, L.; Dalton-Morgan, J.; Batley, J.; Yu, L.; Meng, J.; Li, M. Genome wide analysis of flowering time trait in multiple environments via high-throughput genotyping technique in Brassica napus L. *PLoS ONE* **2015**, *10*, e0119425. [CrossRef] [PubMed]
224. Hou, J.; Long, Y.; Raman, H.; Zou, X.; Wang, J.; Dai, S.; Xiao, Q.; Li, C.; Fan, L.; Liu, B. A Tourist-like MITE insertion in the upstream region of the BnFLC.A10 gene is associated with vernalization requirement in rapeseed (Brassica napus L.). *BMC Plant. Boil.* **2012**, *12*, 238.

225. Luo, Z.; Wang, M.; Long, Y.; Huang, Y.; Shi, L.; Zhang, C.; Liu, X.; Fitt, B.D.; Xiang, J.; Mason, A.S. Incorporating pleiotropic quantitative trait loci in dissection of complex traits: Seed yield in rapeseed as an example. *Theor. Appl. Genet.* **2017**, *130*, 1569–1585. [CrossRef] [PubMed]
226. Hartman, Y.; Hooftman, D.A.; Schranz, M.E.; van Tienderen, P.H. QTL analysis reveals the genetic architecture of domestication traits in Crisphead lettuce. *Genet. Resour. Crop. Evol.* **2013**, *60*, 1487–1500. [CrossRef]
227. Hartman, Y.; Hooftman, D.A.; Uwimana, B.; van de Wiel, C.; Smulders, M.J.; Visser, R.G.; van Tienderen, P.H. Genomic regions in crop–wild hybrids of lettuce are affected differently in different environments: Implications for crop breeding. *Evol. Appl.* **2012**, *5*, 629–640. [CrossRef] [PubMed]
228. Hartman, Y.; Uwimana, B.; Hooftman, D.A.; Schranz, M.E.; Wiel, C.; Smulders, M.J.; Visser, R.G.; Tienderen, P.H. Genomic and environmental selection patterns in two distinct lettuce crop–wild hybrid crosses. *Evol. Appl.* **2013**, *6*, 569–584. [CrossRef] [PubMed]
229. Zhang, L.; Su, W.; Tao, R.; Zhang, W.; Chen, J.; Wu, P.; Yan, C.; Jia, Y.; Larkin, R.M.; Lavelle, D. RNA sequencing provides insights into the evolution of lettuce and the regulation of flavonoid biosynthesis. *Nat. Commun.* **2017**, *8*, 2264. [CrossRef] [PubMed]
230. Javed, N.; Geng, J.; Tahir, M.; McVetty, P.; Li, G.; Duncan, R.W. Identification of QTL influencing seed oil content, fatty acid profile and days to flowering in Brassica napus L. *Euphytica* **2016**, *207*, 191–211. [CrossRef]
231. Rahman, H.; Bennett, R.A.; Kebede, B. Molecular mapping of QTL alleles of Brassica oleracea affecting days to flowering and photosensitivity in spring Brassica napus. *PLoS ONE* **2018**, *13*, e0189723. [CrossRef] [PubMed]
232. Shen, Y.; Xiang, Y.; Xu, E.; Ge, X.; Li, Z. Major Co-localized QTL for Plant Height, Branch Initiation Height, Stem Diameter, and Flowering Time in an Alien Introgression Derived Brassica napus DH Population. *Front. Plant Sci.* **2018**, *9*, 390. [CrossRef] [PubMed]
233. Xu, L.; Hu, K.; Zhang, Z.; Guan, C.; Chen, S.; Hua, W.; Li, J.; Wen, J.; Yi, B.; Shen, J. Genome-wide association study reveals the genetic architecture of flowering time in rapeseed (Brassica napus L.). *DNA Res.* **2015**, *23*, 43–52. [CrossRef] [PubMed]
234. Wang, N.; Chen, B.; Xu, K.; Gao, G.; Li, F.; Qiao, J.; Yan, G.; Li, J.; Li, H.; Wu, X. Association mapping of flowering time QTLs and insight into their contributions to rapeseed growth habits. *Front. Plant Sci.* **2016**, *7*, 338. [CrossRef] [PubMed]
235. Yang, S.; Zhang, B.; Liu, G.; Hong, B.; Xu, J.; Chen, X.; Wang, B.; Wu, Z.; Hou, F.; Yue, X. A comprehensive and precise set of intervarietal substitution lines to identify candidate genes and quantitative trait loci in oilseed rape (Brassica napus L.). *Theor. Appl. Genet.* **2018**, *131*, 2117–2129. [CrossRef] [PubMed]
236. Schiessl, S.; Iniguez-Luy, F.; Qian, W.; Snowdon, R.J. Diverse regulatory factors associate with flowering time and yield responses in winter-type Brassica napus. *BMC Genom.* **2015**, *16*, 737. [CrossRef] [PubMed]
237. Li, X.; Wang, W.; Wang, Z.; Li, K.; Lim, Y.P.; Piao, Z. Construction of chromosome segment substitution lines enables QTL mapping for flowering and morphological traits in Brassica rapa. *Front. Plant Sci.* **2015**, *6*, 432. [CrossRef] [PubMed]
238. Lou, P.; Zhao, J.; Kim, J.S.; Shen, S.; Del Carpio, D.P.; Song, X.; Jin, M.; Vreugdenhil, D.; Wang, X.; Koornneef, M. Quantitative trait loci for flowering time and morphological traits in multiple populations of Brassica rapa. *J. Exp. Bot.* **2007**, *58*, 4005–4016. [CrossRef] [PubMed]
239. Wang, Y.; Zhang, L.; Ji, X.; Yan, J.; Liu, Y.; Lv, X.; Feng, H. Mapping of quantitative trait loci for the bolting trait in Brassica rapa under vernalizing conditions. *Genet. Mol. Res.* **2014**, *13*, 3927–3939. [CrossRef] [PubMed]
240. Liu, Y.; Li, C.; Shi, X.; Feng, H.; Wang, Y. Identification of QTLs with additive, epistatic, and QTL × environment interaction effects for the bolting trait in Brassica rapa L. *Euphytica* **2016**, *210*, 427–439. [CrossRef]
241. Dechaine, J.M.; Brock, M.T.; Weinig, C. QTL architecture of reproductive fitness characters in Brassica rapa. *BMC Plant Boil.* **2014**, *14*, 66. [CrossRef] [PubMed]
242. Shu, J.; Liu, Y.; Zhang, L.; Li, Z.; Fang, Z.; Yang, L.; Zhuang, M.; Zhang, Y.; Lv, H. QTL-seq for rapid identification of candidate genes for flowering time in broccoli× cabbage. *Theor. Appl. Genet.* **2018**, *131*, 917–928. [CrossRef] [PubMed]
243. Hasan, Y.; Briggs, W.; Matschegewski, C.; Ordon, F.; Stützel, H.; Zetzsche, H.; Groen, S.; Uptmoor, R. Quantitative trait loci controlling leaf appearance and curd initiation of cauliflower in relation to temperature. *Theor. Appl. Genet.* **2016**, *129*, 1273–1288. [CrossRef] [PubMed]

244. FitzJohn, R.G.; Armstrong, T.T.; Newstrom-Lloyd, L.E.; Wilton, A.D.; Cochrane, M. Hybridisation within Brassica and allied genera: Evaluation of potential for transgene escape. *Euphytica* **2007**, *158*, 209–230. [CrossRef]
245. Shea, D.J.; Tomaru, Y.; Itabashi, E.; Nakamura, Y.; Miyazaki, T.; Kakizaki, T.; Naher, T.N.; Shimizu, M.; Fujimoto, R.; Fukai, E. The production and characterization of a BoFLC2 introgressed Brassica rapa by repeated backcrossing to an F1. *Breed. Sci.* **2018**, *68*, 316–325. [CrossRef] [PubMed]
246. Lebeda, A.; Doležalová, I.; Křístková, E.; Kitner, M.; Petrželová, I.; Mieslerová, B.; Novotná, A. Wild Lactuca germplasm for lettuce breeding: Current status, gaps and challenges. *Euphytica* **2009**, *170*, 15. [CrossRef]
247. Van Cutsem, P.; Du Jardin, P.; Boutte, C.; Beauwens, T.; Jacqmin, S.; Vekemans, X. Distinction between cultivated and wild chicory gene pools using AFLP markers. *Theor. Appl. Genet.* **2003**, *107*, 713–718. [CrossRef] [PubMed]
248. Hodgins, K.A.; Lai, Z.; Oliveira, L.O.; Still, D.W.; Scascitelli, M.; Barker, M.S.; Kane, N.C.; Dempewolf, H.; Kozik, A.; Kesseli, R.V. Genomics of Compositae crops: Reference transcriptome assemblies and evidence of hybridization with wild relatives. *Mol. Ecol. Resour.* **2014**, *14*, 166–177. [CrossRef] [PubMed]
249. Testone, G.; Mele, G.; Di Giacomo, E.; Gonnella, M.; Renna, M.; Tenore, G.C.; Nicolodi, C.; Frugis, G.; Iannelli, M.A.; Arnesi, G. Insights into the sesquiterpenoid pathway by metabolic profiling and de novo transcriptome assembly of stem-chicory (Cichorium intybus cultigroup "Catalogna"). *Front. Plant Sci.* **2016**, *7*, 1676. [CrossRef] [PubMed]

© 2018 by the authors. Licensee MDPI, Basel, Switzerland. This article is an open access article distributed under the terms and conditions of the Creative Commons Attribution (CC BY) license (http://creativecommons.org/licenses/by/4.0/).

Review

Using Morphogenic Genes to Improve Recovery and Regeneration of Transgenic Plants

Bill Gordon-Kamm *, Nagesh Sardesai, Maren Arling, Keith Lowe, George Hoerster, Scott Betts and Todd Jones

Corteva Agriscience™, Agriculture Division of DowDuPont, Johnston, IA 50131, USA; nagesh.sardesai@corteva.com (N.S.); maren.arling@pioneer.com (M.A.); keith.lowe@pioneer.com (K.L.); george.hoerster@pioneer.com (G.H.); scott.betts@pioneer.com (S.B.); todd.j.jones@pioneer.com (T.J.)
* Correspondence: william.gordon-kamm@pioneer.com; Tel.: 515-535-3243

Received: 23 November 2018; Accepted: 31 January 2019; Published: 11 February 2019

Abstract: Efficient transformation of numerous important crops remains a challenge, due predominantly to our inability to stimulate growth of transgenic cells capable of producing plants. For years, this difficulty has been partially addressed by tissue culture strategies that improve regeneration either through somatic embryogenesis or meristem formation. Identification of genes involved in these developmental processes, designated here as morphogenic genes, provides useful tools in transformation research. In species from eudicots and cereals to gymnosperms, ectopic overexpression of genes involved in either embryo or meristem development has been used to stimulate growth of transgenic plants. However, many of these genes produce pleiotropic deleterious phenotypes. To mitigate this, research has been focusing on ways to take advantage of growth-stimulating morphogenic genes while later restricting or eliminating their expression in the plant. Methods of controlling ectopic overexpression include the use of transient expression, inducible promoters, tissue-specific promoters, and excision of the morphogenic genes. These methods of controlling morphogenic gene expression have been demonstrated in a variety of important crops. Here, we provide a review that highlights how ectopic overexpression of genes involved in morphogenesis has been used to improve transformation efficiencies, which is facilitating transformation of numerous recalcitrant crops. The use of morphogenic genes may help to alleviate one of the bottlenecks currently slowing progress in plant genome modification.

Keywords: transformation; morphogenic; embryogenesis; meristem formation; organogenesis

1. Introduction

Despite progress in crop transformation over the past several decades, efficient production of transgenic plants remains one of the major barriers to crop improvement [1]. There are two components to producing transgenic plants. The first is the ability to introduce and express transgenes (transformation), and the second is the ability to form tissue (typically de novo embryos or shoots) capable of regenerating into a fertile plant. Many plant species (or genotypes) remain difficult to transform and regenerate. Such varieties are referred to as being recalcitrant to transformation and plant regeneration. One of the promising tools helping to reduce this recalcitrance (and thus alleviate the bottleneck) is the use of genes involved in controlling plant growth and development.

Morphogenesis, or the organized spatial development of embryos, tissues, and organs, is a tightly controlled process involving networks of genes acting sequentially or in concert. Within this broad context, the concept of trying to use genes involved in either embryogenesis or meristem maintenance has attracted the attention of plant transformation researchers for many years. Basic research within these two well-defined areas has contributed an ever-expanding number of genes and gene networks involved in embryo development [2] and meristem development [3–6] that we will not attempt to cover here. De novo regeneration of plants typically occurs through either somatic embryogenesis or organogenesis (de novo formation of new meristems or through rearrangement of pre-existing meristems), traditionally manipulated by adjusting auxin/cytokinin ratios in the medium [6–8]. In addition to exogenous hormone manipulation, ectopic overexpression of plant genes that control growth and development has also proven to be useful.

Of course, there are numerous reports where non-plant genes have been used to improve transformation frequencies and/or plant regeneration. Examples include tumor-inducing genes from *Agrobacterium*, such as the isopentyl transferase or *ipt* gene [9,10], *rolC* [11,12], *rolB*, *6B* [13], and *tzs* [14,15], and viral genes that stimulate the plant cell cycle [16]. Further details for non-plant genes are beyond the scope of this review. Instead, we will focus on research that has demonstrated a potentially useful morphogenic growth response due to ectopic overexpression of a plant embryo or meristem gene (morphogenic genes), or has demonstrated a practical benefit for plant transformation or regeneration. Plant transformation and regeneration improvements can be further distinguished by their impact on either improving transformation efficiencies, or improving the regeneration process to recover transgenic plants. Finally, we are making a distinction between reports focused solely on observations of morphogenic responses, and those studies that describe practical methods to improve transformation and/or regeneration. Both types of results are important in terms of their contribution to the field of transformation research, and are described below.

Characterizing morphogenic genes involves phenotypic analysis of knockout mutants and/or transgenic experiments typically involving either ectopic overexpression or downregulation in plants. Often, the two strategies are combined, where introduction of an expression cassette is used to complement a mutant phenotype. Morphogenic responses have been observed using transient expression, constitutive expression, or induction of gene expression by exposure of a stable transgenic event to a chemical ligand. Using a plant morphogenic gene to improve transformation, on the other hand, almost invariably involves limiting the length of time that expression occurs to avoid later pleiotropic effects in regenerated T0 plants and subsequent progeny generations (Figure 1).

2. Phenotypic Responses from Ectopic Overexpression of Morphogenic Genes

The genes involved in embryogenesis, meristem maintenance, and hormone metabolism are numerous and have been studied for many years [2–6]. Within this large and ever-expanding body of literature, there has been a steady stream of reports demonstrating morphogenesis in response to altered expression of these genes. These observations provide the groundwork that inspires new strategies for transformation research, and are discussed below.

Numerous genes mentioned in this review are typically categorized as genes involved in embryogenesis, meristem function, or hormone pathways. However, we have chosen to group these genes based on their practical benefit (or potential) when used for transformation. Therefore, morphogenic genes that have stimulated an embryogenic or meristematic response when overexpressed were grouped into two categories based on the observed growth response: A) those that enhance a pre-existing embryogenic response under conditions (media composition, exogenous hormones, or even the tissue type) that already elicit the growth response, and B) those that produce ectopic somatic embryos or meristems under conditions where such a response is typically not observed (see Table 1 for a list of genes).

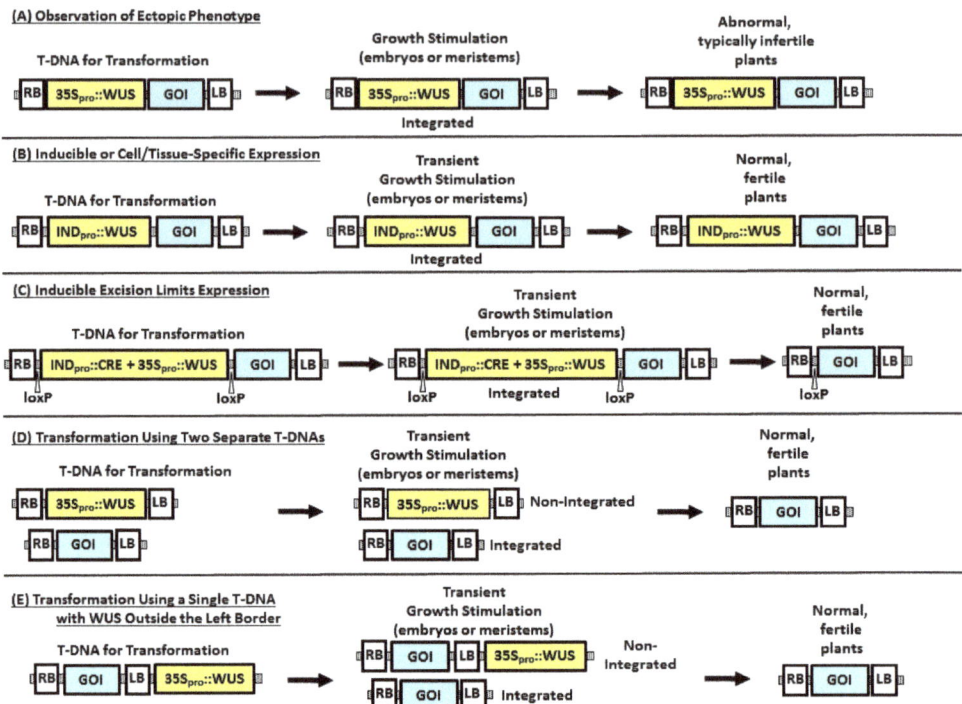

Figure 1. Methods for expression of morphogenic genes in plant transformation. WUS is used to exemplify the morphogenic gene expression cassettes, which could be designed for overexpression of the gene, downregulation of a gene, or combinations of genes, while the box labeled "GOI" (Genes Of Interest) represents trait gene expression cassettes. (**A**) Using a constitutive promoter, such as CaMV35S, to drive expression of a WUS gene results in growth stimulation of cells transformed with the T-DNA, either through somatic embryogenesis or through meristem proliferation. (**B**) Using an inducible promoter to drive expression of WUS will result in growth stimulation only when the plant tissue is exposed to the inducing stimulus (typically a chemical ligand). (**C**) Growth stimulation can also be effectively controlled by using a combination of constitutive expression of WUS and inducible expression of CRE recombinase to remove the WUS expression cassette. (**D**) Transforming the same plant cell with a T-DNA containing the WUS expression cassette and a second T-DNA containing the trait expression cassette will also provide transient growth stimulation sufficient to recover regenerable tissues, such as somatic embryos, without the integration of the morphogenic gene. (**E**) Using a single T-DNA containing the trait, with the WUS expression cassette outside the T-DNA Left Border sequence, higher numbers of the trait-containing T-DNA are introduced relative to the low numbers of "read-through" sequences containing the WUS gene, providing transient growth stimulation without WUS integration. IND_{pro} is used to represent chemically inducible promoters, such as the estradiol-, glucocorticoid-, or tetracycline-responsive promoters in 1-B. Promoters that are induced by physical conditions, such as desiccation (e.g., the RAB17 promoter in [17]), are used to control recombinase-mediated excision (as in 1-C). CRE represents the CRE recombinase expression cassette and loxP are the CRE-recombinase target sites. RB and LB represent the right and left T-DNA border sequences, respectively.

Table 1. Strategies to improve transformation using morphogenic genes.

Strategy	Examples				
	CDS	Promoter for Transgene	Transformed Species	Variety *	Ref.
(A) Enhance pre-existing somatic embryogenic culture response	AtSERK1	35S	Arabidopsis thaliana	Ws	[18]
	CcSERK1 (or RNAi)	35S and Inducible	Coffea canephora	cv. Robusta	[19]
	AtAGL15	35S	A. thaliana	Ws	[20]
	GmAGL15	35S	Glycine max	"Jack"	[21]
	GhAGL15	35S	Gossypium hirsutum	cv. CRI24	[22]
	AtWUS	Inducible	C. canephora	cv. Robusta	[23]
	BnSTM, BoSTM	35S	A. thaliana	Col	[24]
	BnSTM, BoSTM	35S	Brassica napus	cv. Topas	[24]
	BrSTM	35S	A. thaliana	Col	[24]
	BrSTM	35S	B. napus	cv. Topas	[24]
	AtWUS	35S	G. hirsutum	var. Coker 310	[25]
(B) Ectopic formation of somatic embryos or meristems	BnBBM	35S	A. thaliana	Col and C24	[26]
	AtBBM-GR	Inducible	Nicotiana tabacum	Wisconsin 38	[27]
	AtBBM-GR	Inducible	N. tabacum	Petit Havana SR1	[27]
	GmBBM	35S	A. thaliana	not specified	[28]
	TcBBM	35S	Theobroma cacao	Scavina-6 (SCA6)	[29]
	EgBBM	35S	A. thaliana	Col	[30]
	AtEMK	35S	A. thaliana	Col-0	[31]
	AtRKD4	Inducible	Psp (Orchid)	"Sogo Vivian"	[32]
	AtLEC1	35S	A. thaliana	Ws-0	[33]
	CsL1L	35S	Clonorchis sinensis	cv. "Olinda"	[34]
	PaHAP3A	Inducible	Picea abies	cell lines 88 and 61	[35]
	AtFUS3	AtML1	A. thaliana	Col	[36]
	AtLEC2	35S	A. thaliana	Ws-0	[37]
	AtWUS	Inducible	A. thaliana	Col, Ws, Ler	[38]
	AtWUS, AtSTM	Both inducible	A. thaliana	Ler	[39]
	AtWUS	Activated	A. thaliana	Ler	[40]
	AtWUS	Inducible	N. tabacum	cv. Samsun	[41]
	AtWOX5	Inducible	N. tabacum	cv. Samsun	[42]
	ZmKN1	35S	N. tabacum	cv. Xanthi	[43]
	NtKN1	35S	N. tabacum	cv. Samsun	[44]

Table 1. *Cont.*

Strategy	CDS	Examples		Ref.
		Promoter for Transgene	Transformed Species	Variety *	

Strategy	CDS	Promoter for Transgene	Transformed Species	Variety *	Ref.
	AtCUC1, AtCUC2	35S	A. thaliana	Ler	[45]
	AtLEC2	Inducible	A. thaliana	Col-0	[46]
	AtESR1	Inducible	A. thaliana	Ws	[47]
	AtESR2	Inducible	A. thaliana	Ler and Ws	[48]
	AtMPΔ	MP Promoter	A. thaliana	Col-0	[49]
	BnBBM	Inducible	Capsicum annuum	Three hybrids [b]	[50]
	AtBBM	Inducible	A. thaliana	RDL and Ler	[51]
	AtPGA37	Inducible	A. thaliana	Col-0, Ws, Ler	[52]
	AtLEC2	Inducible	T. cacao	var. SCA6	[53]
	AtWOX2 WOX8 WOX9	Inducible	N. tabacum	cv. Samsun	[54]
(C) Restrict morphogenic response to enable recovery of normal plants	BcBBM [a]	35S	Populus tomentosa	not specified	[55]
	ZmBBM/ZmWUS2 [a]	Ubi + NOS	Zea mays	4 Pioneer Inbreds [c]	[17]
	ZmBBM/ZmWUS2 [a]	Ubi + NOS	Oryza sativa	(indica) cv. IRV95	[17]
	ZmBBM/ZmWUS2 [a]	Ubi + NOS	Sorghum bicolor	var. Tx430	[17]
	ZmBBM/ZmWUS2 [a]	Ubi + NOS	Salvia officinalis	var. CP01-1372	[17]
	ZmBBM/ZmWUS2 [a]	Ubi + NOS	Z. mays	public inbred B73	[56]
	ZmBBM/ZmWUS2 [a]	Ubi + NOS	S. bicolor	var. P898012	[56]
	ZmBBM/ZmWUS2	PLTP + AXIG1	Z. mays	maize inbreds [d]	[57]

* "Variety" = ecotype (RDL, Wassilewskija = Ws, Columbia = Col, Landsberg *erecta* = Ler), variety (var.), cultivar (cv.), inbred, or hybrid name; [a] In column labelled "CDS", these included recombinase-mediated excision for removal of morphogenic gene(s); [b] Orchid hybrids Fiesta, Spirit, and Ferrari; [c] Pioneer inbreds PHN46, PH581, PHP38, and PHH56; [d] Public maize inbreds B73, Mo17, and the FFMM line A (Fast Flowering Mini-Maize, line A). Pioneer inbreds PHR03, PH184C, PHH5G, PH1V5T, and PH1V69.

2.1. Enhancing the Somatic Embryogenic Response

In this category, overexpression of the plant gene results in enhanced formation of somatic embryos under in vitro culture conditions in which somatic embryogenesis already occurs (see Table 1, Strategy A). This includes the observation that when *SOMATIC EMBRYOGENESIS RECEPTOR KINASE1* (*SERK1*), a gene normally associated with anther and pollen development, was overexpressed using the Cauliflower Mosaic Virus 35S promoter (CaMV 35S) in *Arabidopsis thaliana*, no changes in plant phenotype were observed, but the embryogenic callus response was improved 3-4-fold over wild-type [18]. This demonstrated that SERK1 stimulated an enhanced somatic embryo response from germinating seedlings placed on media capable of eliciting this response already. Similarly, overexpression of the *Coffea canephora SERK1* gene during the in vitro somatic embryogenesis process enhanced the production of somatic embryos by 2-fold, while silencing the gene dramatically reduced the somatic embryogenesis response [19]. Similar conclusions have been reached in reports in which the *AGAMOUS-LIKE15* (*AGL15*) gene was overexpressed, enhancing the formation of secondary somatic embryos from cultured zygotic embryos in Arabidopsis [20], increasing the number of somatic embryos in soybean cultures [21], and again enhancing production of embryogenic callus in cotton [22]. The similarities between SERK1 and AGL15 overexpression are not surprising, since AGL15 is part of the SERK1 protein complex [58]. Interestingly, SERK1 and SERK3 have been shown to be co-receptors, along with BRASSINOSTEROID INSENSITIVE 1 protein (BRI1), of the brassinosteroid class of plant growth regulators [59,60], and the SERK proteins and BRI1 phosphorylate one another upon brassinosteroid sensing. Based on protein structure, the SERK proteins appear to mediate brassinosteroid signaling across the plasma membrane [61]. This observation makes an intriguing connection between brassinosteroid response and embryogenesis.

Increased embryogenic responses have also been reported using genes more typically associated with meristem formation, as with *A. thaliana WUSCHEL* (*AtWUS*), a key regulator of meristem cell fate [62], or *SHOOT MERISTEMLESS* (*STM*), which is required for proper meristem formation [63]. In transgenic *Coffea canephora* containing an estradiol-inducible *AtWUS* construct, leaf discs placed on estradiol increased somatic embryo formation from a control level of one somatic embryo per leaf segment (non-treated), up to a level of 3–5 somatic embryos per transgenic leaf segment after estradiol exposure [23]. Shortly after, it was reported that constitutive overexpression (using the CaMV 35S promoter) of the *Brassica napus*, *Brassica oleracea*, or *Brassica rapa* homologs of *STM* in *Arabidopsis thaliana* cotyledons placed on auxin-containing medium resulted in an approximately two-fold increase of somatic embryo formation relative to the wild-type control. In transgenic *B. napus* containing the 35S::*BnSTM* construct, a similar two-fold increase was observed in microspore-derived embryogenesis [24]. Similarly, in experiments with the objective of improving transformation methods in cotton (*Gossypium hirsutum* L.), the 35S::*AtWUS* cassette was introduced into hypocotyl segments and a three-fold increase in the formation of somatic embryos was observed [25]. Further, the somatic embryos derived from the WUS treatment (when the *WUS* gene is being overexpressed) produced leaf-like structures but failed to regenerate into plants, likely due to the deleterious effect of WUS ectopic overexpression on subsequent regeneration.

2.2. Ectopic Formation of Somatic Embryos or Meristems

In the second category, overexpression of the plant gene results in direct ectopic formation or spontaneous formation/acquisition of structures resembling embryos (often with embryo characteristics, such as increased oil levels) or meristems in the absence of inductive conditions (see Table 1, Strategy B). In 2002, two important milestone research articles were published that characterized embryonic morphogenesis as a result of ectopic overexpression of either the *Brassica napus BABY BOOM* (*BnBBM*) gene [26] or the *AtWUS* gene [38]. We will first review *BBM* and other genes involved in embryogenesis, and then later turn our attention to genes involved in meristem function. The *BBM* gene, a member of the AP2/ERF superfamily of transcription factors [64], has generated great interest among transformation researchers from the first publication [26]. In these experiments, it was observed that constitutive expression of the Brassica *BBM* gene in Arabidopsis resulted in ectopic somatic embryo formation in vegetative portions of progeny plants, for example in the shoot apex and leaves, and these ectopic somatic embryos could in turn produce plants in the absence of hormones. These results stimulated further research aimed at harnessing BBM-induced somatic embryogenesis to aid in the recovery of transgenic T0 plants.

This first publication was followed by reports in other plant species or using orthologs of the *BnBBM* gene, providing additional insights into how *BBM* worked. Srinivasan et al. [27] investigated the ability of various BBM orthologs to induce embryogenic responses in a less-related species. Spontaneous somatic embryogenesis was not observed using 35S::*AtBBM* in *Nicotiana tabacum* [27]. However, when these authors used a steroid-inducible, post-translationally controlled AtBBM fusion protein (*AtBBM~GR*) regulated by the 35S promoter to create stable lines and evaluated progeny, spontaneous ectopic shoot and root formation was observed upon addition of the inducing ligand dexamethasone (DEX). Further, when hypocotyls were exposed to DEX, somatic embryos could be induced when the growth medium contained either zeatin or benzylaminopurine. While the authors attributed the difference observed between Arabidopsis and tobacco to varying competence in response to the BBM signal, they also pointed out that expression of *BBM* from either Arabidopsis or Brassica in tobacco could produce developmental responses that differ from those observed using the endogenous tobacco *BBM* gene.

In another example of expressing an orthologous *BBM* gene, a constitutively expressed soybean gene (35S::*GmBBM*) was transformed into Arabidopsis [28], and ectopic somatic embryos were observed growing from the cotyledons, the shoot apical meristem, and the hypocotyls of stably transformed plants. Again, differences were observed in the pattern of somatic embryo formation, but, in general, the three studies provide strong evidence that constitutive expression of BBM can result in ectopic somatic embryo formation. Using the genomic clone of *Theobroma cacao BBM* (*TcBBM*) under the control of the 35S promoter, Florez et al. [29] demonstrated that it phenocopied the effects of *AtBBM* in Arabidopsis and stimulated the formation of somatic embryos from Theobroma cotyledons cultured on hormone-free media. Although somatic embryos were formed in cacao using *TcBBM*, constitutive expression prevented normal plant regeneration.

For many species, recovery of transgenic events is not the bottleneck, but instead regeneration of viable T0 plants is inefficient and rate-limiting. For example, when a *BBM* ortholog from oil palm (*Elaeis guineensis*) was cloned into an expression cassette behind the CaMV 35S promoter and then transformed into Arabidopsis, it was observed that cotyledon, leaf, or root segments from stable transgenic events exhibited enhanced rates of shoot formation relative to the wild-type controls [30]. These results are consistent with earlier observations where regeneration was improved through ectopic overexpression of BBM [26]. However, in all these reports, the 35S promoter was used to drive constitutive expression of the transgene and, as a result, no data on recovery of mature fertile plants were presented.

Tsuwamoto et al. [31] using Arabidopsis *EMBRYOMAKER* (*AtEMK*), a gene related to *BBM* (both within the AP2/ERF superfamily) driven by the CaMV 35S promoter in Arabidopsis, produced transgenic progeny that could be phenotypically evaluated. In these experiments, ectopic overexpression of *AtEMK* produced light-green embryo-like structures (possessing morphological and/or biochemical characteristics normally observed in zygotic embryos but lacking the full functionality of being able to develop into a plant) at the tip of cotyledons in 23% of the seedlings. While several of the embryo-like structures developed small features resembling roots and leaves, these outgrowths did not continue to develop. As ectopic overexpression of *AtEMK* resulted in pleiotropic effects, it may be necessary to express *AtEMK* under a regulated system for obtaining normal plantlets from somatic embryos and the embryo-like structures.

Another gene observed to function during early embryo development in Arabidopsis is *RKD4*, a member of the RWP-RK transcription factor family essential for the first asymmetrical division of the zygote to form the two cells that will give rise to the embryo and suspensor [65,66]. Following up on these characterizations of RKD4 function, Mursyanti et al. [32] demonstrated in orchid (the hybrid *Phalaenopsis* "Solo Vivien") that chemical induction of transgenic *RKD4* in leaf tissue resulted in ectopic somatic embryogenesis, a very exciting observation in a species normally reluctant to produce direct somatic embryos (de novo embryos that arise from somatic cells, having the capacity to develop into plants).

Genes normally involved in embryo maturation, such as *LEAFY COTYLEDON1* (*LEC1*), *LEAFY COTYLEDON2* (*LEC2*), and *FUSCA3* (*FUS3*), produce similar morphogenic responses when overexpressed. The first of these genes characterized in Arabidopsis was *LEC1* by Lotan et al. [33], in which a 35S::*AtLEC1* cassette was introduced into Arabidopsis using *Agrobacterium*. Progeny seed were germinated and embryo-like structures were observed in many germinating plantlets. For example, cotyledon-like structures were observed to replace what should have been the first true leaves of the seedling. The embryo-like nature of these tissues was corroborated by other embryo characteristics, such as accumulation of cruciferin-A storage protein and oleosin RNAs. However, despite forming embryo-like structures, no functional ectopic somatic embryos were observed (i.e., embryo formation was incomplete).

In a similar study, a *Citrus sinensis LEC1* paralog called *L1L* (*LEC1-Like*) was constitutively overexpressed in a 35S::*CsL1L* cassette after transformation of "Olinda" sweet orange or "Guoqing No. 1" Satsuma mandarin epicotyls using *Agrobacterium* [34]. In these experiments, the authors observed that the normally recalcitrant epicotyls formed some embryo-like structures after one month, and after another two months on elongation medium formed shoots with aberrant leaves. This suggests that in Citrus, *L1L* overexpression is sufficient to produce functional somatic embryos.

Uddenberg et al. [35] observed that overexpression of the *PaHAP3A* (a *LEC1/L1L* gene from Norway spruce, based on sequence information) did not result in ectopic somatic embryo formation in vegetative tissues. However, when expression was induced during zygotic embryo maturation, ectopic somatic embryos formed on the surface of the zygotic embryos. As noted by Srinivasan et al. [27] for BBM, this suggests that certain cell types may be more receptive to inductive signals, such as that being provided by the PaHAP3A protein [35].

In a variation on the general theme of using the CaMV 35S promoter to drive constitutive expression, Gazzarrini et al. [36] used an epidermal-specific promoter Meristem Layer1 (ML1) from Arabidopsis to drive expression of the Arabidopsis *FUSCA* gene (*AtFUS3*). Consistent with previous *LEC1* results, ectopic overexpression of *AtFUS3* resulted in the formation of cotyledon-like leaves that accumulated storage protein bodies (similar to the cells in an embryo). This observation is similar to phenotypes observed with *LEC1* overexpression.

In experiments similar to those described above, stable transgenic lines were produced by introducing 35S::*AtLEC2* into *lec2-1* and *lec2-5* mutant lines of Arabidopsis, and ectopic somatic embryos formed that were competent to germinate and produce plants [37]. However, the resultant plant phenotypes were aberrant and the authors did not comment on fertility. Nonetheless, it appears that *LEC2* might result in more complete somatic embryo formation, compared to either *LEC1*, *L1L*, or *FUS3*. This is consistent with evidence indicating that LEC2 functions upstream of and activates both *LEC1* and *FUS3* [67], in which case the penetrance of the somatic embryo phenotype might be stronger in plants overexpressing *LEC2*.

Similar to observations in which overexpression of genes involved in meristem initiation and maintenance have increased a pre-existing embryogenic response [19–22,24–26], their ectopic overexpression has also been reported to stimulate the formation of somatic embryos where they would otherwise not be observed.

The first report of a "meristem" gene stimulating embryo formation was by Zuo et al. [38], who obtained an estradiol-induced activation-tagged *pga6-1* mutant line in Arabidopsis that formed somatic embryos from root tips. This was confirmed to be *AtWUS*, which phenocopied the original activation-tagged mutant when expressed using either the estradiol-inducible or 35S promoters. Somatic embryo formation was also observed from a variety of tissues with de novo embryos from root tips being the most common observation. As with the observations by Boutilier et al. [26] with BBM, these results clearly demonstrate that WUS can stimulate the vegetative-to-embryonic transition.

By 2002 it was well-established that the proper expression of numerous genes was essential for morphogenesis. In keeping with this concept, Gallois et al. [39] analyzed the impact that two such genes might have on ectopic meristem initiation, using two different methods to control expression of STM and WUS in Arabidopsis (chemical induction and heat shock, respectively). Based on their observations of treated leaf tissue, the authors hypothesized that STM and WUS expression would produce clusters of cells adjacent to the WUS foci that represented incipient meristems (confirmed by a meristem-specific biomarker). These young ectopic meristems were initiated, but self-perpetuating meristems were not established.

In another interesting paper by Gallois et al. [40], stable transgenic lines were produced in Arabidopsis where WUS expression was activated through either HSP::CRE-mediated excision, or a GAL4-VP16 activation system. Upon either type of WUS activation, unique phenotypes were observed in root tips, and the type of response was dependent on other variables, such as hormone regime or co-expression of another morphogenic transcription factor. For example, when WUS was expressed alone in hormone-free medium, ectopic shoots and leaves were observed. When WUS was expressed in the presence of exogenous auxin (2,4-D), ectopic somatic embryos were formed. Finally, floral structures were observed when WUS was induced along with a constitutively expressed LEAFY gene, a master regulator of floral development [68].

Despite somatic embryogenesis from root tips being consistently observed when WUS expression was induced in Arabidopsis, estradiol induction of AtWUS in Nicotiana tabacum resulted in a direct organogenic response, where the root tips became swollen and developed green shoots rather than somatic embryos [41]. When AtWOX5 (a member of the WUS/WOX gene family expressed in the root tip) was substituted in these experiments, root tip swelling and green shoot formation were again observed [42].

While Gallois et al. [39] focused on the interaction of WUS and STM in Arabidopsis, the impact of the maize STM ortholog KNOTTED1 (KN1) has also been reported. When the maize KN1 gene was constitutively overexpressed (35S::ZmKN1) in Nicotiana tabacum, it resulted in a 3-fold increase in shoot organogenesis, relative to the NPTII-only control [43]. The increase in shoots in the KN1 treatment was obtained with no antibiotic or herbicide selection and no exogenous hormones in the media, and the resultant plants were bushy, with altered leaf morphology and underdeveloped roots.

The overexpression of KN1 (using the CaMV 35S promoter) can also bypass an intermediate callus phase, as reported by Nishimura et al. [44]. Constitutive expression of tobacco KN1 orthologs in Nicotiana tabacum resulted in a range of pleiotropic phenotypes. Transgenic plants containing the NTH20 (a knotted-like homeobox gene) expression cassette occasionally produced ectopic shoot meristems that would develop into small shoots with leaves emerging from the original leaf surface.

Other genes that play a role in shoot meristem formation are the CUP-SHAPED COTYLEDON genes CUC1 and CUC2. Using the Arabidopsis CUC1 and CUC2 genes, Daimon et al. [45] showed that overexpression of these genes under a strong promoter (CaMV 35S) led to the rapid production of adventitious shoots in transgenic calli derived from Arabidopsis hypocotyls. CUC1- and CUC2- overexpressing calli produced an average of 4.8 and 3.3 adventitious shoots per callus, respectively, while the controls produced 0.5 shoots that developed more slowly. In the absence of the phytohormones, no adventitious shoots were formed, indicating that CUC1 and CUC2 function was hormone-dependent.

In the context of using plant genes to improve plant transformation, genes involved in hormone signal transduction (whether receptors or downstream targets) also fall under our 'morphogenic' classification. Two such candidates are the Arabidopsis ENHANCER OF SHOOT REGENERATION genes (ESR1 and ESR2), both identified through mutant screening and demonstrated to be involved in the cytokinin response pathway. As part of this characterization, it was demonstrated that overexpression of both ESR1 and ESR2 conferred cytokinin-independent shoot formation [47,48]. The

Arabidopsis auxin-response gene *MONOPTEROS* (*MP*) is also of interest in this respect. In experiments focusing on the role of the *MP* gene on shoot formation, it was observed that using the endogenous promoter to drive expression of a C-terminally deleted gene referred to as *MP∆* (lacking the domain involved in auxin/IAA interactions) increased the formation of shoot apical meristems from callus [49].

In addition to genes involved in the hormone signaling pathway, levels of hormones can affect the phenotypic response of genes involved in morphogenesis. Transgenic Arabidopsis explants overexpressing *LEC2* under a DEX-inducible system produced somatic embryos in the presence of low auxin concentrations, while increasing the concentration resulted in the production of calli [46].

3. Strategies to Improve Transformation Using Morphogenic Genes

Given the focus of this review, it is easy to see why researchers in plant transformation would view the genes described above as potential tools in the transformation process. However, constitutive and strong expression of these morphogenic genes often caused undesired pleiotropic effects, including reduced fertility. In the examples discussed in this section, constitutive expression of the morphogenic genes (Figure 1A) or inducible expression in stably transformed plants (Figure 1B) were the predominant strategies. To render true utility, an additional step is required: combining optimized expression of the plant morphogenic genes being used with a robust method to limit expression after plant transformation/regeneration has occurred and their utility has expired. Such methods typically demonstrate improved transformation efficiency, enhanced regeneration, or both in a manner that produces healthy, fertile T0 plants. It should also be emphasized that to characterize gain-of-function phenotypes, as in many of the examples above, stable transgenic germplasm was produced using conventional selection methods (for example, after floral-dip transformation in Arabidopsis). As a result, the ectopic phenotype was evaluated in homogeneously transgenic tissues in seed-derived plants. This contrasts with expressing a morphogenic gene in a single cell surrounded by wild-type tissue, as when trying to use a morphogenic gene to recover transgenic events. In this situation, differential expression of the morphogenic gene provides a positive growth advantage or an identifiable phenotype (relative to wild-type cells) that can be used for selection. This is the case in the examples described below (see Table 1, Strategy C).

There are several reports that describe ectopic overexpression of morphogenic genes; however, deleterious pleiotropic phenotypes were observed in plants when constitutively expressed. Studies on how to control the timing and level of expression for these genes through downregulation or elimination have lagged. To date, we can identify four approaches to address this problem in the literature: (i) stimulating the morphogenic growth response through inducible expression of the morphogenic gene (Figure 1B) followed by removal of the inducing ligand to turn off expression, (ii) excision of the plant morphogenic gene (Figure 1C) when no longer needed, (iii) use of a plant promoter that turns off when no longer needed to permit normal growth and reproduction in transgenic plants, and (iv) using Agrobacterium-mediated delivery in a manner that favors transient expression of the morphogenic genes (Figure 1D,E).

Inducible expression has provided a robust method for using morphogenic genes to recover fertile transgenic plants. There are few reports on the transformation of recalcitrant species, and thus the first case presented here of using an inducible morphogenic gene for improving transformation and regeneration deserves emphasis. Pepper varieties (*Capsicum annuum*) have very poor transformation efficiencies and regenerative capacity. Heidmann et al. [50] transformed cotyledon explants of two sweet pepper varieties with a 35S::*BnBBM~GR* construct and cultured the explants on media supplemented with thidiazuron and DEX for 2 months. Emerging shoots were transferred to DEX-free elongation medium for 4 weeks and then pre-rooting medium for a month. Transformation efficiency with the regulated *BBM* expression was >1% compared to 0% with a 35S::*GUS* (*β-glucuronidase*) construct. This is an important step forward in what has historically been a very recalcitrant crop.

Also using the DEX-inducible *AtBBM~GR* system, Lutz and colleagues [51] described a method to obtain fertile, transgenic Arabidopsis plants from leaf cultures. In the absence of auxins, transgenic leaf explants produced prolific shoots in around 4 weeks in the presence of DEX, while explants on medium without the ligand did not produce shoots, and explants from wild-type plants became necrotic, irrespective of whether DEX was added or not. Recovered shoots were further cut into smaller segments and regenerated on medium containing DEX for 3 months. Removing the shoots from the DEX-containing medium allowed for regeneration of plantlets with normal flowering and seed formation over 3 months' time. Fertile transgenic progeny were produced from the collected seeds.

Using the estradiol-inducible system [38], Wang et al. [52] identified *PLANT GROWTH ACTIVATION* genes, such as *PGA37*, which resulted in a vegetative-to-embryogenic transition when overexpressed in Arabidopsis. While the downstream targets have yet to be characterized, *PGA37* was determined to encode a MYB118 transcription factor based on structural similarities within the DNA-binding domain. When *PGA37* was expressed under inducible control, somatic embryos developed from root explants and was associated with increased *LEC1* expression. Estradiol-induced expression of *PGA37* in the presence of auxin produced green-yellowish embryonic calli in 7–10 days and generated somatic embryos upon culturing for 3–5 weeks. Upon removal of estradiol from the medium (thus downregulating *PGA37* expression), the somatic embryos developed into healthy, fertile plantlets. Overexpression of a closely related homolog, *MYB115*, under the estradiol inducible system also led to the formation of somatic embryos from root explants.

When transforming recalcitrant species, such as trees, that in addition have maturation periods running into years, direct production of somatic embryos offers a relatively fast way for genetic manipulation [69]. One such example is use of the morphogenic gene *LEC2* to generate transgenic plants in *Theobroma cacao*, as described by Shires and colleagues [53]. They identified the *TcLEC2* transcription factor sequence and cloned it as a translational fusion with GR (the Glucocorticoid Receptor) driven by a CaMV 35S promoter. This DEX-inducible construct was used to transform cotyledon tissue of the variety Scavina-6 and cultured for about 6 months to screen for transgenic somatic embryos. Consistent with the transformation recalcitrance in *T. cacao*, only one transgenic embryo was recovered. This was proliferated by segmenting the cotyledons into several pieces to produce clonal-transformed somatic embryos. When tissue from these secondary embryos was placed on hormone-free media supplemented with DEX, multiple embryos were formed in 6 days. Levels of DEX up to 50 µM produced the most embryos (403) per 100 explants over a period of 4 months. In this experiment, a single somatic embryo was converted to a transgenic plant that developed normally. Young leaf tissue from the transgenic 35S::*TcLEC2*:GR plant was capable of prolific somatic embryo formation in the presence of DEX after 3 months, providing a promising method to regenerate secondary transgenics from leaf material.

In a recent publication using the estradiol-inducible system to control *WOX* gene expression, two combinations of Arabidopsis-derived *WOX* genes (*WOX2* + *WOX8* or *WOX2* + *WOX9*) were evaluated in the presence of 1 µM 2,4-D for 10 days. Both combinations resulted in substantial plantlet regeneration from *Nicotiana tabacum* leaf pieces, in contrast to the wild-type control where no plantlet regeneration was observed [54].

Excision-based strategies to control morphogenic gene expression are another alternative. The first demonstration of this concept in a dicot was reported in *Populus tomentosa* using *BBM* and FLP-recombinase for excision [55]. These authors designed a T-DNA construct consisting of a single pair of FRT (FLP Recombination Target Sites) flanking both a heat-shock inducible promoter driving FLP recombinase expression and a CaMV 35S promoter driving expression of the *Brassica campestris BBM* gene cassette. Using *Agrobacterium*-mediated transformation, 21 callus cultures of Chinese white poplar were transformed with this T-DNA and cultured on hormone-free medium. Six of the 21 calli developed a total of 12 somatic embryos approximately four weeks after transformation, and half of the somatic embryos germinated to form plantlets that had a dwarf phenotype with small wrinkled leaves when cultured for 60 days. Heat shock treatment at 42 °C for 2 hours led to excision of both the *FLP* recombinase and *BBM* cassettes in four of the six plants, resulting in reappearance of the normal phenotype in regenerated plants.

The first demonstration of using morphogenic genes followed by excision to improve monocot transformation has only recently been reported by Lowe et al. [17]. After *Agrobacterium*-mediated transformation of immature embryos of a normally recalcitrant Pioneer maize inbred, a strongly expressed *ZmBBM* (using the *Zea mays* UBIQUITIN promoter) plus a weakly expressed *ZmWUS2* (using the *Agrobacterium* NOPALINE SYNTHASE, or NOS, promoter) resulted in the stimulation of embryogenic callus. This combination of a weakly expressed *WUS2* gene and a strongly expressed *BBM* gene resulted in high transformation frequencies when immature embryos of the maize inbred were transformed. The callus was placed on dry filter paper for 3 days to stimulate a desiccation-induced maize promoter from an ABA-responsive gene (RAB17) driving *CRE* recombinase, which then efficiently excised all three expression cassettes. After excision of the *CRE*, *WUS2*, and *BBM* transgenes, all that remained in the integrated T-DNA locus were the genes of interest (for example, herbicide resistance and/or a visual marker gene). This also improved transformation efficiency in many other difficult or recalcitrant genotypes (difficult corn inbreds, sorghum, sugarcane, and Indica rice), and permitted transformation of previously non-transformable explants, such as mature embryo sections (starting with mature, dry seed) or leaf segments from 7–14-day-old seedlings. This method is beginning to facilitate enhanced transformation of previously recalcitrant public cereal varieties, such as the maize inbred B73 and the sorghum variety P898012 [56].

More recently, Lowe et al. [57] described an improved transformation system for maize using two new promoters: the maize AXIG1 promoter (auxin-inducible) driving *WUS2* and the maize PLTP promoter (PhosphoLipid Transfer Protein promoter) driving *BBM*. This new configuration of expression cassettes resulted in rapid formation of somatic embryos within 7 days, with germination of these newly formed somatic embryos producing plantlets ready for transplantation into soil and growth in the greenhouse within 21–30 days. Expression of both promoters was so low or confined to specific cell types (or tissues) in the plant that, even without excision of the PLTP::*BBM* and AXIG1::*WUS2* cassettes, the resultant T0 plants were all robust and fertile. This new rapid transformation system has worked in all Pioneer and public inbreds tested, as well as in recalcitrant sorghum and wheat varieties.

A recurring theme in all the transformation methods described above is the necessity of controlling expression of the morphogenic genes in order to recover normal-phenotype plants, through either inducible expression, developmentally regulated expression, or excision. In the case of inducible or developmentally regulated expression of *BBM* and *WUS2*, the morphogenic genes and trait genes are linked within the same construct. Alternatively, constructs designed for excision are larger and more complex, but the morphogenic genes are removed prior to plant regeneration.

However, two additional alternatives exist that avoid both pitfalls. In the first alternative, two T-DNAs are introduced from the same *Agrobacterium* (containing both plasmids). In the second alternative, two separate Agrobacteria are used to introduce different T-DNA plasmids. Using either strategy, the two transgenic loci are later segregated away from each other [70]. The basis for the second approach was illuminated in a report by Florez et al. [29], who demonstrated that transient expression of *TcBBM* delivered by *Agrobacterium* was sufficient to stimulate somatic embryo formation, with no indication of the transgenic *BBM* cassette being detected in the somatic embryos. The authors then speculated that a co-transfection technique could be used to obtain transgenic plants by mixing an *Agrobacterium* strain containing a *BBM* expression cassette along with a trait-containing strain.

One final variation on using *Agrobacterium* to deliver transient morphogenic gene activity relies on a commonly observed characteristic of T-DNA delivery, with many labs over the years reporting on the phenomenon of T-DNA border read-through [71–74]. As the name implies, when the T-DNA is processed in the *Agrobacterium* before delivery, inefficient T-strand processing will produce some percentage of T-DNA molecules that are not terminated precisely at the Left Border (LB) but continue to include sequence beyond the LB. As a means of reducing this type of unwanted read-through, researchers have employed negative selectable markers [72] positioned beyond the LB (outside the T-DNA) to eliminate plant cells that had integrated these sequences. Others have tried to exploit inefficient T-strand processing by positioning a positive marker gene beyond the LB that could be used transiently, producing selectable marker-free transgenic plants [75]. By placing an *Agrobacterium*-derived *IPT* gene outside the LB, many cells would receive a mixture of T-DNAs, with the majority of the T-strands processed properly (only containing the trait), but with a smaller percentage of T-strands not terminated properly at the LB thus containing the flanking *IPT* gene. In these cells, transient *IPT* expression would stimulate cytokinin production and shoot proliferation, and when the processed T-strand integrated and the *IPT*-containing T-strand did not, this would result in the recovery of trait-containing transgenic plants that contain no selectable marker.

Since it has not been reported that shoot proliferation in response to *IPT* expression occurs in maize and other cereal crops, we tested *WUS2* alone or *WUS2* + *BBM* for transient somatic embryo formation and subsequent germination to produce T0 plants by positioning these cassettes outside the LB [76]. This resulted in a simple non-excision method for using *WUS2* and *BBM* for transformation of recalcitrant maize genotypes that exploits inefficiencies of both T-strand processing and integration to allow for rapid transformation of maize while enriching for events with no *WUS2* or *BBM* integration.

For example, *Agrobacterium*-mediated T-DNA delivery into the Pioneer maize inbred PH1V69 is very efficient; however, without *BBM* and *WUS2* expression cassettes present in the T-DNA, it has so far proven impossible to recover transgenic T0 plants. Using inbred PH1V69, transformation was performed using an *Agrobacterium* strain LBA4404 that contained a single T-DNA plasmid [76]. Two versions were tested, with both containing two 'Mock Trait' expression cassettes within the T-DNA, a *Setaria italica*-derived UBIQUITIN promoter driving expression of a green fluorescent protein and an herbicide resistance cassette containing a *Sorghum bicolor* Acetolactate Synthase (ALS) promoter driving expression of *HRA* (a mutant maize *ALS* gene that confers resistance to sulfonylurea herbicides). For the first treatment, two expression cassettes were also positioned outside the LB of the T-DNA, with a PLTP promoter driving *WUS2*, and a PLTP promoter driving expression of *BBM* (*WUS2/BBM*). In a second treatment, only the PLTP::*WUS2* was placed outside the LB (*WUS2* only). *Agrobacterium*-infection, resting, somatic embryo maturation, and regeneration were performed as described in Lowe et al. [57] and T0 plants were analyzed for the presence of the marker genes plus *WUS2* and *BBM* (when applicable) using qPCR. Starting with 196 immature embryos in each treatment, it was observed after analysis that the frequency of T0 plants (relative to the number of starting immature embryos) that were single-copy for marker genes and were negative for *WUS2* and *BBM* (when applicable) was 12.2% and 10.2% for the '*WUS2* only' and '*WUS2/BBM*' treatments, respectively. The percentage of T0 plants that contained either *BBM/WUS2* or *WUS2* alone (depending on the treatment) was 49% and 38%, respectively [76]. This suggests that while 'read-through' copies of *WUS2* or *WUS2/BBM* were clearly having a positive impact in terms of stimulating somatic embryo formation, there was also some unavoidable integration of T-DNA sequences that also carried along the flanking sequence ('backbone' from the *Agrobacterium* T-DNA-containing plasmid). This is expected and, consequently, the method requires PCR screening to identify perfect, single-copy T-DNA events. However, it should also be noted that the frequency of recovering perfect, single-copy events was comparable to that observed for the excision method for this inbred [17]. This represents a viable alternative to excision as a means of creating high-quality transgenic events in recalcitrant monocot crops that do not contain helper genes (in this case *WUS2* and *BBM*).

4. Conclusions

There has been meteoric progress in plant genome modification engendered by CRISPR/*CAS9* (CLUSTERED REGULARLY-INTERSPERSED SHORT PALINDROMIC REPEATS and the *CRISPR-ASSOCIATED 9* gene) over the past half-decade. This explosion has also brought into sharp focus the impediment presented by the state of transformation technology for many crops [1]. For maize, use of the morphogenic genes *WUS2* and *BBM* has mitigated this bottleneck and has been used in-house for several years for all aspects of our genome modification programs. These include particle-gun-mediated creation of mini-chromosomes [77], CRISPR/CAS9-mediated mutagenesis or editing [78,79], and, of course, random *Agrobacterium*-mediated transformation [17,57].

From our experience with other cereal crops [17] and the progress by Mookkan and colleagues in recalcitrant public lines [56], we feel this technology should make all aspects of genome modification accessible to all cereals, with future enhancements continuing to simplify and improve this approach. Similarly, based on observed morphogenic responses in eudicots and gymnosperms, broadening these methods to include more plant species will hopefully continue to erode the barriers that make so many crops inaccessible for genome editing.

In the foreseeable future, however, finding a single solution that works across all crops is unlikely. Different species and even different varieties within the same species will require new combinations of morphogenic triggers (new combinations of genes or varied expression patterns) to produce either somatic embryos or new apical meristems for rapid production of genetically modified plants. Basic research over the past three decades has provided us with a detailed understanding of the genes that control morphogenesis, and the signaling networks that are so critical to meristem and embryo development, with new insights constantly being discovered. These insights will continue to provide the inspiration for testing morphogenic genes (or combinations of genes) and, along with new strategies to control or limit expression, will result in continued improvements that expand the range of plant species amenable to transformation. Hopefully, this will make plant transformation much more efficient, routine, and accessible for all crops of interest, and will alleviate this key bottleneck to crop improvement, enabling CRISPR/CAS-mediated genome modification in many important crops.

Author Contributions: Writing Review & Editing, N.S.; M.A.; K.L.; G.H.; S.B.; T.J.; B.G.-K.

Funding: This research received no external funding.

Conflicts of Interest: The authors declare no conflict of interest.

References

1. Altpeter, F.; Springer, N.M.; Bartley, L.E.; Blechl, A.E.; Brutnell, T.P.; Citovsky, V.; Conrad, L.J.; Gelvin, S.B.; Jackson, D.P.; Kausch, A.P.; et al. Advancing Crop Transformation in the Era of Genome Editing. *Plant Cell* **2016**, *28*, 1510–1520. [CrossRef] [PubMed]
2. Smertenko, A.; Bozhkov, P.V. Somatic embryogenesis: Life and death processes during apical-basal patterning. *J. Exp. Bot.* **2014**, *65*, 1343–1360. [CrossRef] [PubMed]
3. Snipes, S.A.; Rodriguez, K.; DeVries, A.E.; Miyawaki, K.N.; Perales, M.; Xie, M.; Reddy, G.V. Cytokinin stabilizes WUSCHEL by acting on the protein domains required for nuclear enrichment and transcription. *PLOS Genet.* **2018**, *14*, e1007351. [CrossRef] [PubMed]
4. Grienenberger, E.; Fletcher, J.C. Polypeptide signaling molecules in plant development. *Curr. Opin. Plant Biol.* **2015**, *23*, 8–14. [CrossRef] [PubMed]
5. Barton, M.K. Twenty years on: The inner workings of the shoot apical meristem, a developmental dynamo. *Dev. Biol.* **2010**, *341*, 95–113. [CrossRef] [PubMed]
6. Ikeuchi, M.; Ogawa, Y.; Iwase, A.; Sugimoto, K. Plant regeneration: Cellular origins and molecular mechanisms. *Development* **2016**, *143*, 1442–1451. [CrossRef]
7. Skoog, F.; Miller, C.O. Chemical regulation of growth and organ formation in plant tissues cultured. *In Vitro Symp. Soc. Exp. Biol.* **1957**, *11*, 118–130.
8. Murashige, T. Plant propagation through tissue cultures. *Ann. Rev. Plant Physiol.* **1974**, *25*, 135–166. [CrossRef]
9. Smigocki, A.C.; Owens, L.D. Cytokinin gene fused with a strong promoter enhances shoot organogenesis and zeatin levels in transformed plant cells. *Proc. Natl. Acad. Sci. USA* **1998**, *85*, 5131–5135. [CrossRef]
10. Ondřej, M.; Macháčková, I.; Čatský, J.; Eder, J.; Hrouda, M.; Pospíšilová, J.; Synková, H. Potato transformation by T-DNA cytokinin synthesis gene. *Biol. Plant.* **1990**, *32*, 401–406. [CrossRef]
11. Casanova, E.; Valdes, A.E.; Zuker, A.; Frenandez, B.; Vainstein, A.; Trillas, M.I. rolC-transgenic carnation plants: Adventitious organogenesis and levels of endogenous auxin and cytokinins. *Plant Sci.* **2004**, *167*, 551–560. [CrossRef]
12. Gorpenchenko, T.Y.; Kiselev, K.V.; Bulgakov, V.P.; Tchernoded, G.K.; Bragina, E.A.; Khodakovskaya, M.V.; Koren, O.G.; Batygina, T.B.; Zhuravlev, Y.N. The *Agrobacterium rhizogenes rolC*-gene-induced somatic

embryogenesis and shoot organogenesis in *Panax* ginseng transformed calluses. *Planta* **2006**, *223*, 457–467. [CrossRef] [PubMed]
13. Otten, L. The *Agrobacterium* Phenotypic Plasticity (*Plast*) Genes. In *Current Topics in Microbiology and Immunology*; Springer: Berlin/Heidelberg, Germany, 2018.
14. Roeckel, P.; Oancia, T.; Drevet, J.R. Phenotypic alterations and component analysis of seed yield in transgenic *Brassica napus* plants expressing the *tzs* gene. *Physiol. Plant.* **1998**, *102*, 243–249. [CrossRef]
15. Choi, Y.I.; Noh, E.W.; Choi, K.S. Low level expression of prokaryotic tzs gene enhances growth performance of transgenic poplars. *Trees* **2009**, *23*, 7441–7750. [CrossRef]
16. Gordon-Kamm, W.; Dilkes, B.P.; Lowe, K.; Hoerster, G.; Sun, X.; Ross, M.; Church, L.; Bunde, C.; Farrell, J.; Hill, P.; et al. Stimulation of the cell cycle and maize transformation by disruption of the plant retinoblastoma pathway. *Proc. Nat. Acad. Sci. USA* **2002**, *99*, 11975–11980. [CrossRef] [PubMed]
17. Lowe, K.; Wu, E.; Wang, N.; Hoerster, G.; Hastings, C.; Cho, M.J.; Scelonge, C.; Lenderts, B.; Chamberlin, M.; Cushatt, J.; et al. Morphogenic Regulators Baby boom and Wuschel Improve Monocot Transformation. *Plant Cell* **2016**, *28*, 1998–2015. [CrossRef] [PubMed]
18. Hecht, V.; Velle-Calzada, J.P.; Hartog, M.V.; Schmidt, E.D.; Boutilier, K.; Grossniklaus, U.; de Vries, S.C. The Arabidopsis SOMATIC EMBRYOGENESIS RECEPTOR KINASE 1 gene is expressed in developing ovules and embryos and enhances embryogenic competence in culture. *Plant Physiol.* **2001**, *127*, 803–816. [CrossRef]
19. Pérez-Pascual, D.; Jiménez-Guillen, D.; Villanueva-Alonzo, H.; Souza-Perera, R.; Godoy-Hernández, G.; Zúñiga-Aguilar, J.J. Ectopic expression of the *Coffea canephora* SERK1 homolog-induced differential transcription of genes involved in auxin metabolism and in the developmental control of embryogenesis. *Physiol. Plant.* **2018**, *163*, 530–551. [CrossRef]
20. Harding, E.W.; Tang, W.; Nichols, K.W.; Fernandez, D.E.; Perry, S.E. Expression and maintenance of embryogenic potential is enhanced through constitutive expression of AGAMOUS-LIKE15. *Plant Physiol.* **2003**, *133*, 653–663. [CrossRef]
21. Thakare, D.; Tang, W.; Hill, K.; Perry, S.E. The MADS-domain transcriptional regulator AGAMOUS-LIKE15 promotes somatic embryo development in Arabidopsis and soybean. *Plant Physiol.* **2008**, *146*, 1663–1672. [CrossRef]
22. Yang, Z.; Li, C.; Wang, Y.; Zhang, C.; Wu, Z.; Zhang, X.; Liu, C.; Li, F. GhAGL15s, preferentially expressed during somatic embryogenesis, promote embryogenic callus formation in cotton (*Gossypium hirsutum* L.). *Mol. Genet. Genom.* **2014**, *289*, 873–883. [CrossRef] [PubMed]
23. Arroyo-Herrera, A.; Gonzalez, A.K.; Moo, R.C.; Quiroz-Figueroa, F.R.; Loyola-Vargas, V.M.; Rodriguez-Zapata, L.C.; Burgeff D'Hondt, C.; Suárez-Solís, V.M.; Castaño, E. Expression of WUSCHEL in *Coffea canephora* causes ectopic morphogenesis and increases somatic embryogenesis. *Plant Cell Tissue Organ Cult.* **2008**, *94*, 171–180. [CrossRef]
24. Elhiti, M.; Tahir, M.; Gulden, R.H.; Khamiss, K.; Stasolla, C. Modulation of embryo-forming capacity in culture through the expression of Brassica genes involved in the regulation of the shoot apical meristem. *J. Exp. Bot.* **2010**, *61*, 4069–4085. [CrossRef] [PubMed]
25. Bouchabké-Coussa, O.; Obellianne, M.; Linderme, D.; Montes, E.; Maia-Grondard, A.; Vilaine, F.; Pannetier, C. Wuschel overexpression promotes somatic embryogenesis and induces organogenesis in cotton (*Gossypium hirsutum* L.) tissues cultured in vitro. *Plant Cell Rep.* **2013**, *32*, 675–686. [CrossRef] [PubMed]
26. Boutilier, K.; Offringa, R.; Sharma, V.K.; Kieft, H.; Ouellet, T.; Zhang, L.; Hattori, J.; Liu, C.M.; van Lammeren, A.A.; Miki, B.L.; et al. Ectopic expression of BABY BOOM triggers a conversion from vegetative to embryonic growth. *Plant Cell* **2002**, *14*, 1737–1749. [CrossRef] [PubMed]
27. Srinivasan, C.; Liu, Z.; Heidmann, I.; Supena, E.D.; Fukuoka, H.; Joosen, R.; Lambalk, J.; Angenent, G.; Scorza, R.; Custers, J.B.; et al. Heterologous expression of the BABY BOOM AP2/ERF transcription factor enhances the regeneration capacity of tobacco (*Nicotiana tabacum* L.). *Planta* **2007**, *225*, 341–351. [CrossRef]
28. El Ouakfaoui, S.; Schnell, J.; Abdeen, A.; Colville, A.; Labbé, H.; Han, S.; Baum, B.; Laberge, S.; Miki, B. Control of somatic embryogenesis and embryo development by AP2 transcription factors. *Plant Mol. Biol.* **2010**, *74*, 313–326. [CrossRef]
29. 29 Florez, S.L.; Erwin, R.L.; Maximova, S.N.; Guiltinan, M.J.; Curtis, W.R. Enhanced somatic embryogenesis in Theobroma cacao using the homologous BABY BOOM transcription factor. *BMC Plant Biol.* **2015**, *16*, 121. [CrossRef]

30. Morcillo, F.; Gallard, A.; Pillot, M.; Jouannic, S.; Aberlenc-Bertossi, F.; Collin, M.; Verdeil, J.L.; Tregear, J.W. EgAP2-1, an AINTEGUMENTA-like (AIL) gene expressed in meristematic and proliferating tissues of embryos in oil palm. *Planta* **2007**, *226*, 1353–1362. [CrossRef]
31. Tsuwamoto, R.; Yokoi, S.; Takahata, Y. *Arabidopsis EMBRYOMAKER* encoding an AP2 domain transcription factor plays a key role in developmental change from vegetative to embryonic phase. *Plant Mol. Biol.* **2010**, *73*, 481–492. [CrossRef]
32. Mursyanti, E.; Purwantoro, A.; Moeljopawiro, S.; Semiarti, E. Induction of Somatic Embryogenesis through Overexpression of ATRKD4 Genes in *Phalaenopsis* "Sogo Vivien". *Indones. J. Biotechnol.* **2015**, *20*, 42–53. [CrossRef]
33. Lotan, T.; Ohto, M.; Yee, K.M.; West, M.A.; Lo, R.; Kwong, R.W.; Yamagishi, K.; Fischer, R.L.; Goldberg, R.B.; Harada, J.J. Arabidopsis LEAFY COTYLEDON1 is sufficient to induce embryo development in vegetative cells. *Cell* **1998**, *93*, 1195–1205. [CrossRef]
34. Zhu, S.-P.; Wang, J.; Ye, J.-L.; Zhu, A.-D.; Guo, W.-W.; Deng, X.-X. Isolation and characterization of LEAFY COTYLEDON 1-LIKE gene related to embryogenic competence in *Citrus sinensis*. *Plant Cell Tiss. Organ Cult.* **2014**, *119*, 1–13. [CrossRef]
35. Uddenberg, D.; Abrahamsson, M.; von Arnold, S. Overexpression of PaHAP3A stimulates differentiation of ectopic embryos from maturing somatic embryos of Norway spruce. *Tree Genet. Genomes* **2016**, *12*, 18. [CrossRef]
36. Gazzarrini, S.; Tsuchiya, Y.; Lumba, S.; Okamoto, M.; McCourt, P. The transcription factor FUSCA3 controls developmental timing in Arabidopsis through the hormones gibberellin and abscisic acid. *Dev. Cell* **2004**, *7*, 373–385. [CrossRef] [PubMed]
37. Stone, S.L.; Kwong, L.W.; Yee, K.M.; Pelletier, J.; Lepiniec, L.; Fischer, R.L.; Goldberg, R.B.; Harada, J.J. LEAFY COTYLEDON2 encodes a B3 domain transcription factor that induces embryo development. *Proc. Natl. Acad. Sci. USA* **2001**, *98*, 11806–11811. [CrossRef]
38. Zuo, J.; Niu, Q.W.; Frugis, G.; Chua, N.H. The WUSCHEL gene promotes vegetative-to-embryonic transition in Arabidopsis. *Plant J.* **2002**, *30*, 349–359. [CrossRef]
39. Gallois, J.L.; Woodward, C.; Reddy, G.V.; Sablowski, R. Combined SHOOT MERISTEMLESS and WUSCHEL trigger ectopic organogenesis in Arabidopsis. *Development* **2002**, *129*, 3207–3217.
40. Gallois, J.L.; Nora, F.R.; Mizukami, Y.; Sablowski, R. WUSCHEL induces shoot stem cell activity and developmental plasticity in the root meristem. *Genes Dev.* **2004**, *18*, 375–380. [CrossRef]
41. Rashid, S.Z.; Yamaji, N.; Kyo, M. Shoot formation from root tip region: A developmental alteration by WUS in transgenic tobacco. *Plant Cell Rep.* **2007**, *26*, 1449–1455. [CrossRef]
42. Rashid, S.Z.; Kyo, M. Ectopic Expression of WOX5 Dramatically Alters Root-tip Morphology in Transgenic Tobacco. *Transgenic Plant J.* **2009**, *3*, 92–96.
43. Luo, K.; Zheng, X.; Chen, Y.; Xiao, Y.; Zhao, D.; McAvoy, R.; Pei, Y.; Li, Y. The maize Knotted1 gene is an effective positive selectable marker gene for Agrobacterium-mediated tobacco transformation. *Plant Cell Rep.* **2006**, *25*, 403–409. [CrossRef] [PubMed]
44. Nishimura, A.; Tamaoki, M.; Sakamoto, T.; Matsuoka, M. Over-Expression of Tobacco knotted 1-Type ClassI Homeobox Genes Alters Various Leaf Morphology. *Plant Cell Physiol.* **2000**, *41*, 583–590. [CrossRef] [PubMed]
45. Daimon, Y.; Takabe, K.; Tasaka, M. The *CUP-SHAPED COTYLEDON* genes promote adventitious shoot formation on calli. *Plant Cell. Physiol.* **2003**, *44*, 113–121. [CrossRef] [PubMed]
46. Wójcikowska, B.; Jaskóła, K.; Gąsiorek, P.; Meus, M.; Nowak, K.; Gaj, M.D. LEAFY COTYLEDON2 (LEC2) promotes embryogenic induction in somatic tissues of Arabidopsis, via YUCCA-mediated auxin biosynthesis. *Planta* **2013**, *238*, 425–440. [CrossRef] [PubMed]
47. Banno, H.; Ikeda, Y.; Niu, Q.W.; Chua, N.H. Overexpression of Arabidopsis ESR1 induces initiation of shoot regeneration. *Plant Cell* **2001**, *13*, 2609–2618. [CrossRef]
48. Ikeda, Y.; Banno, H.; Niu, Q.W.; Howell, S.H.; Chua, N.H. The ENHANCER OF SHOOT REGENERATION 2 gene in Arabidopsis regulates CUP-SHAPED COTYLEDON 1 at the transcriptional level and controls cotyledon development. *Plant Cell Physiol.* **2006**, *47*, 1443–1456. [CrossRef]
49. Ckurshumova, W.; Smirnova, T.; Marcos, D.; Zayed, Y.; Berleth, T. Irrepressible MONOPTEROS/ARF5 promotes de novo shoot formation. *New Phytol.* **2014**, *204*, 556–566. [CrossRef]
50. Heidmann, I.; de Lange, B.; Lambalk, J.; Angenent, G.C.; Boutilier, K. Efficient sweet pepper transformation mediated by the BABY BOOM transcription factor. *Plant Cell Rep.* **2011**, *30*, 1107–1115. [CrossRef]

51. Lutz, K.A.; Martin, C.; Khairzada, S.; Maliga, P. Steroid-inducible BABY BOOM system for development of fertile Arabidopsis thaliana plants after prolonged tissue culture. *Plant Cell Rep.* **2015**, *34*, 1849–1856. [CrossRef] [PubMed]
52. Wang, X.; Niu, Q.-W.; Teng, C.; Li, C.; Mu, J.; Chua, N.-H.; Zuo, J. Overexpression of *PGA37/MYB118* and *MYB115* promotes vegetative-to-embryonic transition in *Arabidopsis*. *Cell Res.* **2009**, *19*, 224–235. [CrossRef] [PubMed]
53. Shires, M.E.; Florez, S.L.; Lai, T.S.; Curtis, W.R. Inducible somatic embryogenesis in *Theobroma cacao* achieved using the DEX-activatable transcription factor-glucocorticoid receptor fusion. *Biotechnol. Lett.* **2017**, *39*, 1747–1755. [CrossRef] [PubMed]
54. Kyo, M.; Maida, K.; Nishioka, Y.; Matsui, K. Coexpression of WUSCHEL related homeobox (WOX) 2 with WOX8 or WOX9 promotes regeneration from leaf segments and free cells in *Nicotiana tabacum* L. *Plant Biotechnol.* **2018**, *35*, 23–30. [CrossRef]
55. Deng, W.; Li, Z.; Luo, K.; Yang, Y. A novel method for induction of plant regeneration via somatic embryogenesis. *Plant Sci.* **2009**, *177*, 43–48. [CrossRef]
56. Mookkan, M.; Nelson-Vasilchik, K.; Hague, J.; Zhang, Z.J.; Kausch, A.P. Selectable marker independent transformation of recalcitrant maize inbred B73 and sorghum P898012 mediated by morphogenic regulators BABY BOOM and WUSCHEL2. *Plant Cell Rep.* **2017**, *36*, 1477–1491. [CrossRef] [PubMed]
57. Lowe, K.; La Rota, M.; Hoerster, G.; Hastings, C.; Wang, N.; Chamberlin, M.; Wu, E.; Jones, T.; Gordon-Kamm, W. Rapid genotype "independent" *Zea mays* L. (maize) transformation via direct somatic embryogenesis. *In Vitro Cell Dev. Biol. Plant.* **2018**, *54*, 240–252. [CrossRef] [PubMed]
58. Karlova, R.; Boeren, S.; Russinova, E.; Aker, J.; Vervoort, J.; de Vries, S. The Arabidopsis SOMATIC EMBRYOGENESIS RECEPTOR-LIKE KINASE1 protein complex includes brassinosteroid-insensitive1. *Plant Cell* **2006**, *18*, 626–638. [CrossRef] [PubMed]
59. Gou, X.; Yin, H.; He, K.; Du, J.; Yi, J.; Xu, S.; Lin, H.; Clouse, S.D. Genetic Evidence for an Indispensable Role of Somatic Embryogenesis Receptor Kinases in Brassinosteroid Signaling. *PLOS Genet.* **2012**, *8*, e1002452. [CrossRef]
60. Santiago, J.; Henzler, C.; Hothorn, M. Molecular mechanism for plant steroid receptor activation by somatic embryogenesis co-receptor kinases. *Science* **2013**, *341*, 889–892. [CrossRef]
61. Hohmann, U.; Lau, K.; Hothorn, M. The structural basis of ligand perception and signal activation by receptor kinases. *Ann. Rev. Plant Biol.* **2017**, *68*, 109–137. [CrossRef]
62. Laux, T.; Mayer, K.F.; Berger, J.; Jurgens, G. The *WUSCHEL* gene is required for shoot and floral meristem integrity in Arabidopsis. *Development* **1996**, *122*, 87–96. [PubMed]
63. Long, J.A.; Moan, E.I.; Medford, J.I.; Barton, M.K. A member of the KNOTTED class of homeodomain proteins encoded by the *STM* gene of Arabidopsis. *Nature* **1996**, *379*, 66–69. [CrossRef] [PubMed]
64. Licausi, F.; Ohme-Takagi, M.; Perata, P. APETALA 2/Ethylene Responsive Factor (AP 2/ERF) transcription factors: Mediators of stress responses and developmental programs. *New Phytol.* **2013**, *199*, 639–649. [CrossRef] [PubMed]
65. Waki, T.; Kiki, T.; Watanabe, R. The Arabidopsis RWP-RK Protein RKD4 Triggers Gene Expression and Pattern Formation in Early Embryogenesis. *Curr. Biol.* **2011**, *21*, 1277–1281. [CrossRef] [PubMed]
66. Jeong, S.; Palmer, T.M.; Lukowitz, W. The RWP-RK Factor GROUNDED Promotes Embryonic Polarity by Facilitating YODA MAP Kinase Signaling. *Curr. Biol.* **2011**, *21*, 1268–1276. [CrossRef] [PubMed]
67. Santos Mendoza, M.; Dubreucq, B.; Miquel, M.; Caboche, M.; Lepiniec, L. LEAFY COTYLEDON 2 activation is sufficient to trigger the accumulation of oil and seed specific mRNAs in Arabidopsis leaves. *FEBS Lett.* **2005**, *579*, 4666–4670. [CrossRef] [PubMed]
68. Weigel, D.; Alvarez, J.; Smyth, D.R.; Yanofsky, M.F.; Meyerowitz, E.M. LEAFY controls floral meristem identity in Arabidopsis. *Cell* **1992**, *69*, 843–859. [CrossRef]
69. Nagle, M.; Déjardin, A.; Pilate, G.; Strauss, S.H. Opportunities for innovation in genetic transformation of forest trees. *Front. Plant Sci.* **2018**, *9*, 1443. [CrossRef]
70. Miller, M.; Tagliani, L.; Wang, N.; Berka, B.; Bidney, D.; Zhao, Z.-Y. High Efficiency Transgene Segregation in Co-Transformed Maize Plants using an Agrobacterium Tumefaciens 2 T-DNA Binary System. *Transgenic Res.* **2002**, *11*, 381–396. [CrossRef]
71. Kononov, M.E.; Bassuner, B.; Gelvin, S.B. Integration of T-DNA binary vector 'backbone' sequences into the tobacco genome: Evidence for multiple complex patterns of integration. *Plant J.* **1997**, *11*, 945–957. [CrossRef]

72. Hanson, B.; Engler, D.; Moy, Y.; Newman, B.; Ralston, E.; Gutterson, N. A simple method to enrich an Agrobacterium-transformed population for plants containing only T-DNA sequences. *Plant J.* **1999**, *19*, 727–734. [CrossRef] [PubMed]
73. Kim, S.R.; Lee, J.; Jun, S.H.; Park, S.; Kang, H.G.; Kwon, S.; An, G. Transgene structures in TDNA-inserted rice plants. *Plant Mol. Biol.* **2003**, *52*, 761–773. [CrossRef] [PubMed]
74. Podevin, N.; De Buck, S.; De Wilde, C.; Depicker, A. Insights into recognition of the T-DNA border repeats as termination sites for T-strand synthesis by Agrobacterium tumefaciens. *Transgenic Res.* **2006**, *15*, 557–571. [CrossRef] [PubMed]
75. Richael, C.; Kalyeava, M.; Chretien, R.C.; Rommens, C.M. Cytokinin vectors mediate marker-free and backbone-free plant transformation. *Transgenic Res.* **2008**, *17*, 905–917. [CrossRef] [PubMed]
76. Lowe, K.; Hoerster, G.; Anand, A.; Arling, M.; Wang, N.; McBride, K.; Gordon-Kamm, W. Transient expression of morphogenic genes positioned outside the T-DNA borders results in rapid formation of somatic embryos and fertile transgenic cereal plants. Unpublished; manuscript in preparation.
77. Ananiev, E.V.; Wu, C.; Chamberlin, M.A.; Svitashev, S.; Schwartz, C.; Gordon-Kamm, W.; Tingey, S. Artificial chromosome formation in maize (*Zea mays* L.). *Chromosoma* **2009**, *118*, 157–177. [CrossRef] [PubMed]
78. Svitashev, S.; Young, J.K.; Schwartz, C.; Gao, H.; Falco, S.C.; Cigan, A.M. Targeted Mutagenesis, Precise Gene Editing, and Site-Specific Gene Insertion in Maize Using Cas9 and Guide RNA. *Plant Physiol.* **2015**, *169*, 931–945. [CrossRef] [PubMed]
79. Svitashev, S.; Schwartz, C.; Lenderts, B.; Young, J.K.; Cigan, A.M. Genome editing in maize directed by CRISPR–Cas9 ribonucleoprotein complexes. *Nat. Commun.* **2016**, *7*, 13274. [CrossRef] [PubMed]

 © 2019 by the authors. Licensee MDPI, Basel, Switzerland. This article is an open access article distributed under the terms and conditions of the Creative Commons Attribution (CC BY) license (http://creativecommons.org/licenses/by/4.0/).

Article

Inheritance and Genetic Mapping of the Reduced Height (*Rht18*) Gene in Wheat

Nathan P. Grant [1], Amita Mohan [1], Devinder Sandhu [2,*] and Kulvinder S. Gill [1,*]

1. Department of Crop and Soil Sciences, Washington State University—Pullman, 277 Johnson Hall, PO Box 646420, Pullman, WA 99164-6420, USA; ngrant@wsu.edu (N.P.G); amitamohan@wsu.edu (A.M.)
2. USDA-ARS Salinity Lab., 450 W. Big Springs Rd., Riverside, CA 92507, USA
* Correspondence: devinder.sandhu@ars.usda.gov (D.S.); ksgill@wsu.edu (K.S.G.); Tel.: +951-369-4832 (D.S.); +509-335-4666 (K.S.G.)

Received: 12 June 2018; Accepted: 11 July 2018; Published: 15 July 2018

Abstract: Short-statured plants revolutionized agriculture during the 1960s due to their ability to resist lodging, increased their response to fertilizers, and improved partitioning of assimilates which led to yield gains. Of more than 21 reduced-height (*Rht*) genes reported in wheat, only three—*Rht-B1b*, *Rht-D1b*, and *Rht8*—were extensively used in wheat breeding programs. The remaining reduced height mutants have not been utilized in breeding programs due to the lack of characterization. In the present study, we determined the inheritance of *Rht18* and developed a genetic linkage map of the region containing *Rht18*. The height distribution of the F_2 population was skewed towards the mutant parent, indicating that the dwarf allele (*Rht18*) is semi-dominant over the tall allele (*rht18*). *Rht18* was mapped on chromosome *6A* between markers barc146 and cfd190 with a genetic distance of 26.2 and 17.3 cM, respectively. In addition to plant height, agronomically important traits, like awns and tiller numbers, were also studied in the bi-parental population. Although the average tiller number was very similar in both parents, the F_2 population displayed a normal distribution for tiller number with the majority of plants having phenotype similar to the parents. Transgressive segregation was observed for plant height and tiller number in F_2 population. This study enabled us to select a semi-dwarf line with superior agronomic characteristics that could be utilized in a breeding program. The identification of SSRs associated with *Rht18* may improve breeders' effectiveness in selecting desired semi-dwarf lines for developing new wheat cultivars.

Keywords: *Rht18*; reduced height; wheat; semi-dwarf; linkage map

1. Introduction

The Green Revolution, in the mid-twentieth century, brought about advancements in agriculture that are still in practice to date. The introduction of semi-dwarf varieties that are more responsive to changing agriculture practices like response to fertilizers was pivotal in bringing the green revolution by increasing cereal production to meet the population demands particularly in developing countries like China, India, Brazil, and Egypt [1]. Two genotypes, Norin10 {*Rht1* (*Rht-B1b*) and *Rht2* (*Rht-D1b*)} and Akakomugi (*Rht8*), were first incorporated into breeding programs to introduce the semi-dwarf genes in wheat cultivars in the United States and Italy [2,3].

The development of semi-dwarf cultivars can be attributed to a shorter yet stronger culm that accommodates high yields and prevents lodging [4,5]. Of the 21 wheat mutants reported to be associated with height reduction, only *Rht-B1b* and/or *Rht-D1b*, *Rht8* and *Rht12* have been characterized in detail [6,7]. *Rht-B1* and *Rht-D1* are two homoeologous genes present on B and D genomes in hexaploid wheat that code for DELLA proteins, which suppress gibberellin (GA)-responsive growth [8]. Normally, GA regulates binding of the GA insensitive dwarf 1 (GID1) receptor protein with DELLA proteins and promotes their degradation. Mutant alleles, *Rht-B1b* and *Rht-D1b*, produce DELLA proteins that do not bind GID1 resulting in growth inhibition due to insensitivity to GA [8]. Similarly, modulation in GA synthesis or signaling is known to be involved in reducing plant height in different species. Studies in *Arabidopsis*, maize [6], rice [9,10], and barley [11], suggest that GA affects the inter-nodal elongation and thus alters plant height.

Height reduction in present day cultivars of wheat is achieved mainly by *Rht-B1b* and/or *Rht-D1b*, accounting for ~95% of the cultivated wheat around the world [2]. Of the other 19 height mutants reported in wheat, only *Rht8* has been used in some European wheat cultivars. The rest have not been utilized either because of the lack of genetic characterization or mapping information. The limited genetic variability in semi-dwarf lines used in breeding programs is becoming a bottleneck for further wheat improvement, due to the association of some negative effects with the *Rht-B1b* and *Rht-D1b* genes, particularly under abiotic stresses or changing environmental conditions [12]. Currently used semi-dwarf wheat lines are defective from the perspective of GA, which plays an important role in the growth and development of the plant. These genotypes display a significant effect on early seedling growth. Specifically, coleoptile length, first leaf elongation, seedling emergence, and plant height reduction have been reported in the genotypes carrying *Rht-B1b* and *Rht-D1b* compared to tall parents [6,13]. The GA-responsive *Rht12* and *Rht13* were reported to reduce plant height with no adverse effect on the coleoptile and root trait during the seedling stage [14,15]. *Rht12* delayed ear emergence, reducing flag leaf length and grain size, while *Rht13* adversely affected the 1000 kernel weight and flag leaf length. Initially classified as GA-responsive, *Rht8* was reported to be involved in reduced sensitivity to brassinosteroids that resulted in reducing plant height [16]. The 17 other reduced-height mutants have not been fully characterized. *Rht18* was found to be GA-sensitive and was identified as a possible reduced height mutant candidate for future breeding programs [4,17]. In durum wheat, *Rht18* was previously mapped to the short arm of chromosome *6A* at the same locus as *Rht14* and *Rht16* [17,18]. Applications of exogenous gibberellins (GA$_3$) restored plant height and other agronomic traits of *Rht18* dwarf lines to the wild-type levels, indicating that *Rht18* dwarf mutants are impaired in GA biosynthesis [19]. In this investigation, we have mapped *Rht18* to chromosome *6A* using a cross between a pre-green revolution tall line (Indian) devoid of any know height reducing genes and *Rht18* mutant Icaro. The transfer of the *Rht18* allele into bread wheat and the selection of potential semi-dwarf lines with good agronomic characteristics can be useful for wheat breeding programs.

2. Results and Discussion

2.1. Plant Height of F_2 and $F_{2:3}$ Progenies

The plant height of the F_2 population was recorded under controlled environmental conditions in a greenhouse along with parental lines Indian and Icaro. The height of the tall parent Indian and dwarf mutant parent Icaro averaged 86 ± 2.82 cm (Mean ± S.E.) and 44 ± 1.02 cm (Mean ± S.E.), respectively (Figure 1). Of the 94 F_2 plants, approximately 55 were within 10 cm of the Icaro height range. Only four of the plants in the F_2 population had a phenotype similar to Indian (86 ± 10 cm). Three of the originally-sown plants did not grow to maturity. This is expected as sterility is often associated with the incompatibility among the tetraploid and hexaploid crosses [20,21]. The F_2 population had a height distribution skewed towards the parent Icaro (Figure 1). The skewed distribution towards reduced height parents was also reported in the *Rht3* F_2 mapping population [22]. This distribution suggests that the mutant phenotype is dominant, as only a few plants had the tall phenotype. Interestingly, a few F_2 plants were taller than the tall parent and many were shorter than the dwarf parent. The height distribution pattern suggests that additional modifier genes might be involved in regulating plant height. Plant height is known to be a complex trait regulated by interaction and interplay among major and minor genes [23]. The transgressive segregation observed for plant height might be due to epistatic gene actions [23]. Transgressive segregation was reported earlier in wheat for several agronomic traits, including plant height [24], grain yield and its components [25], heading date, and vernalization requirement [26]. In a previous study involving *Rht8*, transgressive segregants were observed for longer peduncles and grains per spike with no significant change in spike length, spikelet number, or number of fertile tillers [27]. Additionally, no significant effect was observed on roots, while a slight decrease in coleoptile length occurred. Partitioning of dry matter to ears was increased at anthesis, however, dry weight of stems and above-ground biomass, including ears, decreased [27].

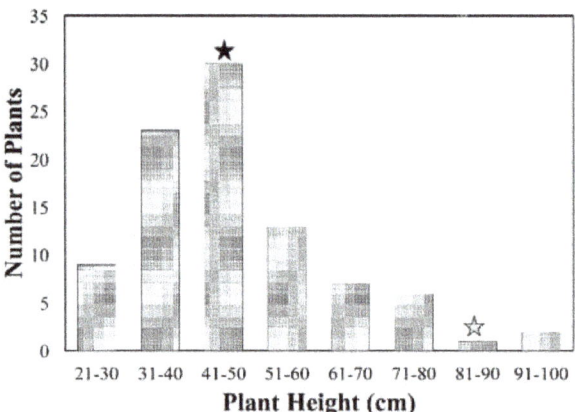

Figure 1. Height distribution in the F_2 population. Plant height was grouped into 10 cm series. The star represents plant height of tall or dwarf mutant parent. The average plant height of Indian is recorded 86 cm (from eight plants) and Icaro as 44 cm (from six plants).

Forty seeds from each individual F_2 plant representing the $F_{2:3}$ progenies were sown in the field the following summer to evaluate the genotypes of the F_2 plants, as it was difficult to classify plants into distinct categories in F_2. The $F_{2:3}$ population showed segregation for plant height (Figure 2), with 14 progenies classified as homozygous short, one as homozygous tall, and 54 were classified as heterozygous. Highly significant effects were found for the plant height (Table 1). For the $F_{2:3}$ population the height was found to be on average taller than the F_2, possibly due to the photoperiod effect in the field. As seen in the F_2 generation, we observed some very dwarf and some very tall plants in $F_{2:3}$ progenies (Figure 2), indicating the role of additional modifier genes in transgressive segregation.

Table 1. Analysis of variance (ANOVA) of plant height for the $F_{2:3}$ population.

Source	DF	SS	MS	F Value	Pr > F
Model	76	155,138.4	2041.29	14.49	<0.0001
Error	648	9131.879	140.91		
Corrected total	724	246,452.3			

Figure 2. Plant height distribution among $F_{2:3}$ families. (**A**) Indian and Icaro; and (**B–H**) different $F_{2:3}$ families.

2.2. Spike Morphology

Along with the plant height, the F_2 population also segregated for awn-less/short or long, black awns. Among the parents, Indian spikes were awn-less and Icaro spikes bear long black awns (Figure 3). Among the F_2 plants, 55 plants had awns and 36 plants were awn-less. We have also observed a difference in spike morphology among the F_2 and $F_{2:3}$ plants (Figure 3). The Indian spike is long and had loose spikelets, while the Icaro head is small with compact spikelets (Figure 3). We have observed plants with Indian-type heads with awns and Icaro-type heads without awns (Figure 3).

Figure 3. Spike morphology of parents and $F_{2:3}$ families in the mapping population. (**A**) Indian; (**B**) Icaro; and (**C–H**) different $F_{2:3}$ progenies.

2.3. Tiller Number

The F$_2$ population displayed a range for the number of tillers per plant ranging from three tillers per plant to 28 tillers per plant (Figure 4). Fifty-five percent of plants have tiller numbers ranging from 9 to 15 per plant, resembling the average for both Indian and Icaro, which were approximately 11 and 12 tillers per plant, respectively. The highest tillering plants were usually dwarf and sterile or contained only a few seeds in a spike. This might be due to incompatibility between the two genotypes. The higher or lower number of tillers compared to the average of both the parents might be due to multigenic nature of the trait. Extreme dwarf plants were sterile and did not set seeds. Further, of the 91 F$_2$ lines used for F$_{2:3}$ field evaluation, only 75 plants produced seed. This is expected for a hexaploid and tetraploid cross due to pollen viability issues restricting the seed set [28]. Among the plants that set seeds, some had good seed sets while others only contributed a few per plant.

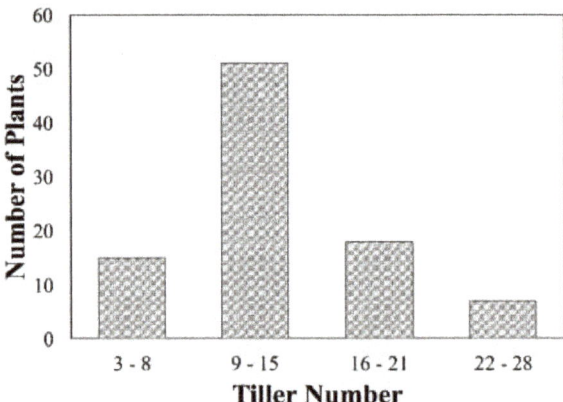

Figure 4. Tiller number distribution in the F$_2$ population. The average number of tillers for parents Indian and Icaro were 11 and 12, respectively.

The variation observed in spike morphology was not associated with the height phenotype each plant displayed (data not shown). The seed weight did not correlate with plant height. Tiller number in F$_2$ plants did not associate with the seed weight or number of seeds harvested at maturity (data not shown). The 100 seed weight for Indian and Icaro were 3.43 g and 3.69 g, respectively. Among the F$_{2:3}$ families, the short families had an average 100 seed weight of 2.8 g while the tall families had 3.1 g. The height mutation in wheat is reported to have affected the seed weight compared to the tall counterpart. Of the studied reduced height mutants, *Rht12* reduces the grain weight more compared to *Rht-B1b*, *Rht-B1c*, and *Rht8* [29]. The reduction in grain weight might be due to the adverse effect of *Rht18* on grain size [27,30]. In fields conditions, the tiller number per plant was difficult to measure, hence, was not recorded for the F$_{2:3}$ plants. The F$_{2:3}$ families were also evaluated in the field for their agronomic characteristics to identify the agronomically important plant to be utilized in hexaploid wheat breeding. We have selected one line (line 29) based on plant height, stem strength, and spike morphology. More detailed agronomic and molecular analysis will be performed on the selected line to determine its suitability for utilization in a breeding program.

2.4. Genetic Mapping of Rht18

In order to map the gene on a wheat chromosome, over 700 SSR markers [31] were used to screen parents Indian and Icaro. Of these, 154 markers showed polymorphism between the parents and were used to genotype the population. The *Rht18* gene was mapped to the short arm of chromosome 6A and was flanked by barc146 and cfd190 (Figure 5). The SSR marker cfd190 was placed at a distance of 17.3 cM away from *Rht18*. Previously, barc003, a marker from the short arm of the chromosome *6A*, was mapped 25.1 cM away from *Rht18* in durum wheat [17,32]. Earlier, *Rht18* was mapped on chromosome *6A* between barc118 and IWA4371 using recombinant inbred lines (RILs) in durum wheat [18]. The mapping location of *Rht18* in our study is consistent with the previous map position [17,18]. Recently, several independent single nucleotide variants in the *GA2oxA9* gene located on chromosome 6A were associated with the *Rht18* mutant phenotype [33]. *GS20xA9* is predicted to encode GA 2-oxidase, which reduces the amount of bioactive GA (GA1).

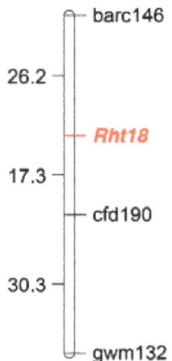

Figure 5. Genetic linkage map showing the position of *Rht18* on chromosome *6A*. Genetic distances are shown in centiMorgans (cM).

Reduced-height genes in wheat have been imperative to the agronomic success of the crop. The resulting yield increases have been credited to the improved structure of the plant that responded better to the agronomic practices in use today. The semi-dwarf phenotype increases resistance to lodging along with increasing the number of grains per plants. Incorporating additional reduced-height genes into breeding programs could help contribute to the diversification of the genotype. Considering climate change and the demand for food security, incorporating additional dwarfing genes into the germplasm and evaluating their agronomic worth might help to address the wheat productivity under a changing climate. As the photoperiod and the background of a genotype affect height, a marker close to the gene may assist in easy and precise selection of the locus. Thus, identification of SSR markers closer to the *Rht18* locus may assist breeders in early identification of dwarfing lines for breeding populations. Further, conducting the genomic and agronomic characterization of this mutant gene may become instrumental in developing a better dwarfing system in wheat. Additionally, we have identified a semi-dwarf line from $F_{2:3}$ families with superior agronomic characters that might have potential to be used in wheat breeding to incorporate the gene into the hexaploid background of Pacific Northwest region.

3. Materials and Methods

3.1. Plant Materials

The dwarf parent, Icaro (tetraploid; 4×) (*Rht18*; PI 503555), was originally derived in 1987 in Italy from fast-neutron treatment of cv. Anhinga (PI 428455). The tall line Indian (hexaploid; 6×) (CI 4489), was developed at the University of Idaho, Idaho before 1915. As the tall parent is released before the introduction of reduced height genes, we presumed that cv. Indian would be devoid of the *Rht18* allele in the background. Both the germplasms were procured from GRIN [34].

3.2. F₁ and Plant Growing Conditions

The F_1 produced by crossing Indian as the female parent and Icaro as the male parent was self-pollinated, and 120 F_2 seeds were collected. The F_2 mapping population was grown at the plant growth facility, Washington State University, under controlled conditions of 16 h days (22 °C) and 8 h (18 °C) nights. For ease of genotyping, 94 randomly-selected F_2 plants were selected for further analysis. Forty seeds of each F_2 plant were grown in three-foot rows at the Spillman Agronomy Farm, Pullman, WA for phenotypic screening. Four rows were planted in each plot with a row-to-row and plot-to-plot spacing of one foot. Each row represented progeny of a single F_2 plant. The seeds were planted mechanically using four planter drills and the plants were grown until maturity using the standard regional agricultural practices with no irrigation.

3.3. Phenotypic Screening

The phenotypic data for height, awns, tiller number, and seed weight was collected for the F_2 and $F_{2:3}$ populations. The plant height was recorded at maturity to the nearest cm excluding the awns. The population was characterized into tall, intermediate, and dwarf based on the plant height at maturity. Tiller numbers were counted manually per plant and seed weight was measured for each individual plant.

3.4. DNA Isolation and Genotyping

Young leaf tissue of F_2 plants was collected in 96-well DNA extraction plates. Four, 2-cm long leaf segments were clipped and lyophilize for three days. The lyophilized tissue was used for DNA isolation using a modified SDS extraction method [35]. The DNA was diluted to a final concentration of 25 ng/μL. Primer sequence information for simple sequence repeat (SSR) markers were obtained from GrainGenes website [36].

Over 700 SSR markers were first screened for polymorphism between the parental genotypes. The PCR was performed in 12 μL reaction volume containing 1× NEB reaction buffer, 200 μM of dNTPs, 2.5 mM MgCl₂, 0.05 μM forward primer, 0.25 μM reverse primer, 0.2 μM M13 forward-labeled primer, and 1U homemade Taq polymerase. For multiplexing, the M13 sequence was fluorescently labeled separately with FAM, HEX, NED, and PET dyes. The amplification of SSR loci was performed using the protocol consisted of 94 °C/4 min for initial denaturation, followed by 37 cycles (94 °C/30 s, 60 °C/45 s, 72 °C/60 s), with final extension at 72 °C/10 min. The amplification products were separated using ABI DNA analyzer 3100 (Applied Biosystems Inc., Carlsbad, CA, USA). Alleles were sized relative to internal size standard (cassual445 labeled with Dy630) using GeneMarker software (Softgenetics, State College, PA, USA). MapMaker 2.0 was used to construct the genetic linkage map using the Kosambi mapping function [37,38].

Author Contributions: N.P.G., A.M. and D.S. carried out the research study, analyzed and interpreted the data, and drafted the manuscript. N.P.G. and A.M. collected the phenotypic data. K.S.G., D.S. and A.M. conceived and designed the study. Authors read and approved the final manuscript.

Funding: Financial assistance from NSF BREAD-0965533 to support Nathan Grant's undergraduate research and for conducting the experiment is acknowledged.

Acknowledgments: We thank Deven R. See and the Western Regional Small Grains Genotyping Laboratory (WRSGGL), Pullman, WA, for providing the genotyping facility.

Conflicts of Interest: The authors declare no conflict of interest. Mention of trade names or commercial products in this publication is solely for the purpose of providing specific information and does not imply recommendation or endorsement by the U.S. Department of Agriculture.

References

1. Germplasm Resources Information Network. Available online: https://www.ars-grin.gov/ (accessed on 12 July 2018).
2. Hedden, P. The genes of the Green Revolution. *Trends Genet.* **2003**, *19*, 5–9. [CrossRef]
3. Borojevic, K.; Borojevic, K. The transfer and history of "reduced height genes" (Rht) in wheat from Japan to Europe. *J. Hered.* **2005**, *96*, 455–459. [CrossRef] [PubMed]
4. Ellis, M.H.; Rebetzke, G.J.; Chandler, P.; Bonnett, D.; Spielmeyer, W.; Richards, R.A. The effect of different height reducing genes on the early growth of wheat. *Funct. Plant Biol.* **2004**, *31*, 583–589. [CrossRef]
5. Chapman, S.C.; Mathews, K.L.; Trethowan, R.M.; Singh, R.P. Relationships between height and yield in near-isogenic spring wheats that contrast for major reduced height genes. *Euphytica* **2007**, *157*, 391–397. [CrossRef]
6. Peng, J.; Richards, D.E.; Hartley, N.M.; Murphy, G.P.; Devos, K.M.; Flintham, J.E.; Beales, J.; Fish, L.J.; Worland, A.J.; Pelica, F.; et al. 'Green revolution' genes encode mutant gibberellin response modulators. *Nature* **1999**, *400*, 256–261. [CrossRef] [PubMed]
7. Ellis, M.H.; Rebetzke, G.J.; Azanza, F.; Richards, R.A.; Spielmeyer, W. Molecular mapping of gibberellin-responsive dwarfing genes in bread wheat. *Theor. Appl. Genet.* **2005**, *111*, 423–430. [CrossRef] [PubMed]
8. Pearce, S.; Saville, R.; Vaughan, S.P.; Chandler, P.M.; Wilhelm, E.P.; Sparks, C.A.; Al-Kaff, N.; Korolev, A.; Boulton, M.I.; Phillips, A.L.; et al. Molecular characterization of *Rht-1* dwarfing genes in hexaploid wheat. *Plant Physiol.* **2011**, *157*, 1820–1831. [CrossRef] [PubMed]
9. Ogawa, M.; Kusano, T.; Katsumi, M.; Sano, H. Rice gibberellin-insensitive gene homolog, *OsGAI*, encodes a nuclear-localized protein capable of gene activation at transcriptional level. *Gene* **2000**, *245*, 21–29. [CrossRef]
10. Ikeda, A.; Ueguchi-Tanaka, M.; Sonoda, Y.; Kitano, H.; Koshioka, M.; Futsuhara, Y.; Matsuoka, M.; Yamaguchi, J. *Slender* rice, a constitutive gibberellin response mutant, is caused by a null mutation of the *SLR1* gene, an ortholog of the height-regulating gene *GAI/RGA/RHT/D8*. *Plant Cell* **2001**, *13*, 999–1010. [CrossRef] [PubMed]
11. Chandler, P.M.; Marion-Poll, A.; Ellis, M.; Gubler, F. Mutants at the *Slender1* locus of barley cv Himalaya. Molecular and physiological characterization. *Plant Physiol.* **2002**, *129*, 181–190. [CrossRef] [PubMed]
12. Tolmay, V.L. Resistance to biotic and abiotic stress in the Triticeae. *Hereditas* **2001**, *135*, 239–242. [CrossRef] [PubMed]
13. Schillinger, W.F.; Donaldson, E.; Allan, R.E.; Jones, S.S. Winter wheat seedling emergence from deep sowing depths. *Agron. J.* **1998**, *90*, 582–586. [CrossRef]
14. Chen, L.; Phillips, A.L.; Condon, A.G.; Parry, M.A.; Hu, Y.G. GA-responsive dwarfing gene *Rht12* affects the developmental and agronomic traits in common bread wheat. *PLoS ONE* **2013**, *8*, e62285. [CrossRef] [PubMed]
15. Wang, Y.; Chen, L.; Du, Y.; Yang, Z.; Condon, A.G.; Hu, Y.-G. Genetic effect of dwarfing gene *Rht13* compared with *Rht-D1b* on plant height and some agronomic traits in common wheat (*Triticum aestivum* L.). *Field Crops Res.* **2014**, *162*, 39–47. [CrossRef]
16. Gasperini, D.; Greenland, A.; Hedden, P.; Dreos, R.; Harwood, W.; Griffiths, S. Genetic and physiological analysis of *Rht8* in bread wheat: An alternative source of semi-dwarfism with a reduced sensitivity to brassinosteroids. *J. Exp. Bot.* **2012**, *63*, 4419–4436. [CrossRef] [PubMed]
17. Watanabe, N. Genetic collection and development of near-isogenic lines in durum wheat. *Vavilov J. Genet. Breed.* **2008**, *12*, 636–643.
18. Vikhe, P.; Patil, R.; Chavan, A.; Oak, M.; Tamhankar, S. Mapping gibberellin-sensitive dwarfing locus *Rht18* in durum wheat and development of SSR and SNP markers for selection in breeding. *Mol. Breed.* **2017**, *37*, 28. [CrossRef]
19. Yang, Z.Y.; Liu, C.Y.; Du, Y.Y.; Chen, L.; Chen, Y.F.; Hu, Y.G. Dwarfing gene *Rht18* from tetraploid wheat responds to exogenous GA3 in hexaploid wheat. *Cereal Res. Commun.* **2017**, *45*, 23–34. [CrossRef]
20. Padmanaban, S.; Zhang, P.; Hare, R.A.; Sutherland, M.W.; Martin, A. Pentaploid wheat hybrids: Applications, characterisation, and challenges. *Front. Plant Sci.* **2017**, *8*, 358. [CrossRef] [PubMed]
21. Martin, A.; Simpfendorfer, S.; Hare, R.A.; Eberhard, F.S.; Sutherland, M.W. Retention of D genome chromosomes in pentaploid wheat crosses. *Heredity* **2011**, *107*, 315–319. [CrossRef] [PubMed]

22. Navarro, C.; Yang, Y.; Mohan, A.; Grant, N.; Gill, K.S.; Sandhu, D. Microsatellites based genetic linkage map of the *Rht3* locus in bread wheat. *Mol. Plant Breed.* **2014**, *5*, 43–46. [CrossRef]
23. Rieseberg, L.H.; Archer, M.A.; Wayne, R.K. Transgressive segregation, adaptation and speciation. *Heredity* **1999**, *83*, 363–372. [CrossRef] [PubMed]
24. Bhatt, G.M. Inheritance of heading date, plant height, and kernel weight in two spring wheat Crosses. *Crop Sci.* **1972**, *12*, 95–98. [CrossRef]
25. El-Hennawy, M.A.; Abdalla, A.F.; Shafey, S.A.; Al-Ashkar, I.M. Production of doubled haploid wheat lines (*Triticum aestivum* L.) using anther culture technique. *Ann. Agric. Sci.* **2011**, *56*, 63–72. [CrossRef]
26. Shindo, C.; Tsujimoto, H.; Sasakuma, T. Segregation analysis of heading traits in hexaploid wheat utilizing recombinant inbred lines. *Heredity* **2003**, *90*, 56–63. [CrossRef] [PubMed]
27. Yang, Z.; Zheng, J.; Liu, C.; Wang, Y.; Condon, A.G.; Chen, Y.; Hu, Y.-G. Effects of the GA-responsive dwarfing gene *Rht18* from tetraploid wheat on agronomic traits of common wheat. *Field Crops Res.* **2015**, *183*, 92–101. [CrossRef]
28. Bhagyalakshmi, K.; Vinod, K.K.; Kumar, M.; Arumugachamy, S.; Prabhakaran, A.; Raveendran, T.S. Interspecific hybrids from wild x cultivated *Triticum* crosses-a study on the cytological behavior and molecular relations. *J. Crop Sci. Biotechnol.* **2008**, *11*, 257–262.
29. Casebow, R.; Hadley, C.; Uppal, R.; Addisu, M.; Loddo, S.; Kowalski, A.; Griffiths, S.; Gooding, M. Reduced height (*Rht*) alleles affect wheat grain quality. *PLoS ONE* **2016**, *11*, e0156056. [CrossRef] [PubMed]
30. Tang, T. *Physiological and Genetic Studies of an Alternative Semi-Dwarfing Gene Rht18 in Wheat*; University of Tasmania: Tasmania, Australia, 2015.
31. Somers, D.J.; Isaac, P.; Edwards, K. A high-density microsatellite consensus map for bread wheat (*Triticum aestivum* L.). *Theor. Appl. Genet.* **2004**, *109*, 1105–1114. [CrossRef] [PubMed]
32. Haque, M.A.; Martinek, P.; Watanabe, N.; Kuboyama, T. Genetic mapping of gibberellic acid-sensitive genes for semi-dwarfism in durum wheat. *Cereal Res. Commun.* **2011**, *39*, 171–178. [CrossRef]
33. Ford, B.; Foo, E.; Sharwood, R.E.; Karafiatova, M.; Vrána, J.; MacMillan, C.; Nichols, D.S.; Steuernagel, B.; Uauy, C.; Doležel, J.; et al. Rht18 semi-dwarfism in wheat is due to increased expression of GA 2-oxidaseA9 and lower GA Content. *Plant Physiol.* **2018**. [CrossRef] [PubMed]
34. Gerritsen, V.B. All things dwarfed and beautiful. Available online: https://web.expasy.org/spotlight/pdf/sptlt070.pdf (accessed on 12 July 2018).
35. Sandhu, D.; Champoux, J.A.; Bondareva, S.N.; Gill, K.S. Identification and physical localization of useful genes and markers to a major gene-rich region on wheat group *1S* chromosomes. *Genetics* **2001**, *157*, 1735–1747. [PubMed]
36. GrainGenes. Available online: https://wheat.pw.usda.gov/cgi-bin/GG3/browse.cgi?class=marker;begin=8251 (accessed on 12 July 2018).
37. Lander, E.S.; Green, P.; Abrahamson, J.; Barlow, A.; Daly, M.J.; Lincoln, S.E.; Newburg, L. MAPMAKER: An interactive computer package for constructing primary genetic linkage maps of experimental and natural populations. *Genomics* **1987**, *1*, 174–181. [CrossRef]
38. Kosambi, D.D. The estimation of map distance from recombination values. *Ann. Eugen.* **1944**, *12*, 172–175. [CrossRef]

© 2018 by the authors. Licensee MDPI, Basel, Switzerland. This article is an open access article distributed under the terms and conditions of the Creative Commons Attribution (CC BY) license (http://creativecommons.org/licenses/by/4.0/).

Communication

Two Rye Genes Responsible for Abnormal Development of Wheat–Rye Hybrids Are Linked in the Vicinity of an Evolutionary Translocation on Chromosome 6R

Natalia V. Tsvetkova [1,2], Natalia D. Tikhenko [2,3], Bernd Hackauf [4] and Anatoly V. Voylokov [2,*]

1. Department of Genetics and Biotechnology, St. Petersburg State University, Universiteskaya nab.7/9, St. Petersburg 199034, Russia; ntsvetkova@mail.ru
2. Vavilov Institute of General Genetics Russian Academy of Sciences, St. Petersburg Branch, Universiteskaya nab.7/9, St. Petersburg 199034, Russia; tikhenko@mail.ru
3. Leibnitz Institute of Plant Genetics and Crop Plant Research, Corrensstrasse 3, Stadt Seeland OT, D-06466 Gatersleben, Germany
4. Julius Kühn-Institut, Institute for Breeding Research on Agricultural Crops, Rudolf-Schick-Platz 3a, D-18190 Sanitz, Germany; bernd.hackauf@julius-kuehn.de
* Correspondence: av_voylokov@mail.ru

Received: 25 May 2018; Accepted: 6 July 2018; Published: 10 July 2018

Abstract: The post-zygotic reproductive isolation (RI) in plants is frequently based on the negative interaction of the parental genes involved in plant development. Of special interest is the study of such types of interactions in crop plants, because of the importance of distant hybridization in plant breeding. This study is devoted to map rye genes that are incompatible with wheat, determining the development of the shoot apical meristem in wheat–rye hybrids. Linkage analysis of microsatellite loci, as well as genes of embryo lethality (*Eml-R1*) and hybrid dwarfness (*Hdw-R1*) was carried out in hybrids of Chinese Spring wheat with recombinant inbred lines as well as interline rye hybrids. *Eml-R1* and *Hdw-R1* could be mapped proximal and distal of two closely linked EST-SSR markers, *Xgrm902* and *Xgrm959*, on rye chromosome 6R. Both rye genes are located on a segment of chromosome 6R that contains a breakpoint of evolutionary translocation between the ancestral chromosomes of homeologous groups 6 and 3. The obtained results are discussed in relation to genes interacting in developmental pathways as a class of causal genes of RI.

Keywords: wheat-rye hybrids; genes of reproductive isolation; stem apical meristem; molecular marker

---

1. Introduction

Post-zygotic incompatibility in plants is often expressed in an autoimmune reaction (tissue necrosis) and disturbances of plant development, leading to a decrease in the viability or death of hybrids. Immunity and ontogenesis are controlled by many interacting genes. The interaction of genes is the basis of the canonical scheme (Bateson–Dobzhansky–Muller model), explaining the emergence and functioning of post-zygotic incompatibility. In cultivated and wild plants, a large number of genes controlling hybrid incompatibility in accordance with the classic two-locus scheme have been described [1]. The complementary interaction of incompatible alleles is established in interspecific and intraspecific hybrids. Phenotypes associated with hybrid necrosis resemble those elicited in response to various abiotic and biotic stresses [2]. However, while substantial progress has been achieved to uncover the molecular mechanisms by which disease resistance is achieved [2], the molecular mechanisms of hybrid incompatibility connected with the disturbances of development have not been studied in detail. The interaction of incompatible alleles of wheat (*Eml-A1*) and rye (*Eml-R1*) genes lead to ungerminating hybrid seeds. Mature seeds have normal endosperm, but hybrid embryos may be lacking, or are varying in size from small to normal, and are comprising of dead tissues, undifferentiated in the region of shoot apical meristem (SAM) [3]. Another rye mutation (*Hdw-R1b*) affects the shoot apical meristem in wheat–rye hybrids at the transition from vegetative to reproductive phases of development [4]. As a result, a phenotype develops similar to grass-clump dwarfs found in some intraspecific hybrids of bread wheat [5], and in hybrids of its progenitor species [6,7]. These wheat–rye hybrids stop development at the tillering stage, having three to five short tillers, and die within two months. Scanning electron microscopy showed that the apices of dwarf plants to that time do not reach the double-ridge stage, which is reminiscent of apices of young seedlings. The elucidation of the molecular control of developmental disorders in wheat–rye hybrids complements similar studies in related cereals, and should help unravel their causal function in the evolution of isolation mechanisms. The recently published draft genome sequence of rye [8] together with the high-quality reference genome sequence of barley [9] provide invaluable genomic resources to precisely characterize *Eml-R1* and *Hdw-R1* by their position in the rye genome. Here, we report on the identification of expressed sequence tag (EST) derived microsatellite markers, which are linked to both genes. Comparative mapping allowed to integrate the target interval in the barley genome sequence as a prerequisite for fine mapping and subsequent positional cloning.

2. Results

2.1. The Linkage of Mutant Genes with Molecular Markers

The locus of hybrid embryo lethality, *Eml-R1*, has been localized previously on chromosome 6R [10,11] based on the linkage to the co-segregating markers *Xgwm1103/Xgwm732*. In the present study, we have mapped *Eml-R1* distal of *Xgwm1103/Xgwm732*, and closely linked to *Xgrm959* and *Xgrm902* in hybrids of Chinese Spring (CS) wheat, and a set of rye recombinant inbred lines (RILs) originating from the cross between L2 × L7 (Figure 1). We were able to integrate three additional microsatellite markers—*Xgrm173*, *Xgrm130*, and *Xgwm751*—in this linkage group on chromosome 6R. The loci *Xgrm173* and *Xgrm959*–*Xgrm902* carry different alleles in lines L6 and V1. This enabled testing the linkage of *Hdw-R1* with these markers in a CS × F1 (L6 × V1) cross. The markers revealed linkage to *Hdw-R1*. This gene, compared with *Eml-R1*, is located distal of the linked markers *Xgrm959*–*Xgrm902* (Figure 1b). Thus, both genes of hybrid incompatibility may be inherited together. Their joint segregation was validated in three hybrid populations.

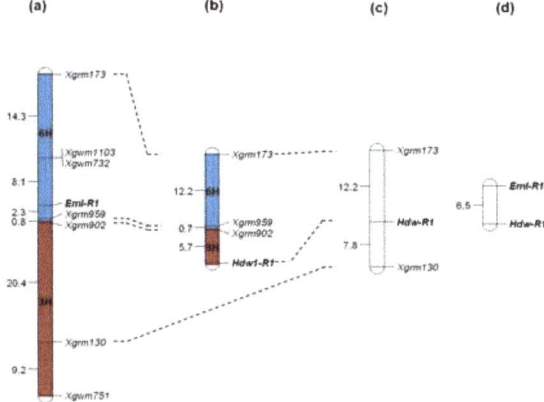

Figure 1. Genetic maps of *Eml-R1* and *Hdw-R1* on rye chromosome 6RL and their relationships with the homeologous barley chromosomes. Linkage maps were established for different wheat–rye hybrids: (**a**) CS × recombinant inbred lines (L2 × L7), n = 74; (**b**) CS × F1 (L6 × V1), n = 230; and (**c,d**) CS × F1 (V1 × L2), n = 230. Recombination frequency is shown in %.

2.2. Comparative Mapping

Each of the four EST-derived SSR markers *Xgrm173*, *Xgrm959*, *Xgrm902*, and *Xgrm130* could be integrated in the draft of the rye genome sequence (Table 1). The length of the corresponding Lo7 contigs varies between 362–11,532 bp; two of these contigs have been mapped on chromosome 6R. Sequence similarity searches in the barley genome sequence revealed that the orthologs of *Xgrm173* and *Xgrm959*, both flanking *Eml-R1*, are residing on chromosome 6H, while the *Hdw-R1* flanking markers *Xgrm902* and *Xgrm130* correspond to segments on chromosome 3H (Table 1, Figure 1).

Table 1. Integration of the 6R markers derived from rye EST sequences in the rye and barley genome sequences.

Marker [a]	BLASTN Query [a]	Rye				Barley			
		Subject	Expect	Chr [b]	pos	Subject	s_start	s_end	Expect
GRM0173	Sce_Assembly02_c6346	Lo7_v2_contig_257767	1×10^{-100}	0R		chr6H	570137757	570137532	6×10^{-59}
GRM0959	Sce_Assembly02_c87163	Lo7_v2_contig_4801	3×10^{-145}	6R	108.7	chr6H	579212782	579213063	9×10^{-43}
GRM0902	Sce_Assembly02_c81266	Lo7_v2_contig_1427427	1×10^{-104}	0R		chr3H	19821332	19821201	2×10^{-12}
GRM0130	Sce_Assembly02_c4514	Lo7_v2_contig_126444	5×10^{-167}	6R	124.6	chr3H	674285409	674285721	5×10^{-122}

[a] According to Martis et al., 2013; [b] according to Bauer et al., 2017.

2.3. Joint Segregation Analysis of Hybrid Dwarfness and Embryo Lethality

For segregation analysis, the germinating seeds of three populations with dihybrid segregations CS × F1 (V1 × L2), CS × F1 (L2 × V1), and CS × F1 (L2 × V10) were divided into five phenotypic classes (Table 1).

The ratio of seeds with normal embryos to seeds with abnormal embryos (with undifferentiated embryos and without embryos) corresponds to the expected monohybrid segregation in all of the studied hybrids. The segregation for the gene of hybrid dwarfness [4] can be observed only in plants grown from the seeds with normal alive embryos. In theory, the ratio of dwarf to normal plants depends on the linkage of genes *Eml-R1* and *Hdw-R1*. If these genes are not linked (segregate independently), we will observe monohybrid gametic segregation 1 (*Hdw-R1a*):1 (*Hdw-R1b*). In reality, we observed a case that was attributed to the tight linkage of the studied genes, which follows from their map position (Figure 1a,b). A large fraction of dwarf plants and a significantly smaller fraction of normal plants were found in the progeny of each cross (Table 2). Gametes, producing the normal wheat–rye hybrids, appear as a result of crossing over between these linked genes in the meiosis of the rye parent. With this assumption, we calculated a recombination frequency between the rye genes *Eml-R1* and *Hdw-R1* as a frequency of normal hybrid plants (rye recombinant gametes) among all of the hybrid plants (the all rye gametes). This frequency is equal to 6.5 ± 1.7% (Figure 1d), 3.2 ± 1.6%, and 9.0 ± 2.6% in the studied dihybrid cross combinations CS × F1 (V1 × L2), CS × F1 (L2 × V1), and CS × F1 (L2 × V10), respectively. As a mean value, one may consider 6.3 ± 1.1%, which was calculated for the pooled sample. A recombination rate between genes *Xgwm173* and *Hdw-R1* calculated in hybrids with monohybrid and dihybrid segregations do not differ to a large extent, and are equal to 17.9% and 12.2%, correspondingly. It is important to note that the frequency of seeds with normal differentiated embryos, but that are not capable of germination, varied significantly between these hybrids: 6.6%, 15.5%, and 69.5%. The appearance of this phenotypic class in segregation and its variable frequency are attributable to environmental variation. The fertilization and embryo development under distant hybridization are very sensitive to variability in temperature, humidity, and mineral supply. For this reason, the development of embryos carrying the normal *Eml-R1a* allele may be disturbed before seed maturity, and such embryos die. The mutant *Eml-R1b* allele expresses far earlier, and its expression leads to the development of morphologically distinct embryos that also die before maturity.

Table 2. Segregation for hybrid dwarfness (*Hdw-R1*) and embryo lethality (*Eml-R1*) in crosses of Chinese Spring wheat with F1 interline rye hybrids.

Hybrid Combination	Seeds with Normal Embryo (*Eml-R1a*)			Seeds with Abnormal Embryo (*Eml-R1b*)		χ^2 1 *Eml-R1a*: 1 *Eml-R1b*
	Alive		Dead	Undifferentiated Embryo	Without Embryo	
	Dwarf	Normal				
CS × F1(V1 × L2)	215	15	34 (6.6) *	244	10	1.19
CS × F1(L2 × V1)	121	4	51 (15.5) *	145	7	1.76
CS × F1(L2 × V10)	111	11	228 (65.9) *	321	37	0.09

* Percentage is shown in brackets.

3. Discussion

In an initial attempt, we have mapped *Eml-R1* on chromosome 6RL based on linkage to the genomic wheat microsatellite markers *Xgwm1103*/*Xgwm732* [10,11]. In the present study, we describe for the first time that *Eml-R1* is linked to another gene controlling post-zygotic reproductive isolation between wheat and rye, *Hdw-R1*. Furthermore, the integration of EST-derived rye SSR marker enabled comparative mapping, and revealed that both genes are located on an interstitial region on chromosome 6RL, covering a previously reported 3L/6L translocation breakpoint [12,13]. While *Eml-R1* is residing on a segment that is homeologous with barley chromosome 6H, *Hdw-R1* maps distally from *Eml-R1* on a 6RL segment that is homeologous with the long arm of 3H. The 3L/6L translocation breakpoint is located within a 0.8-cM interval defined by *Xgrm959* and *Xgrm902*, respectively. The localization of both genes on different ancestral segments shaping the modern chromosome 6R is further supported by the genomic wheat microsatellite loci *Xgwm751*, *Xgwm1103*, and *Xgwm732*. The co-segregating microsatellite markers *Xgwm1103*/*Xgwm732* in wheat are located on chromosomes 6A [14] and 6D [15]. In contrast, marker *Xgwm751* was mapped in wheat close to the centromere on chromosome arm 3AL, and on the long arm of 3B [16]. The progress achieved in the present study concerning the localization of *Eml-R1* and *Hdw-R1* with respect to the position of translocation breakpoints is important in terms of understanding the mechanisms of reproductive isolation and the evolution of the rye karyotype. In *Saccharomyces cerevisiae*, chromosomal rearrangements have been identified as a major mechanism to reproductively isolate different strains [17]. In plants, knowledge of chromosomal rearrangements is still scarce, and their importance for speciation is controversial discussed [18]. Recently, 4L/5L translocation breakpoints have been comprehensively described at the molecular level as two hotspots of chromosomal rearrangements that have been reused during Triticeae evolution [19]. *Eml-R1* and *Hdw-R1* reside at the 3L/6L translocation in rye, and highlight that the fitness of wheat/rye hybrids can be genetically affected via embryogenesis or the vegetative development, respectively. As a consequence, both genes might have contributed to the speciation of rye, which diverged from *Triticum aestivum* approximately three to four million years ago [20]. The molecular genetic control of hybrid inviability in plants is yet not well understood [21]. The natural genetic diversity of rye inbred lines in the *Eml-R1* as well as *Hdw-R1* genes and a sophisticated phenotyping system based on test-crosses with wheat enables a forward genetics approach to isolate genes involved in the reproductive isolation of Triticeae species. With the recent availability of a draft genome sequence of rye [8] and a high-quality reference genome sequence of barley [9], the positional cloning of both genes has now become a feasible task in the large and complex rye genome.

There are numerous examples of hybrid incompatibility manifesting itself as an arrest of plant development at different stages [1]. They include embryo and seedling lethality, failure to transition from vegetative to reproductive stages of development, or forming reproductive organs. Some of the described examples closely resemble the expression of known mutant genes controlling plant development through SAM maintenance and function. The death of hybrid plants at different stages of development was frequently connected with the necrosis of tissues, suggesting an autoimmune reaction. Thus, it is not easy to find the true cause of hybrid incompatibility. One key to solve the problem is an approach based on the identification of candidate genes. That approach allows the unraveling of complex hierarchical relationships of genes performing different functions, using the knowledge of the functions of the interacting candidate genes. Thus, the identification of the corresponding genes is critical for understanding the molecular mechanisms of complementary negative interactions in hybrids. Transcriptional analysis of the incompatibility of hybrids of tetraploid wheat and wild diploid relatives illustrates well the need for identification of the causal genes of incompatibility [6,7,22,23]. The authors describe the changes in the transcription activity of the hundreds of genes at shoot apices in hybrids between tetraploid wheat (*Triticum turgidum* ssp. *durum*, AABB genome) variety Langdon, and two wheat diploid relatives, *Aegilops tauschii* (DD genome) and *Aegilops umbellata* (UU genome). F1 hybrids of wheat with some accessions of both wild relatives show one of two developmental abnormalities: severe growth abortion (SGA), which may be considered as seedling lethality, and grass-clump dwarfness/hybrid necrosis. Both have some features closely resembling the morphological expression of the wheat–rye dwarfness. Lethal at a three-leaf stage and temperature independent SGA connected with the down-regulation of numerous transcription factors, including the KNOTTED1-like homeobox gene, maintaining SAM, and the cell cycle-related genes functioning in SAM and leaf primordia. The temperature-dependent grass-clump dwarfness is explained by the down-regulation of the APETALA-like MADS box genes, known as flowering promoters, and by increased miR156 transcription, leading to a reduced level of target mRNA of *SPL* genes (Squamosa promoter binding protein-box transcription factors), some of which promote tillering. It is very interesting that the grass-clump dwarf phenotype is characteristic of hybrids, growing under normal temperature conditions. The same hybrids at low temperature express a typical autoimmune response connecting with the repression of cell division. Transcription factors, such as small RNA, are capable of physically interacting with target DNA and RNA, correspondingly, and their genes can be considered as the most likely for the role of candidate genes. It is worth noting that the cited authors revealed the presence of compatible and non-compatible genotypes in all of the parents, but carried out the segregation analysis and mapped only one of the genes (*Net2*) in *Aegilops tauschii*. We found only incompatible alleles in Chinese Spring bread wheat and both types of alleles in the rye inbred lines. The wheat gene *Eml-A1*, which is complementary to rye incompatible allele *Eml-R1b* in the expression of hybrid embryo lethality, was mapped on the distal part of the long arm of chromosome 6A with the aid of the deletion lines of CS [24]. One would expect that *Eml* genes may have orthologs on the chromosomes of homeologous group 6 in different species of the tribe Triticeae, including one with sequenced genomes. The comprehensive study of our, and the cited, examples of genome incompatibility would resolve the subject under discussion. Namely, would the genes of developmental pathways be considered as a separate class of plant genes that is capable of serving as causal genes for reproductive isolation? The progress in this direction is closely connected with genome sequencing in species of tribe Triticeae. Rye now does not limit the comparative studies, owing to new genomic resources [8].

4. Materials and Methods

4.1. Plant Material

For segregation analysis, the wheat–rye hybrid seeds were produced in two types of crosses. The female parent in both cases was bread wheat of the Chinese Spring variety, as the male parents were interline rye F1 hybrids, or the set of 74 L2 × L7 RILs were used. It was shown previously that rye lines L6 and L7 carry normal (compatible) alleles in both studied genes (*Eml-R1a* and *Hdw-R1a*). Line L2 carries the incompatible allele *Eml-R1b*, which leads to embryo lethality in hybrids with Chinese Spring wheat. Lines V1 and V10 carry the incompatible allele of hybrid dwarfness *Hdw-R1b*. To produce wheat–rye hybrid seeds, wheat spikes were emasculated 1–2 days before anthesis, and pollinated 2–4 days later with freshly collected rye pollen. Wheat plants were pollinated by pollen collected from individual plants of corresponding F1 hybrids, or each of 74 RIL plants.

4.2. Phenotyping and Genotyping

Mature wheat–rye caryopses were soaked in water, and 3–4 days later, the embryos were classified as normal (completely differentiated) or abnormal (undifferentiated or without embryos). To study the segregation for dwarfness, the seeds with normal alive embryos were sown in the soil, and one month old plants were differentiated as either dwarf or normal. DNA was isolated from the leaves of grown plants using the CTAB method [25]. For each hybrid combination, the polymorphic microsatellite markers were selected on the basis of preliminary screening, and data for the linkage of *Eml-R1* with two co-segregating markers *Xgwm1103/Xgwm732* on chromosome 6R [8,9]. In all, three wheat microsatellites (*Xgwm1103/Xgwm732*, and *Xgwm751*) and four rye ones (*Xgrm173*, *Xgrm959*, *Xgrm902*, and *Xgrm130*) were used for mapping. Segregation for each marker corresponded to the expected gametic ratio of 1:1 ($p > 0.05$) in hybrids of wheat with RILs and with rye F1 L6 × V1. The segregation for the markers *Xgrm173* and *Xgrm130*, which were studied in hybrids CS × (V1 × L2), differed to a large extent from the monohybrid ratio. In this hybrid, markers may be studied only in segregating dwarf plants and rare recombinant plants with normal phenotypes. Information for the used wheat microsatellite (GWM) belongs to the Institute of Plant Genetics and Crop Plant Research (Gatersleben, Germany). A set of EST-derived rye microsatellites (GRM) were used according to Martis et al. [12]. Electrophoresis was performed in 6% denaturing polyacrylamide gel on an automatic laser fluorescent sequencer ALFexpress II (Amersham-Pharmacia-Biotech, Amersham, UK). The sizes of the fragments were calculated using the program Fragment Analyser 1.02 (Amersham-Pharmacia-Biotech) by comparison with internal standards of known size.

4.3. Linkage Map Construction and Comparative Mapping

Segregation for the studied genes and markers were tested by χ^2. Linkage groups for different hybrids were built by MultiPoint3.3 (MultiQTL Ltd., Institute of Evolution, Haifa, Israel, http://www.multiqtl.com). A recombination frequency in percent was used as a measure of the genetic distance. For comparison purposes, RF per single meiosis was calculated for RILs. To identify the orthologous *Eml-R1* and *Hdw-R1* segments in the genomes of barley and rye, rye EST assemblies representing the GRM markers [12] were compared against masked barley pseudomolecules [9] as well as rye whole genome shotgun contigs v2 [8] using BLASTN and the IPK barley and rye blast server (http://webblast.ipk-gatersleben.de/).

Author Contributions: A.V.V. designed the study and wrote the manuscript; N.V.T. produced the plant material, and performed the microsatellite and phenotype analysis; and N.D.T. coordinated and validated the phenotype analysis. B.H. performed comparative mapping and revised the manuscript.

Funding: This study was supported by the RF state budget (project "Genetics and breeding of rye on the base of natural hereditary diversity"—AAAA-A16-116111610177-3).

Conflicts of Interest: The authors declare no conflict of interest. The founding sponsors had no role in the design of the study; in the collection, analyses, or interpretation of data; in the writing of the manuscript; or in the decision to publish the results.

References

1. Voilokov, A.V.; Tikhenko, N.D. Genetics of postzygotic reproductive isolation in plants. *Russ. J. Genet.* **2009**, *45*, 637–650. [CrossRef]
2. Bomblies, K.; Weigel, D. Hybrid necrosis: Autoimmunity as a potential gene-flow barrier in plant species. *Nat. Rev. Genet.* **2007**, *8*, 382–393. [CrossRef] [PubMed]
3. Tikhenko, N.; Rutten, T.; Voylokov, A.; Houben, A. Analysis of hybrid lethality in F1 wheat-rye hybrid embryos. *Euphytica* **2008**, *159*, 367–375. [CrossRef]
4. Tikhenko, N.; Rutten, T.; Tsvetkova, N.; Voylokov, A.; Börner, A. Hybrid dwarfness in crosses between wheat (*Triticum aestivum* L.) and rye (*Secale cereale* L.): A new look at an old phenomenon. *Plant Biol.* **2015**, *17*, 320–326. [CrossRef] [PubMed]
5. Worland, A.J.; Law, C.N. The genetics of hybrid dwarfing in wheat. *Z. Pflanzenzücht.* **1980**, *85*, 28–39.
6. Okado, M.; Yoshida, K.; Takumi, S. Hybrid incompatibilities in interspecific crosses between tetraploid wheat and its wild diploid relative *Aegilops umbellulata*. *Plant Mol. Biol.* **2017**, *95*, 625–645. [CrossRef] [PubMed]
7. Matsuda, R.; Masaru Lenisa, J.-C.; Sakaguchi, K.; Ohno, R.; Yoshida, K.; Takumi, S. Global gene expression profiling related to temperature-sensitive growth abnormalities in interspecific crosses between tetraploid wheat and *Aegilops tauschii*. *PLoS ONE* **2017**, *12*, e0176497. [CrossRef] [PubMed]
8. Bauer, E.; Schmutzer, T.; Barilar, I.; Mascher, M.; Gundlach, H.; Martis, M.M.; Twardziok, S.O.; Hackauf, B.; Gordillo, A.; Wilde, P.; et al. Towards a whole-genome sequence for rye (*Secale cereale* L.). *Plant J.* **2017**, *89*, 853–869. [CrossRef] [PubMed]
9. Mascher, M.; Gundlach, H.; Himmelbach, A.; Beier, S.; Twardziok, S.O.; Wicker, T.; Radchuk, V.; Dockter, C.; Hedley, P.E.; Russell, J.; et al. A chromosome conformation capture ordered sequence of the barley genome. *Nature* **2017**, *544*, 427–433. [CrossRef] [PubMed]
10. Tikhenko, N.; Tsvetkova, N.; Priyatkina, S.; Voylokov, A.; Börner, A. Gene mutations in rye causing embryo lethality in hybrids with wheat: Allelism and chromosomal localization. *Biol. Plant.* **2011**, *55*, 448–452. [CrossRef]
11. Tikhenko, N.D.; Tsvetkova, N.V.; Lyholay, A.N.; Voylokov, A.V. Identification of complementary genes of hybrid lethality in crosses of bread wheat and rye. Results and prospects. *Russ. J. Genet.* **2017**, *7*, 153–158. [CrossRef]
12. Devos, K.M.; Atkinson, M.D.; Chinoy, C.N.; Francis, H.A.; Harcourt, R.L.; Koebner, R.M.; Liu, C.J.; Masojć, P.; Xie, D.X.; Gale, M.D. Chromosomal rearrangements in the rye genome relative to that of wheat. *Theor. Appl. Genet.* **1993**, *85*, 673–680. [CrossRef] [PubMed]
13. Martis, M.M.; Zhou, R.; Haseneyer, G.; Schmutzer, T.; Vrána, J.; Kubaláková, M.; König, S.; Kugler, K.G.; Scholz, U.; Hackauf, B.; et al. Reticulate evolution of the rye genome. *Plant Cell* **2013**, *25*, 3685–3698. [CrossRef] [PubMed]
14. Kosellek, C.; Pillen, K.; Nelson, J.C.; Weber, W.E.; Saal, B. Inheritance of field resistance to *Septoria tritici* blotch in the wheat doubled-haploid population Solitär × Mazurka. *Euphytica* **2013**, *194*, 161–176. [CrossRef]
15. Ganal, M.W.; Röder, M.S. Microsatellite and SNP Markers in Wheat Breeding. In *Genomics-Assisted Crop Improvement*; Varshney, R.K., Tuberosa, R., Eds.; Springer: Dordrecht, The Netherlands, 2007; pp. 1–24.
16. Marone, D.; Laidò, G.; Gadaleta, A.; Colasuonno, P.; Ficco, D.B.; Giancaspro, A.; Cattivelli, L. A high-density consensus map of A and B wheat genomes. *Theor. Appl. Genet.* **2012**, *125*, 1619–1638. [CrossRef] [PubMed]
17. Hou, J.; Friedrich, A.; De Montigny, J.; Schacherer, J. Chromosomal Rearrangements as a Major Mechanism in the Onset of Reproductive Isolation in *Saccharomyces cerevisiae*. *Curr. Biol.* **2014**, *24*, 1153–1159. [CrossRef] [PubMed]
18. Rieseberg, L.H. Chromosomal rearrangements and speciation. *Trends Ecol. Evol.* **2001**, *16*, 351–358. [CrossRef]
19. Li, W.; Challa, G.S.; Zhu, H.; Wie, W. Recurrence of Chromosome Rearrangements and Reuse of DNA Breakpoints in the Evolution of the Triticeae Genomes. *G3 (Bethesda)* **2016**, *6*, 3837–3847. [CrossRef] [PubMed]

20. Middleton, C.P.; Senerchia, N.; Stein, N.; Akhunov, E.D.; Keller, B.; Wicker, T.; Kilian, B. Sequencing of chloroplast genomes from wheat, barley, rye and their relatives provides a detailed insight into the evolution of the Triticeae tribe. *PLoS ONE* **2014**, *9*, e85761. [CrossRef] [PubMed]
21. Chen, C.; E, Z.; Lin, H.X. Evolution and Molecular Control of Hybrid Incompatibility in Plants. *Front. Plant Sci.* **2016**, *7*, 1208. [CrossRef] [PubMed]
22. Mizuno, N.; Shitsukawa, N.; Hosogi, N.; Park, P.; Takumi, S. Autoimmune response and repression of mitotic cell division occur in inter-specific crosses between tetraploid wheat and *Aegilops tauschii* Coss. that show low temperature-induced hybrid necrosis. *Plant J.* **2011**, *68*, 114–128. [CrossRef] [PubMed]
23. Hatano, H.; Mizuno, N.; Matsuda, R.; Shitsukawa, N.; Park, P.; Takumi, S. Disfunction of mitotic cell division at shoot apices triggered severe growth abortion in interspecific hybrids between tetraploid wheat and *Aegilopstauschii*. *New Phytol.* **2012**, *194*, 1143–1154. [CrossRef] [PubMed]
24. Tikhenko, N.; Poursarebani, N.; Rutten, T.; Schnurbusch, T.; Börner, A. Embryo lethality in wheat-rye hybrids: Dosage effect and deletion bin mapping of the responsible wheat locus. *Biol. Plant.* **2017**, *61*, 342–348. [CrossRef]
25. Stein, N.; Herren, G.; Keller, B. A new DNA extraction method for high-throughput marker analysis in a large-genome species such as *Triticum aestivum*. *Plant Breed.* **2001**, *120*, 354–356. [CrossRef]

© 2018 by the authors. Licensee MDPI, Basel, Switzerland. This article is an open access article distributed under the terms and conditions of the Creative Commons Attribution (CC BY) license (http://creativecommons.org/licenses/by/4.0/).

MDPI
St. Alban-Anlage 66
4052 Basel
Switzerland
Tel. +41 61 683 77 34
Fax +41 61 302 89 18
www.mdpi.com

Plants Editorial Office
E-mail: plants@mdpi.com
www.mdpi.com/journal/plants

www.ingramcontent.com/pod-product-compliance
Lightning Source LLC
LaVergne TN
LVHW071942080526
838202LV00064B/6658